Road Transportation: Planning and Management

Road Transportation: Planning and Management

Edited by Jacob Rowe

STATES
ACADEMIC PRESS
www.statesacademicpress.com

States Academic Press,
109 South 5th Street,
Brooklyn, NY 11249, USA

ISBN: 978-1-63989-474-1

Cataloging-in-Publication Data

Road transportation : planning and management / edited by Jacob Rowe.
p. cm.
Includes bibliographical references and index.
ISBN 978-1-63989-474-1
1. Transportation, Automotive. 2. Transportation, Automotive--Planning.
3. Transportation, Automotive--Management. 4. Transportation--Planning.
5. Transportation--Management. I. Rowe, Jacob.
HE5611 .R63 2022
388.31--dc23

For information on all States Academic Press publications
visit our website at www.statesacademicpress.com

STATES
ACADEMIC PRESS

Contents

Preface

Road transportation is the transit of goods and people on roads. Cargo is transported by trucking companies, and passengers are transported via mass transit. The transportation of goods depends on the local infrastructure, the distance the goods are to be transported by road, the kind of goods transported, and the volume and weight of an individual shipment. Major roads today are primarily made of concrete, while local roads are made of asphalt. The design of modern pavements is such that these can support heavier vehicular loads and faster speeds. All roadways have signs, signals or pavement markings to control traffic. This book elucidates the various aspects in the planning and management of road transportation in a multidisciplinary manner. It presents researches and studies performed by experts across the globe. With state-of-the-art inputs by acclaimed experts of this field, this book targets students and professionals.

Significant researches are present in this book. Intensive efforts have been employed by authors to make this book an outstanding discourse. This book contains the enlightening chapters which have been written on the basis of significant researches done by the experts.

Finally, I would also like to thank all the members involved in this book for being a team and meeting all the deadlines for the submission of their respective works. I would also like to thank my friends and family for being supportive in my efforts.

Editor

1

Group Gap Acceptance: A New Method to Analyze Driver Behavior and Estimate the Critical Gap at Multilane Roundabouts

Khaled Shaaban and Hassan Hamad

Department of Civil Engineering, Qatar University, P.O. Box 2713, Doha, Qatar

Correspondence should be addressed to Khaled Shaaban; kshaaban@qu.edu.qa

Academic Editor: Irena I. Otković

The operations of multilane roundabouts, especially three-lane roundabouts, are unique and more complicated than any other type of roundabouts. This study aims to analyze driver behavior and estimate the critical gap at three-lane roundabouts. Video data were collected at two roundabouts. The analysis identified a pattern of group gap acceptance, where vehicles entering the roundabout from different lanes moved in groups during the same gap. In this case, the decision of vehicles entering from outside lanes greatly depended on the gap acceptance decision of vehicles in the inside lane. Analysis showed that the vast majority of the vehicles accept the gap in groups and the critical gap was estimated accordingly. The study provides a new explanation for the operation at multilane roundabouts. The use of this simple method is recommended when estimating critical gaps for multilane roundabouts.

1. Introduction

Roundabouts are popular in many countries, especially in the case of lower volumes when compared to signalized intersections as they reduce queuing and delays at the approaches. Yield-controlled entries of roundabouts oblige vehicles to decide to either reject or accept a gap in the circulating flow. The decision making in most cases depends on the driver's behavior. Roundabout capacity is greatly affected by the gap acceptance behavior of drivers. The Highway Capacity Manual (HCM) provides a method to calculate the capacity of roundabouts, and it involves a number of inputs. One of the main inputs is the critical gap [1]. This value is used in the capacity model to estimate the capacity of existing facilities without the need to perform field measurements, assess the level of service, and predict the capacity at future locations. Furthermore, this value is needed in simulation and traffic modeling of roundabouts for design and research projects.

A gap is an opening in the circulating flow that circulates around the island of the roundabout. As illustrated in Figure 1, gap vehicles are the vehicles that travel in the circulating lanes and create the gaps. The decision vehicle is the vehicle at the entry of the roundabout that takes the decision, whether rejecting or accepting a gap. The gap closes on the left side and reopens on the right side as shown in Figure 1. Rejecting a gap happens when the driver of the decision vehicle finds the gap small enough that s/he refuses to merge with the circulating flow and remains stopped at the entry point. Accepting a gap is when the driver of the decision vehicle merges with the circulating flow from the vehicle's position at the entry of the roundabout as soon as the gap is large enough not to cause crashes or severe conflicts.

The critical gap is the smallest gap that a driver is assumed to accept [2]. The critical gap is an important parameter that affects the capacity and delay of roundabouts. Monitoring and evaluating of capacity values at roundabouts are a difficult matter due to the high dependence of these parameters on the drivers' behavior [1]. This study aims to investigate the gap acceptance behavior and to determine the critical gap for high-volume three-lane roundabouts. These types of roundabouts are unique, and their operation is more complicated than other types of roundabouts. Furthermore, previous studies have not thoroughly addressed these types of roundabouts to the best of our knowledge.

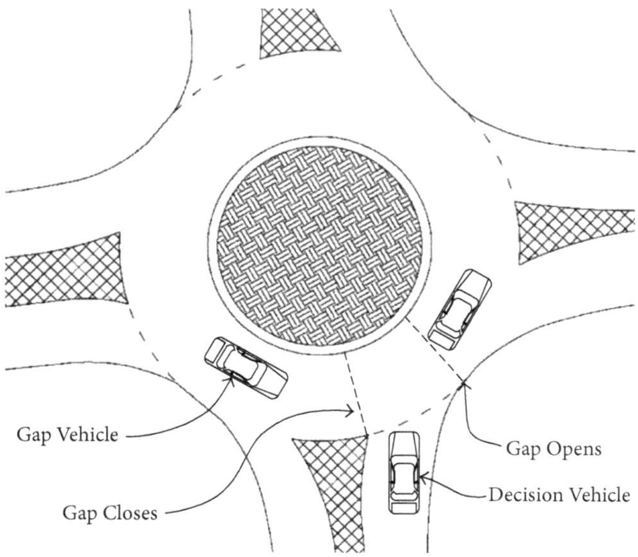

FIGURE 1: Gap acceptance definition.

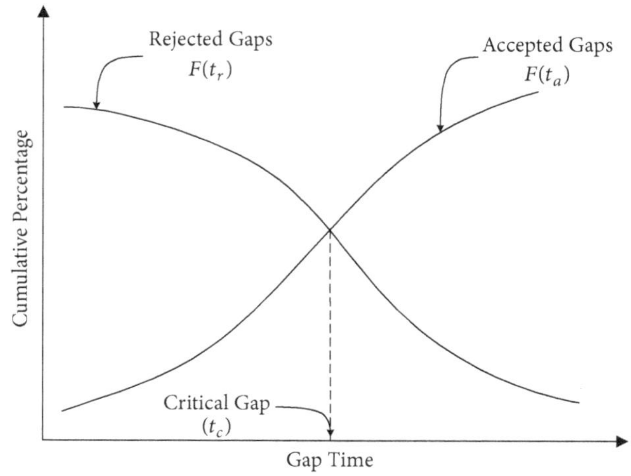

FIGURE 2: Critical gap based on Raff's method.

2. Literature Review

Numerous studies have investigated critical gaps. Raff [9] developed the earliest and the most popular method of evaluating the critical lag. He defined the critical lag as the lag for which the number of accepted lags shorter than it is similar to the number of rejected lags longer than it. A graphical model was developed in which two cumulative distribution curves related to the number of accepted and rejected gaps intersect to provide the critical lag value. The original Raff's procedure estimated critical lags on the basis of lags accepted and rejected. Miller corrected the Raff's method by including the entire gap data instead of lags only [10]. This new method is known as the modified Raff method. As shown in Figure 2, the modified Raff method determines the critical gap value (t_c), graphically, by finding the intersection point between the two functions:

$$1 - F(t_r), F(t_a), \tag{1}$$

where t_a and t_r are accepted and rejected gap times and $F(t_a)$ and $F(t_r)$ are cumulative distribution functions (CDFs) of accepted and rejected gaps.

The intersection point between the cumulative distribution curves corresponds to the critical gap value on the horizontal axis. This value is called the critical gap because most vehicles will more likely to accept a gap of a value larger than them and reject a gap of a value smaller than them. Previous studies showed that the value of t_c is affected by the existing traffic volumes at which this value has been evaluated [11, 12].

Different studies estimated the critical gap values obtained from different analysis methods and then compared them to different international standards or other studies. A summary of the critical gap values determined by previous studies is shown in Table 1. The table contains the analysis methods, location, and types of roundabouts. In summary, most studies addressed only one- and two-lane roundabouts. In summary, to our knowledge, limited studies investigated or developed a methodology to study the critical gap value for three-lane roundabouts.

Guo [7] compared three methods for the estimation of the critical gap for roundabouts, including Ashworth's method, Raff's method, and the maximum likelihood method to calculate the critical gap at roundabouts. He provided justification for recommending some of the methods over the others in practical applications. As per Guo, the modified Raff's method and the maximum likelihood method are considered easy to use and practical. On the other hand, Ashworth's method will result in a larger result because it only uses values of the accepted gap. Tupper et al. [13] compared four methods of calculating the critical gap for unsignalized T-intersections in terms of the ease of use and the use of data. Raff and the cumulative acceptance methods were identified as the most computationally simple followed closely by the fit maximization method. The equilibrium of probabilities method was the most computationally demanding. The Raff, equilibrium of probabilities, and fit maximization methods utilized both the accepted and rejected gap data, requiring a smaller sample size. The cumulative acceptance method utilized only the accepted gap data requiring a larger sample size for meaningful results. It was therefore decided to use the modified Raff's method in this study because it is easy to use, requires a small sample size, and produces reasonably accurate results.

3. Methods

3.1. Group Gap Acceptance. Vehicle interactions at the roundabouts vary depending on the number of circulating lanes. As the number of lanes increases, the complexity of the vehicular interactions increases. Three-lane roundabouts have complicated interaction patterns. Any two-gap vehicles in the circulating lanes regardless of the lane they are occupying can create an acceptable gap. Therefore, there should be 14 interaction cases as illustrated in Figure 3. If the decision

TABLE 1: Summary of past studies.

Author(s)	Year	Analysis method	Critical gap (s)	Type of roundabout	City, country
Flannery and Datta [3]	2007	Maximum likelihood method	3.94	1-lane	Florida and Maryland, USA
Xu and Tian [4]	2008	Maximum likelihood method	4.8		California, USA
Mensah et al. [5]	2010	Raff's method	2.55		Maryland, USA
Fitzpatrick et al. [6]	2013	Raff's method	2.2		Massachusetts, USA
Xu and Tian [4]	2008	Maximum likelihood method	L: 4.70/ R: 4.40	2-lane	California, USA
Guo [7]	2010	Raff's method	2.91		Dalian, China
		Revised Raff's method	2.78		
		Maximum likelihood method	2.65		
		Ashworth method	3.2		
Kusuma and Koutsopoulos [8]	2011	Maximum likelihood method	3.58		Stockholm, Sweden

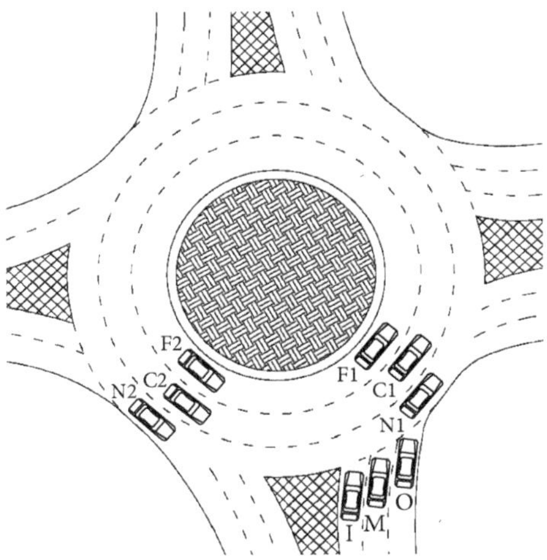

Case	Vehicle Interactions
1	Vehicle I with Vehicles F1 & F2
2	Vehicle I with Vehicles F1 & C2
3	Vehicle I with Vehicles F1 & N2
4	Vehicle I with Vehicles C1 & F2
5	Vehicle I with Vehicles C1 & C2
6	Vehicle I with Vehicles C1 & N2
7	Vehicle I with Vehicles N1 & F2
8	Vehicle I with Vehicles N1 & C2
9	Vehicle I with Vehicles N1 & N2
10	Vehicle M with Vehicles C1 & C2
11	Vehicle M with Vehicles C1 & N2
12	Vehicle M with Vehicles N1 & C2
13	Vehicle M with Vehicles N1 & N2
14	Vehicle O with Vehicles N1 & N2

F: Far
C: Center
N: Near

I: Inside
M: Middle
O: Outside

FIGURE 3: Traditional interaction cases.

vehicle is in the inside lane, cases 1 through 9 are applicable covering the path of the decision vehicle and all the possible paths vehicles combinations in the circulating lanes. If the gap vehicle is in the middle lane of the approach, cases 10 through 13 are applicable. For the outside lane, only one case is applicable.

Vehicles were observed to accept gaps in groups. The group gap acceptance behavior is based on the actual mechanics of accepting or rejecting gaps observed in the field at the two study locations. Vehicles in the middle and outside lanes, follow vehicles, take advantage of the gap for the inside vehicle to enter the roundabout. As shown in Figure 4, seven cases were observed (A to G) in the field. Cases A, B, and C are based on the vehicle in the inside lane accepting a gap. Case A occurs when all lanes are occupied. In this case, the vehicle in the inside lane accepts the gap. At the same time, the follow vehicles in middle and outside lanes accept the same gap by following the vehicle in the inside lane. Case B occurs if the vehicle in the outside lane does not accept the gap with the follow vehicle in the middle lane. Case C occurs when the vehicle in the middle lane rejects a gap whereas the vehicles in the inside and outside lanes accept it. Case D occurs in case of no vehicles in the inside lane. In this case, the vehicle in the middle lane accepts a gap, and the follow vehicle in the outside lane accepts the same gap. The remaining cases involve only one decision vehicle. Cases E, F, and G occur if the decision vehicles accept the gap individually.

3.2. Data Collection. Video footage was used to collect the data at two high-volume three-lane roundabouts in the city of Doha, Qatar. The first roundabout is located at the intersection of Al Gharrafa Road and Al Maszhabiya Street. The second roundabout is located at the intersection of Haloul Street and Mesaimeer Road. The two selected roundabouts are similar in geometry. Both represent the typical configuration of three-lane roundabouts in the city of Doha. The circulation

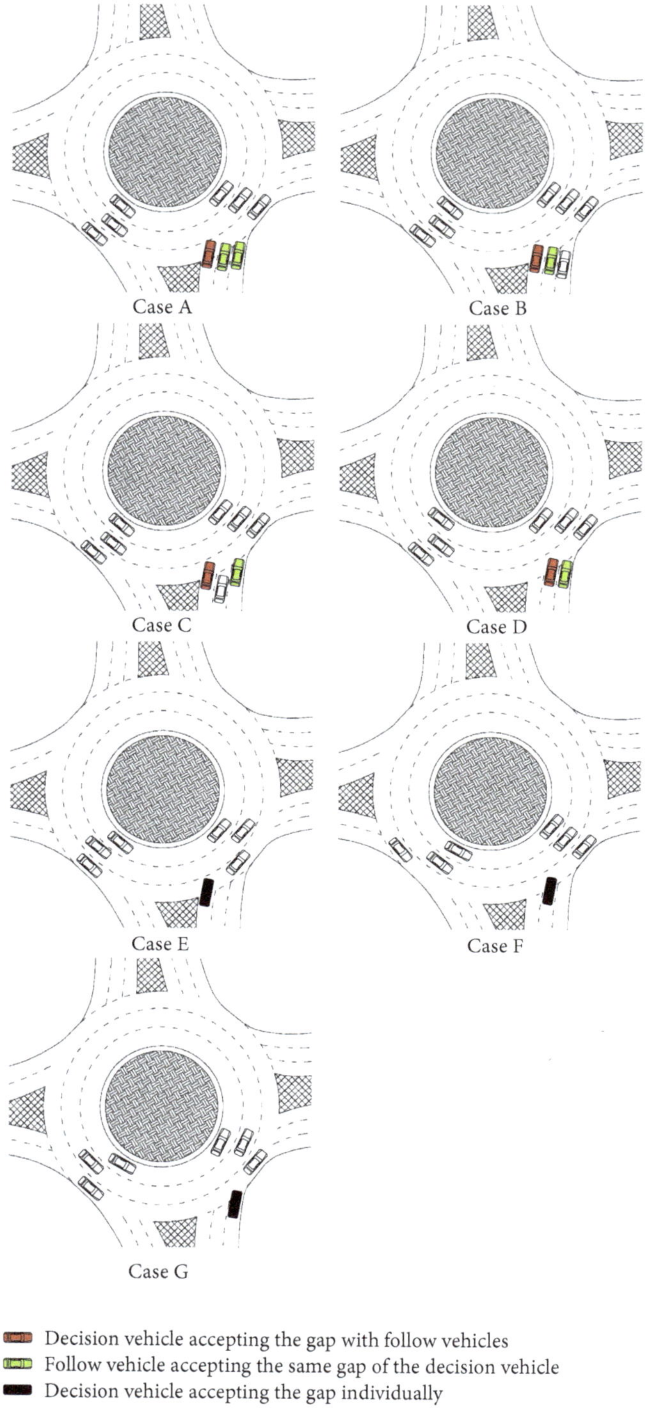

FIGURE 4: Interaction cases based on group gap acceptance.

inside the roundabouts is counterclockwise in Qatar. Vehicles traveling in the circulating lanes around the middle circular island have the right-of-way, where the vehicles at the yield-controlled entrances have to find a gap in order to merge with the circulating flow. Both roundabouts were selected because of the heavy traffic volumes and consistent entry flow to ensure a large sample size. Cameras were installed strategically at different angles at the studied approach to ensure a clear view without the obstruction of large-sized vehicles. Data were collected for the northbound approaches at both intersections as shown in Figure 5. A total of 15 hours of footage was captured covering the peak and off-peak hours. Data collection was performed from 6:00 a.m. to 9:00 p.m. in clear weather conditions.

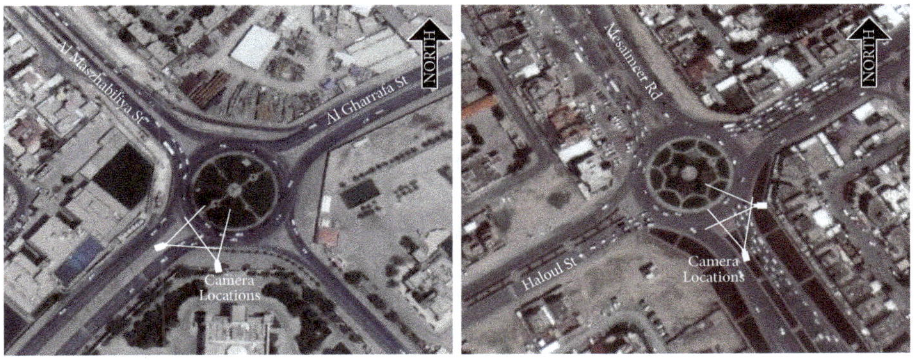

FIGURE 5: Aerial photo for roundabouts and camera locations.

3.3. Gap Measurement. The video data were analyzed frame by frame with an accuracy of 0.03 s based on a frame rate of 30 frames per second using Forevid analysis software to determine the time of each accepted and rejected gap. In Raff's method, the critical gap is estimated by either considering only gaps [12] or by combining gaps and lags together [14, 15]. This study considered only gaps and did not consider any lags similar to Brilon et al. [15]. In this study, only decision vehicles that come to a full stop at the yield line were considered. This condition was not a problem because of the relatively large circulating volumes. The majority of the vehicles entering the roundabout stopped first before proceeding. A gap is formed in the circulating lanes by having two-gap vehicles following each other in the same or a different lane with enough gap that can fit a decision vehicle. The gap is considered accepted when the decision vehicle's driver takes the decision to proceed into a gap in the circulating flow moving from the state of full stop at the approach. On the other hand, a gap is considered rejected when the driver of the decision vehicle decides not to proceed in the case of a gap in the circulation flow regardless of its size. For the vehicles entering the roundabout, direction, vehicle type, and departure time were recorded. For the vehicles circulating in the roundabout, the passing time at specified lines across the road and direction were recorded. This information was used to identify the different interaction cases and to determine the accepted and rejected gaps for the vehicles.

3.4. Vehicle Types. In this study, vehicles were classified under three categories: passenger vehicles, medium vehicles, and heavy vehicles. Each of the used classes is basically a group of vehicles that are approximately similar in dimensions and performance. The passenger vehicles group includes sedans, sport-utility vehicles, pickup trucks, and vans. Medium vehicles group includes single-unit two-axle trucks, small recreational vehicles, minibuses, and ambulances. Heavy vehicles group includes large buses, trailers of all sizes, and dump trucks. The summary of the collected data is presented in Table 2. Most of the vehicles in the sample (95.47%) were passenger vehicles.

3.5. Interaction Cases. The vehicular interaction cases are not similar amongst the different types of roundabouts because the number of lanes in each type of the roundabouts dictates the complexity of the interaction cases. Three-lane roundabouts have complex interaction cases, as explained earlier, and, due to that, new interaction cases were developed in a way that reflects the driver decision-taking more accurately. Some of the cases rarely occurred such that there are not enough data points to perform the analysis. Only the cases that have an abundance of data points have been analyzed. The total number of observed interactions exceeded 4,500 interactions over the duration of the data collection for both locations. As indicated in Table 2, case A was the most occurring case with 74.98% of all cases.

4. Results

4.1. Overall Critical Gap. Using the modified Raff's method, the overall critical gap, defined as the intersection between the CDFs of the rejected and accepted gaps at which the probability of rejecting or accepting, was estimated for all the data points. This value was estimated by equalizing the sigmodal functions of the cumulative distribution plots. The overall critical gap for the two roundabouts was 2.40 seconds. Figure 6 shows the CDFs that yielded the critical gap value of the two three-lane roundabouts.

4.2. Critical Gap for the Interaction Cases. Most of the interaction cases observed were case A, where the vehicles at the roundabout entry proceed together. The critical gap value for case A was 2.45 s, which is close to the overall critical gap value. A summary of the overall critical gap, the critical gaps of the different interaction cases, and vehicle types are listed in Table 3. Figure 7 shows the graphical representation of the data based on the modified Raff's definition to obtain the critical gap for each vehicular interaction case.

4.3. Critical Gap for the Vehicle Types. It was found that the critical gap values for passenger, medium, and heavy vehicles are in an ascending order. The passenger vehicles had a critical gap value of 2.39 s compared to 2.53 s and 3.03 s for medium and heavy vehicles, respectively. The number of data points for medium and heavy vehicles shows that more data points are required to obtain more accurate results. Figure 7 shows the graphical representation of modified Raff's method for the different vehicle types.

TABLE 2: Summary of data collected.

Location		Al Gharrafa St. roundabout		Haloul St. roundabout		Combined	
Classification	Type	Count	Percentage	Count	Percentage	Count	Percentage
Gap	Accepted	1025	44.55%	1112	46.66%	2137	45.62%
	Rejected	1276	55.45%	1271	53.34%	2547	54.38%
Vehicle type	Passenger vehicle	2210	96.05%	2262	94.92%	4472	95.47%
	Medium vehicle	62	2.69%	63	2.64%	125	2.67%
	Heavy vehicle	29	1.26%	58	2.43%	87	1.86%
Group gap	A	1860	80.83%	1652	69.32%	3512	74.98%
	B	118	5.13%	302	12.67%	420	8.97%
	C	205	8.91%	180	7.55%	385	8.22%
	D	23	1.00%	4	0.17%	27	0.58%
Individual gap	E	87	3.78%	239	10.03%	326	6.96%
	F	6	0.26%	5	0.21%	11	0.23%
	G	2	0.09%	1	0.04%	3	0.06%
Total		2301	100%	2383	100%	4684	100%

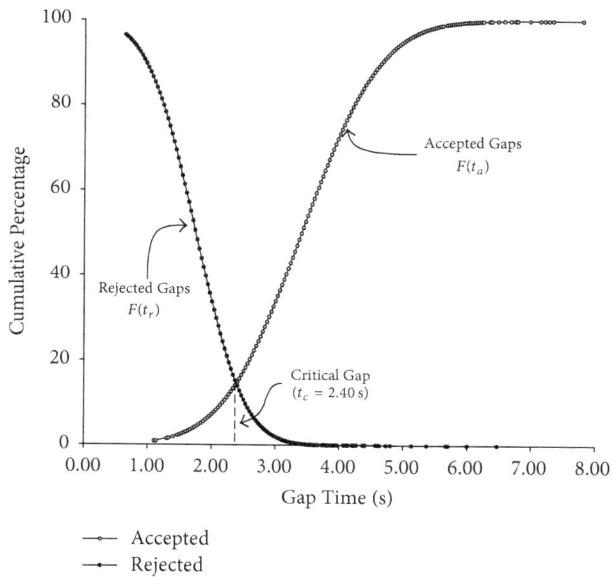

FIGURE 6: Overall critical gap.

5. Conclusions

The presented method, group gap acceptance, was used to analyze driver behavior and estimate the critical gap for three-lane roundabouts. Data were collected at two roundabouts in the city of Doha, Qatar. More than 4,500 interactions were recorded. Data were classified based on the vehicle type and interaction between vehicles. Field results indicated that the gap acceptance decision for vehicles entering from the inside lane affects the decision of vehicles entering from the outside lanes. In this behavior, named group gap acceptance, the vehicles in the outside lanes follow the vehicle in the inside lane and enter the roundabout during the same gap.

The overall critical gap value was 2.40 s. The critical gap for passenger vehicles was the lowest (2.39 s) compared to medium (2.53 s) and heavy vehicles (3.03 s). Nine driver interaction cases were identified (A–G). Most of the interactions (92.75%) involved more than one vehicle entering the roundabout and moving together in groups. Case A, three vehicles in a group, was the dominant interaction case, accounting for more than 74.98% of the data points collected from two locations, with a critical gap value of 2.45 s.

The presented method, group gap acceptance, was used to analyze driver behavior and estimate the critical gap for three-lane roundabouts. The critical gap for three-lane roundabouts, but the value of the overall critical gap, seems to be much lower than that obtained for one- and two-lane roundabouts [3–8]. Considering that driver behavior is a major contributor to the operational performance of roundabouts, the results may be an indication of aggressive driver behavior, which was identified in prior studies conducted

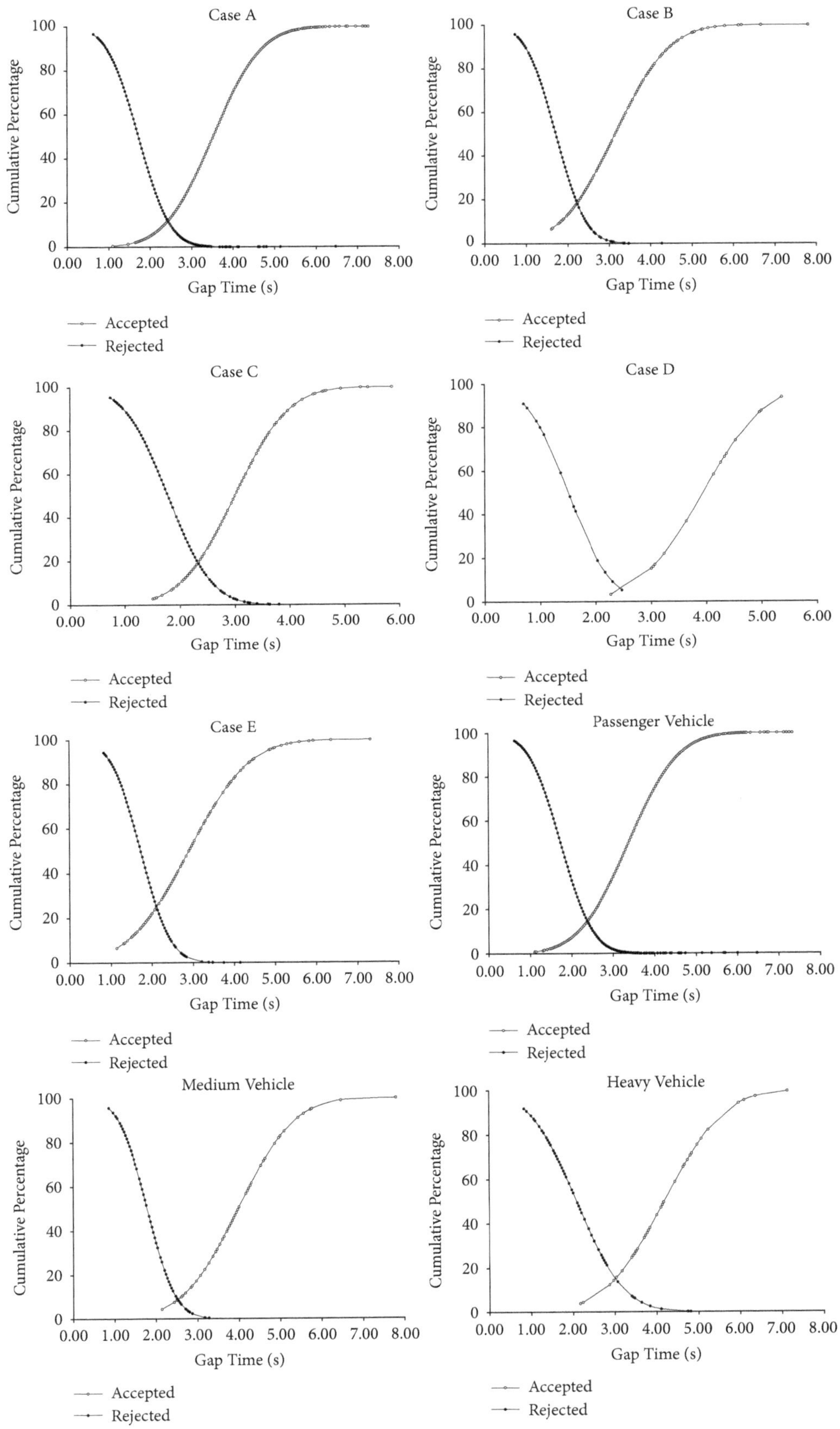

Figure 7: Critical gap for different cases.

Table 3: Summary of critical gap values.

Classification	Type	Critical gap (s)
Overall		2.40
Vehicle types	Passenger vehicle	2.39
	Medium vehicle	2.53
	Heavy vehicle	3.03
Interaction cases	A	2.45
	B	2.22
	C	2.33
	D	2.47
	E	2.11
	F	Not enough data
	G	Not enough data

in Qatar [16–18]. Such behavior may be caused by drivers becoming frustrated at heavily congested roundabouts and entering the roundabout when there is an insufficient gap.

The values obtained in this study can be used as a baseline for critical gap values of three-lane roundabouts in Qatar and other countries in the region, such as Saudi Arabia, United Arab Emirates, Bahrain, Kuwait, and Oman. These values can be used in numerous applications. For example, they can be input in capacity-prediction models and used in the simulation of traffic flow at multilane roundabouts to replicate real-life conditions. The method developed in this study can also be employed to estimate critical gap values for different types of multilane roundabouts in other regions.

The current study has a number of limitations. It did not consider the effect of pedestrians or cyclists on critical gap values because no pedestrians or cyclists were observed at the study locations, as expected throughout this region [19]. Furthermore, this study did not consider the effects of the geometry of the roundabout. Some roundabouts have an elliptical shape, whereas others have a steep slope, with circulating lanes, three legs, or slip ramps. In the current study, both roundabouts were circular, flat, and had four legs. Furthermore, the approaches to the roundabout did not feature slip ramps. In addition, the vehicular interaction cases in the study were based on allowable legal movements. Driving behaviors considered illegal, such as a vehicle entering the far circulating lane from the outside lane, were not included, as such behaviors, besides being careless, high-risk maneuvers, are in violation of traffic law and do not depict acceptable gap acceptance maneuvers.

Disclosure

The statements made herein are solely the responsibility of the authors.

Conflicts of Interest

The authors declare that there are no conflicts of interest regarding the publication of this paper.

Acknowledgments

This publication was made possible by the Qatar Research Fund (a member of Qatar Foundation) through an NPRP award [NPRP 4-1170-2-456].

References

[1] C: National Research and B: Transportation Research, *HCM 2010 : highway capacity manual*, Transportation Research Board, Washington, Wash, USA, 2010.

[2] R. J. Troutbeck, “Estimating the mean critical gap,” *Transportation Research Record*, vol. 2461, pp. 76–84, 2014.

[3] A. Flannery and T. Datta, “Operational performance measures of American roundabouts,” *Transportation Research Record*, vol. 1572, pp. 68–75, 1997.

[4] F. Xu and Z. Z. Tian, “Driver behavior and gap-acceptance characteristics at roundabouts in California,” *Transportation Research Record*, vol. 2071, no. 1, pp. 117–124, 2008.

[5] S. Mensah, S. Eshragh, and A. Faghri, “A critical gap analysis of modern roundabouts,” in *Proceedings of the 89th Annual Meeting of the Transportation Research Board*, Washington, Wash, USA, January 2010.

[6] C. Fitzpatrick, D. Abrams, Y. Tang, and M. Knodler, “Spatial and temporal analysis of driver gap acceptance behavior at modern roundabouts,” *Transportation Research Record*, no. 2388, pp. 14–20, 2013.

[7] R. Guo, “Estimating critical gap of roundabouts by different methods,” in *Proceedings of the 6th Advanced Forum on Transportation of China (AFTC 2010)*, pp. 84–89, Beijing, China, 2010.

[8] A. Kusuma and H. N. Koutsopoulos, “Critical gap analysis of dual lane roundabouts,” pp. 709–717.

[9] M. S. Raff, “A volume warrant for urban stop signs,” 1950.

[10] A. J. Miller, “A note on the analysis of gap—acceptance in traffic,” *Journal of the Royal Statistical Society*, vol. 23, no. 1, pp. 66–73, 1974.

[11] A. Miller, “Nine estimators of gap-acceptance parameters,” in *Bulletin Transport Section, Civil Engineering*, vol. 5, 1971.

[12] W. Brilon, R. Koenig, and R. J. Troutbeck, “Useful estimation procedures for critical gaps,” *Transportation Research Part A: Policy and Practice*, vol. 33, no. 3-4, pp. 161–186, 1999.

[13] S. Tupper, M. A. Knodler Jr, C. Fitzpatrick, and D. S. Hurwitz, “Estimating Critical Gap-A Comparison of Methodologies Using a Robust, Real-World Data Set,” in *Proceedings of the Transportation Research Board 92nd Annual Meeting*, 5019, 13 pages, 2013.

[14] R. Ashalatha and S. Chandra, “Critical gap through clearing behavior of drivers at unsignalised intersections,” *KSCE Journal of Civil Engineering*, vol. 15, no. 8, pp. 1427–1434, 2011.

[15] W. Brilon, R. Koenig, and R. J. Troutbeck, “Useful estimation procedures for critical gaps,” *Transportation Research Part A: Policy and Practice*, vol. 33, no. 3, pp. 161–186, 1999.

[16] K. Shaaban, J. S. Wood, and V. V. Gayah, “Investigating driver behavior at minor-street stop-controlled intersections in Qatar,” *Transportation Research Record*, vol. 2663, pp. 109–116, 2017.

[17] A. Tageldin, T. Sayed, and K. Shaaban, “Comparison of time-

proximity and evasive action conflict measures case studies from five cities," *Transportation Research Record*, vol. 2661, pp. 19–29, 2017.

[18] K. Shaaban and A. Pande, "Evaluation of red-light camera enforcement using traffic violations," *Journal of Traffic and Transportation Engineering (English Edition)*, vol. 5, no. 1, pp. 66–72, 2018.

[19] K. Shaaban, D. Muley, and A. Mohammed, "Analysis of illegal pedestrian crossing behavior on a major divided arterial road," *Transportation Research Part F: Traffic Psychology and Behaviour*, vol. 54, pp. 124–137, 2018.

2

A New Method Based on Field Strength for Road Infrastructure Risk Assessment

Yi Li and Yuren Chen

The Key Laboratory of Road and Traffic Engineering, Ministry of Education, College of Transportation Engineering, Tongji University, No. 4800 Cao'an Hwy., Shanghai 201804, China

Correspondence should be addressed to Yi Li; smilesusanly@163.com

Guest Editor: Xiaobo Qu

Because road infrastructures have significant impact on driving safety, their risk levels need to be evaluated dynamically according to drivers' perception. To achieve this, this paper proposes two field strength models to quantify the impact of road infrastructures on drivers. First, road infrastructures are classified into two types (continuous and discrete). Then, two field strength models for these types are proposed. Continuous field strength model describes the impact of long-belt-shape infrastructure by differential and integral methods. Discrete field strength model describes the static and dynamic characteristics of infrastructures. This model includes four parameters: mass of vehicles, mass of infrastructures, warning level, and kinetic energy of road infrastructures. The field strength is a relative concept, which changes with vehicle state. At the end of this paper, risk assessment principles are listed to clarify the nature of road infrastructure risk evaluation. A workflow of risk assessment and a case study are presented to illustrate the application of this novel method. The result of this study shows that ① the field strength is positively related to its risk level; ② the distribution of road infrastructure risks explains driver behaviour correctly; ③ drivers tend to keep driving in low-risk area. These findings help to explain the impact mechanism of road infrastructures on drivers, which can be applied in AI-based driving assistance system in the future.

1. Introduction

Road infrastructures are basic components of traffic environment. China has built more than 40,000km highway in the last five years. In such circumstance, a large number of roadside facilities need to be assessed. They are not only the structure of road alignment, but also the guidance for drivers. Past studies have proposed a lot of models and systems for road infrastructure assessment.

(1) Studies on the Relationship between Road Infrastructure and Driver's Visual Perception. Drivers do not capture infrastructure information through direct contact. More than 80% of such information is obtained from visual perception [1, 2]. Plenty of experiments have been done to verify it. It showed that different traffic environment would result in different driving intention and behaviour [3]. Road line ratio from driver's view was collected through field tests by Victor [4]. The results showed that it would decrease with the climb of driver's vision pressure. Wang et al. (2006) regarded the road infrastructure information as a series of stimulation. If the stimulation density from the infrastructure was in a proper range, then driver behaviour would be safe. On the other hand, some eye movement criteria (spot distribution, spot strength, etc.) were used to evaluate the visual burden caused by small radius and small angle of road alignment [5]. The results indicated that sharp radius would result in large visual burden for drivers. Different sweeping duration on road condition also showed some certain impact on lane keeping behaviour [6].

Other researchers focused on identifying such influence at intersections where the infrastructures were more various. Corresponding models were built to describe the relationship between real-world experimental data and driver's inner pressure [7]. Werneke and Vollrath [8] did field and simulated tests at intersections, and they concluded that poor planning of infrastructures near intersections would lead to severe vehicle collisions. Similar tests were conducted on mountain highway. The cross analysis between driver's vision pressure and psychological tension confirmed the importance of proper road shoulder width and access management [9–11]. Besides, drivers' speed adaption behaviour was proved to vary with different road infrastructures and traffic complexity conditions. Complex infrastructure information would significantly result in frequent speed adaption behaviour [12].

Based on such tests and analysis, it can be seen that scientific infrastructure arrangement and proper infrastructure information are important for driving safety [13, 14].

(1) Studies on Driver's Vision Pressure Field. As a part of "Human-Vehicle-Road" circle, road infrastructures will influence the whole driving environment. Therefore, the risk of road infrastructure should be assessed dynamically from the view of its user–driver. The risk assessment is only valid when the interaction between infrastructure and driver does exist. Based on such understanding, driver's vision pressure field theory is proposed as the extension of artificial potential field theory [15]. In his theory, there was a potential field in which objects would attract and exclude the moving objects, and this was the source of movement. This theory assumed that objects were pushed under virtual force. It has been widely applied in solving robot path planning problem. Similarly, the moving vehicles can be regarded as a kind of movement in a potential field composed of road infrastructures. Then, many pressure field models were built according to this theory: ① discrete pressure field model [16] explained driver's decision pattern; ② vehicle path planning model combined pressure field theory and elastic band theory to predict vehicle's path (Thomas and Thorsten 2008); ③ driver's workload theory helped to estimate the impact of other traffic on drivers [17]. With the development of road scanning technology, more infrastructure information and indexes are available for precise analysis [18]. Moreover, the pressure field was simulated to visualize its impact on vehicle movement [19]. These studies helped to quantify the impact of road infrastructures on drivers, and such impact was also known as driving safety field proposed by Wang et al. [20, 21]. In driving safety field, road infrastructures were the sources of static potential field that had an important role in collision warning algorithm. Based on this, a dynamic 3D virtual hazard potential field model was built to calculate the field energy of infrastructures for driving decision assistance [22, 23]. It can be seen that the whole road environment is a physical field, and the road infrastructures build the framework of it (Ni 2013).

Past studies mainly focus on four aspects: ① the relationship between pressure field and real road environment; ② what is the physical rule of this pressure field; ③ how to explain such field phenomenon from drivers' view; ④ how to apply this theory in road risk assessment. Therefore, driver's vision pressure field still needs a deep study on the interaction between road infrastructures and drivers. This paper will focus on its application on risk assessment of road infrastructure.

(3) Studies on Road Infrastructure Risk Assessment Methods. Road infrastructure assessment is a major topic in traffic safety. In India, planning and building safe road infrastructure have been proposed as a high priority [24].

To reduce the negative influence of road infrastructures, the project MARVin in Austria built a database which included accident data and road parameters (radius, gradient, etc.). The risk of road infrastructure was assessed through comparison between tested road and nonaccident road. Two roads shared similar parameters, but they had different speed limits or weather conditions. Then high-crash-risk road sections or infrastructures can be spotted [25]. A similar method was used in risk model proposed by Appleton [26]. By comparison with benchmark road in New Zealand, the researcher got personal risk and collective risk scores. The score was the product of risk level and feature extent. Zhang and Hu [27] built a modified Bayesian network model to assess the facility risks on freeways, in which traffic data or field survey data were applied to calibrate parameters of the model. The results showed that smaller model result resulted in higher facility risk, and the threshold was 0.5. Bayesian network was also applied in multirisk assessment of road infrastructure system, and fully probabilistic approach was adopted to integrate multirisk interactions at both hazard and fragility levels [28].

Road infrastructure assessment is also a part of macroscopic road management, which focuses on road-user risk caused by road infrastructure. CEDR-project developed a risk assessment framework to obtain the risk of pavement, structure, and drainage in network level for most of the European road administrations [29]. Specifically, systematic risk model described the relationship among road infrastructure, vehicle, and driver. Multidimensional probability distribution quantified cause-and-effect chain of the risk impact of infrastructure on vehicle [30]. Other studies took the concept of infrastructure risk as collision risk. No collision means no risk [31]. Besides, improper facility was also a main reason for unsafe behaviour, such as speeding [32], distraction [33], and run-off-road crashes [34, 35].

Recently, intelligent vehicles which can collect vehicle movement and roadside information have been widely developed. With such technology, a series of experiments were conducted in northern Virginia. The results indicated that wide shoulder had significant impact on driver behaviour change [36]. Severe accidents would increase with improper slopes, bends, or pavement surface conditions for HGV (Heavy Goods Vehicles) [37].

It can be seen that past studies on risk assessment of road infrastructure mainly focused on performance comparison and systematic assessment. They presented general comments, but they ignored the interaction between infrastructures and users (e.g., drivers) [38]. Recently, Scott-Parker et al. [39] have tested drivers stress in response to different infrastructures, but they did not explain its impact mechanism. To overcome this weakness, this paper regards drivers as assessors of road infrastructures. Their subjective visual pressure is the key to evaluate the risk levels of nearby facilities. Corresponding field strength models and risk assessment methods are discussed in detail in this paper. This helps to quantify the relationship between driver behaviour and road infrastructures, and it also presents a better understanding on the nature of road infrastructure risk assessment. The concept

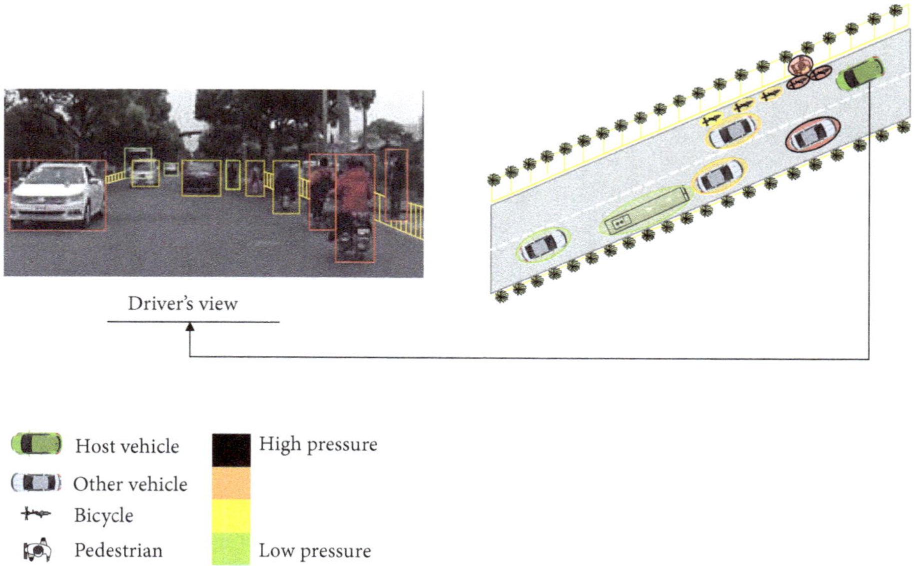

FIGURE 1: Conceptual image of pressure field from driver's view.

of pressure sources and field strength models may become the foundation of risk assessment system.

This paper is organized as follows. Section 1 introduces the achievements and weaknesses of past studies on driver's vision pressure field and road infrastructure risk assessment methods. Section 2 classifies some common road infrastructures based on pressure sources. Section 3 describes the field strength model of two road infrastructure types. Section 4 presents some assessment principles and a case study. Section 5 concludes the achievements and limits of this paper.

2. Classification of Road Infrastructure Based on Pressure Sources

In a complex traffic system, road infrastructure is a critical part. It does not have direct contact with vehicles, but it has influence on driver behaviour. In this paper, we define this mechanism as pressure field, which is not visible but measurable. Figure 1 shows a conceptual image of pressure field from driver's view.

The chance of direct collision between vehicle and road infrastructure is small; however driver behaviour continuously changes under the restrictions of nearby road infrastructures (including road geometry, road facilities, and other road users). This means that the pressure field is similar to gravity field, electric field, and magnetic field, which can force the vehicles to accelerate, decelerate, or change lane. If the risk level of road infrastructure is too high, it will form a stressful road environment for drivers, which may lead to unsafe driver behaviour. Therefore, a scientific risk assessment method for road infrastructure is necessary to understand and quantify such impact.

As the basis of road infrastructure assessment, we first need to classify some common road infrastructures. The classification is based on physical features of road infrastructures. Although other factors like weather, visibility, time of day, etc. may also have some impact on driver behaviour [40, 41], we eliminate these factors by experiments only on sunny days during 8:00 to 15:00 with good visibility in this research.

There are various shapes of infrastructures on road, but most of them are continuous type and discrete type. Continuous road infrastructures have continuous impact on the vehicles during the driving process. They are mostly long-belt-shape facilities standing or attaching to the road surface. Discrete road infrastructures are arranged separately on road. Although there are many kinds of discrete infrastructures, they can be mainly divided into two types: independent and centralized. Typical independent discrete infrastructures are traffic signs or signals. Centralized discrete infrastructures consist of a number of independent discrete infrastructures, such as disperse barriers. Tables 1 and 2 show some typical continuous and discrete road infrastructures, respectively.

The road infrastructures have two kinds of effect on vehicles: exclusion and attraction. The exclusion effect forces the vehicles to keep a certain distance from pressure sources. The attraction effect, however, leads the vehicles to move towards the pressure sources. Table 3 shows the pressure effect of some typical pressure sources.

Some elements in Table 3 are listed in both exclusion and attraction columns because they show different impacts on different vehicle types. For example, bus lane has attraction effect on buses in rush hour. However, it shows exclusive effect on other cars. Besides, in off-peak hours it becomes an ordinary lane, and it has no restriction at all. Meanwhile, some elements only have attraction effect on some vehicles in some specific circumstances. For example, emergency parking lane has no effect on normal vehicles, but it has great attraction effect on out-of-control vehicles.

Based on the above classification and the discussion, we will build two models to quantify the impact of these two road infrastructure types on drivers.

TABLE 1: Typical continuous road infrastructure.

Name	Figure	Name	Figure	Illustration
Central Separation Belt	 	Double amber lines		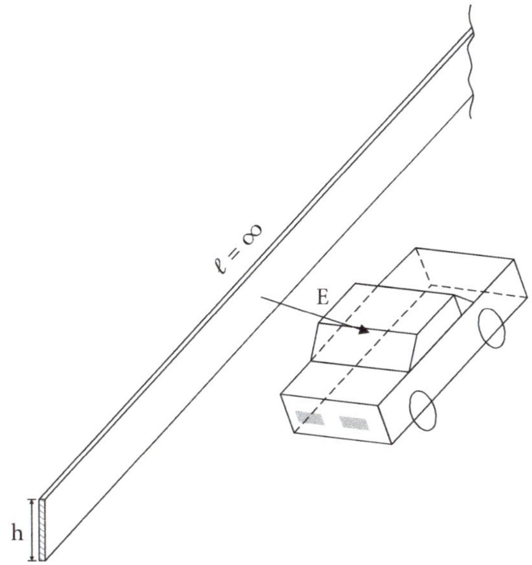
		W-Beam Barrier		
Continuous barrier	 	Single lane marking		

Table 2: Typical discrete road infrastructure.

Name	Figure	Name	Figure	Illustration
Disperse barrier		Prohibitory sign		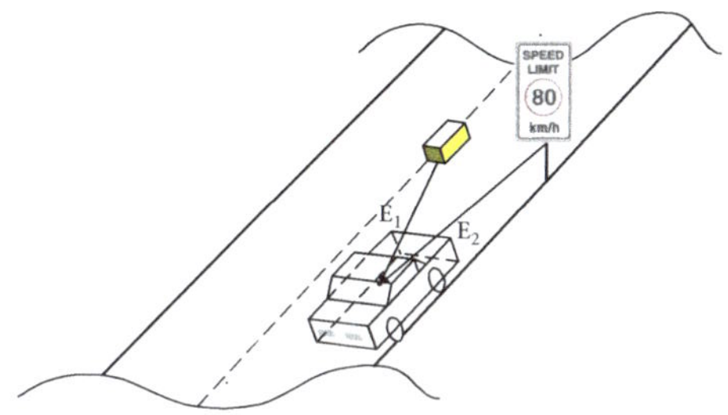
Accommodation lane sign		Warning sign		
Traffic signal		Directional sign		
Nearby vehicle/bicycle/pedestrian			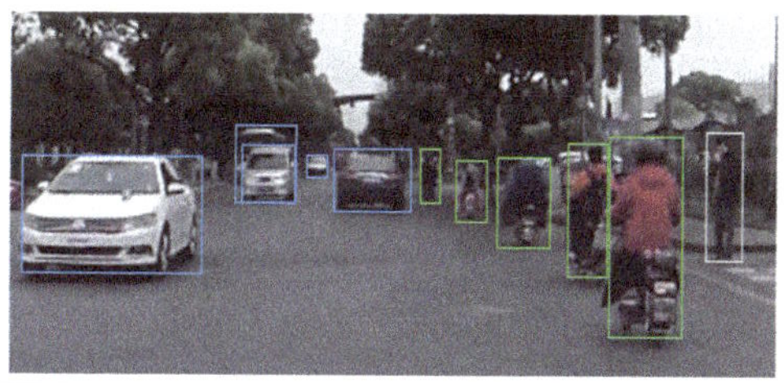	

TABLE 3: The pressure effect of typical pressure sources.

Pressure sources	Pressure effect	
	Exclusion	Attraction
Lane marking	Prohibitory marking Warning marking	Amber dash marking[b] Lane dash marking[b]
Road sign	Prohibitory sign Warning sign Accommodation lane sign[a] Intersection guide line[a] Traffic signal (red/yellow)	Directional sign Intersection waiting area[b] Accommodation lane sign[a] Intersection guide line[a] Traffic signal (green)
Barrier	Guardrail Crash barrier Working zone	Escape lane[b] Emergency parking lane[b]

Note. a. The element shows different pressure effect under different circumstances; b. the element only is shown as a certain kind of pressure effect in some circumstances.

3. Field Strength Model of Different Road Infrastructure Types

3.1. Continuous Road Infrastructure Field Strength Model. To simplify the calculation, the host vehicles is represented by a rectangle object ($a \times b \times c$). The origin of the coordinate system is the geometrical center. The positive Y-axis points to the direction of vehicle movement. Figure 2 shows the simplified vehicle model and coordinate system.

Because a continuous road infrastructure always has a relative long dimension compared to other dimensions (see Figure 2), the differential analysis can be used to select the microsource unit of such item (ds: microunit, width=dz, length=∞). Figure 3 shows the impact of field strength of microsource unit on vehicle in z plane.

Then, the corresponding field strength of this microsource unit can be calculated as follows.

$$dE = \frac{\vartheta}{\alpha \cdot l} dz = \frac{\tau}{\alpha \cdot \pi \cdot \left\| \overrightarrow{x} + \overrightarrow{y} + \overrightarrow{z} \right\|_2} dz \quad (1)$$

where dE is the field strength of microsource unit ds; dz is the width of microsource unit ds; α is the reliability parameter of continuous field source, which consists of four levels: 1–rigid, 2–semirigid, 3–flexible, and 4–traversable.

l is the boundary length of field range. When the infrastructure is in the center of road, the field range is centrosymmetric distributed around pressure sources. When it locates at roadside, the range covers a 1/4 circle area around pressure sources.

τ is the characteristic parameter of continuous field source, which shows the protection intensity of field source. The value of this parameter depends on the "Specification for Design of Highway Safety Facilities" (JTG D81-2017) in China. To cover different levels of guardrail and lane marking, τ is divided into eight levels, shown in Table 4.

According to Figures 3 and 4, the microfield source is infinitely small along z-axis, so the vehicle is parallel to the microfield source along y-axis. If the infrastructure locates

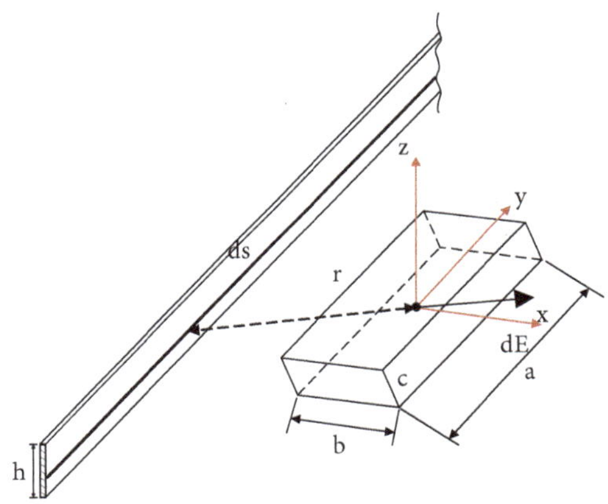

FIGURE 2: Simplified vehicle and coordinate system.

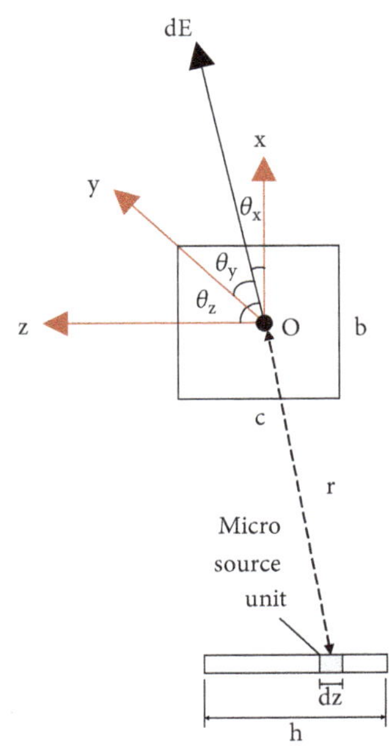

dE: the field strength of micro source unit *ds*

FIGURE 3: Field strength of microsource unit.

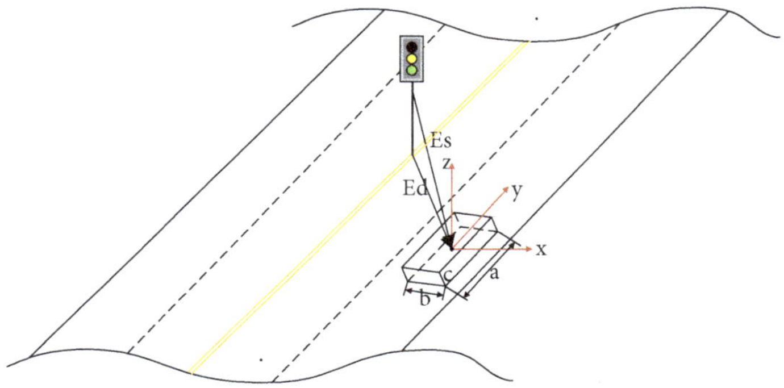

FIGURE 4: Static field strength (Es) and dynamic field strength (Ed) of independent pressure source.

TABLE 4: The level of characteristic parameter of continuous field source.

Protection intensity	τ	Protection intensity	τ
Dash lane marking	1	A/A$_m$-class protection	5
White lane marking	2	SB/SB$_m$-class protection	6
Amber lane marking	3	SA/SA$_m$-class protection	7
B-class protection	4	SS-class protection	8

in the center of road, the field strength along x-axis can be calculated as follows.

$$dE_x = \frac{\tau \cdot dz}{\alpha \cdot l} \cdot \cos\theta_x = \frac{\tau \cdot x/\sqrt{x^2+z^2}}{\alpha \cdot \pi \cdot \sqrt{(|x| - b/2)^2 + (|z| - b/2 \cdot z/x)^2}} dz = \frac{\tau}{\alpha\pi} \cdot \frac{|x| - b/2}{(|x| - b/2)^2 + (|z| - b/2 \cdot z/x)^2} dz \quad (2)$$

Then the whole field strength function can be obtained as follows.

$$E_x = \int_{-c/2}^{h-c/2} \frac{\tau}{\alpha\pi} \cdot \frac{|x| - b/2}{(|x| - b/2)^2 + (|z| - b/2 \cdot z/x)^2} dz \quad (3)$$

This function shows that the field strength along x-axis will change with the distance between the vehicle and infrastructure.

3.2. Discrete Road Infrastructure Field Strength Model. Compared with continuous infrastructures, discrete infrastructures not only appear occasionally but also have instruction and guidance information. The field strength of a discrete infrastructure consists of two parts: static field strength and dynamic field strength (see Figures 4 and 5).

The overall field strength can be obtained as follows.

$$\overrightarrow{E_{dis}} = \overrightarrow{E_s} + \overrightarrow{E_d} \quad (4)$$

where $\overrightarrow{E_{dis}}$ is the overall field strength of a discrete infrastructure pressure source; $\overrightarrow{E_s}$ is the static field strength of a discrete infrastructure pressure source; $\overrightarrow{E_d}$ is the dynamic field strength of a discrete infrastructure pressure source.

The concept of static field strength is similar to the field strength of continuous sources. It reflects the inherent characteristics of an independent pressure source. The distribution of its field intensity is within the circle (in the middle of road) or semicircle (on the side of road) around the pressure sources. Figure 6 shows the decomposition of static field strength.

In Figure 6, the mass center is the original point of field strength. Its coordinate is as follows.

$$\begin{aligned} \overline{x} &= \frac{\iiint xu(x,y,z)\,dv}{m} = \frac{\iiint x\,dv}{V} \\ \overline{y} &= \frac{\iiint yu(x,y,z)\,dv}{m} = \frac{\iiint y\,dv}{V} \\ \overline{z} &= \frac{\iiint zu(x,y,z)\,dv}{m} = \frac{\iiint z\,dv}{V} \end{aligned} \quad (5)$$

where $(\overline{x}, \overline{y}, \overline{z})$ is the coordinate of mass center; V is the volume of field source; m is the mass of field source.

The distance (r_1) between the field source and vehicle is as follows.

$$r_1 = \sqrt{\left(\sqrt{\overline{x}^2 + \overline{y}^2} + \frac{b}{2\cos\theta_x}\right)^2 + \left(|\overline{z}| + \frac{b\cot\theta_z}{2\cos\theta_x}\right)^2} \quad (6)$$

Then the static field strength can be obtained:

$$E_s = \begin{cases} \dfrac{\beta \cdot m_{\text{source}}}{r_1^2}, & \text{independent pressure source} \\ \displaystyle\sum_{i=1}^{n} \frac{\beta_i \cdot m_i}{r_1^2}, & \text{centralized pressure source,} \end{cases} \quad (7)$$

where m_{souce} is the inertial characteristics of field source, which can be quantified by mass; β is the warning level of field source, which is divided into three levels; see Table 5.

Compared with static field strength, dynamic field strength shows guidance effect on traffic flow, so it covers only a specific direction of traffic flow. In reference to point-charge field theory in physics, discrete pressure source and

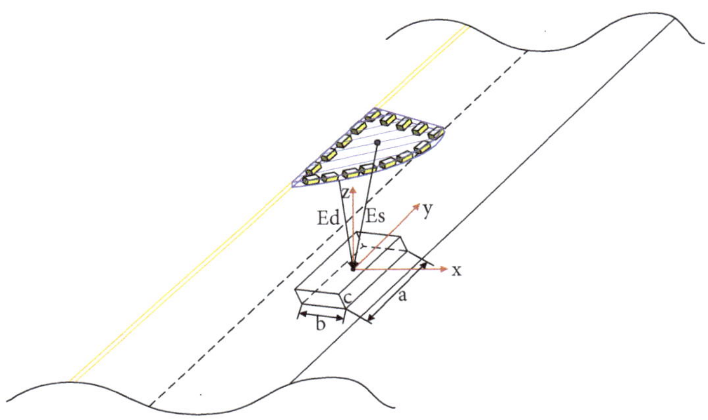

FIGURE 5: Static field strength (Es) and dynamic field strength (Ed) of centralized pressure source.

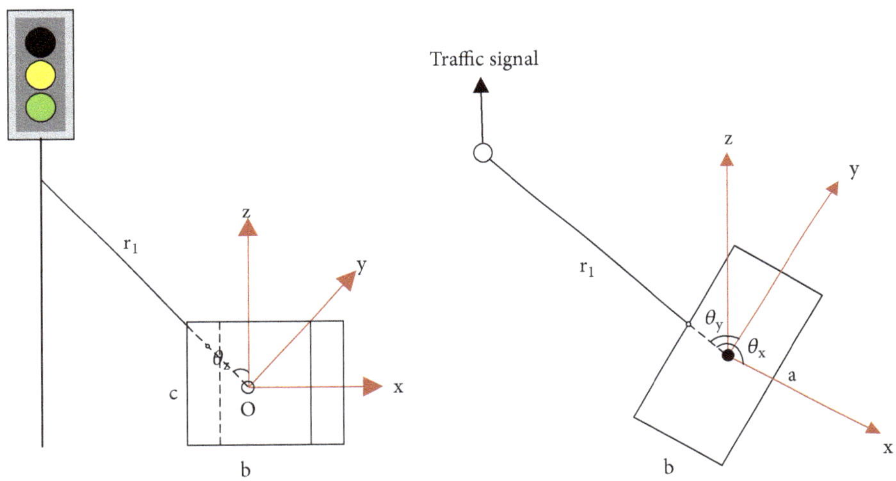

(a) (b)

FIGURE 6: Static field strength decomposition. ((a) longitudinal, (b) overlook).

TABLE 5: Warning level of field source.

Level	Infrastructure
1	Warning sign/Accommodation lane sign/Directional sign
2	Prohibitory sign/Traffic signal
3	Disperse barrier

vehicle can be regarded as a couple of energetic bodies, so the dynamic field strength is as follows.

$$E_d = \begin{cases} \dfrac{\beta \cdot \zeta}{{r_2}^2}, & \text{independent pressure source} \\ \dfrac{\beta_{cen} \cdot \zeta}{{r_2}^2}, & \text{centralized pressure source} \end{cases} \quad (8)$$

where r_2 is the vertical distance from vehicle to infrastructure, $r_2 = \sqrt{x^2 + y^2}$; β_{cen} is the warning level of the group field source; ζ is the warning kinetic energy of field source, which is the required kinetic energy for host vehicle. For example, in car-following situation $\zeta = (1/2)m_{front\ car}v^2_{front\ car}$; in speed limit situation $\zeta = (1/2)m_{\text{host}}v^2_{limit}$.

4. Risk Assessment Principles and Case Study

In this section, the above models are utilized to assess the risk of road infrastructures with a case study.

First, some risk assessment principles need to be declared. (In this paper, risk source is equal to pressure source.)

(1) Variability of Risk Source Type under Different Scenes. In Section 2, we classify the road infrastructures into two types. In the case of multisource combination, the types of some field sources change with their directional text and time limit. This leads to an opposite risk assessment result (see the example of bus lane in Section 2). Therefore, before assessing the road infrastructure risk, we should check the pressure source type and state first.

TABLE 6: Indexes of pressure sources.

No	Infrastructure	Pressure effect	Indexes	Illustration
1	Continuous barrier	Exclusion	τ=5 α=2 h=2.5m	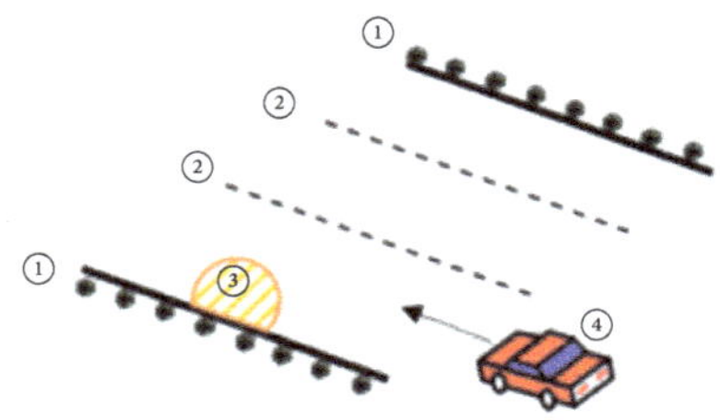
2	Lane marking	Attraction	Dash line	
3	Discrete barrier	Exclusion	m_{source}=225kg h=0.7m	
4	Host vehicle	/	m_{host} = 1500kg b=1.8m c=2m v_{host}=40km/h	
*	Lane width	/	3.5m	

(2) Traffic-Rule-Based Pressure Source Priority. Traffic rules and guidance clarify the right-of-way and drivers' behaviour standard. Similarly, we divide the pressure source priority into four levels. Among these priority levels, driver's safety and vehicle collision prevention are the most important. The specific priority levels are described as follows.

Priority 1: Fixed road infrastructure in driver's view.

Priority 2: Prohibitory sign/line or signal.

Priority 3: Warning sign/line or signal.

Priority 4: Traversable line, directional sign/line or signal.

In the process of pressure combination, the priority of each pressure source needs to be evaluated. First, combine the field strength of Priority 1. Then, add the combination results of lower priorities on the previous result. This forms a priority-based level system, which is easy for intelligent detection and analysis in the future.

(3) The Relativity of Road Infrastructure Risk. There are various risk sources (pressure sources) in road environment and different drivers would perceive different risk levels in multisource field. Generally, such characteristic can be explained in two aspects.

① When facing with the same pressure source, drivers will perceive different priority according to their vehicle types. For example, some road infrastructures only restrict the passage of truck, so they have high priority for truck drivers and low priority for vehicle drivers.

② When the same driver drives the same vehicle along the road several times, the risk level and combination of road infrastructures vary with vehicle position and vehicle state.

Therefore, the road infrastructure risk is a changeable and relative factor. It needs to be analyzed and calculated specifically in each scenario.

Next, according to the principles discussed above, we put forward an example (a section on Cao'an Hwy., Shanghai, China) with continuous and discrete road infrastructures to show the workflow of risk assessment based on the models proposed above. The indexes of pressure sources are listed in Table 6.

To explain the risk assessment clearly, we draw the assessment workflow with the example data in Figure 7. Each view and vehicle state of current moment during driving process is a series of data input for the assessment.

(i) Firstly, each infrastructure in driver's view is classified according to its physical features (continuous or discrete), and its pressure indexes are also collected with vehicle state data. (In this paper, we assume that all indexes of road infrastructures are digitalized and stored in a database, which can be collected intelligently on road.)

(ii) Secondly, based on the vehicle and pressure source data, we can grade the priority level (e.g., P1 and P4) of each infrastructure.

(iii) Thirdly, the field strength of each priority is obtained based on the above results of classification and individual field strength calculation (P1: 11.32~175.68, P4: -14.14~-4.46). Moreover, the overall field strength of the whole road infrastructure environment in driver's view can be obtained (E: -2.15~171.22).

(iv) Then, the risk of each infrastructure needs to be assessed according to a risk level guide which can be customized to meet road authorities' needs. In this particular example, four risk levels are provided in Table 7. If the risk level of infrastructure is below the minimum threshold, then the infrastructure is regarded as a safe one. If the field strength is within the other three intervals, corresponding risk level can be calibrated.

(v) Finally, we can get the risk assessment result of current view of road infrastructure.

In this example, the discrete barrier on roadside shows high risk for drivers on the nearest lane. Continuous barriers are safe for driving, and the dash lane markings have attractive effect on vehicle movement.

To show the impact of risk assessment results on driver behaviour, corresponding vehicle trajectories (pink dots and white arrow) are drawn on the field strength diagram. Figure 8 shows the test data of 20 drivers and average vehicle trajectory in this situation. It can be seen that all drivers chose to change lane in order to keep away from the high field strength area. They always kept their trajectories in low/no risk area. However, these trajectories were not the same because each driver had his/her own vision perception and risk assessment result. If some drivers got high pressure

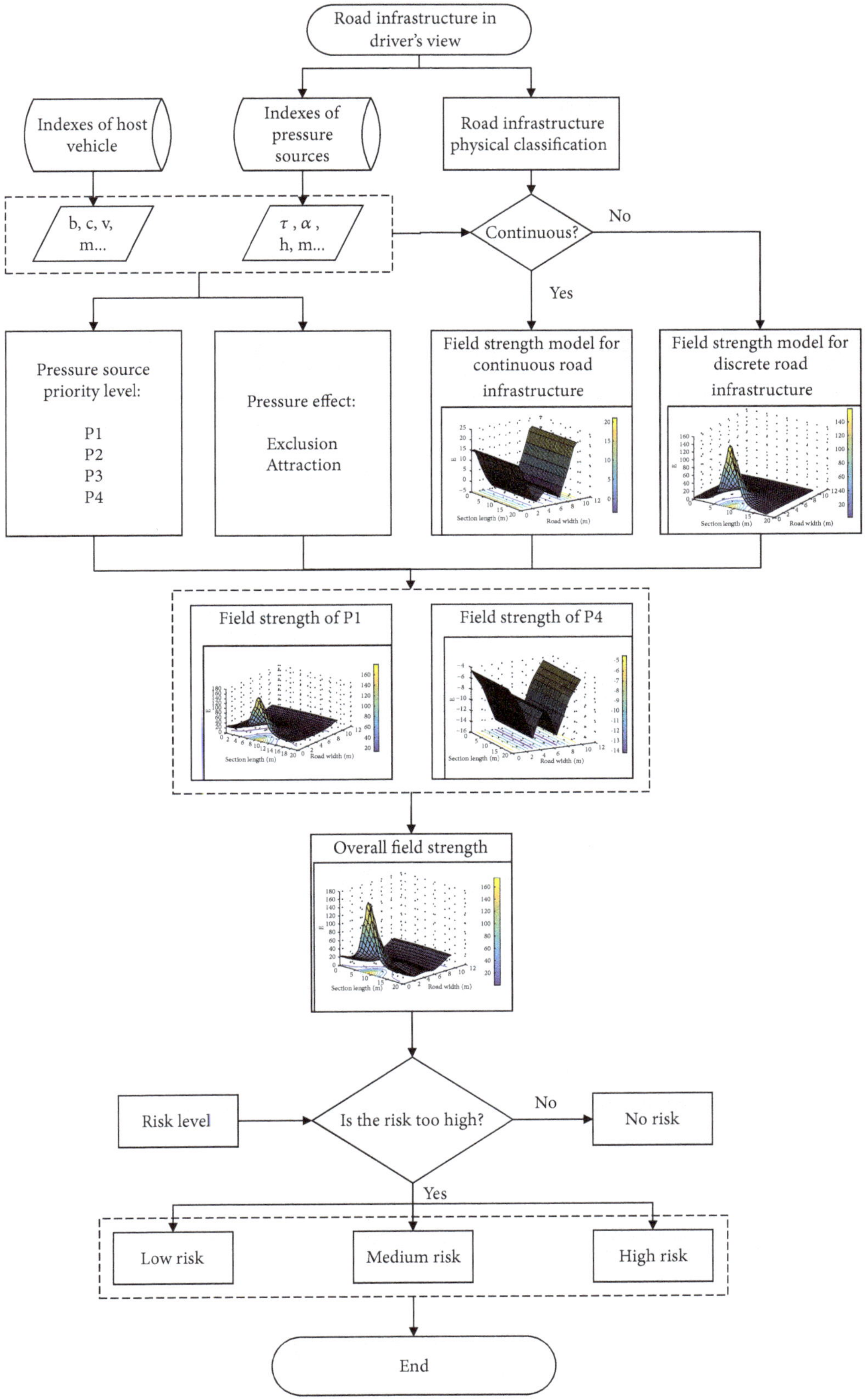

FIGURE 7: Workflow of road infrastructure risk assessment with test data.

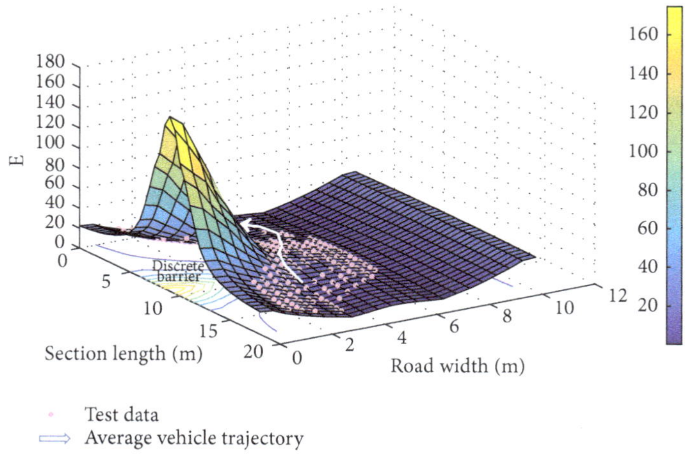

FIGURE 8: Vehicle trajectories in 3D field strength diagram.

TABLE 7: Risk levels of road infrastructure (case study).

Risk level	Field strength	Explanation
No risk	E<60	The infrastructure is good enough for safe driving.
Low risk	60⩽E<100	The infrastructure is safe for most drivers, but it is risky for a small part of drivers in harsh weather.
Medium risk	100⩽E<140	The infrastructure is quite safe under normal condition, but it is risky for a large part of drivers in harsh weather.
High risk	E⩾140	The infrastructure is risky for most drivers, and it may lead to severe traffic problem.

earlier than others, then their psychology index (such as heart rate) would force them to react more quickly [42]. Although drivers' reactions are different, the average behaviour shows that ① the field strength model can quantify the road infrastructure risk correctly; ② driver behaviour will change with the risk levels of nearby road infrastructures, so it verifies the necessary of this assessment study and the importance of road infrastructure planning.

5. Conclusion

This paper presents a new method and novel field strength models for road infrastructure risk assessment. Past risk assessment methods are mainly based on the physical indexes of road infrastructures. Driver's perception and its impact mechanism are neglected in such methods, so they are static and inflexible. The risk assessment method we present here takes driver's subjective visual perception of road infrastructures as an important factor. This driver-vision-based method helps to quantify such process, which overcomes the weakness of traditional methods. In other words, our method provides a dynamic and specific way to measure the risk impact of road infrastructures on driver behaviour. The main contributions of this research are as follows. The classification method of typical road infrastructures is provided. All infrastructures are components of road environment, which can be quantified by field strength. Because different kinds of road infrastructures have different pressure impact on drivers, we build two field strength models. Continuous field strength model describes the persistent impact on driving behaviour. Discrete field strength model shows the static and dynamic impact of the infrastructures on drivers. Based on the above models and analysis, three risk assessment principles show the nature of risk source evaluation. Finally, the workflow of risk assessment is presented with a case study, and corresponding risk levels are listed and explained.

However, there are still some limits in this research. Firstly, we assume that all characteristics of road infrastructures have been intelligently collected. This may be possible for some advanced autopilot vehicles or connected vehicles, but it is hard for traditional vehicles or platforms. Secondly, as each frame of recorded videos from drivers' view needs to be analyzed, the amount of calculation is huge. Therefore, high-speed calculation, transmission, and distributed procession will be thresholds for online analysis. Thirdly, we only discuss the overall risk level of road infrastructures for all drivers. Nevertheless, this risk assessment method is based on the field strength captured from driver's view, so different drivers may have different driving preference and decisions during the trip, which may lead to some deviation among drivers. Fourthly, we mainly divide common road infrastructures into continuous type and discrete type, but there are still various infrastructures and features to be analyzed for precise assessment in the future. Besides, the influence of bad weather and poor visibility is not considered in the pressure calculation in this research, which needs to be analyzed with more test data under various weather conditions. In the following studies, a more elaborate and dynamic risk evaluation system is necessary for personalized service.

Conflicts of Interest

The authors declare that they have no conflicts of interest regarding the publication of this paper.

Acknowledgments

This work was supported by Zhejiang Provincial Communication Department [Grant no. 2015J07].

References

[1] S. Sivaraman and M. M. Trivedi, "Looking at vehicles on the road: a survey of vision-based vehicle detection, tracking, and behavior analysis," *IEEE Transactions on Intelligent Transportation Systems*, vol. 14, no. 4, pp. 1773–1795, 2013.

[2] B. Yu and Y. Chen, "Driving Comfort Evaluation of Urban Road from Driver's Visual Perception," in *Proceedings of the 15th COTA International Conference of Transportation Professionals*, pp. 2570–2579, Beijing, China.

[3] T. Deng, K. Yang, Y. Li, and H. Yan, "Where Does the Driver Look? Top-Down-Based Saliency Detection in a Traffic Driving Environment," *IEEE Transactions on Intelligent Transportation Systems*, vol. 17, no. 7, pp. 2051–2062, 2016.

[4] T. W. Victor, J. L. Harbluk, and J. A. Engström, "Sensitivity of eye-movement measures to in-vehicle task difficulty," *Transportation Research Part F: Traffic Psychology and Behaviour*, vol. 8, no. 2, pp. 167–190, 2005.

[5] X. Pan, Q. Fang, and H. Jiang, "Driving Visual Demand-based Safety Evaluation of Mountainous Highway Horizontal Curve," *Journal of Tongji University. Natural Science*, vol. 38, no. 12, pp. 1763–1766, 2010.

[6] W. J. Horrey and C. D. Wickens, "Focal and Ambient Visual Contributions and Driver Visual Scanning in Lane Keeping and Hazard Detection," *Proceedings of the Human Factors and Ergonomics Society Annual Meeting*, vol. 48, no. 19, pp. 2325–2329, 2016.

[7] Z. H. Khattak, M. D. Fontaine, and R. A. Boateng, "Evaluating the impact of adaptive signal control technology on driver stress and behavior using real-world experimental data," *Transportation Research Part F: Traffic Psychology and Behaviour*, vol. 58, pp. 133–144, 2018.

[8] J. Werneke and M. Vollrath, "What does the driver look at? the influence of intersection characteristics on attention allocation and driving behavior," *Accident Analysis & Prevention*, vol. 45, pp. 610–619, 2012.

[9] Y. Yun-xing, C. Fang, and Z. Yong-fu, "Research on Visual Behavior Characteristics of Drivers in Downhill Sections of the Mountainous Expressway," *Highway*, vol. 1, pp. 132–137, 2018.

[10] L. Zhong, X. Zhao, H. Ding et al., "Experimental Research on Safety Impact of Inside Shoulder Width on Drivers without Speed Limitation," *Journal of Highway & Transportation Research Development*, 2015.

[11] K. Dixon and L. Brown, "Assessing how drivers of through vehicles react to driveway activity," *Transportation Research Record*, no. 2404, pp. 77–84, 2014.

[12] O. Oviedo-Trespalacios, M. M. Haque, M. King, and S. Washington, "Effects of road infrastructure and traffic complexity in speed adaptation behaviour of distracted drivers," *Accident Analysis & Prevention*, vol. 101, pp. 67–77, 2017.

[13] M. S. Young, J. M. Mahfoud, N. A. Stanton, P. M. Salmon, D. P. Jenkins, and G. H. Walker, "Conflicts of interest: the implications of roadside advertising for driver attention," *Transportation Research Part F: Traffic Psychology and Behaviour*, vol. 12, no. 5, pp. 381–388, 2009.

[14] S. Bendak and K. Al-Saleh, "The role of roadside advertising signs in distracting drivers," *International Journal of Industrial Ergonomics*, vol. 40, no. 3, pp. 233–236, 2010.

[15] O. Khatib, J. Warren, V. De Sapio, and L. Sentis, "Human-Like Motion from Physiologically-Based Potential Field," in *Building the Information Society*, vol. 156 of *IFIP International Federation for Information Processing*, pp. 747-748, Springer US, Boston, MA, 2004.

[16] Y. Sha and Q. Shi, "Pedestrian Simulation Model based on Discrete Potential Field," *Highway Engineering*, vol. 34, no. 2, pp. 153–156, 2009.

[17] E. Teh, S. Jamson, and O. Carsten, "Mind the gap: Drivers underestimate the impact of the behaviour of other traffic on their workload," *Applied Ergonomics*, vol. 67, pp. 125–132, 2018.

[18] T. Sattel, T. Hesse, and C. Sondermann-Wölke, "Experimental study of evasion maneuvers with a potential field-based assistance function for collision avoidance," *VDI-Berichte*, no. 2009, pp. 475–493, 2008.

[19] H. Wang and T. Wu, "New vector field microcosmic model for traffic flow," *China Journal of Highway and Transport*, vol. 16, pp. 99–102, 2003 (Chinese).

[20] J. Wang, J. Wu, X. Zheng, D. Ni, and K. Li, "Driving safety field theory modeling and its application in pre-collision warning system," *Transportation Research Part C: Emerging Technologies*, vol. 72, pp. 306–324, 2016.

[21] J. Wang, J. Wu, and Y. Li, "Concept, Principle and Modeling of Driving Risk Field Based on Driver-vehicle-road Interaction," *China Journal of Highway and Transport*, vol. 29, pp. 105–114, 2016.

[22] J. Ji, H. Peng, W. Chi et al., "Decision Making for Vehiel Collision Avoidance Based on Dangerous Potential Field," *SAE-China Congress & Exhibition*, vol. 2, 2015.

[23] J. Ji, P. Ji, and H. Peng, "Design of 3D Virtual Dangerous Potential Field for Vehicle Active Collision Avoidance," *Automotive Engineering Magazine*, vol. 38, no. 9, pp. 1065–1071, 2016.

[24] R. V. Ponnaluri and Y. D. Santhi, "Road crash history and risk groups in India: Need for new initiatives and safety policies," *Transportation Research Record*, no. 2114, pp. 64–71, 2009.

[25] S. Peter, P. Maurer, and R. Stütz, *MARVIN–Model for Assessing Risks of Road Infrastructure*, Arsenal Research, Vienna, Austria, 2007.

[26] I. Appleton, "Road Infrastructure Safety Assessment," in *Proceedings of the 4th IRTAD Conference*, Seoul, Korea, 2009.

[27] D. Zhang and X. Hu, "Risk Assessment of Traffic Facility on Freeway Based on a Modified Bayesian Network Model," *Journal of Transport Information and Safety*, vol. 34, no. 5, pp. 102–107, 2016.

[28] P. Gehl, *Bayesian Networks for The Multi-Risk Assessment of Road Infrastructure*, UCL (University College London), 2017.

[29] A. Weninger-Vycudil, B. Brozek, R. Spielhofer, C. Britton, and M. Oldfield, "Integration of cross asset risk assessment into road infrastructure asset management," in *Proceedings of the IABSE Conference, Geneva 2015: Structural Engineering: Providing Solutions to Global Challenges*, pp. 1010–1016, Switzerland, September 2015.

[30] J. Bald, K. Stumpf, and T. Wallrabenstein, "Systematic risk analysis for safety assessments of road systems," *Journal of the Acoustical Society of America*, vol. 123, no. 5, pp. 3466-3466, 2008.

[31] S. SUGIURA, Y. KANAMORI, A. TAKAGI, F. KURAUCHI, and H. MORIMOTO, "Development of integrated asset management technique of road infrastructure based on risk evaluation," *Journal of Japan Society of Civil Engineers, Ser. F4 (Construction and Management)*, vol. 67, no. 4, pp. I_103–I_112, 2011.

[32] M. S. Nemmang, R. Rahman, M. M. Rohani et al., "Analysis of Speeding Behaviour During Approaching the U-Turn Facility Road Segment Based On Driving Simulation Test," *Matec Web of Conferences*, vol. 103, p. 08008, 2017.

[33] G. K. Kountouriotis and N. Merat, "Leading to distraction: Driver distraction, lead car, and road environment," *Accident Analysis & Prevention*, vol. 89, pp. 22–30, 2016.

[34] S. Matena, A. Hegewald, W. Louwerse et al., "Reduction of head-on collisions and run-off-the-road-accident. Report D 3.2 of the RiPCORD-iSEREST project (Road Infrastructure Safety Protection - Core-Research and Development for Road Safety in Europe; Increasing safety and reliability of secondary road)," 2009.

[35] M. Tziotis, V. Pyta, N. Mabbott et al., "Road safety engineering risk assessment part 10: rural run-off-road crashes," *Ran Off Road Crashes*, 2010.

[36] J. Schorr, S. H. Hamdar, and C. Silverstein, "Measuring the safety impact of road infrastructure systems on driver behavior: Vehicle instrumentation and real world driving experiment," *Journal of Intelligent Transportation Systems: Technology, Planning, and Operations*, vol. 21, no. 5, pp. 364–374, 2017.

[37] M. Gothié, V. Cerezo, and F. Conche, "Relationship between road infrastructure characteristics and HGV accidents," in *Proceedings of the International Conference on Heavy Vehicles, HVTT10: 10th International Symposium on Heavy Vehicle Transportation Technologies*, pp. 319–331, 2008.

[38] R. Jobanputra and M. Vanderschuren, "Micro-simulation modelling of the impact of infrastructure provision and vehicle and pedestrian behaviour on road crash risk," in *Proceedings of the 3rd International Conference on Road Safety and Simulation*, 2011.

[39] B. Scott-Parker, C. Jones, and J. Tucker, "Driver stress in response to infrastructure and other road users: simulator research informing an innovative approach to improving road safety," in *Proceedings of the Australasian Road Safety Conference 2016*, Canberra, ACT, Australia, 2016.

[40] S. H. Hamdar, L. Qin, and A. Talebpour, "Weather and road geometry impact on longitudinal driving behavior: Exploratory analysis using an empirically supported acceleration modeling framework," *Transportation Research Part C: Emerging Technologies*, vol. 67, pp. 193–213, 2016.

[41] K. McCann and M. D. Fontaine, "Assessing driver speed choice in fog with the use of visibility data from road weather information systems," *Transportation Research Record*, vol. 2551, pp. 90–99, 2016.

[42] Yi Li, Yuren Chen, and Fan Wang, "The Impact of Traffic Environmental Vision Pressure on Driver Behaviour," *Journal of Advanced Transportation*, vol. 2018, Article ID 4941605, 12 pages, 2018.

3

Effects of Safety Facilities on Driver Distance Perception in Expressway Tunnels

Leyu Wei, Jinliang Xu, Xingli Jia, Xiaodong Zhang, and Haoru Li

School of Highway, Chang'an University, Middle Section of South 2 Ring Rd., Beilin District, Xi'an, Shaanxi 710064, China

Correspondence should be addressed to Jinliang Xu; xujinliang@chd.edu.cn

Academic Editor: Andrea D'Ariano

We investigated the effects of four safety facilities in expressway tunnels—information boards, flashing lights, human-voice broadcasts, and siren broadcasts—on driver distance perception by questionnaire surveys and field experiments. Results from a survey questionnaire given to 436 drivers indicated that each of the facilities, except the human-voice broadcast, was perceived to increase the driving safety. Consistently, results from field experiments involving 150 participants in China's Xingshuliang Tunnel indicated that information boards, flashing lights, and siren broadcasts increased the distance perception accuracy of drivers, while human-voice broadcasts decreased this accuracy. The results of human-voice broadcasts may be due to the fact that drivers could not catch and understand the information they heard from human-voice broadcasts while driving in tunnels. This research can assist engineers in identifying the effective safety facilities in tunnels and provide a basis for prioritizing the implementation of these facilities, ultimately increasing driver distance perception accuracy and decreasing rear-end collisions.

1. Introduction

Expressway tunnels play an enormous role in transportation systems to make travelling convenient and fast; however, the constantly increasing tunnel accidents become a big problem for the safety of expressway operation. It is generally accepted that the frequency of rear-end collisions is the highest among the tunnel accidents that lead to large fatalities, casualties, and economic damage. In Norway's one-way, two-lane expressway tunnels, Amundsen and Ranes [1] found that approximately 62% of the traffic accidents involved rear-end collisions. Similarly, in China's expressway tunnels, Ma et al. [2] found that almost 57.5% of the 134 traffic accidents involved rear-end collisions. The Handbook of Tunnel Fire Safety [3] and Singapore accident statistics [4] also confirmed that rear-end collisions were the most commonly occurring accidents in expressway tunnels.

Several factors contribute to the occurrence of a rear-end collision in an expressway tunnel, and human perception is the key factor. Driving in a dark and enclosed tunnel can be monotonous and may lead to physical fatigue and cognitive distraction [5]. In addition, many drivers tend to travel more closely to the vehicle ahead of them, perceiving a longer distance between vehicles in tunnels [6]. When an emergency arises, those travelling too close to the lead vehicle may not have time to react appropriately, resulting in a rear-end collision. To effectively reduce rear-end collisions in expressway tunnels, it is important to learn more about the role of human perception and, in particular, driver distance perception.

Various facilities or devices including flashing lights, information boards, cable broadcasts, pavement markings, speed bumps, and delineators have been used to improve driver's perceive accuracy in tunnel environments [7]. In China, detailed specifications effectively guide engineers in the design of conventional expressway facilities and devices [8, 9]; however, such specifications are lacking for expressway tunnels. Instead, Chinese engineers must rely upon their prior project experience and published research when designing expressway tunnel facilities and devices. An enhanced understanding of driver distance perception and the associated effects of various safety facilities in tunnels would provide a basis for engineering design.

Previous studies investigating driver distance perception commonly use survey questionnaire or interview methods to reflect a driver's subjective experience. Considering a driver's sensory, cognitive, physical, and functional affordance, Ronchi and Nilsson [10] surveyed 62 drivers regarding the design and function of information signs and flashing lights of tunnels and recommended that any sign text should be displayed in an amber color to better capture a driver's attention. They also assessed the effect of acoustic systems in road tunnels and suggested that human-voice broadcasts should be avoided because a driver may not hear or distinguish messages in a noisy tunnel environment. However, their study of acoustic systems was mainly based on theoretical analysis and lacked experiment evidence. Further surveys are needed to confirm the effects of human-voice broadcast and siren broadcasts. Dudek and Ullman [11] compared the effect of flashing messages, flashing lines, and alternating lines on a variable message sign based on the reading time, response time, and level of comprehension reported by 260 Texas drivers. They found that the average reading time was significantly higher for flashing messages. Other studies [12, 13] have reported that a driver's reading comprehension can be increased with clear semantic sign messages and proper training.

As an alternative to survey questionnaires or interviews, select studies have investigated the effects of safety facilities using field experiments conducted in real tunnels or driving simulators. During the experiments, specialized devices were used to measure dynamic driver data such as heart rate, eye movement, myoelectricity, skin temperature, and respiration depth. In addition, driving behaviors such as driving speed, acceleration, distance to the nearest wall, and overtaking rate can be measured [14]. For example, Manser and Hancock [15] used a driving simulator to determine that wall pattern design significantly affects a driver's speed control and perception.

Some debate exists regarding the ability of driving simulators to accurately replicate driving experience and associated driver behavior in real tunnels. Select studies found that driver performance was similar in virtual and real environments [16–18], while other studies reported a significant difference in driver behavior in these two environments [19, 20]. For example, driving speed was higher and vehicle distance to the nearest tunnel wall was much lower in simulated tunnels as compared with real tunnels. This phenomenon may be attributable to a driver's lower stress level in a driving simulator or their feelings of dizziness and nausea when using the virtual visual equipment [21]. This ongoing debate regarding the accuracy of driving simulators suggests that field experiments conducted in real tunnels are more robust.

Previous research has shown that safety facilities in tunnels significantly affect a driver's speed perception and concentration [22, 23], but few studies have considered the relationship between driver distance perception and these facilities. According to statistics, rear-end collision accidents in tunnels are mainly caused by passenger cars and trucks, accounting for 42.18% and 56.25% of tunnel accidents, respectively [24]. It is important for both truck drivers and passenger car drivers to estimate distance right and maintain a safety distance from vehicles in front of them. In addition, information boards, flashing lights, human-voice broadcasts, and siren broadcasts are the most common safety facilities in China's expressway tunnels. Information boards and flashing lights primarily affect a driver's visual perception, and cable broadcasts using a human voice or siren affect a driver's audio perception. The research on the effects of these four facilities is very important and necessary for road traffic safety. In this study, we investigated the effects of four types of safety facilities of expressway tunnels—information boards, flashing lights, human-voice broadcasts, and siren broadcasts—on distance perception of passenger car drivers and truck drivers. To support this study's objective, a survey questionnaire was administered to 436 drivers at two expressway service areas. Complementing this qualitative data, field experiments involving 150 participants were conducted in the Xingshuliang Tunnel along the Yanxi Expressway in the Shaanxi Province of China. The results of this study can assist engineers in identifying the most effective safety facilities in tunnels and provide a basis for prioritizing their implementation, ultimately increasing driver distance perception accuracy and decreasing rear-end collisions.

2. Experimental Method

2.1. Questionnaire Survey. A questionnaire survey was conducted in Yaozhou and Huangling service areas along the Yanxi Expressway. Vehicle drivers and occupants possessing a valid driver's license were asked to independently complete questionnaires. When completing the questionnaire, drivers were first asked to provide basic demographic information (e.g., gender and age) as well as driving experience. Next, drivers were asked to assess the effects of the four safety facilities on their driving behavior using three indicators (e.g., increasing their attention, decelerating, or maintaining space between vehicles). Using a five-point Likert scale, a rank of "1" indicated a very negative effect on driving behavior, while a rank of "5" indicated a very positive effect on driving behavior. Higher rankings suggest a more effective facility [21]. The original Chinese version of the questionnaire was published on the website https://www.wjx.cn/jq/26197255.aspx, and the translated English version of the questionnaire could be found on the website https://www.wjx.cn/jq/26193643.aspx.

The minimum sample size required in this study was determined as follows:

$$N \geq \left(\frac{SK}{E}\right)^2 \tag{1}$$

where N is the sample size; S is the sample standard deviation; K is the z-score for a specified probability; and E is the allowable error. Assuming that S=0.5, K=1.96 (for a 95% confidence level), and E=5%, a minimum of 384 questionnaires was required.

In 2017, the statistics by Chinese ministry of public security showed that female drivers account for only 25% of the car drivers. Besides, few of the truck drivers are females because trucking is an intensive work that most of the female drivers could not handle. Considering the small proportion of the female drivers in China, fewer female drivers were investigated in both questionnaire surveys and field experiments.

TABLE 1: Installation setups in 5 experimental conditions.

Condition	Information Board	Flashing Light	Human-voice Broadcast	Siren Broadcast
1	×	×	×	×
2	√	×	×	×
3	×	√	×	×
4	×	×	√	×
5	×	×	×	√

Note: "√" means "activated"; "×" means "deactivated".

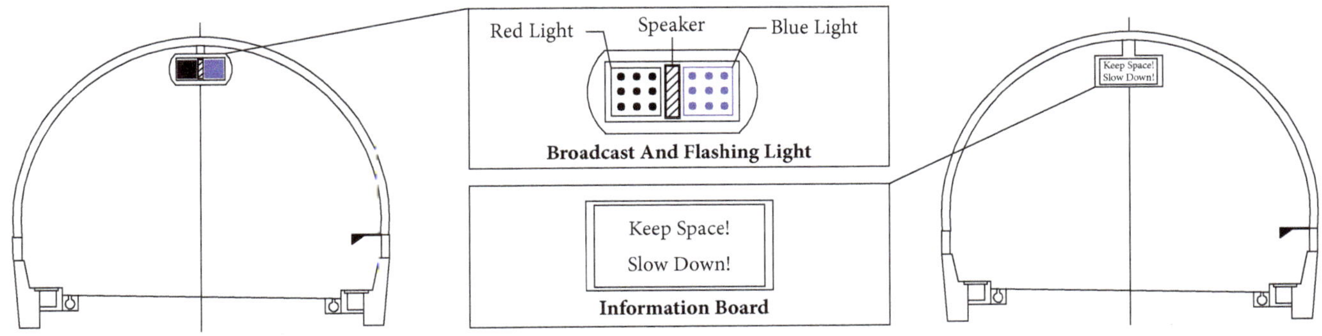

FIGURE 1: Safety facilities in the Xingshuliang Tunnel.

2.2. Field Experiment. To complement the qualitative survey questionnaire analysis, field experiments were conducted in the Xingshuliang Tunnel along the Yanxi Expressway in Shaanxi Province of China. The Xingshuliang Tunnel, which is located between the Yaozhou and Huangling service areas, is a one-tube three-lane tunnel that is 1100 m long. Owing to the low traffic volume of the experimental road section, the traffic flow during field experiments could be considered as the free flow.

Drivers stopping at either the Yaozhou or Huangling service areas were invited to participate in the field experiments. Car drivers and truck drivers were invited to participate. Prior to participating, drivers were informed that they may be asked several simple questions regarding expressway tunnel facilities, but the specific study purpose was not disclosed.

During the experiments, the Xingshuliang Tunnel Monitoring Center, under our guidance, applied five different experimental conditions in the tunnel. As shown in Table 1, no devices were activated in condition 1, and only one type of facility was activated in the other four conditions. A driver would only participate in one experiment of the 5 conditions at a time. To ensure that the type of safety facilities is the only variable in this experiment, it is necessary to keep the characteristics of participates in each group homogeneous. Therefore, participates were divided into different groups according to their gender and age in order to ensure the demographic composition of each group similar. All of the field experiments were conducted during low traffic time period at the low volume road section to minimize the influence of road traffic flows on participants in the real environment.

Figure 1 shows the safety facilities in the tunnel. The information boards were located every 1 km and displayed "Keep Space! Slow Down!" when activated. The flashing lights and cable broadcasts were jointly mounted every 300 m. The red/blue flashing lights had a flash frequency of 180 times/min when activated. The cable broadcast system of the tunnel offered a human-voice or siren model. When the human-voice model was activated, a female voice broadcasted "Please slow down! Please keep space!"

Figure 2 shows the experimental setup used to determine driver distance perception accuracy. Two sets of flags (one yellow and two blue) were positioned outside and inside the tunnel. The distance between the yellow and nearest blue flag was 100 m and that between the blue flags was 60 m, representing a safe distance between vehicles at 60 km/h. To reduce the confounding effects of speed on a driver's visual perception during the experiments, each driver was required to travel at 60 km/h [15, 22]. When approaching the yellow flag, drivers were asked to estimate the distance between the two blue flags.

3. Data Analysis Method

3.1. Questionnaire Survey. To determine whether drivers subjectively perceived a safety facility to have a positive or negative effect on driver distance perception, survey questionnaire responses were analyzed using a one-sample t-test and nonparametric Wilcoxon sign test, with the neutral rank of 3 as the baseline. The t-tests would be applied when normality assumptions were satisfied otherwise the Wilcoxon sign test would be used. Rejection of the null hypothesis confirmed that devices ranked <3 and >3 negatively and positively affected driver behavior, respectively.

To better understand the effects of demographics and driving experience on the subjective device rankings, a

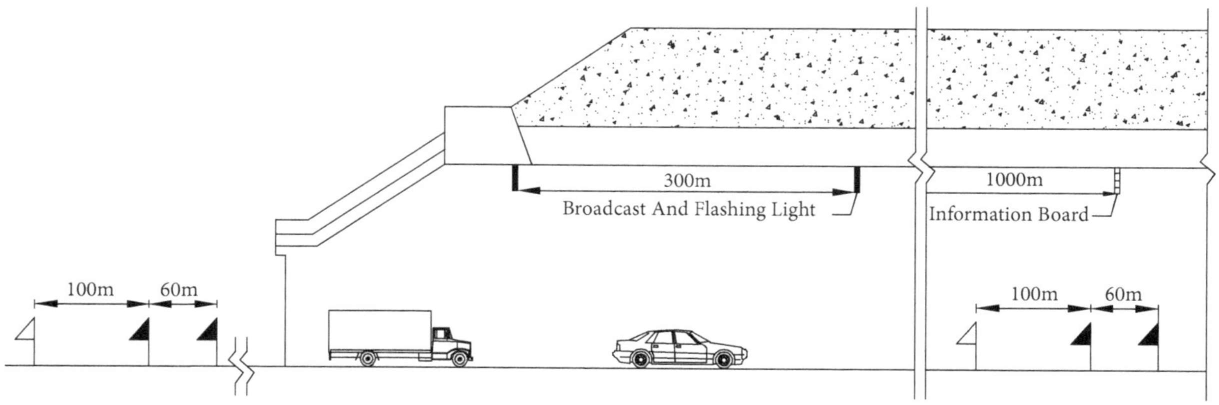

FIGURE 2: Driver distance perception measurement in the Xingshuliang Tunnel.

series of regressions were performed using an ordered logit regression model appropriate for the five-point Likert scale based data. For ordinal responses, ordered probit or logit models yield consistent and efficient estimates [25, 26]. In an ordered logit model, the probability of a response having a Likert rank j is determined as follows [27]:

$$\Pr(y_i > j) = \frac{\exp(X_i\beta' - \tau_i)}{1 + \exp(X_i\beta' - \tau_i)} \quad j = 1, \ldots, 5 \qquad (2)$$

where X_i is a $(k\times 1)$ vector of observed nonrandom explanatory variables, β is a $(k\times 1)$ vector of unknown parameters to be estimated, and τ_i is a $(k\times 1)$ vector of unknown cut points to be estimated. The model parameters (β) and cut points (τi) are estimated using maximum likelihood methods [27]. When estimating an ordinal regression, a test of the parallel lines assumption was needed to verify that the slope coefficients did not vary over different alternatives [28]. If this assumption was violated, a generalized or partially constrained generalized ordered logit model would have been estimated instead [29, 30].

3.2. Field Experiment. In this study, the difference between the actual and estimated distance reflects a driver's distance perception accuracy. This error can be defined as follows:

$$e = |S_p - S_r|, \quad (S_p > 0,\ S_r > 0) \qquad (3)$$

where S_p is the perceived distance (m) and S_r is the real distance (60 m).

Then, a driver's distance perception accuracy can be defined as follows:

$$\alpha = \frac{e}{S_r} = \frac{|S_p - S_r|}{S_r} = \left|\frac{S_p}{S_r} - 1\right|, \quad (S_p > 0,\ S_r > 0) \qquad (4)$$

Driver distance perception accuracy is optimal when α=0. To introduce the effects of safety facilities on driver distance perception, individual accuracy values can be averaged for each set of drivers exposed to a particular safety device as follows:

$$\overline{\alpha} = \frac{1}{n}\sum_{i=1}^{n} \alpha(\mathrm{i}) \qquad (5)$$

where i is the number of drivers exposed to one of the five experimental conditions.

To account for the differences in visibility and noise levels, we compared field measurements from outside and inside the tunnel and determined efficiency values for individual safety devices as follows:

$$\beta = \frac{1}{n}\sum_{i=1}^{n} \left(\frac{S_{P1}(i)}{S_{p2}(i)} \cdot \frac{S_{r1}(i)}{S_{r2}(i)} \right) \qquad (6)$$

where S_{p1}(i) and S_{p2}(i) are the perceived distances for driver i and S_{r1}(i) and S_{r2}(i) are the real distances (60 m) for driver i inside and outside the tunnel, respectively. A higher β value suggests higher safety device performance.

4. Results and Discussion

4.1. Questionnaire Survey. In total, 461 drivers at the Yaozhou and Huangling service areas were recruited to fill the questionnaire. However, some respondents did not answer all questions on the paper, and these incomplete questionnaires were excluded from the final sample. Therefore, only a total of 436 valid responses were considered for the data analysis (> 384).

Of the 436 respondents, 83.03% were males and 16.97% were females. Most respondents were aged between 26 and 35 yr (35.09%) or 36 and 45 yr (30.27%). Respondents aged between 18 and 25, 46 and 55, and 56 and 65 yr accounted for 10.55%, 15.83%, and 8.26% of the sample, respectively. With respect to driving experience, most respondents reported driving for >6 yr (51.33%). Respondents who reported driving for <1, 1–3, and 4–6 yr accounted for 6.67%, 28%, and 14% of the sample, respectively.

Uniquely, cable broadcasts in tunnels are deactivated in most of the time and only activated for special situations or emergencies like severe weather events, traffic accidents, and

TABLE 2: Subjective effects of safety facilities on driver behavior.

Safety facility	Driver behavior	Very negative 1	Slightly negative 2	Neutral 3	Slightly positive 4	Very positive 5	Mean	Standard Deviation
Information boards	Increase attention	4.36	5.96	8.72	56.19	24.77	3.91	0.98
	Decelerate	4.13	8.49	9.63	24.54	53.21	4.14	1.15
	Keep space	10.32	7.11	11.47	44.04	27.06	3.70	1.23
Flashing lights	Increase attention	2.06	1.61	22.71	35.09	38.53	4.06	0.93
	Decelerate	2.98	4.13	18.81	37.39	36.70	4.01	0.99
	Keep space	6.42	13.07	25.00	22.48	33.03	3.63	1.24
Human-voice broadcasts*	Increase attention	0.00	7.57	5.28	46.33	40.83	4.20	0.85
	Decelerate	0.00	5.28	4.36	36.70	53.67	4.39	0.80
	Keep space	28.21	26.61	13.53	20.64	11.01	2.60	1.37
Siren broadcasts*	Increase attention	2.52	19.27	1.83	26.38	50.00	4.02	1.23
	Decelerate	3.67	3.21	10.32	34.86	47.94	4.20	1.00
	Keep space	15.60	22.71	11.01	34.17	16.51	3.13	1.35

*Note: 42.2% of drivers had never heard cable broadcasts and based responses on expectation not experiences. Analyses excluding these 42.2% of participants showed a similar pattern to those reported in Table 2.

road maintenance. As such, 42.2% of those drivers had never heard cable broadcasts in tunnels and filled the questionnaire based on their expectation instead of experience. Statistics tests indicated that the results of respondents who have experienced the cable broadcasts were similar to that of all respondents.

Table 2 summarizes the subjective effects of the four safety facilities on driver behavior in tunnels based on the five-point Likert scale. It is shown that all of the questionnaire responses followed abnormal distribution, so the Wilcoxon sign tests were applied to all cases. All of the Wilcoxon sign tests rejected the null hypothesis that the mean ranks were equal to the neutral rank of 3 at the 99% confidence level.

According to the survey, drivers considered that all of the four facilities had a positive effect (>3) on their willingness to increase attention and decelerate their vehicle. It was also believed that each of the facilities was useful (>3) for drivers to maintain the distance, except the human-voice broadcasts (<3). Additional Kruskal-Wallis tests were applied to specifically compare the subjective perception of the safety facilities. The asymptotic significance of "increase attention" function, "decelerate" function, and "keep space" function were all 0 (<0.05). It can be inferred that people generally considered that the most helpful safety facility for increasing attention (4.20) and slowing down (4.39) was the human-voice broadcast, and the most effective safety facility for keeping a safety distance (4.14) between vehicles was the information board.

For information boards, the mean value of decelerating was the highest in all functions. Respondents reflected that the messages on the information boards could usually remind them of the speed limit and encourage them to apply their brakes to avoid a fine. Human-voice and siren broadcasts in tunnels were reportedly most effective for encouraging drivers to decelerate (4.39 and 4.20, respectively) and increase their attention (4.20 and 4.02, respectively). Most notably, human-voice broadcasts had a negative effect (<3) on encouraging drivers to maintain adequate spacing between vehicles. When cable broadcasts were activated, most drivers realized that a serious event has occurred, and they would slow down and increase their attention accordingly. At the same time, they were anxious to exit the tunnel and followed the lead vehicle more closely. Flashing lights, which create a strong visual stimulation for drivers in dark tunnels, were most effective for encouraging drivers to increase their attention (4.06) and decelerate (4.01). These findings suggest that flashing lights can effectively evoke immediate reactions that do not require complex thought (e.g., applying brakes or increasing attention) but are less effective for complex reactions such as estimating and responding to distance changes between vehicles.

Regarding any confounding effects on the subjective facility rankings, results of the ordered logit model indicated no age or driving experience effects for any of the reported driving behaviors. Gender, however, affected several of the reported driving behaviors. Table 3 summarizes these model results. Compared to female respondents, male respondents were more likely to increase attention but less likely to decelerate when exposed to information boards. Comparatively, female respondents were more likely to maintain adequate distance when exposed to flashing lights, while males were unaffected. Cable broadcasts had a positive and negative effect on distance maintenance for males and females, respectively. These collective results suggested that male respondents were more confident in their driving skills, more rational when facing complex situations, and less

Table 3: Ordered logit model results regarding gender effects on driver behavior.

Driver behavior	Information boards		Flashing lights		Human-voice broadcasts		Siren broadcasts	
	Coefficient	*p*	Coefficient	*p*	Coefficient	*p*	Coefficient	*p*
Increase attention	-0.275	0.009	-0.413	<0.001	-0.403	<0.001	-0.423	<0.001
Decelerate	-0.297	0.005	0.228	0.040	-0.421	<0.001	-0.452	<0.001
Keep space	-0.427	<0.001	0.236	0.032	0.317	0.005	-0.210	0.047

Table 4: Distance perception accuracy (α) and device efficiency (β) statistics.

Driver group	*N*	α		β	
		Mean	Standard deviation	Mean	Standard deviation
1 Control Group	30	0.16	0.0903	0.63	0.1322
2 Information Board Group	30	0.09	0.0775	0.78	0.1911
3 Flashing Light Group	30	0.12	0.0955	0.76	0.1710
4 Human-voice Broadcast Group	30	0.25	0.1025	0.52	0.1640
5 Siren Broadcast Group	30	0.13	0.0947	0.73	0.1516

Table 5: ANOVA of distance perception accuracy among driver groups.

	Sum of Squares	df	Mean Square	F	Sig.
Between Groups	.478	4	.120	13.983	.000
Within Groups	1.240	145	.009	—	—
Total	1.718	149	—	—	—

affected by safety facilities. Female drivers were more cautious and less confident in their driving skills, less calm under pressure, and more affected by safety facilities.

4.2. Field Experiment. Results of the field experiments were based on the behaviors of 150 drivers (90 car and 60 truck drivers). Thirty drivers (18 car and 12 truck drivers) were exposed to one of the five experimental conditions. The sample was predominantly male, comprising only 16 female passenger cars (17.7%) and no female truck drivers. As mentioned before, the number of female drivers is much less than male drivers in China. In addition, only few of the females work for truck transportation since truck driving is a labor-intensive work for them. The few female drivers we encountered at the expressway service areas often rejected the request to participate in experiments owing to their security concerns. The age of participants ranged from 18 to 59 yr (36.64±12.78 yr), and the average driving experience was 7.38±3.81 yr.

Table 4 summarizes the mean and standard deviation values for a driver's distance perception accuracy (α) and for an individual safety facility's efficiency (β) when exposed to individual safety facilities. The mean accuracies for drivers exposed to information boards, flashing lights, and siren broadcasts (groups 2, 3, and 5, respectively) were lower (α=0 is optimal) than the mean accuracy for drivers exposed to no devices (group 1). It can assume that information boards, flashing lights, and siren broadcasts were effective for improving driver distance perception in tunnels. To verify this assumption, further comparisons of distance perception accuracy between groups were needed. Besides, human-voice broadcasts decreased distance perception accuracy by almost 56.25%. During the field experiments, several drivers complained that they could not hear the message of human-voice broadcast clearly, and a few drivers even opened car windows to hear it.

To further confirm the effects of safety facilities on driver distance perception, ANOVA (the Analysis of Variance) was used to compare driver groups 1–5. The homogeneity test of variances proved that ANOVA was appropriate and effective ($p=0.52>0.05$). The results of ANOVA are summarized in Table 5. There existed significant difference of distance perception accuracy between these groups ($p<0.05$). To further study the significant level of safety facilities, LSD (least-significant difference) were made to calculate multiple comparisons and the results are shown in Table 6.

By comparison between group 1 and groups 2-5, a driver's distance perception accuracy was significantly different when exposed to information boards and human-voice broadcasts. According to the Letter Marking Method, the distance perception accuracy of information boards, flashing lights, and siren broadcasts wasclassified as the first level, the distance perception accuracy without devices was classified as the second level, and the distance perception accuracy of human-voice broadcast was classified as the third level. The order of effectiveness of tunnel safety facilities was consistent with results in Table 4.

Generally, information boards provide clear and simple message to guide drivers, and they do not create tense atmosphere as broadcasts do. Compared with information boards, siren broadcast is usually loud and harsh to catch people's

TABLE 6: Multiple comparisons of distance perception accuracy among driver groups.

Group I	Group J	Mean Difference (I-J)	Std. Error	Sig.	95% Confidence Interval	
					Lower Bound	Upper Bound
1*	2	.069444445	.023877258	.004	.02225201	.11663688
	3	.040555556	.023877258	.092	-.00663688	.08774799
	4	-.094444444	.023877258	.000	-.14163688	-.04725201
	5	.031666667	.023877258	.187	-.01552577	.07885910
2*	1	-.069444445	.023877258	.004	-.11663688	-.02225201
	3	-.028888889	.023877258	.228	-.07608132	.01830355
	4	-.163888889	.023877258	.000	-.21108132	-.11669645
	5	-.037777778	.023877258	.116	-.08497021	.00941466
3*	1	-.040555556	.023877258	.092	-.08774799	.00663688
	2	.028888889	.023877258	.228	-.01830355	.07608132
	4	-.135000000	.023877258	.000	-.18219244	-.08780756
	5	-.008888889	.023877258	.710	-.05608132	.03830355
4*	1	.094444444	.023877258	.000	.04725201	.14163688
	2	.163888889	.023877258	.000	.11669645	.21108132
	3	.135000000	.023877258	.000	.08780756	.18219244
	5	.126111111	.023877258	.000	.07891868	.17330355
5*	1	-.031666667	.023877258	.187	-.07885910	.01552577
	2	.037777778	.023877258	.116	-.00941466	.08497021
	3	.008888889	.023877258	.710	-.03830355	.05608132
	4	-.126111111	.023877258	.000	-.17330355	-.07891868

*Note: 1: control group, 2: information board group, 3: flashing light group, 4: human-voice broadcast group, and 5: siren broadcast group.

attention, but it does not transmit explicit instructions for drivers. Human-voice broadcasts contain specific directions; however, it is hard to clearly catch the words and follow them. Therefore, the information board performs better in improving distance perception ability, and cable broadcast in tunnels is not always helpful for drivers to drive safely.

The noise inside tunnels is normally 10~20 dB larger than outside tunnels, and the reverberation time is probably 5~10 s [31]. For long tunnels, when play cable broadcast, tunnel walls will strengthen the noise level and reverb effects more than usual [32]. To solve this problem, the decibel level and arrangement distance of broadcast should be further studied to decrease echo and reverberation effect in tunnels. Recently, advanced technology of wireless broadcast applying in expressway tunnels has been developed to achieve the goal that people could listen to specific radio in vehicles [33, 34]. It seems that wireless tunnel radio in vehicles would totally solve the problem of echo and reverberation and make it possible to transit more information without clear sound concern. Some other problems will appear, such as that drivers can only hear wireless broadcast instructions only if their car radios are turned on.

An individual safety device's efficiency (β), which accounted for differences in visibility and noise levels outside and inside the tunnel, varied based on the driver. Figure 3 shows the distribution of β values across the sample. For drivers exposed to information boards, flashing lights, siren broadcasts, no facilities, and human-voice broadcasts, the range of β values was 0.450–1.256, 0.356–1.167, 0.356–1.000, 0.381–0.857, and 0.296–0.857, respectively. Compared with outside the tunnel, driver distance perception inside the tunnel improved when exposed to information boards, flashing lights, and siren broadcasts. Conversely, driver distance perception declined when exposed to human-voice broadcasts inside the tunnel as compared with outside the tunnel. Nearly 13% of the β values for information boards and flashing lights were larger than 1, indicating that driver distance perception was better inside rather than outside the tunnel.

The results of field experiments indicated that the human-voice broadcast did little contribution on distance perception, and the results of questionnaire survey also showed that people generally believed that the human-voice broadcast against them to keep space from vehicles ahead.

One limitation should be noted regarding this study's findings. Driving speed has been shown to significantly affect a driver's perceived distance; higher vehicle speeds result in lower perceived distances [17, 23, 24]. As such, drivers participating in this study's field experiments were required to drive at a speed of 60 km/h. Because it was difficult for drivers to maintain a constant speed of 60 km/h, the perceived distances obtained in this study may include small errors.

5. Conclusion

Information boards, flashing lights, and cable broadcasts are the most common safety facilities used in expressway tunnels. Despite their prevalence, few studies have confirmed the effects of these facilities on driver distance perception and associated rear-end collision rates in tunnels. In this study, we investigated the effects of four safety facilities—information

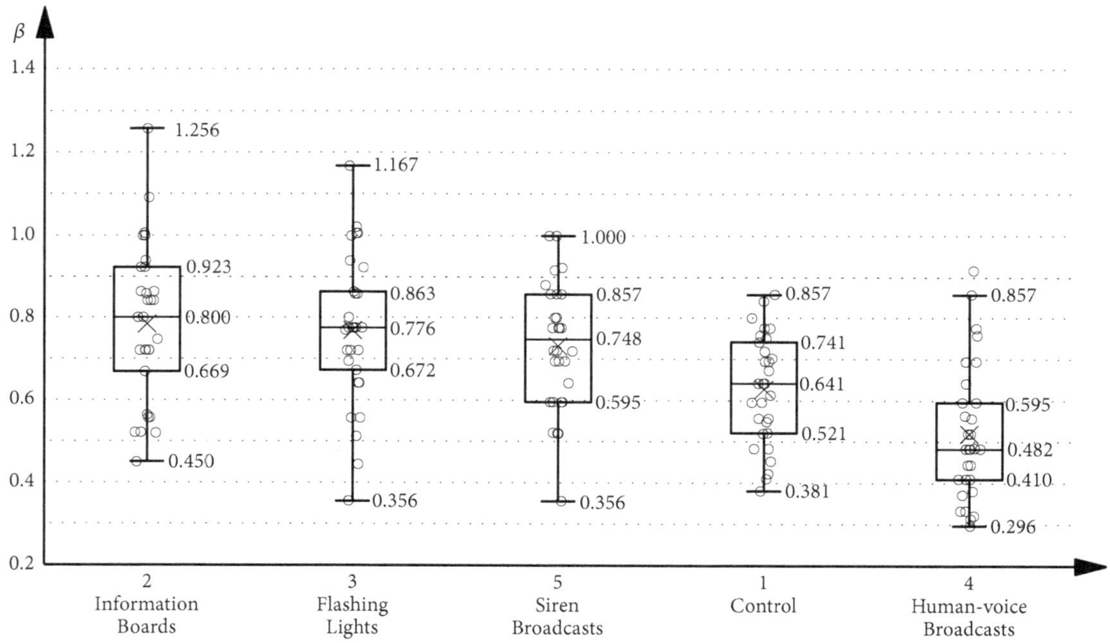

FIGURE 3: Sample distribution of facility efficiency (β).

boards, flashing lights, human-voice broadcasts, and siren broadcasts—on driver distance perception in expressway tunnels. A survey questionnaire was given to 436 drivers at two service areas along the Yanxi Expressway in Shaanxi Province of China. In addition, field experiments involving 150 participants were conducted in the Xingshuliang Tunnel along the Yanxi Expressway.

Results from the questionnaire survey indicated that drivers believed that information boards, flashing lights, and siren broadcasts were most effective (in descending order) in encouraging drivers to increase attention, decelerate, and maintain safe distance between vehicles. Human-voice broadcasts were positive in encouraging drivers to increase attention and decelerate but had a negative effect on distance maintenance of drivers. It could be explained that the information of human-voice broadcast in expressway tunnels was difficult for drivers to catch and understand. In this study, the questionnaire survey was mainly focused on the perceived effects of four safety facilities; however, how much better or worse with each safety facility to reduce crashes needs further study.

Based on the statistical analysis of the distance perception accuracy (α), the study of field experiments came to a conclusion that information boards, flashing lights, and siren broadcasts increased the distance perception accuracy of drivers, while human-voice broadcasts decreased this accuracy. There was no significant difference of distance perception accuracy between information boards, flashing lights, and siren broadcasts, and the distance perception accuracy of these three facilities was significant higher than without facilities. According to the calculation of individual safety device efficiency (β), for each safety facility in tunnels, the distance perception of information boards, flashing lights, and siren broadcasts performed better inside tunnels than outside tunnels.

The results of this study substantially contribute to the state of knowledge regarding the effects of safety facilities on driver distance perception. These findings can assist engineers in identifying the most effective safety facilities in tunnels and provide a basis for prioritizing their implementation, ultimately increasing driver distance perception accuracy and decreasing rear-end collisions.

Some of the differences of the results obtained from the questionnaire surveys and the field experiments may be contributed to the difference between the subjective judgement and objective behaviors. Some of the small differences of the results related to the cable broadcasts obtained in this study and other research may also be due to the 42.2% of responders who had never experienced cable broadcast in expressway tunnels. Also, the sample size of drivers, especially female drivers, was not large enough to represent the reactions of drivers in China; the real effects of each safety facility may be better or worse than the results in this research.

Despite these contributions, additional research is required in a number of areas. Future research focused on driver distance perception in tunnel facilities should consider the effects of alternative technologies (e.g., wireless broadcast systems) to improve human-voice broadcasts, the combined use of various safety devices (in this study, we considered each facility singularly), and alternate messaging and installation arrangements. A related study considering driver speed perception in tunnels could be combined with this study's findings to ultimately improve road safety. Besides, the effects of safety facilities of tunnels under special situations or emergencies should be further studied.

Conflicts of Interest

The authors declare that there are no conflicts of interest regarding the publication of this paper.

Acknowledgments

This research was supported in part by the National Key Research and Development Program of China (no. 2016YFC0802208) and the Natural Science Foundation of Shaanxi Province (no. 2017JQ5122).

References

[1] F. H. Amundsen and G. Ranes, "Studies on traffic accidents in Norwegian road tunnels," *Tunnelling and Underground Space Technology*, vol. 15, no. 1, pp. 3–11, 2000.

[2] Z. Ma, C. Shao, and S. Zhang, "Characteristics of traffic accidents in Chinese freeway tunnels," *Tunneling Underground Space Technology Incorporating Trenchless Technology Research*, vol. 24, no. 3, pp. 350–355, 2009.

[3] A. Beard and R. Carvel, *The Handbook of Tunnel Fire Safety*, Thomas Telford Publishing, 2005.

[4] Q. Meng and X. Qu, "Estimation of rear-end vehicle crash frequencies in urban road tunnels," *Accident; Analysis And Prevention*, vol. 48, no. 5, pp. 254–263, 2012.

[5] F. H. Amundsen, "Studies of driver behaviour in Norwegian road tunnels," *Tunnelling and Underground Space Technology*, vol. 9, no. 1, pp. 9–15, 1994.

[6] S. Bassan, "Overview of traffic safety aspects and design in road tunnels," *Iatss Research*, vol. 40, no. 1, pp. 35–46, 2016.

[7] Federal Highway Administration, "Manual on Uniform Traffic Control Devices for Streets and Highways," 2009.

[8] Ministry of Communications of PRC, *Guidelines for Design of Highway Safety Facilities: JTG D81-2006*, China Communications Press, 2006.

[9] Ministry of Communications of PRC, *Specification for Layout of Highway Traffic Signs and Markings: JTG D82-2009*, China Communications Press, 2009.

[10] E. Ronchi and D. Nilsson, *Traffic Information Signs, Color Scheme of Emergency Exit Portals and Acoustic Systems for Road Tunnel Emergency Evacuations*, Department of Fire Safety Engineering Lund University Sweden, 2013.

[11] C. Dudek and G. Ullman, "Flashing Messages, Flashing Lines, and Alternating One Line on Changeable Message Signs," *Transportation Research Record Journal of the Transportation Research Board*, vol. 1803, no. 1, pp. 94–101, 2002.

[12] Y.-K. Ou and Y.-C. Liu, "Effects of sign design features and training on comprehension of traffic signs in Taiwanese and Vietnamese user groups," *International Journal of Industrial Ergonomics*, vol. 42, no. 1, pp. 1–7, 2012.

[13] D. Shinar and M. Vogelzang, "Comprehension of traffic signs with symbolic versus text displays," *Transportation Research Part F Traffic Psychology and Behaviour*, vol. 18, no. 4, pp. 72–82, 2013.

[14] World Road Association, *Human Factors and Road Tunnel Safety regarding Users*, 2008.

[15] M. P. Manser and P. A. Hancock, "The influence of perceptual speed regulation on speed perception, choice, and control: Tunnel wall characteristics and influences," *Accident Analysis & Prevention*, vol. 39, no. 1, pp. 69–78, 2007.

[16] M. Lidström, "Using advanced driving simulator as design tool in road tunnel design," *Transportation Research Record Journal of the Transportation Research Board*, vol. 1615, no. 1, pp. 51–55, 1998.

[17] L. Domenichini, F. La Torre, D. Vangi, A. Virga, and V. Branzi, "Influence of the lighting system on the driver's behavior in road tunnels: A driving simulator study," *Journal of Transportation Safety & Security*, vol. 9, no. 2, pp. 216–238, 2017.

[18] F. Bella, "Driving simulator for speed research on two-lane rural roads," *Accident Analysis & Prevention*, vol. 40, no. 3, pp. 1078–1087, 2008.

[19] J. Törnros, "Driving behavior in a real and a simulated road tunnel–a validation study," *Accident Analysis & Prevention*, vol. 30, no. 4, pp. 497–503, 1998.

[20] A. Calvi and F. D'Amico, "A study of the effects of road tunnel on driver behavior and road safety using driving simulator," *Advances in Transportation Studies*, 2013.

[21] E. Ronchi, D. Nilsson, S. Kojić et al., "A Virtual Reality Experiment on Flashing Lights at Emergency Exit Portals for Road Tunnel Evacuation," *Fire Technology*, vol. 52, no. 3, pp. 623–647, 2016.

[22] X. Jia, M. Li, and J. Li, "Study on Effects of Safety Facilities on driver's Speed Perception and Concentration in Expressway Tunnel," *Journal of Residuals Science & Technology*, vol. 17, no. 7, pp. 139.1-139.5, 2016.

[23] H. Wan, Z. Du, B. Ran, and M. Wang, "Speed control method for highway tunnel safety based on visual illusion," *Transportation Research Record Journal of the Transportation Research Board*, vol. 2485, pp. 1–7, 2015.

[24] Y. Dai, C. An, and Z. Liao, "Characteristics of Highway Tunnel accidents," *Computer and Communications*, vol. 28, no. 2, pp. 101–106, 2010.

[25] S. Yasmin, N. Eluru, C. R. Bhat, and R. Tay, "A latent segmentation based generalized ordered logit model to examine factors influencing driver injury severity," *Analytic Methods in Accident Research*, vol. 1, pp. 23–38, 2014.

[26] C. S. Duncan, A. J. Khattak, and F. M. Council, "Applying the ordered probit model to injury severity in truck-passenger car rear-end collisions," *Transportation Research Record Journal of the Transportation Research Board*, vol. 1635, no. 1, pp. 63–71, 1998.

[27] J. S. Long, "Regression models for categorical and limited dependent variables," *Technometrics*, vol. 40, no. 1, pp. 80-81, 2006.

[28] A. Schreiner, P. Bergmans, P. Cherubin et al., "Effects of Drivers' Actions on Severity of Emergency Vehicle Collisions," *Transportation Research Record Journal of the Transportation Research Board*, vol. 2318, no. 1, pp. 90–97, 2018.

[29] S. Anowar, S. Yasmin, and R. Tay, "Factors Influencing the Severity of Intersection Crashes in Bangladesh," *Asian Transport Studies*, vol. 3, no. 2, pp. 1–12, 2014.

[30] B. G. Witmer and P. B. Kline, "Judging perceived and traversed distance in virtual environments," *Presence*, vol. 7, no. 2, pp. 144–167, 1998.

[31] S. Li-Ming and S. Yang, "The theoretical research on traffic noise in highway tunnels," in *Proceedings of the 2011 IEEE 3rd International Conference on Communication Software and Networks*, pp. 746–749, IEEE, 2011.

[32] L. N. Yang and B. M. Shield, "Development of a ray tracing computer model for the prediction of the sound field in long enclosures," *Journal of Sound and Vibration*, vol. 229, no. 1, pp. 133–146, 2000.

[33] Z. Sun and I. F. Akyildiz, "Channel modeling of wireless networks in tunnels," in *Proceedings of the IEEE GLOBECOM 2008 - 2008 IEEE Global Telecommunications Conference*, pp. 158–162, IEEE, 2008.

[34] Z. Sun and I. F. Akyildiz, "Channel modeling and analysis for wireless networks in underground mines and road tunnels," *IEEE Transactions on Communications*, vol. 58, no. 6, pp. 1758–1768, 2010.

4

Decision-Support Framework for Selecting the Optimal Road Toll Collection System

Marina Milenković,[1] Draženko Glavić,[1] and Miloš N. Mladenović[2]

[1] *University of Belgrade, Faculty of Transport and Traffic Engineering, V. Stepe 305, 11000 Belgrade, Serbia*
[2] *Aalto University, Department of Built Environment, Otakaari 4, 02150 Espoo, Finland*

Correspondence should be addressed to Marina Milenković; marina.milenkovic@sf.bg.ac.rs

Academic Editor: Jose E. Naranjo

One of the central decision-making questions in planning road tolling is the selection of the optimal toll collection system (TCS). The question of TCS selection arises in the situation when the existing TCS is to be upgraded or when TCS is selected for a newly constructed road. Considering that there are multiple TCS available nowadays, with their particular advantages and disadvantages, and that there is a range of often conflicting criteria for TCS selection, this decision-making issue belongs to the group of multicriteria decision-making (MCDM) problems. The MCDM-based methodology used in this research integrates Strengths-Weakness-Opportunities-Threat (SWOT) analysis and Fuzzy Preference Ranking Organization Method for Enrichment Evaluations (F-PROMETHEE). The expert-based decision-support framework includes a procedure for defining evaluation criteria and their weights, scoring of alternatives, and sensitivity analysis. Presented decision-support framework is tested with fourteen toll systems. Results indicate that the best-ranked TCS is the dedicated short-range communication multilane free flow. Decision-support framework is developed for transferability to different contexts, where local features can be taken into account by choosing specific alternatives, criteria, and criteria values. Finally, this development opens up opportunities for further analysis of criteria values and considerations of user attitudes in road pricing scheme planning.

1. Introduction

Road pricing scheme represents the direct charges collected for the road use, such as distance or time-based fees, and urban congestion pricing [1–5]. These charges often depend on the vehicle type, number, and configuration of vehicle axles, vehicle dimensions, trailer types, and vehicle emission class. In general, road pricing has two purposes. One purpose of road pricing is demand management, while the other purpose is a mechanism for financing road infrastructure construction, operation, and maintenance. With its strong origins in the economic theory, road pricing has been advocated as an efficient policy to account for externalities and infrastructural funding. Despite its sound purposes, road pricing is a typical example of a theoretically well-developed transport policy that has been shown difficult to implement [6]. Since road pricing is a type of a policy that radically shifts distribution of costs and benefits, this policy is often value-laden and politely charged. One of the crucial aspects in planning road pricing scheme is public acceptability of road pricing overall. A range of previous research indicates the low public acceptability of road pricing, pointing out to the infringement of freedom and fairness [7–10]. Along with the questions of public acceptability, another crucial issue with road pricing is political acceptability and support [11–13]. For example, politicians often perceive road pricing as complicated new charge for something that is free [12]. Moreover, political unacceptance of road pricing might originate from preference for policies with highly visible benefits and distributed and low-visibility costs [13, 14]. Consequently, planning road pricing scheme often requires balancing contradicting objectives that planners and infrastructure managers might have with respect to users.

An important part in the delicate process of road pricing planning is the selection of tolling system, as an important ancillary asset. Just as different optimal tool schemes relate to different policy objectives [15], toll collection system (TCS) directly affects the user's experience and consequently acceptance of road pricing scheme. Additional challenge in the decision-making process emerges from the fact that the choice of a particular TCS assumes the operational responsibility for the next 10 to 20 years. This long-term responsibility for TCS asset is particularly important for agencies with large road pricing systems, often responsible for large-scale investment into a single TCS. In practice, the agency is facing a choice between a multitude of TCS currently in use. Nowadays, TCS used worldwide include manual and automatic coin machine (ACM), vignettes, dedicated short-range communication (DSRC) with barriers, DSRC multilane free flow (MLFF), barcode, radiofrequency identification (RFID) with barriers, RFID open road tolling (ORT), global navigation satellite system/cellular network or vehicle positioning system (GNSS/CN or VPS), automated number plate recognition system (ANPR), infrared, tachograph, smart card, and smartphones (Figure 1). These systems differ in their performance, if one considers, for example, accuracy, costs, or interoperability. Bearing in mind the many and diverse interests at stake, each TCS has a range of advantages and drawbacks from the standpoint of road manager and user. For example, different TCS respond differently to the need to maximize the toll revenue, which is often an important criterion for road managers. On the contrary, different TCS cater differently to the user needs for reducing delays and costs or improving level of service (LOS) and safety.

Taking into account the importance of TCS in the effectiveness of the road pricing scheme, as well as large-scale investment and conflicting criteria related to a multitude of alternative TCS, transport planners face a challenging question—which TCS should we select? Previously, only one study has dealt with the issue of selecting the optimal TCS [16]. This study from India considered five alternative TCS including barcode-based ETC, RFID-based ETC, ANPR ETC, VPS ETC, and active infrared system. A multicriteria decision-making method fuzzy VIKOR was used, concluding that the ETC system based on RFID technology was the best of all considered options. Thus, with this gap in the previous research, the objective of this paper is the development of a decision-support framework for selecting the optimal TCS. The next section will present an expert-based methodological framework centered on Fuzzy Preference Ranking Organization Method for Enrichment Evaluations (F-PROMETHEE) multicriteria decision analysis technique. Moreover, the methodological framework integrates Strengths-Weakness-Opportunities-Threat (SWOT) analysis. Decision-support framework is tested with fourteen toll systems, in the context of the newly constructed motorway. Following the results from the implemented decision-support framework (DSF), the discussion of implications is provided, including recommendations for future research.

2. Methodology

Considering the multitude of conflicting criteria stemming from both planners and users' perspective, as well as a range of advantages and disadvantages that each TCS has, decision-making is not intuitive. Lack of intuitiveness originates from the fact that TCS selection problem cannot be easily organized due to large number of interdependencies and is thus not easily specified in advance. As TCS selection does not have a conventional mathematical formulation, the choice of the optimal TCS belongs to the category of semistructured decision problems [17]. The challenge of semistructured problem is contrasted with the fact that agency often has additional requirements. These agency's requirements might include the need that comparison of alternatives is analytically based and that there is cross-referencing to the relevant sources of information, transferability among locations or agencies, and communication of evaluation results in a multifaceted form [18]. On the contrary, transport planners and road managers often have extensive experience with their transport system. Consequently, this expert knowledge can be an essential resource for aiding the selection of the optimal TCS.

Considering the semistructured nature of the TCS selection problem and the extensive expert knowledge about the transport system, there is an opportunity for a DSF based on multicriteria decision-making (MCDM) methodology. MCDM refers to the process of making decisions between a number of alternatives by defining the decision criteria and their weights. The application of MCDM results in the ranking of alternatives, from the most to the least favorable, thus allowing comparison of alternatives. Some common methods for MCDA in transport applications include the technique for order of preference by similarity to ideal solutions (TOPSIS) [19, 20], analytic hierarchy process (AHP) [21–23], grey evaluation method (GE) [24, 25], simple additive weighting (SAW) [26], and elimination and choice expressing reality ELECTRA [27]. In particular, MCDM methods are often used for formulating DSFs for optimal technological system selection [28–36].

This DSF centers on Preference Ranking Organization Method for Enrichment Evaluations (PROMETHEE), as the outranking method. When contrasted with other multicriteria analysis methods, PROMETHEE performs better in dealing with the problems having a large number of alternatives [37–41]. On the contrary, the capability to deal with a large number of alternatives is not the case with the AHP method [42]. In practice, PROMETHEE is a group of methods, where PROMETHEE I is used for partial ranking of alternatives and PROMETHEE II is used for complete ranking of alternatives [37]. In order to overcome the uncertainty in the process of the selecting optimal toll collection system, we utilize the extension of the PROMETHEE with fuzzy numbers (F-PROMETHEE). As a fuzzy subset of real numbers, fuzzy numbers are included to deal with subjective dilemma that comes from using linguistic variables for problem representation [43]. As a result, the use of fuzzy numbers allows consideration of vagueness inherent in this decision-making context through the user friendliness from linguistic evaluations. The integration of PROMETHEE I and II, as well as other elements into a DSF, is presented in the following section.

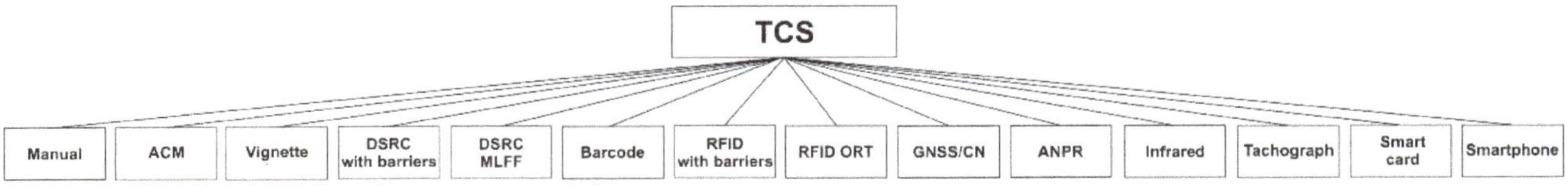

FIGURE 1: Toll collection systems.

FIGURE 2: Proposed decision-support framework.

2.1. Decision-Support Framework. The DSF involves a procedure, presented in the Figure 2, to be organized as an expert-based workshop. The figure shows subprocesses as well as outputs from each subprocess. Involvement of stakeholders in the decision-making process has been especially highlighted before [44] and is thus an important aspect of the proposed DSF. As Figure 2 shows, first step includes the description of the MCDM problem with a definition of the TCS goals. During workshops, the authors first defined a set of alternatives, as TCS used worldwide. Then, SWOT analysis, including a discussion about strengths, weakness, opportunities, and threats of each alternative, was conducted to help experts get better understanding of the alternative. In general, groups of experts should be between four and six people, to allow sufficient diversity of opinions while still having possibility for simultaneous discussion. In addition, by bringing in experts with different roles and backgrounds, discussion allows for complementary knowledge exchange in a constructive setting framed by SWOT. In this particular workshop, four tolling experts from the fields of transport planning, traffic operations, transport policy, and transport economics have participated. This combination has allowed for a sufficient diversity of opinion to emerge, while still being able to have a common understanding on some of the core issues. SWOT-related processes are colored blue in Figure 2. TSC characteristics: in the workshop, the experts discussed the MCDM task, tolling options, definition of criteria and their weights. Next, sets of the criteria were defined, as well as the criteria weights. The values of the criteria weights were determined using subjective fuzzy evaluations by experts. Then, the experts scored alternatives per each criterion using subjective fuzzy evaluations. Processes using fuzzy logic are colored green in Figure 2. The F-PROMETHEE was used for ranking the alternatives. While implementing the PROMETHEE, the preference functions were defined for each of the considered criteria, as well as the values of the parameters for the selected preference functions. This method determined the values of the positive and negative outranking flows, as well as the net outranking flow, which led to the ranking of the alternatives. Following the alternative ranking, the sensitivity analysis was conducted in order to determine the stability interval for each of the considered TCS alternatives. Processes that rely on PROMETHEE are colored red in Figure 2. The following subsections elaborate on each of the DSF subprocesses.

2.1.1. Establishing Decision Context. The first subprocess involves the definition of TCS goal and all the possible alternative TCS, as well as the criteria list for evaluating the alternatives. The definition of the criteria is preceded with a SWOT analysis of the TCS alternatives. SWOT analysis is a structured planning method that helps evaluate the strengths, weaknesses, opportunities, and threats of each TCS, as follows:

Strengths: characteristics (relative) of the toll alternative that give it an advantage over other alternatives

Weaknesses: characteristics (relative) of the toll alternative that represent disadvantage relative to other alternatives

Opportunities: characteristics (absolute) of the toll alternative that could be used for its advantage during implementation

Threats: characteristics (absolute) of the toll alternative that could be a cause of issues during implementation.

In this DSF, the objective of SWOT is to enable experts to better understand alternatives and their scores and as well define criteria. SWOT is selected as a technique that has minimum structure, thus not overburdening the discussion with its process but allows focus on substance. Having a divergent method is particularly important in this subprocess, because it allows expert knowledge acquisition from both explicit and implicit knowledge, as well as from knowledge about transport system from different aspects. Procedurally, as experts are discussing strengths, weaknesses, opportunities, and threats of different TCS alternatives, they are exchanging knowledge and identifying issues and potential criteria to be discussed, selected, and categorized. Consequently, although SWOT analysis is an output in itself, the process of getting to this output is important for knowledge development among the group of experts. As a result, this DSF subprocess results in refinement of alternatives, their scores, and criteria, as well as the objective per criterion, i.e., minimization or maximization.

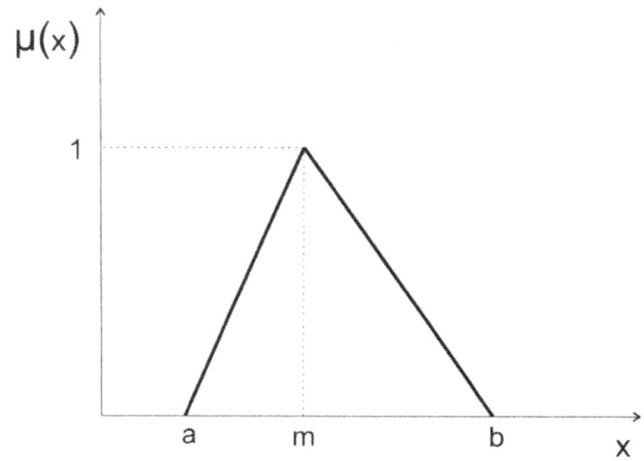

FIGURE 3: Graphical presentation of a triangular fuzzy number.

2.1.2. Determination of Criteria Weight. Determination of the weights is the second subprocess in this DSF. SWOT analysis is used to assist the process of criteria weighting by deepening knowledge of toll alternative characteristics. The relevant linguistic terms are defined in order to assess the criteria individually, instead of using pairwise comparisons. These determined linguistic terms are than expressed by triangular fuzzy numbers. Linguistic definitions for the criteria weight and their corresponding fuzzy numbers are shown in Table 1. Graphical interpretation of a triangular fuzzy number is presented in Figure 3, where a is the lower and b is the upper limit, while m is the point where membership function μ is equal to one.

Experts evaluate the criteria weights in the form of linguistic terms. In order to define the weight factor of each criterion, each expert is presented with a list of criteria defined in the previous step. Then, linguistic expert evaluations are converted into related fuzzy numbers. Next, the importance of each criterion (i), taking into account the estimates of all experts (k), is determined as follows:

$$W_i = (a_i, m_i, b_i) \longrightarrow$$
$$W_i = \left[\min(a_{ik}), \frac{\sum_{i=1}^{k} m_{ik}}{k}, \max(b_{ik})\right] \tag{1}$$

The fuzzy numbers were defuzzified by using the distance minimization method suggested by Asady and Zendehman [58] as given below:

$$W_{\text{defuzzified}} = m + \frac{b-a}{4} \tag{2}$$

Defuzzified values are then normalized as follows:

$$W_{\text{normalized}} = \frac{w_i}{\sum_{i=1}^{n} w_i}, \tag{3}$$

where

TABLE 1: Linguistic terms and corresponding fuzzy numbers.

Linguistic terms		Triangular fuzzy number
For determination of criteria weight	For scoring alternative per criteria	
Very Low (VL)	Very Weak (VW)	(0.00, 0.00, 0.15)
Low (L)	Weak (W)	(0.00, 0.15, 0.30)
Medium Low (ML)	Medium Weak (MW)	(0.15, 0.30, 0.50)
Medium (M)	Medium (M)	(0.30, 0.50, 0.70)
Medium High (MH)	Medium High (MH)	(0.50, 0.70, 0.85)
High (H)	High (H)	(0.70, 0.85, 1,00)
Very high (VH)	Very high (VH)	(0.85, 1.00, 1.00)

i denotes number of criteria ($i = 1, 2, \ldots, n$);

k denotes number of decision-makers ($k = 1, 2, \ldots, p$);

j denotes number of alternatives ($j = 1, 2, \ldots, m$);

a_{ik} denotes left threshold of fuzzy number estimated by expert k for criteria i;

m_{ik} denotes midpoint of fuzzy number estimated by expert k for criteria i;

b_{ik} denotes right threshold of fuzzy number estimated by expert k for criteria i;

W_i denotes weight of criteria i.

2.1.3. Scoring of Alternatives. In addition to assessing the weight of the criteria, tolling experts are responsible for scoring alternatives, also partially relying on information provided from SWOT analysis. In order to evaluate the performance of each alternative on the basis of the defined criteria, the expert opinions are collected as linguistic terms, defined in Table 1. Due to the fact that determination of analytical scoring is not possible for such a large number of criteria and alternatives, linguistic scale is used. Linguistic expressions are then transformed into the corresponding fuzzy numbers (Table 1). The values of the fuzzy numbers boundaries are the same as those during the activity of criteria weighting. The estimates of all experts were taken into account using (1). The obtained fuzzy numbers were than defuzzified using (2) and normalized using (3).

2.1.4. Evaluation of Alternatives. The detailed procedure for evaluation of alternatives using PROMETHEE II method consists of five steps [see [38, 39]]. The subprocess starts with the determination of deviations based on pairwise comparisons. This step is then followed by applying a relevant preference function for each criterion defined in subprocess 2 above. Step three involves calculating global preference index, while stop four involves calculating positive and negative outranking flows for each alternative and partial ranking. The subprocess ends in step five, with the calculation of net outranking flow for each alternative and with complete ranking [38, 39].

PROMETHEE is based on pairwise comparisons and the specificity of this method is preference function. Its rule is that difference in values for each pair of alternatives leads to the range from zero to one, per criterion. In this way, normalization of values and obtaining information about the preferences of each alternative in relation to each, according to each of the criteria, are achieved.

The PROMETHEE I partial ranking is based on the computation of two preference flows. Phi+ (positive flow) is a measure of the strength of the alternative, while Phi– (negative flow) is a measure of the weakness of the alternative. PROMETHEE I partial ranking allows incomparability, which usually happens when one alternative is good in a subset of criteria in which another one is weak and vice versa [59]. The possibility of incompatibility is excluded by introducing the characteristic of net outranking flow which denotes the difference between the positive and negative outranking flows. Thus, alternative a is preferred to alternative b in the PROMETHEE II ranking only if it is preferred to b according to the net preference flow. In this particular case, $aP^{\mathrm{II}}b$ only if $\Phi(a) > \Phi(a)$ [60]. Thus, PROMETHEE II provides a complete ranking of the alternatives from the best to the worst ones [59].

2.1.5. Sensitivity Analysis. As the last subprocess, sensitivity analysis is essential for determining the changes of alternative ranking due to changes in the value of particular criteria weight. As part of PROMETHEE, weight sensitivity analysis can be done using Walking Weights or Stability Intervals analysis [60]. For this DSF, we have implemented Stability Intervals, which shows how the Phi score and the PROMETHEE II ranking vary as a function of the weight of a criterion and identifies the interval of stability of the first place and complete ranking.

3. Application of the Decision-Support Framework

Decision-support framework is tested with fourteen toll systems, in the context of the newly constructed motorway. DSF has been implemented in a workshop setting, with participation by four tolling experts from the fields of transport planning, traffic operations, transport policy, and transport economics. Experts' number has been decided based on the general recommendations that group activity is usually most effective in groups of four to six people.

3.1. Description of the Alternative TCS. Toll collection has changed historically from manual toll collection to various forms of electronic toll collection. The development of digital

TABLE 2: Description of the alternative TCS.

TCS	Description of the alternative TCS
Manual	This system relies on cash payment for motorway usage which is conducted manually by tolling staff. Driver has to stop her vehicle at the toll station and pay to officer who is responsible for it. Officer determines the amount to be paid on the basis of the characteristics and classification of vehicles, and the distance traveled by the vehicle. Since paying takes some amount of time, queuing is usual for this tolling type [45].
ACM	It is a machine with a slot for inserting coins or paper money. Driver has to stop her vehicle and pay a certain amount of money on the basis of the classification of vehicles, and the distance traveled by the vehicle. This technology has better service time when compared to the manual TCS, consequently resulting in lower delays.
Vignette	It is a sticker which, once bought, pays the toll for a specific time period. There are vignettes which pay the toll in the duration of several days, week, month or a year. The vignette must be applied to the car windshield. The so-called electronic vignettes are not in the form of stickers, but the number of the car's registration plate is inserted into the system for the days the vignette is paid for. The control of the participants is performed with the help of the database, traditional stopping on the roadside and penalizing the violators.
DSRC with barriers	It is a type of non-contact TCS where the users do not have contact with toll booths or cashiers. Users can choose bills to be sent once a month or choose a prepaid model. A vehicle does not have to stop at the toll plaza; it only needs to reduce its speed in order to establish the contact and recognition through the on-board unit (OBU) to pass. In the meantime, all data are recorded in the central toll system of the operator.
DSRC MLFF	In this system, antennas are set on gantries, on particular locations (e.g., between two interchanges) along the highway with road pricing. Antennas detect traffic flow and record the use on the OBU, as well as in the central toll collection system of the operator for further processing and payment. The technology used in MLFF system is designed so that the vehicles can maintain their speed and change lanes (including emergency stop lanes) while passing below the collection portal.
Barcode -based	This system is a subcategory of ETC. Within this system, the barcode is a sticker applied to the windshield of the vehicle which is read by a laser scanner while the vehicle passes through the toll gate [46].
RFID with barriers	This system contains the OBU or sticker installed on the front windshield of the vehicle. At toll gate, either prepaid or postpaid system is read by a RFID frequency reader, [16]. RFID reader is a device used to communicate with an RFID tag, via radio waves [47]. A vehicle does not have to stop at the toll gate; it only needs to reduce its speed in order to establish the contact and recognition so that system can let the vehicle pass.
RFID ORT	This system contains the OBU installed on the front of the vehicle. At the toll gate, this system is read by a RFID frequency reader [16]. The technology used in RFID ORT system is designed so that the vehicles can maintain their speed and change lanes (including emergency stop lanes) while passing below the collection portal.
GNSS/CN	It includes a global navigation satellite system incorporated with the cellular networks. This TCS functions using a global positioning system unit (GPS/Galileo/Glonass) installed on the OBU which stores the route coordinates of the vehicle and sends the transaction information to toll collection central system through global mobile communication system (GSM/3G/4G) [48–50].
ANPR	It uses a stationary camera for recording and identifying the registration plate numbers of vehicles passing through the toll gates. The detected registration plates are paired in the database and the toll is deducted (a particular amount of money is taken). If the recorded plate number is not read accurately or not found in the database, an enforcement violation alarm is generated to alert the authorities. In this manner, two issues are simultaneously solved – identifying the vehicles for toll collection and issuing/recording the violation enforcement alert [51, 52].
Infrared	This system is similar to DSRC with barriers TCS; the only difference is that it has an active infrared unit installed in the vehicle which contains all the information [53–55].
Tachograph -based	It records the mileage driven by the user through an OBU connected electronically to the vehicle's odometer, measuring the vehicle's mileage. A tachograph system which is in place in New Zealand requires manual, rather than electronic, data collection [56].
Smart card	It represents a memory card in which the details of a particular person and certain amount of money are stored. Smart cards must be recharged with some amount of money and whenever a person wants to pay the toll, she needs to insert her smart card and deduct an amount of money. The smart card based toll collection system functions on the basis of contact communication between the smart card and the reader [57].
Smartphones	It is not matured TCS and it is in stages of constant development. An example of mobile and Smartphone ETC integration is the m-Toll, Ptoll, GeoToll, and PayTollo projects. This TCS relies on the use of smartphone capabilities, such as WiFi, 4G, GSNN, and NFC connection to authenticate, validate and charge road users without the need for any additional hardware for the end user.

Table 3: SWOT matrix for alternative TCS.

TCS	Strengths	Weaknesses	Opportunities	Threats
Manual	Backup tolling option for OBU non-equipped motorway users; No user investment	Low LOS, Long queues, High Opex; Increased emissions; Health and safety risks for toll operators	-	Thefts
ACM	Additional tolling option on toll plazas	Stopping, small coin values	-	No exact amount of coins
Vignette	Stimulating the everyday users to use motorways; Decrease of noise; Decrease of air pollution; Savings in infrastructure costs	Applicability; Not fair pricing for foreigners and occasional users; Many stickers on the windshield visually degraded vehicles; High enforcement costs; The difficulties of controlling all vehicles	Temporary tolls; Additional toll system for certain vehicle categories	Acceptance of a wider driving population; Frequent control of the driver; Being considerably more expensive for occasional users; No interoperability
DSRC				
DSRC with barriers	Payment without stopping; No queuing; Decrease of VOC; Decrease of TTC; Reduced environmental degradation; Pay in accordance with travelled kilometres; Large number of DSRC OBUs currently in operation	Inefficiency for occasional users without OBU; High capital expenditure and maintenance cost of tolling infrastructure; Difficulty in modifying the tolled network once implemented	Easy transition from existing manual system to ETC; Ability to provide/support other value added services through the OBU; Interoperability	Low % of users with OBUs
DSRC MLFF	Payment without decreasing the speed; No delays; Without increase of VOC and TTC; Increase of traffic safety; Minimum of environmental degradation; Pay in accordance with travelled kilometres; High reliability & performance, low signal interference; Large number of DSRC OBUs currently in operation; User friendly	Inefficiency for occasional users without OBU; High enforcement costs; The difficulties of controlling vehicles without OBU. Necessity to install road-side infrastructure (gantries) along the road; High capital expenditure and maintenance cost of tolling infrastructure; Difficulty to modify the tolled network once implemented	Increasingly common technology; Efficiency for drivers and toll operators; Ability to provide/support other value added services through the OBU; Interoperability	Avoiding payment of tolls; Being less profitable in low traffic volume roads
Barcode-based	Technology which does not require complex system; Pay in accordance with travelled kilometres	Unreliability; Lower accuracy during bad weather; Lack of flexibility; Low rate of reading data; Low passenger comfort	-	Theft of barcode label
Infrared	Payment without stopping; No queuing; Decrease of VOC; Decrease of TTC; Less environmental degradation; Pay in accordance with travelled kilometres; High rate of reading data; High reliability; Accurate in all-weather condition	The problem of interference with other signals; High Opex and Capex	Easy transition from existing manual to calm active infrared system	The problem of interference with the frequencies of other devices; The lack of interoperability
RFID				
RFID with barriers	Payment without stopping; No queuing; Decrease of VOC; Decrease of TTC; Less environmental degradation; Pay in accordance with travelled kilometres; High rate of reading data	High Opex and Capex; The problem of interference with the frequencies of other devices; Inefficiency for occasional users without OBU	Increasingly common technology; Efficiency for drivers and toll operators; Low RFID sticker/OBU price	The angle of object installation and compatibility having an important role in the reliability and punctuality of these systems
RFID ORT	Payment without stopping; No queuing; Decrease of VOC; Decrease of TTC; Less environmental degradation; Pay in accordance with travelled kilometres; High rate of reading data	The problem of interference with the frequencies of other devices; Inefficiency for occasional users without OBU	Increasingly common technology; Efficiency for drivers and toll operators; Low RFID sticker/OBU price	The angle of object installation and compatibility having an important role in the reliability and of punctuality these systems

Table 3: Continued.

TCS	Strengths	Weaknesses	Opportunities	Threats
Smart card	Reduced queuing time; Small decrease of VOC; and TTC; Environmental degradation	Payment with stopping; Reliability	Reduced number of toll employees	An insufficient number of Smart card users;
ANPR	Payment without stopping; No queuing; Decrease of VOC; Decrease of TTC; Less environmental degradation; No OBU; Ability to deliver cost savings for automatic handling processing, subject to fine tuning and secondary processing methods used; Most successful when combined with other technologies; No performance restrictions regarding vehicle speeds	Requires good quality license plates; Susceptibility to poor lighting and weather conditions; Access to needed vehicle data of foreigners vehicle database; Cost of manual checking that can increase operational costs; Suitability for supporting relatively simple charging policies	Continuous improvements in video camera quality; Always being required for enforcement; Increased overall safety	Thefts and toll avoiding; Lack of standardisation of license plates; Manual verification needed for full effectiveness subject to level of "tuning"
GNSS/CN	Payment without stopping; Payment without decreasing the speed; No delays and queuing; Without increase of VOC and TTC; Minimum of environmental degradation; Pay in accordance with travelled kilometres; Increase of traffic safety; Flexibility to define and modify what is to be charged and how it is to be charged; Once installed, less costly to maintain; Easily expandable to other roads and regions	Privacy of the user's movement; High Opex and Capex; Inefficiency for occasional users without OBU; High enforcement costs; Less used and mature technology than other technologies; Accuracy errors in certain sections of the tolled network, such as parallel free roads and intersections; Additional roadside devices that need to be installed	Route guidance; Increased overall and traffic safety; Ability to provide/support other necessary to be pursued value-added services through the OBU; traffic information, speed control	Interoperability; Gaining the public support regarding the mobility privacy policy
Tachograph-based	Absence of privacy and data protection issues; Investment in tolling infrastructure relatively independent of the tolled network within the area; Low toll collection costs; Maintenance of tolling infrastructure is limited to cross-border control stations	Rigidity in defining and modifying what is to be charged and how it is to be charged within the tolled area; Low accuracy of the tolling technology (±4%); Complex and costly on-board unit; No interoperability; High start-up costs of cross-border control stations; Not commonly used technology	-	Additional position services which cannot be provided through the OBU; No interoperability
Smartphones	Flexibility to define and modify what is to be charged and how it is to be charged; Little physical tolling roadside infrastructure investment; No need for in-vehicle device or costly enforcement infrastructure; User-friendly interface; Interoperability with other tolling technologies; Low maintenance costs	Device battery duration; Less used and mature technology than other options; Ability to become obsoletely quickly, given technological developments in the sector; Variable proliferation of mobile and smart phones in different areas; Some areas not having appropriate GSM coverage; No proven data about accuracy on certain sections of the tolled network, such as parallel free roads and intersections; Not being able to classify vehicles; No standards currently available	Continuous technological improvements in position mobile phone and smartphone industry; Possibility of integrating toll payment with other user services; interoperability	Data protection issue in relation to the cellular network used to track user position

TABLE 4: Groups of criteria and description of criteria.

Criteria group (group of criteria)	Criteria	Description
(1) Technical operation	Interoperability	Interoperability with other toll collection systems inside a country or in other countries.
	Flexibility	Ability of a technology to be used for other purposes (paying for parking, gas, or congestion charge, as well as for navigation, traffic control, vehicle theft prevention etc.).
	Efficiency	The capacity per lane as capability of a system to register and identify the number of vehicles at the same time, based on data rate and system complexity.
	Reliability/accuracy	The ability of a system to correctly identify vehicle's identity and payment options, especially in poor weather conditions.
	Enforcement	The amount of resources engaged for enforcement. Share (percentage) of those who avoid payment.
(2) Traffic operations	Level of service	The impact of the system on the level of service of traffic flow (speed, volume/capacity ratio).
	Traffic safety	The impact of the system on the traffic safety.
	Number of stops	The number of stops during the tolling.
	Delays	The increase in travel time that each of the systems causes.
(3) Financial	Average cost for user	Value that the user has to pay for using the system.
	Ratio of revenues/costs	The value of the ratio of revenues/costs for each system.
	Capex	Value of capital costs, as the size of the capital expenditures of each system.
	Opex	Value of operational costs, as the size of operation expenditures of each system.
(4) Environmental	Visual impact of defacing ambient environment	How each system visually defaces the ambient environment.
	Air pollution	To what extent the system contributes to air pollution.
	Noise	To what extent the system contributes to noise pollution.
(5) Socio-economic	Fair pricing	Whether the system is fair to all users. Payment for what is used.
	User friendliness	Easiness of using the system.
	Aesthetic effect on the vehicle	How the equipment that each system requires affects the aesthetic effect of the vehicle.
	Possibility of theft and fraud	Possibility of system's theft and fraud in general.

technology has been extremely dynamic in recent years and has led to the rapid application of numerous technologies in toll collection. Short description of alternative TCS is given in Table 2.

3.2. SWOT Analysis of Toll Alternatives. Results of SWOT analysis conducted during the workshop are represented in Table 3. Resulting SWOT matrix has assisted experts in further DSF steps, when defining the criteria and scoring the alternative per criterion.

3.3. Criteria List and Groups. During the workshop, experts have decided on the list of criteria which were then grouped. As Table 4 shows, the groups of criteria have been divided into technical operation, traffic operation, financial, environmental, and socioeconomic criteria. Each group contains three to five criteria that cover all characteristics of TCS alternatives, while total number of criteria is twenty. Table 4 also provides a brief description of each criterion.

3.4. Criteria Weights. During the workshop, experts have also assigned criteria weights using the fuzzy-based process described in Section 2. Table 5 provides the evaluations of four experts participating in the study expressed by linguistic terms. The final defuzzified and normalized values of criteria weight are shown in Table 6.

3.5. Scoring of Alternatives and Base Matrix of Input Data. Experts were scoring alternatives per criterion, but due to large number of criteria and alternatives, Table 7 shows an example of scoring for one criterion, namely "interoperability". Defuzzified values of alternatives' scores per criterion estimated by experts are shown in Table 8.

Table 9 shows the type of particular criteria, which preference function was selected for each criterion, and what the values of each parameter for each preference function are.

3.6. PROMETHEE I and II Alternative Ranking. The PROMETHEE Table (Table 10) displays the Phi, Phi+, and

TABLE 5: The importance of criteria according to experts expressed by linguistic terms.

Group of criteria	Criteria	DM1	DM2	DM3	DM4
(1) Technical operation	Interoperability	MH	H	H	MH
	Flexibility	M	ML	H	H
	Efficiency	VH	H	VH	VH
	Reliability	VH	VH	VH	VH
	Enforcement	MH	ML	H	M
(2) Traffic operations	Level of service	VH	MH	H	H
	Traffic safety	H	M	H	H
	Number of stops	H	M	H	H
	Delays	H	H	H	H
(3) Financial	Average cost for user	H	MH	MH	MH
	Revenue/cost ratio for road agency	ML	M	VH	VH
	Capex	M	MH	H	VH
	Opex	ML	ML	H	VH
(4) Environmental	Visual degradation of road environ.	L	ML	ML	M
	Air pollution	M	ML	MH	MH
	Noise	L	L	MH	MH
(5) Socio-economic	Fair pricing	MH	M	H	H
	User friendliness	MH	MH	MH	MH
	Aesthetic effect on the vehicle	ML	M	M	ML
	Possibility of theft and fraud	H	H	H	H

TABLE 6: The values of criteria weights.

	Fuzzy Weight (W_i)	$W_{\text{defuzzified}}$	$W_{\text{normalized}}$
Interoperability	(0.50, 0.73, 1.00)	0.85	0.05
Flexibility	(0.15, 0.60, 1.00)	0.81	0.05
Efficiency	(0.65, 0.95, 1.00)	1.04	0.06
Reliability	(0.80, 1.00, 1.00)	1.05	0.06
Enforcement	(0.15, 0.56, 1.00)	0.78	0.05
Level of service	(0.50, 0.81, 1.00)	0.94	0.06
Traffic safety	(0.30, 0.73, 1.00)	0.90	0.06
Number of stops	(0.30, 0.73, 1.00)	0.90	0.06
Delays	(0.65, 0.80, 1.00)	0.89	0.05
Average cost for user	(0.50, 0.69, 1.00)	0.81	0.05
Revenue/cost ratio for road agency	(0.15, 0.70, 1.00)	0.91	0.06
Capex	(0.30, 0.74, 1.00)	0.91	0.06
Opex	(0.15, 0.60, 1.00)	0.81	0.05
Visual degradation of road environment	(0.00, 0.31, 0.65)	0.48	0.03
Air pollution	(0.15, 0.53, 0.80)	0.69	0.04
Noise	(0.00, 0.40, 0.80)	0.60	0.04
Fair pricing	(0.30, 0.69, 1.00)	0.86	0.05
User friendliness	(0.50, 0.65, 0.80)	0.73	0.04
Aesthetic effect on the vehicle	(0.15, 0.40, 0.65)	0.53	0.03
Possibility of theft and fraud	(0.65, 0.80, 1.00)	0.89	0.05

Phi− scores. The alternatives are ranked according to the PROMETHEE II complete ranking.

Analyzing the flows given in Table 10 it can be concluded that two best-ranked tolling options according to PROMETHEE I partial ranking and according to PROMETHEE II complete ranking are DSRC MLFF (Phi = 0.0425) and RFID ORT (Phi = 0.0454). They are followed by the alternatives with slightly lower values of net flow (GNSS/CN and Smartphone), while the other alternatives have significantly poorer values of positive, negative, and net flow. The lowest ranked alternative is the Manual TCS.

TABLE 7: The opinion of decision makers for each alternative on the basis of criterion "Interoperability".

Criteria	Alternatives (TCS)	DM1	DM2	DM3	DM4
Interoperability (C1)	Manual	W	M	MW	W
	ACM	W	W	MW	W
	Vignette	W	VW	MW	W
	DSRC with barriers	H	MH	M	MW
	DSRC MLFF	H	MH	MH	M
	RFID with barriers	W	MW	M	MW
	RFID ORT	W	MW	MH	M
	Barcode	VW	W	W	W
	GNSS/CN	VH	VH	VH	H
	ANPR	H	M	MH	M
	Infrared	W	MW	MW	W
	Tachograph	VW	W	W	W
	Smart card	VW	W	M	M
	Smartphone	H	MH	H	VH

3.7. Sensitivity Analysis. Table 11 shows the sensitivity analysis results, as the possible range of criteria weights values needed to keep the complete ranking unchanged or for the first place to remain unchanged.

4. Discussion and Conclusion

Planning road pricing scheme is a delicate and often politically charged process, highly dependent upon user and political acceptability for a successful implementation. In this context, toll collection system is crucial for effectiveness of road pricing scheme, both from the standpoint of road managers and from the standpoint of road users. Moreover, toll collection system installation is usually a large-scale investment. Thus, the selection of the optimal toll collection system is a very important decision in the process of planning a road pricing scheme. However, decision-makers are often faced with a choice between a range of alternatives, which often have conflicting advantages and disadvantages. Considering the significance of the toll collection system selection, this research provides a decision-support framework. The discussion here will focus on elaborating the application of this decision-support framework under the assumption of inter- and intraorganizational learning.

In addition to the described framework, results include a systematized analysis of fourteen toll collection systems using SWOT analysis, including their technical and operation characteristics. From the SWOT matrix, one can observe that this analysis itself provides decision-makers with a deeper understanding of advantages and limitations of toll collection systems. Here, it is important to highlight those features that are common to many of toll collection systems. One could then easily identify groups of technologies based on their effect on the traffic flow. For example, manual tolling certainly belongs to the group of technologies that require full stopping for the road pricing transaction to occur. On the other hand, technology such as ANPR with barriers might require deceleration (approximately 30–50 km/h), while GNSS-CN does not require any reduction in vehicle's velocity. Nonetheless, despite its usefulness to organizational learning, SWOT analysis itself is insufficient to help decision-makers select the optimal toll collection system.

Results indicate that the DSRC MLFF, GSNN-CN, and RFID ORT are grouped in the top of ranking. In particular, DSRC MLFF toll system has a narrow advantage over RFID ORT toll technology, while the GSNN-CN technology achieved somewhat worse value for financial criteria (Capex and Opex). Furthermore, smartphone technology got something worse score for traffic criteria (traffic safety). However, when the final ranking is compared to the ranking obtained in the study of Vats et al. (2014), some differences can be noticed. Vats et al. (2014) determined that the optimal alternative for India was the ETC system based on RFID technology, while ANPR and VPS occupied the second and third place, respectively. It should be noted that the research presented here includes additional technologies which were not included in the study by Vats et al. (2014). Moreover, the study presented here contains a larger number of criteria, as well as a greater number of experts from various fields that participated in the evaluation.

In addition to the ranking results, sensitivity analysis provides several other conclusions. First, regarding sensitivity to changes in the technical criteria, one can conclude that the first place is stable in a range from 0%–8.84% up to 0%–100% depending on the criterion. Second, regarding sensitivity to changes in the traffic operations criteria, one can conclude that the first place is stable in a range from 2.89%–15.25% up to 0%–100% depending on the particular criterion. Third, regarding sensitivity to changes in the financial criteria, one can conclude that the ranking of the first place remains stable with increase within 0.00%–8.02% for criterion "average cost for user", within 1.39%–100% for criterion "revenue/cost ratio", and within 7.73%–27.62% for criterion "Capex and Opex", respectively. Fourth, with respect to changes in the environmental criteria, the ranking of the first place is stable in the range up to 100%, depending on the criterion. Lastly, regarding sensitivity to changes in the socioeconomic group of criteria, the ranking of the first place is stable in range

Table 8: Defuzzified alternatives' scores matrix.

Criteria	Manual	ACM	Vignette	DSRC with barriers	DSRC MLFF	RFID with barriers	RFID ORT	Barcode	GNSS/CN	ANPR	Infrared	Tacho-graph	Smart card	Smart-phone
Interoperability	0.06	0.04	0.04	0.10	0.11	0.06	0.08	0.02	0.14	0.10	0.05	0.02	0.06	0.12
Flexibility	0.01	0.03	0.00	0.11	0.11	0.09	0.09	0.02	0.11	0.10	0.08	0.05	0.08	0.12
Efficiency	0.03	0.04	0.04	0.09	0.11	0.08	0.11	0.05	0.11	0.05	0.09	0.06	0.04	0.10
Reliability	0.09	0.06	0.07	0.08	0.07	0.07	0.07	0.05	0.09	0.05	0.07	0.06	0.08	0.08
Enforcement	0.02	0.03	0.13	0.01	0.10	0.01	0.10	0.07	0.10	0.08	0.09	0.10	0.05	0.09
Level of service	0.01	0.03	0.04	0.08	0.10	0.08	0.10	0.06	0.10	0.10	0.08	0.08	0.04	0.09
Traffic safety	0.03	0.03	0.07	0.07	0.09	0.07	0.09	0.08	0.10	0.10	0.08	0.09	0.03	0.06
Number of stops	0.12	0.12	0.11	0.06	0.04	0.06	0.04	0.08	0.04	0.06	0.07	0.06	0.11	0.05
Delays	0.15	0.13	0.13	0.06	0.03	0.06	0.03	0.08	0.03	0.04	0.07	0.04	0.10	0.03
Average cost for user	0.08	0.09	0.12	0.06	0.06	0.06	0.05	0.07	0.07	0.05	0.07	0.08	0.09	0.05
Revenue/cost ratio	0.03	0.06	0.07	0.08	0.09	0.08	0.09	0.07	0.07	0.08	0.07	0.07	0.06	0.09
Capex	0.07	0.06	0.05	0.08	0.07	0.08	0.06	0.08	0.09	0.07	0.08	0.07	0.06	0.07
Opex	0.10	0.07	0.09	0.06	0.07	0.06	0.07	0.06	0.08	0.07	0.06	0.07	0.06	0.07
Visual degrad. of road environ.	0.10	0.09	0.05	0.09	0.05	0.09	0.05	0.09	0.05	0.05	0.09	0.05	0.08	0.05
Air pollution	0.12	0.10	0.08	0.07	0.04	0.07	0.04	0.09	0.04	0.06	0.07	0.06	0.10	0.05
Noise	0.10	0.08	0.07	0.07	0.07	0.07	0.07	0.08	0.07	0.07	0.07	0.07	0.08	0.07
Fair pricing	0.08	0.06	0.01	0.08	0.08	0.08	0.08	0.08	0.08	0.08	0.08	0.08	0.08	0.08
User friendliness	0.01	0.03	0.02	0.08	0.11	0.08	0.10	0.06	0.11	0.09	0.08	0.07	0.05	0.11
Aesthetic effect on the vehicle	0.02	0.02	0.17	0.07	0.07	0.07	0.08	0.13	0.08	0.02	0.08	0.09	0.05	0.05
Possibility of theft and fraud	0.13	0.11	0.12	0.03	0.05	0.03	0.05	0.11	0.04	0.08	0.07	0.09	0.07	0.05

TABLE 9: Preference parameters for alternative ranking.

Criteria	The nature of criteria (Min/Max)	Weight	Preference function	Values of parameter p
Interoperability	max	0.05	V-shape	0.12
Flexibility	max	0.05	V-shape	0.12
Efficiency	max	0.06	V-shape	0.08
Reliability	max	0.06	V-shape	0.04
Enforcement	min	0.05	V-shape	0.12
Level of service	max	0.06	V-shape	0.09
Traffic safety	max	0.06	V-shape	0.07
Number of stops	min	0.06	V-shape	0.08
Delays	min	0.05	V-shape	0.12
Average cost for user	min	0.05	V-shape	0.07
Revenue/cost ratio	max	0.06	V-shape	0.06
Capex	min	0.06	V-shape	0.04
Opex	min	0.05	V-shape	0.04
Visual degradation of road environ	min	0.03	V-shape	0.05
Air pollution	min	0.04	V-shape	0.08
Noise	min	0.04	V-shape	0.03
Fair pricing	max	0.05	V-shape	0.07
User friendliness	max	0.04	V-shape	0.10
Aesthetic effect on the vehicle	min	0.03	V-shape	0.15
Possibility of theft and fraud	min	0.05	V-shape	0.10

TABLE 10: Ranking of the toll alternatives using PROMETHEE method.

Rank	Multicriteria flows	Phi	Phi+	Phi−
1	DSRC MLFF	0.2348	0.2773	0.0425
2	RFID ORT	0.2297	0.2752	0.0454
3	Smartphone	0.2200	0.2708	0.0509
4	GNSS/CN	0.2180	0.2961	0.0781
5	DSRC with barriers	0.1506	0.2277	0.0770
6	ANPR	0.1148	0.2142	0.0995
7	RFID with barriers	0.0995	0.1957	0.0962
8	Infrared	0.0172	0.1448	0.1277
9	Tachograph	−0.0148	0.1384	0.1532
10	Smart card	−0.1296	0.1163	0.2459
11	Barcode	−0.1718	0.0861	0.2579
12	ACM	−0.2721	0.0747	0.3467
13	Vignette	−0.3146	0.0835	0.3981
14	Manual	−0.3816	0.0850	0.4666

from 0.00%–12.06% up to 0%–100%, depending on the criterion.

Here, it is important to highlight the importance of defining and weighting criteria for evaluation of alternative TCS. The suggested decision-making process presented in this study indicates that agency's staff should incorporate their knowledge of specific expert areas along with the agency's objectives. One way this knowledge is incorporated is by defining criteria as well as by grouping them. With this in mind, the presented decision-support framework provides an extensive list of possible criteria for evaluation. These criteria and their group have to be based on the long-term goals that specific transport agency has for its overall system. For example, an agency that is responsible for dominantly regional and intercity motorway network might have different objectives from an agency that is dominantly responsible for suburban or urban motorway tolling. In addition, expert knowledge is particularly important in determining criteria weights. Special attention should be paid to determining criteria weight since they have a crucial role in the analytical procedure for determining the final ranking of alternatives.

In addition to the elements of the proposed decision-support framework above, it is important to highlight also the aspects of the process itself, in the context of organizational learning. In particular, special attention should be paid to the expert knowledge acquisition. The decision-support process

Table 11: Stability intervals matrix for criteria weight regarding the complete ranking and the first place.

Criteria	Complete ranking remains unchanged (%)	First place remains unchanged (%)
Interoperability	3.18–6.03	3.18–10.57
Efficiency	0.00–7.36	0.00–100.00
Flexibility	2.86–8.32	2.25–18.44
Reliability/accuracy	1.63–6.68	0.00–8.84
Enforcement	2.86–7.26	0.00–13.96
Level of service	0.00–7.53	0.00–100.00
Traffic safety	3.97–6.30	2.89–15.24
Number of stops	0.00–7.36	0.00–100.00
An increase in travel time	0.00–15.10	0.00–100.00
Average cost for user	4.39–8.02	0.00–8.02
Ratio of revenues/costs	5.48–23.90	1.39–100.00
Capex	5.65–7.73	2.98–7.73
Opex	4.30–10.10	0.00–27.62
Visual impact of defacing ambient environment	1.25–6.48	0.00–100.00
Air pollution	0.00–5.39	0.00–100.00
Noise	0.00–14.18	0.00–100.00
Fair payment	0.00–10.56	0.00–100.00
Easy to use	0.00–11.97	0.00–100.00
Aesthetic effect on the vehicle	2.10–7.20	0.00–12.06
Possibility of theft and fraud	0.00–6.71	0.00–17.80

itself should aim at maximizing the performance of experts' decisions by minimizing human cognitive errors. In addition, the process should be able to generate several alternative plans by making changes in the decision criteria weights, which would allow experts to evaluate impacts of those changes in a matter of exploratory what-if analysis. One example of such exploratory analysis would be completely excluding a group of criteria. Such exploratory analysis has particular potential in the case of the existence of TCS in the particular area, where some aspects, such as interoperability, might be worthy of a more focused investigation.

It is important to highlight here that this is ultimately a decision-support and not a decision-making framework. Thus, joint discussion between experts during the workshop is an important aspect of supporting the decision-making process. Such a discussion is ultimately necessary for transparent and uniform presentation of alternative decisions. Moreover, experts should be encouraged from the start to understand the decision-making process as an iterative learning experience, where mutually agreed alternative is sought as the ultimate goal. Thus, approaching decision-making process as an iterative one allows for a range of perspectives and agency's objectives to be taken into account. Here, the process can benefit from experts with various backgrounds, such as planning, economics, operations, or maintenance. However, as the proposed decision-support framework uses fuzzy logic, subjective uncertainty can be filtered out for increasing coherence in decision-making.

The methodological framework presented here is transferable, as it can be applied to different case studies, while adopting to local specifics. As objective implementation of MCDM procedure is of the utmost importance, special attention should be paid to identifying relevant experts to join the workshop. However, bearing in mind the need to select one alternative, there is a range of requirements that an agency needs to take into account in its decision-making procedures. The framework presented here can help in organizing road toll planning workshops by providing analysis based and transparent comparison of alternatives. Moreover, approaching these workshops as an opportunity to enhance inter- and intraorganizational learning, the framework presented here can be used as a communication medium.

On the contrary, one has to recognize that a large number of alternatives and criteria are also a disadvantage from the perspective of workload scope. In addition, for many of the alternatives, it is difficult to determine their scores per criterion in an analytical manner. Thus, the lack of analytically determined data in the scoring procedure for each of the alternatives per criterion is a limitation of this effort. Including as wide as possible range of experts can help in overcoming the challenge of missing alternative data. However, since group dynamics in workshops cannot be properly maintained with a significant number of participants, decision-makers will have to seek other solutions. Thus, one potential direction for future research is investigating the potential of using microsimulation techniques for evaluating some toll collection system effects, such as environmental pollution. In addition, when selecting the optimal toll collection system, it is also extremely significant to consider the attitudes of the highway users. In order to achieve this, agency can conduct a survey to determine the users' attitudes towards particular toll collection systems. Surveying users would provide information about their expectations, which would contribute to the TCS ranking methodology from the users' point of view. Finally, exploration of implications for the development of emerging self-driving vehicle technology in relation to different existing toll collection technologies is an important future effort.

Conflicts of Interest

All the authors declare that there are no conflicts of interest regarding the publication of this paper.

References

[1] E. Deakin, "Toll roads: a new direction for US highways?" *Built Environment*, vol. 15, no. 3-4, pp. 185–194, 1989.

[2] S. Ison, "Pricing road space: Back to the future? the cambridge experience," *Transport Reviews*, vol. 16, no. 2, pp. 109–126, 1996.

[3] R. Arnold, V. C. Smith, J. Q. Doan et al., "Reducing congestion and funding transportation using road pricing in Europe and Singapore," Report No. FHWA-PL-10-030, Federal Highway Administration, Washington, D.C., USA, 2010.

[4] D. Glavić, M. Milenković, A. Trpković, M. Vidas, and M. N. Mladenović, "Assessing sustainability of road tolling technologies," in *Proceedings of the nternational Congress on Transport Infrastructure AND Systems - TIS Roma, 803-810*, Roma, Italy, 2017.

[5] D. Glavic, M. Mladenovic, T. Luttinen, S. Cicevic, and A. Trifunovic, "Road to price: User perspectives on road pricing in transition country," *Transportation Research Part A: Policy and Practice*, vol. 105, pp. 79–94, 2017.

[6] C. H. Sørensen, K. Isaksson, J. Macmillen, J. Åkerman, and F. Kressler, "Strategies to manage barriers in policy formation and implementation of road pricing packages," *Transportation Research Part A: Policy and Practice*, vol. 60, pp. 40–52, 2014.

[7] C. Jakobsson, S. Fujii, and T. Gärling, "Determinants of private car users' acceptance of road pricing," *Transport Policy*, vol. 7, no. 2, pp. 153–158, 2000.

[8] S. Jaensirisak, M. Wardman, and A. D. May, "Explaining variations in public acceptability of road pricing schemes," *Journal of Transport Economics and Policy*, vol. 39, no. 2, pp. 127–153, 2005.

[9] J. Schade and M. Baum, "Reactance or acceptance? Reactions towards the introduction of road pricing," *Transportation Research Part A: Policy and Practice*, vol. 41, no. 1, pp. 41–48, 2007.

[10] L. Eriksson, J. Garvill, and A. M. Nordlund, "Acceptability of single and combined transport policy measures: the importance of environmental and policy specific beliefs," *Transportation Research Part A: Policy and Practice*, vol. 42, no. 8, pp. 1117–1128, 2008.

[11] S. Pahaut and C. Sikow, "History of thought and prospects for road pricing," *Transport Policy*, vol. 13, no. 2, pp. 173–176, 2006.

[12] D. King, M. Manville, and D. Shoup, "The political calculus of congestion pricing," *Transport Policy*, vol. 14, no. 2, pp. 111–123, 2007.

[13] C. G. Chorus, J. A. Annema, N. Mouter, and B. van Wee, "Modeling politicians' preferences for road pricing policies: A regret-based and utilitarian perspective," *Transport Policy*, vol. 18, no. 6, pp. 856–861, 2011.

[14] F. Oberholzer-Gee and H. Weck-Hannemann, "Pricing road use: Politico-economic and fairness considerations," *Transportation Research Part D: Transport and Environment*, vol. 7, no. 5, pp. 357–371, 2002.

[15] H. Li, M. C. J. Bliemer, and P. H. L. Bovy, "Network reliability-based optimal toll design," *Journal of Advanced Transportation*, vol. 42, no. 3, pp. 311–332, 2008.

[16] S. Vats, G. Vats, R. Vaish, and V. Kumar, "Selection of optimal electronic toll collection system for India: A subjective-fuzzy decision making approach," *Applied Soft Computing*, vol. 21, pp. 444–452, 2014.

[17] C. Ragsdale, *Spreadsheet Modeling and Decision Analysis: A Practical Introduction to Business Analytics*, Nelson Education, 2014.

[18] M. M. Abbas, M. N. Mladenovic, S. Ganta, Y. Kasaraneni, and C. C. McGhee, "Development and use of critical functional requirements for controller upgrade decisions," *Transportation Research Record*, no. 2355, pp. 83–92, 2013.

[19] D. Teodorović, "Multicriteria ranking of air shuttle alternatives," *Transportation Research Part B: Methodological*, vol. 19, no. 1, pp. 63–72, 1985.

[20] S. Chen, Y. Leng, B. Mao, and S. Liu, "Integrated weight-based multi-criteria evaluation on transfer in large transport terminals: A case study of the Beijing South Railway Station," *Transportation Research Part A: Policy and Practice*, vol. 66, no. 1, pp. 13–26, 2014.

[21] S. Yedla and R. M. Shrestha, "Multi-criteria approach for the selection of alternative options for environmentally sustainable transport system in Delhi," *Transportation Research Part A: Policy and Practice*, vol. 37, no. 8, pp. 717–729, 2003.

[22] A. Zubaryeva, C. Thiel, N. Zaccarelli, E. Barbone, and A. Mercier, "Spatial multi-criteria assessment of potential lead markets for electrified vehicles in Europe," *Transportation Research Part A: Policy and Practice*, vol. 46, no. 9, pp. 1477–1489, 2012.

[23] M. Kumru and P. Y. Kumru, "Analytic hierarchy process application in selecting the mode of transport for a logistics company," *Journal of Advanced Transportation*, vol. 48, no. 8, pp. 974–999, 2014.

[24] J. Li, Z. Q. Yue, and S. C. Wong, "Performance evaluation of signalized urban intersections under mixed traffic conditions by gray system theory," *Journal of Transportation Engineering*, vol. 130, no. 1, pp. 113–121, 2004.

[25] T.-Y. Pai, K. Hanaki, H.-H. Ho, and C.-M. Hsieh, "Using grey system theory to evaluate transportation effects on air quality trends in Japan," *Transportation Research Part D: Transport and Environment*, vol. 12, no. 3, pp. 158–166, 2007.

[26] M. Jakimavičius and M. Burinskiene, "A GIS and multi-criteria-based analysisand rankingof transportation zonesof Vilniuscity," *Technological and Economic Development of Economy*, vol. 15, no. 1, pp. 39–48, 2009.

[27] M. Rogers and M. P. Bruen, "Using ELECTRE to rank options withina highway environmental appraisal-two case studies," *Civil Engineering Systems*, vol. 13, no. 3, pp. 203–221, 1996.

[28] L. F. A. M. Gomes, "Multicriteria ranking of urban transportation system alternatives," *Journal of Advanced Transportation*, vol. 23, no. 1, pp. 43–52, 1989.

[29] H. Oeltjenbruns, W. J. Kolarik, and R. Schnadt-Kirschner,

"Strategic planning in manufacturing systems - AHP application to an equipment replacement decision," *International Journal of Production Economics*, vol. 38, no. 2-3, pp. 189–197, 1995.

[30] O. Kulak, "A decision support system for fuzzy multi-attribute selection of material handling equipments," *Expert Systems with Applications*, vol. 29, no. 2, pp. 310–319, 2005.

[31] M. Dağdeviren, "Decision making in equipment selection: an integrated approach with AHP and PROMETHEE," *Journal of Intelligent Manufacturing*, vol. 19, no. 4, pp. 397–406, 2008.

[32] M. Mladenovic and M. Abbas, "A guide to effective Adaptive Traffic Control Systems," *Traffic, Engineering and Control*, vol. 53, no. 2, pp. 52–54, 2012.

[33] M. N. Mladenovic and M. M. Abbas, "Improved decision-support making for selecting future traffic signal controllers using expert-knowledge acquisition," in *Proceedings of the 2013 16th International IEEE Conference on Intelligent Transportation Systems - (ITSC 2013)*, pp. 732–735, The Hague, Netherlands, October 2013.

[34] M. N. Mladenovic, M. M. Abbas, S. Gopinath, and G. Faddoul, "Decision-Support System for Assessment of Alternative Signal Controllers using Expert Knowledge Acquisition," in *Proceedings of the 92nd Annual Meeting of Transportation Research Board*, Washington, D.C., USA, 2013.

[35] M. N. Mladenović, A. Stevanović, I. Kosonen, and D. Glavić, "Adaptive Traffic Control Systems: Guidelines for Development of Functional Requirements," in *Proceedings of the Mobil. TUM 2015-International Scientific Conference on Mobility and Transport*, Munich, Germany, 2015.

[36] M. N. Mladenovic, K. Mangaroska, and M. M. Abbas, "Decision support system for planning traffic operations assets," *Journal of Infrastructure Systems*, vol. 23, no. 3, Article ID 05017001, 2017.

[37] J. P. Brans, "Lingenierie de la decision. Elaboration dinstrumentsdaide a la decision. Methode PROMETHEE," in *Laide a la Decision: Nature*, R. Nadeau and M. Landry, Eds., pp. 183–214, Instruments et Perspectives Davenir, de Universite Laval, Quebec, Canada, 1982.

[38] J.-P. Brans, P. Vincke, and B. Mareschal, "How to select and how to rank projects: the PROMETHEE method," *European Journal of Operational Research*, vol. 24, no. 2, pp. 228–238, 1986.

[39] M. Behzadian, R. B. Kazemzadeh, A. Albadvi, and M. Aghdasi, "PROMETHEE: a comprehensive literature review on methodologies and applications," *European Journal of Operational Research*, vol. 200, no. 1, pp. 198–215, 2010.

[40] J. D. McMillan, W. S. Adney, J. R. Mielenz, and K. T. Klasson, "Development of a multicriteria assessment model for ranking biomass feedstock collection and transportation systems," in *Proceedings of the 27th Symposium on Biotechnology for Fuels and Chemicals*, pp. 71–87, Humana Press, 2006.

[41] H. Safaei Mohamadabadi, G. Tichkowsky, and A. Kumar, "Development of a multi-criteria assessment model for ranking of renewable and non-renewable transportation fuel vehicles," *Energy*, vol. 34, no. 1, pp. 112–125, 2009.

[42] A. Jahan and K. Edwards, "Multi-criteria Decision Analysis for Supporting the Selection of Engineering Materials in Product Design," *Butterworth-Heinemann*, 2013.

[43] H. Zimmermann, *Fuzzy Set Theory—and Its Applications*, Kluwer Academic, Dordrecht, The Netherlands, 2nd edition, 1992.

[44] C. Macharis, A. De Witte, and J. Ampe, "The multi-actor, multi-criteria analysis methodology (MAMCA) for the evaluation of transport projects: Theory and practice," *Journal of Advanced Transportation*, vol. 43, no. 2, pp. 183–202, 2009.

[45] D. A. Hensher, "Electronic toll collection," *Transportation Research Part A: General*, vol. 25, no. 1, pp. 9–16, 1991.

[46] P. Sharma and V. Sharma, "Electronic toll collection technologies: A state of art review," *International Journal of Advanced Research in Computer Science and Software Engineering*, vol. 4, no. 7, pp. 621–625, 2014.

[47] V. Surendran, K. S. Vignesh, S. Baburaj, V. S. Vishnu, S. R. Krishnaveni, and K. S. Sowmya, "RFID Based Automated Toll Collection System," *International Journal of Computer Science and Information Technology & Security (IJCSITS)*, vol. 6, no. 1, pp. 582–585, 2016.

[48] P. Blythe, "RFID for road tolling, road-use pricing and vehicle access control," *IEE Colloquium (Digest)*, no. 123, pp. 67–82, 1999.

[49] I. Catling, "Road user charging using vehicle positioning systems," in *Proceedings of the Tenth International Conference on Road Transport Information and Control*, pp. 126–130, London, UK.

[50] G. Charpentier and G. Fremont, "The ETC system for HGV on motorways in Germany: first lessons after system opening," in *Proceedings of the European Transport Conference (ETC)*, Association for European Transport, Strasbourg, France, 2003.

[51] P. A. Sorensen and B. D. Taylor, *Review and Synthesis of Road-Use Metering and Charging Systems*, Transportation Research Board, Washington, D.C, 2005.

[52] M. Bibaritsch and C. Egeler, "GO Maut-Enforcement: The enforcement system of the Austrian heavy goods vehicle toll," in *Proceedings of the 13th World Congress on Intelligent Transport Systems and Services, ITS 2006*, gbr, October 2006.

[53] W.-Y. Shieh, W.-H. Lee, S.-L. Tung, and C.-D. Ho, "A novel architecture for multilane-free-flow electronic-toll-collection systems in the millimeter-wave range," *IEEE Transactions on Intelligent Transportation Systems*, vol. 6, no. 3, pp. 294–301, 2005.

[54] M. Staudinger and E. A. Mulka, "Electronic vehicle identification using active infrared light transmission," in *ITS America 14th Annual Meeting and Exposition*, the Crossroads: Integrating Mobility Safety and Security, ITS America 14th Annual Meeting and Exposition, 2004.

[55] S. Tropartz, E. Hoerber, and K. Gruener, "Experiences and results from vehicle classification using infrared overhead laser sensors at toll plazas in New York City," in *Proceedings of the 1999 IEEE/IEEJ/JSAI International Conference on Intelligent Transportation Systems*, pp. 686–691, October 1999.

[56] European Parliament's Committee on Transport and Tourism, *Technology options for the European Electronic Toll Service*, European Union, 2014.

[57] V. Sridhar and M. Nagendra, "Smart Card Based Toll Gate Automated System," *International Journal of Advanced Research in Computer Engineering & Technology (IJARCET)*, vol. 1, no. 5, pp. 203–212, 2012.

[58] B. Asady and A. Zendehnam, "Ranking fuzzy numbers by distance minimization," *Applied Mathematical Modelling*, vol. 31, no. 11, pp. 2589–2598, 2007.

[59] C. Macharis, J. P. Brans, and B. Mareschal, "The GDSS promethee procedure," *Journal of Decision Systems*, vol. 7, no. 4, pp. 283–307, 1998.

[60] *Visual PROMETHEE 1.4 Manual*, http://www.promethee-gaia.net/assets/vpmanual.pdf.

5

Characterizing Heterogeneity in Drivers' Merging Maneuvers Using Two-Step Cluster Analysis

Gen Li[1] and Lu Sun[2]

[1]*School of Transportation, Southeast University, Nanjing 210096, China*
[2]*Department of Operations Research and Financial Engineering, Princeton University, 229 Sherrerd Hall, Princeton, NJ 08544, USA*

Correspondence should be addressed to Gen Li; gilg4226307@aliyun.com

Academic Editor: Ludovic Leclercq

In order to investigate the heterogeneity in merging behaviors on freeways, a novel data mining tool, called two-step cluster analysis, is applied to the merging maneuvers (namely, initial speed, merging speed, and merging position). Merging maneuvers of 370 drivers collected from the NGSIM dataset are automatically and optimally segmented into four clusters (Early Merging Drivers at High Speed, Early Merging Drivers at Low Speed, Late Merging Drivers at Low Speed, and Late Merging Drivers at High Speed) by the two-step cluster analysis. Hypothesis test confirms the significant differences in merging maneuvers between different clusters. The clustered data are used to find the best corresponding fitting distributions. Seven distributions (Normal, Log-normal, Student's t, Logistic, Log-Logistic, Gamma, and Weibull) are considered for each cluster and the Kolmogorov-Smirnov test statics are used to select the best fitted distributions. It is found that merging drivers may merge either early or late, under congestion or uncongested traffic condition. Further analysis of merging durations shows that Late Merging Drivers use significantly shorter time than Early Merging Drivers to finish the merging maneuver, no matter if they are at high or at low speed. Hypothesis test of accepted lead gaps and lag gaps indicate that merging drivers are more sensitive to the lag gaps under congestion. The proposed method can automatically identify the heterogeneity in merging drivers and the results obtained in this paper can be used to enhance the accuracy of the merge behavior models in microscopic simulation software.

1. Introduction

Car following and lane changing are two fundamental tasks conducted by drivers and the car-following model and lane-changing model are two of the most significant submodels in microscopic traffic simulation models. However, lane-changing models have not been given as much attention as car-following models [1–4].

Lane changes can be divided into two types: discretionary lane changes and mandatory lane changes. Merging behavior is one kind of typical mandatory lane changes when vehicles have to move from an on-ramp to the main road. Merging behavior has been confirmed to have impact influence on the traffic operations, and it may trigger traffic congestions and breakdowns [5, 6].

Recently, several studies criticized the ability of lane-changing models to represent reality due to the ignorance of driver heterogeneity [7–10] (Hoogendoorn et al. 2006). Several studies have investigated the heterogeneity among drivers using macroscopic traffic models [11–15]. Other studies investigated the driver heterogeneity during car-following process [9, 10]. It is believed that accommodating the heterogeneity within the driver population is inevitable for producing a more precise car-following model (Kim et al. 2013). However, heterogeneity in lane-changing models has not received much emphasis in the literature.

To investigate the heterogeneity in merging behaviors, this study proposes a new data mining tool, called two-step cluster analysis, combining cluster analysis to capture the heterogeneity in merging behaviors by dividing merging into some homogeneous components. The paper is organized as follows. The next section will provide a literature review on the existing studies followed by Section 3, which describes the NGSIM data used in this paper. Section 4 gives the methodology. Results and discussions are presented in Section 5. Finally, the conclusions are presented in Section 6.

2. Literature Review

Gap acceptance was considered to be the most important part in existing lane-changing models [17–21] (Lee 2006). It is assumed that a driver makes a lane change when both the lead and lag gaps in the target lane are larger than the so-called critical gap. Different definitions of critical gap were used in different studies. Herman and Weiss [22] assumed an exponential distribution for critical gaps, Drew et al. [23] assumed Log-normal distribution, and Miller [24] assumed a normal distribution. Gipps [25] is believed to be the first to use gap acceptance theory to develop a comprehensive framework for lane-changing model. Hidas [26] and Wang [27] applied similar principle to build freeway mandatory lane-changing models. The framework of Gipps' model has also been use in several microscopic traffic simulation software [28–30]. In VISSIM, for a mandatory lane change, the critical gap depends on the acceptable maximum deceleration for the lane-changing driver and his presumed follower. In CORSIM, ten different driver types can be defined with variable gap acceptance values [28] and each gap acceptance decision is independent considering the current available gap and the personal gap acceptance value. In gap acceptance models using critical gaps, the most basic assumption is that a driver will accept the adjacent gap only if both the lead and lag gaps are larger than the critical gap [31] (Ma and Andreasson 2007; Hastie et al. 2001; Martin et al. 2012). However, this assumption is often criticized as it is often inconsistent with the real world observations that vehicles still take lane changes when only the lead or lag gap or even none of them are larger than the critical gap [2, 32, 33]. To overcome this deficiency, a binary logit model was built by Kita (1993). Weng and Meng [34] and Marczak et al. [32] used the same kind of model to predict merging decision in short-term work zone merging areas and to compare the gap acceptance of merging decision at two sites, respectively.

It was claimed that the existing lane-changing models could not reflect the real traffic behavior under congestion [26, 35] (Ahmed et al. 1996). Thus, "forced" and "cooperative" lane-changing models were proposed to describe distinctive behaviors of vehicles under congestion [18, 26] (Ahmed et al. 1996; Hidas 2005; Lee 2006; Rao 2006). A framework for merging behavior with latent plans was proposed by Choudhury et al. (2007). In this paper, three plans, normal merge, merge with courtesy and forced merge, were considered to be latent in the model. However, only accepted gaps were considered and rejected gaps were ignored, and some of the estimated coefficients in the model were not significant [32].

In recent years, driver heterogeneity has been considered as a key element in driver behavior and investigated in several macroscopic traffic studies [12, 14, 36–39] (Sun et al. 2001). In microscopic traffic models, heterogeneity in car-following models was investigated by deriving the joint distribution of model coefficients depending on an empirical basis [7–10] (Hoogendoorn et al. 2006). However, driver heterogeneity in lane-changing models has not drawn similar attention, and no studies try to investigate the heterogeneity among lane-changing drivers.

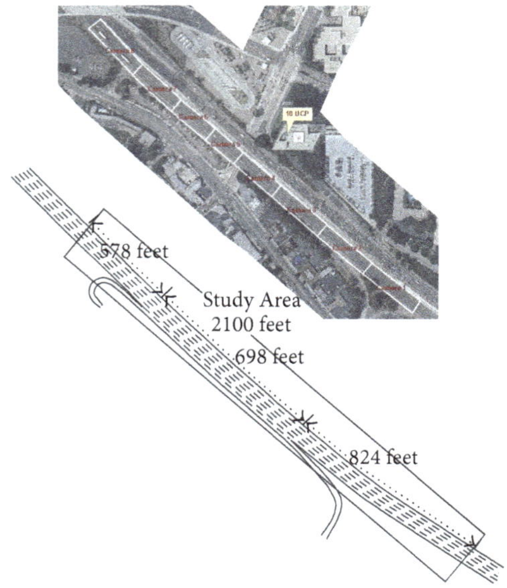

Figure 1: US Highway 101 study corridor from NGSIM [16].

Clustering is a widely used technique in data mining applications for identifying patterns in datasets. However, there are few studies that apply cluster analysis to identifying heterogeneity in drivers. The only related study was conducted by Hill et al. [40], in which 46 participant drivers were classified into 4 clusters based their behaviors in driving experiments using *k-means* clustering method. Two-step cluster analysis is a statistical method that can automatically identify similar clusters of people or objects within data sets [41]. Two-step cluster analysis is also considered more reliable and accurate than traditional clustering methods such as the *k*-means clustering algorithm [42].

A methodology framework than combines two-step cluster analysis; hypothesis test and probability distribution fitting are proposed in this paper. The two-step cluster analysis can naturally and automatically segment the merging drivers into optimal number of clusters. The hypothesis test can identify the significant differences between the clusters and capture the heterogeneity in merging behaviors. At last different probability distribution functions of related parameters can be identified through analyzing the clusters.

3. Data Preparation

In order to investigate the merging behaviors in the weaving sections, the NGSIM vehicle trajectory data collected on a segment of southbound US Highway 101 (Hollywood Freeway) in Los Angeles, CA, are chosen in this paper [43]. Figure 1 shows the site for US Highway 101. This US-101 section is 640 meters long and has five main lanes and one auxiliary lane. The vehicle trajectories were collected from 7:50 a.m. to 8:35 a.m. on June 15, 2005. The road section was covered by eight cameras and the dataset was updated at a resolution of 10 frames per second [16].

In our research, we focused on the merging behavior. The NGSIM database distinguishes the on-ramp, auxiliary lane, and main lane clearly. Thus, it is easy for us to collect the trajectories of 399 merging vehicles. However, the previous study [31, 44] shows that the original trajectory data seem to contain some noise and errors. The velocities and acceleration provided by the NGSIM cannot be directly used. Thus, we applied the following data smoothing techniques proposed by Thiemann et al. [44]:

(1) The velocities and acceleration of vehicles are directly estimated from the longitudinal positions.

(2) The locations (both local lateral and longitudinal coordinates), velocities, and acceleration of vehicles are smoothed by the symmetric exponential moving average filter (sEMA) proposed by Thiemann et al. [44] to decrease measurement errors in the data. The smoothing times of sEMA method are set as the suggested values for the US Highway 101 data set in Thiemann et al. [44].

Although the random errors can be reduced by the smoothing process, there are still some errors in the data. Thus, the following heuristic rules are applied to filter the data sets:

(1) Filter out the data with motorcycle or heavy vehicle to focus on merge behaviors of cars.

(2) Filter out the data if the merging vehicle cross the lane line before entering the auxiliary lane.

(3) Filter out the data if the merging vehicle cross the lane line at the position further than the length of auxiliary lane

After filtering and smoothing the trajectory data, we conduct a further consistency checking to insure the consistency and accuracy of all the data subsets from US-101 dataset.

First, we check the frame ID, number of frames, vehicle length, vehicle width, and vehicle type and find that the frame IDs, the total numbers of the frames, global time, and the vehicle length and width of each vehicle in US-101 dataset are found to be consistent. However, through a searching process, we found that the local coordinates and global coordinates of the three sub-datasets are inconsistent. We search several data points that have the same global coordinates among the three data subsets of US-101 datasets and found that the three data subsets of US-101 dataset are consistent in local x, but inconsistent in local y. As this paper is focused on merging behavior, we use the beginning point of the auxiliary lane, for example. The local y for the beginning points of the auxiliary lane for three data sets are 615.402 ft, 655.675 ft, and 650.172 ft, respectively. It means the upstream edge in data set 1 is at 40.273 ft in data set 2 and 34.770 ft in data set 3. Thus, the three datasets are unified by using the local coordinates of data set 2.

After smoothing, filtering, and consistency checking, 370 trajectories are kept. After getting the trajectories, the parameters of merge maneuvers and gap acceptance can be interpolated. In this paper, merge maneuvers include merging speed, merging position, and merging duration. The parameters of gap acceptance include the accepted gaps, lead gaps, and lag gaps.

For each identified merging vehicle, the merging speed, merging position, accepted gap, accepted lead gap, and lag gap are collected when the front center of the merging vehicle crossed the lane markers.

To extract the merging durations, the method used to determine the start and end point of discrepancy lane-changing behavior in Wang et al. [1] is used in this paper. For each merging vehicle, the start time is taken as the first time when the merging vehicle has the lateral velocity larger than 0.2 m/s and the end time is taken as the first time when the merging vehicle has the lateral velocity less than 0.2 m/s after crossing the lane line.

4. Methodology

4.1. Two-Step Cluster Analysis. The two-step cluster analysis is a scalable cluster analysis algorithm that was designed to manage large datasets. The analysis has two steps: preclustering and clustering.

In the preclustering step, all the cases in the data are scanned and the log-likelihood distance between them is measured to determine whether they are going to form preclusters based on some threshold distance criterion [45].

The log-likelihood distance is a probability based distance. The distance between two clusters is related to the decrease in log-likelihood as they are combined into one cluster. The distance between clusters j and s is defined as follows:

$$d(i,j) = \varepsilon_i + \varepsilon_j - \varepsilon_{\langle i,j\rangle} \tag{1}$$

$$\varepsilon_i = -n_i\left(\sum_{k=1}^{K}\frac{1}{2}\log\left(\hat{\sigma}_{ik}^2 + \hat{\sigma}_k^2\right)\right) \tag{2}$$

$$\varepsilon_j = -n_j\left(\sum_{k=1}^{K}\frac{1}{2}\log\left(\hat{\sigma}_{jk}^2 + \hat{\sigma}_k^2\right)\right) \tag{3}$$

$$\varepsilon_{\langle i,j\rangle} = -n_{\langle i,j\rangle}\left(\sum_{k=1}^{K}\frac{1}{2}\log\left(\hat{\sigma}_{i,jk}^2 + \hat{\sigma}_k^2\right)\right), \tag{4}$$

where $d(i,j)$ is the log-likelihood distance between clusters i and j; $\langle i,j\rangle$ represents the cluster formed by merging clusters i and j; K is the total number of continuous variables; $\hat{\sigma}_k^2$ is the estimated variance of the continuous variable k for the entire dataset; $\hat{\sigma}_{ik}^2$, $\hat{\sigma}_{jk}^2$, and $\hat{\sigma}_{i,jk}^2$ are the estimated variances of the continuous variable k in clusters i, j and $\langle i,j\rangle$.

In the second step, the subclusters resulting from the preclustering step are clustered into the optimal number of clusters using an agglomerative clustering algorithm. The subclusters are regarded as single cases and merged into one cluster by satisfying the minimum distance in (1) until all data are placed within a single cluster.

If a desired number of clusters is not predetermined, the algorithm can automatically determine the optimal number of cluster using Schwarz's Bayesian Criterion (BIC) or the Akaike Information Criterion (AIC) [45]:

$$\begin{aligned} \mathrm{AIC}(J) &= -2\sum_{j=1}^{J}\varepsilon_j + 2m_J \\ \mathrm{BIC}(J) &= -2\sum_{j=1}^{J}\varepsilon_j + m_J\log(N), \end{aligned} \tag{5}$$

where J is the number of clusters; K is the total number of continuous variables clusters; N is the total number of observations; $m_J = 2KJ$.

The two-step analysis is conducted using SPSS 22.0 in this paper.

4.2. Hypothesis Test for Difference between Two Medians. To statistically investigate if there are significant differences between different clusters, the Mann–Whitney U test is conducted in this paper.

Normally, Student's t-test is used to determine if the means of two sets of data are different from each other. However Student's t-test is most commonly applied when the test statistic would follow a Normal distribution if the value of a scaling term in the test statistic was known [46]. It has been shown in Section 4 that the distributions of merging maneuver parameters do not all follow Normal distribution. Thus, Mann–Whitney U test, which is a distribution-free test and has been used in previous study [47], is performed in this paper. It is assumed in Mann–Whitney U test that the populations must be continuous and their probability density functions must have the same shape and spread, which are somewhat less restrictive than the assumptions needed for Student's t-test [47].

4.3. Best Probability Distribution. The clustered data are used to find the best corresponding fitting distributions using EasyFit 5.4 [48, 49]. To best investigate the distributions of the merging parameters, 7 distributions (Normal, Log-normal, Student's t, Logistic, Log-Logistic, Gamma, and Weibull) are considered for each cluster and the Kolmogorov-Smirnov test is used to select the best fitted distribution. The Kolmogorov-Smirnov test statistic is defined by the greatest vertical distance at any point between the two cumulative probability distributions. The smaller the Kolmogorov-Smirnov static is and the bigger the p value is, the better the distribution can describe the sample.

4.3.1. Normal Distribution. A random variable X has Normal distribution with the location parameter μ and scale parameter σ^2, if its density reads the following:

$$f\left(x \mid \mu, \sigma^2\right) = \frac{1}{\sqrt{2\pi\sigma^2}} \exp\left(-\frac{(x-\mu)^2}{2\sigma^2}\right) \quad (6)$$

The Normal distribution is usually denoted as $X \sim N(\mu, \sigma^2)$.

4.3.2. Log-Normal. A random variable X has Log-normal distribution if its logarithm is normally distributed with location parameter μ and scale parameter σ^2, which is usually denoted as $\ln(X) \sim N(\mu, \sigma^2)$

4.3.3. Student's t. A random variable Y follows a skew-t distribution with location the parameter μ and scale parameter σ^2 and degrees of freedom ν if it has the following representation:

$$Y = \mu + \sigma\frac{X}{W}$$
$$X \sim N\left(\mu, \sigma^2\right), \quad (7)$$
$$W \sim \Gamma\left(\frac{\nu}{2}, \frac{\nu}{2}\right),$$

where X is a standard Normal distribution and Γ is the Gamma function.

4.3.4. Logistic. A random variable X follows a logistic distribution if its cumulative distribution function is the logistic function. It resembles the Normal distribution in shape but has heavier tails (higher kurtosis). The probability density function of the logistic distribution is given by the following:

$$f(x \mid \mu, s) = \frac{e^{-(x-\mu)/s}}{s\left(1 + e^{-(x-\mu)/s}\right)^2}, \quad (8)$$

where μ is the location parameter and s is the scale parameter. The logistic distribution is usually denoted as $X \sim \text{Logistic}(\mu, s)$.

4.3.5. Log-Logistic. A random variable X has log-logistic distribution if its logarithm has logistic distribution with the location parameter μ and scale parameter s, which is usually denoted as $\ln(X) \sim \text{Logistic}(\mu, s)$

4.3.6. Gamma. A random variable X follows Gamma distribution with the shape parameter μ and scale parameter k if its density reads the following:

$$f(x \mid k, \theta) = \frac{x^{k-1}e^{-x/\theta}}{\theta^k\Gamma(k)} \quad (9)$$

Gamma distribution is usually denoted as $X \sim \Gamma(\theta, k)$.

4.3.7. Weibull. The probability density function of a Weibull random variable is

$$f(x \mid \lambda, k) = \begin{cases} \frac{k}{\lambda}\left(\frac{x}{\lambda}\right)^{k-1} e^{-(x/\lambda)^k} & x \geq 0 \\ 0 & x < 0, \end{cases} \quad (10)$$

where $k > 0$ is the shape parameter and $\lambda > 0$ is the scale parameter. Weibull distribution is usually denoted as $X \sim \text{Weibull}(k, \lambda)$.

5. Cluster Results and Discussions

5.1. Clustering Results. To investigate the heterogeneity among merging drivers, the two-step cluster analysis is applied to segment the data of merging maneuvers. We choose the initial speed, merging speed, and merging position for cluster analysis. The initials speed is defined as the speed when the merging vehicle enters the auxiliary lane.

TABLE 1: Results of cluster analysis.

Parameters	Variables	Cluster 1	Cluster 2	Cluster 3	Cluster 4
proportion		33.6%	26%	24%	16.4%
mean	Initial Speed	14.547	10.481	13.427	16.768
	Merging Speed	14.186	9.478	9.950	15.613
	Merging Position	45.81	37.09	147.52	194.89
median	Initial Speed	14.167	10.936	13.576	16.663
	Merging Speed	14.082	10.104	10.525	14.921
	Merging Position	44.00	32.11	145.48	190.48
standard deviation	Initial Speed	1.566	2.035	2.069	1.763
	Merging Speed	1.630	2.366	2.511	2.943
	Merging Position	21.46	25.95	49.19	47.89
Skewness	Initial Speed	0.91	−1.12	−0.05	0.30
	Merging Speed	0.52	−0.95	−0.69	0.08
	Merging Position	0.68	1.34	0.28	0.79

It can more or less reflect the expected speeds of merging drivers. The merging speed is defined as the speed when the front center of merging vehicle crosses the lane line. It can be linked to the speed adjustment behavior on the auxiliary lane. And the merging position is defined as the position of the vehicle when the front center of merging vehicle crosses the lane line. All the three maneuvers are indicators of aggressiveness and can reflect drivers' personal characteristics.

The optimal number of clusters is automatically determined by the two-step cluster analysis. The optimal number is both 4 by using AIC and BIC criteria. Table 1 shows the results of the cluster analysis. Figure 2 shows the 3D scatter plots and 2D scatter plots of the clustering results.

For comparison, a K-means clustering approach was also performed and the cluster number was also set at 4. Figure 3 shows the 3D scatter plots and 2D scatter plots of the K-means clustering results. One can find that the K-means clustering cannot well separate the initial speed and merging speed and the results of the two-step cluster analysis show better separation results in all three variables.

From Table 1, one can observe that merging drivers in each cluster is 33.6%, 26%, 24%, and 16.4%, respectively. It means no cluster dominated the most prevailing merging drivers.

From Table 1, the mean merging positions of clusters 1, 2, 3, and 4 are 45.81, 37.09, 147.52, and 194.89 m, respectively. Both the initial speed and merging speed have the lowest values in cluster 2 and largest values in cluster 4. The merging speeds are smaller than the initial speed in all clusters. It is interesting to find that mean values of merging speeds of clusters 2 and 3 are similar. However, the mean values of initial speeds of clusters 2 and 3 are 10.481 m/s and 13.427 m/s, respectively, the difference of which is much larger than that of merging speeds. It indicates that the merging drivers in cluster 2 have lower expecting speed than cluster 3 and they might have prepared for merging before they get on the auxiliary lane.

One can further observe that drivers of cluster 1 and cluster 2 both merge very early (i.e., 45.81 for cluster 1 and 37.09 m for state 2), corresponding to the so-called Early Merging Drivers. However, there are essential differences between the two clusters in terms of mean values of initial speed and merging speed. The mean values of initial speed and merging speed in cluster 2 are much smaller than those cluster 1, indicating that the traffic condition of cluster 2 is better than cluster 1. Clearly, most merging drivers choose to merge as soon as possible at both low and high speeds. Drivers in clusters 1 and 2 are called Early Merging Drivers at High Speed and Early Merging Drivers at Low Speed.

It is interesting to find that cluster 3 consists of merging drivers that have the largest speed reduction during the merging process (i.e., from 13.427 to 9.950 m/s). It means that drivers in cluster 3 have much higher speeds than the mainline traffic speeds when they enter the auxiliary lane and they use the auxiliary lane to reduce the speed difference between their speeds and mainline traffic speeds. Thus, the mean merging position of cluster 3 is much bigger than that of clusters 1 and 2. Drivers in cluster 3 are called Late Merging Drivers at Low Speed.

Cluster 4 contains the merging drivers that merge latest in the auxiliary lane. Previous studies have stated that drivers merge late because they cannot find a gap larger than their critical gap under congestion. However, it is interesting to find that the mean values of initial speeds and merging speeds are also the largest among the four clusters, indicating that cluster 4 has the best traffic condition. It means that drivers in cluster 4 merge late probably because they use the auxiliary lane to get further downstream of the mainline rather than that they cannot find gaps larger than critical gaps. Drivers in cluster 4 are called Late Merging Drivers at High Speed.

5.2. Hypothesis Tests between Different Clusters. To statistically investigate if there are significant differences in the median values of merging durations and gap acceptance between different clusters, two-sample Mann–Whitney U Tests are carried out. The obtained p values are given in Table 2 and the significance level is set as 0.1.

From Table 2, one can find that most p values are 0.000 except for the merging speed when testing clusters 2 and 3. It means that the two-step cluster analysis can well distinguish

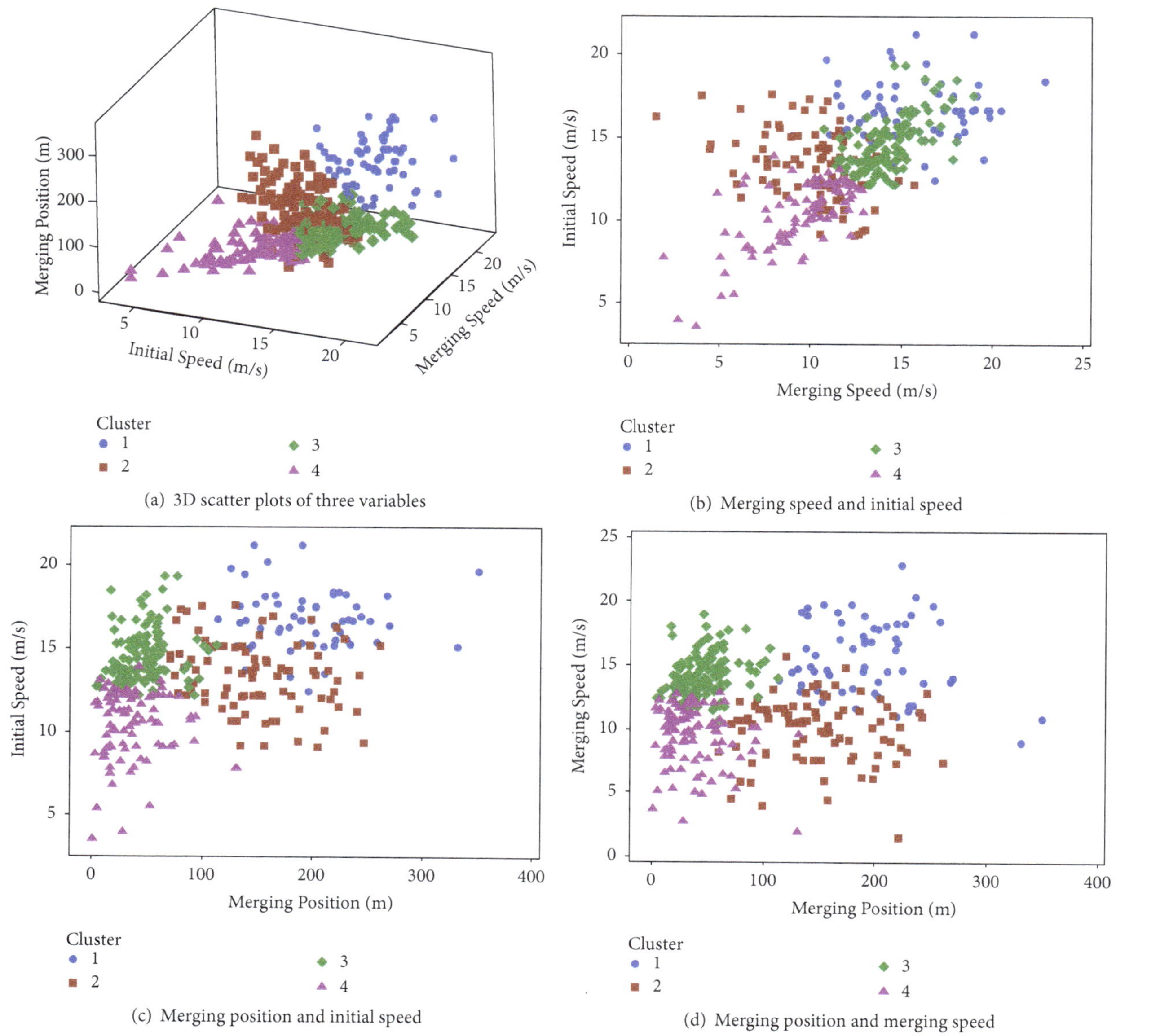

(a) 3D scatter plots of three variables

(b) Merging speed and initial speed

(c) Merging position and initial speed

(d) Merging position and merging speed

FIGURE 2: 3D scatter plots and 2D scatter plots of two-step clustering result.

TABLE 2: Mann-Whitney U test results of merging maneuvers.

Cluster	Initial Speed	Merging Speed	Merging Position
Clusters 1 & 2	0.000	0.000	0.000
Clusters 1 & 3	0.000	0.000	0.000
Clusters 1 & 4	0.000	0.000	0.000
Clusters 2 & 3	0.000	0.197	0.000
Clusters 2 & 4	0.000	0.000	0.000
Clusters 3 & 4	0.000	0.000	0.000

different merging drivers and identify the heterogeneity among merging drivers based on the merging maneuver parameters. Clusters 2 and 3 have similar merging speed probably because the traffic condition are similar. However the differences in initial speed and merging position indicates that different drivers do behave differently under the same traffic condition.

5.3. Probability Distribution Fitting Based on Clustering Results. To best investigate the heterogeneity in merging behaviors, the clustering results are used to develop the mixture models of merging maneuvers. 7 distributions (Normal, Log-normal, Student's t, Logistic, Log-Logistic, Gamma, and Weibull) are considered for each cluster and the Kolmogorov-Smirnov test is used to select the best fitted distributions. Table 3 shows best fitted distributions and corresponding K-S Test statistics, critical values, and the corresponding p values. The probability distributions of each merge maneuver parameters for different traffic densities are provided in Figure 2.

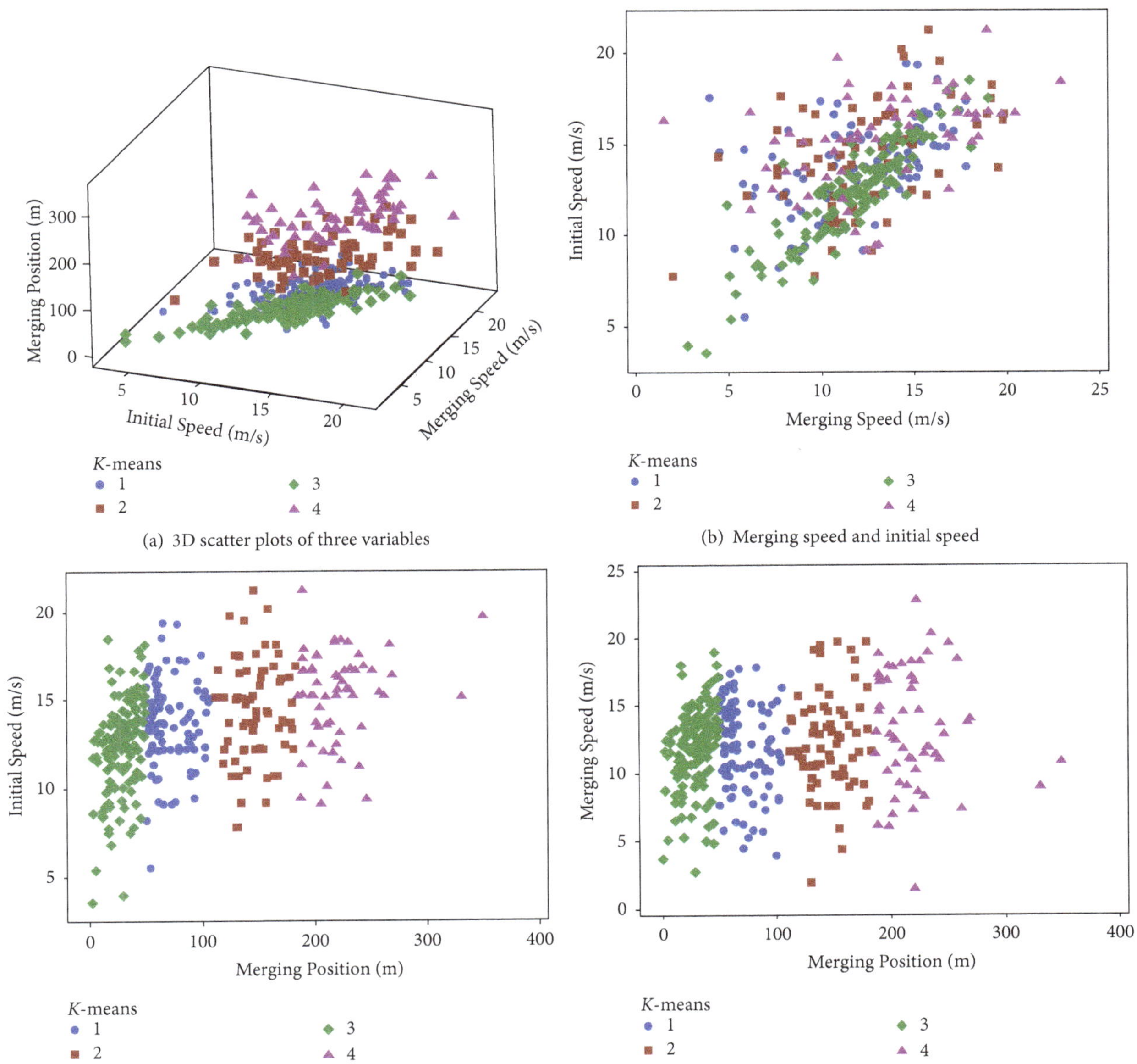

(a) 3D scatter plots of three variables

(b) Merging speed and initial speed

(c) Merging position and initial speed

(d) Merging position and merging speed

FIGURE 3: 3D scatter plots and 2D scatter plots of *K*-means clustering result.

One can observe from Table 3 and Figure 4 that the best fitted distributions of initial speed and merging speed are all Normal distributions except cluster 1. The best fitted distributions of initial speed and merging speed for cluster 1 are Gamma and Log-Logistic, respectively. The best distributions of merging position for 4 clusters are Gamma, Gamma, Log-normal, and Normal, respectively. The different distribution functions and different parameters and shapes of the fitted distributions show significant heterogeneities among the merging drivers across different clusters.

5.4. Analysis of Merging Durations and Gap Acceptance Based on Clustering Results. To further investigate the heterogeneity in different clusters, a comprehensive analysis of the merging durations and gap acceptance is given in this section. Figure 3 provides the cumulative curves of merging durations and gap acceptance parameters. The basic statics are provided in Table 4. One can observe from Table 4 and Figure 5 that cluster 2 has the largest mean value of merging duration and cluster 1 has the second largest value. The merging duration seems to be similar for clusters 3 and 4, both of which are smaller than that of clusters 1 and 2. One can also observe that the accepted gaps, lead gaps, and lag gaps have similar trends in statics and shapes. Clusters 1 and 4 have bigger mean values of three parameters than clusters 2 and 3, indicating that the traffic conditions of clusters 1 and 4 are better than those of clusters 2 and 3. Besides, the mean and median values of accepted lag gaps are much larger than accepted lead gaps in clusters 2 and 4. However they are similar in clusters 1 and 4.

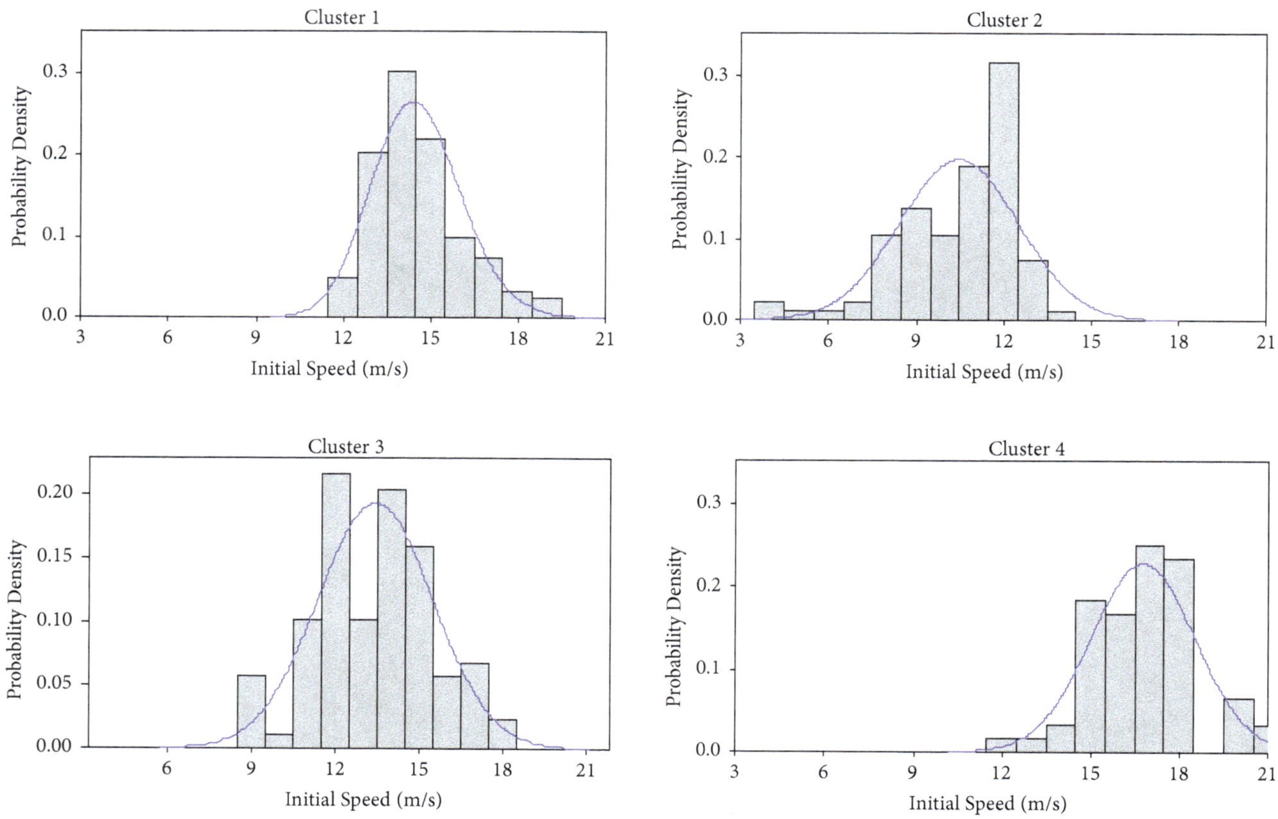

(a)

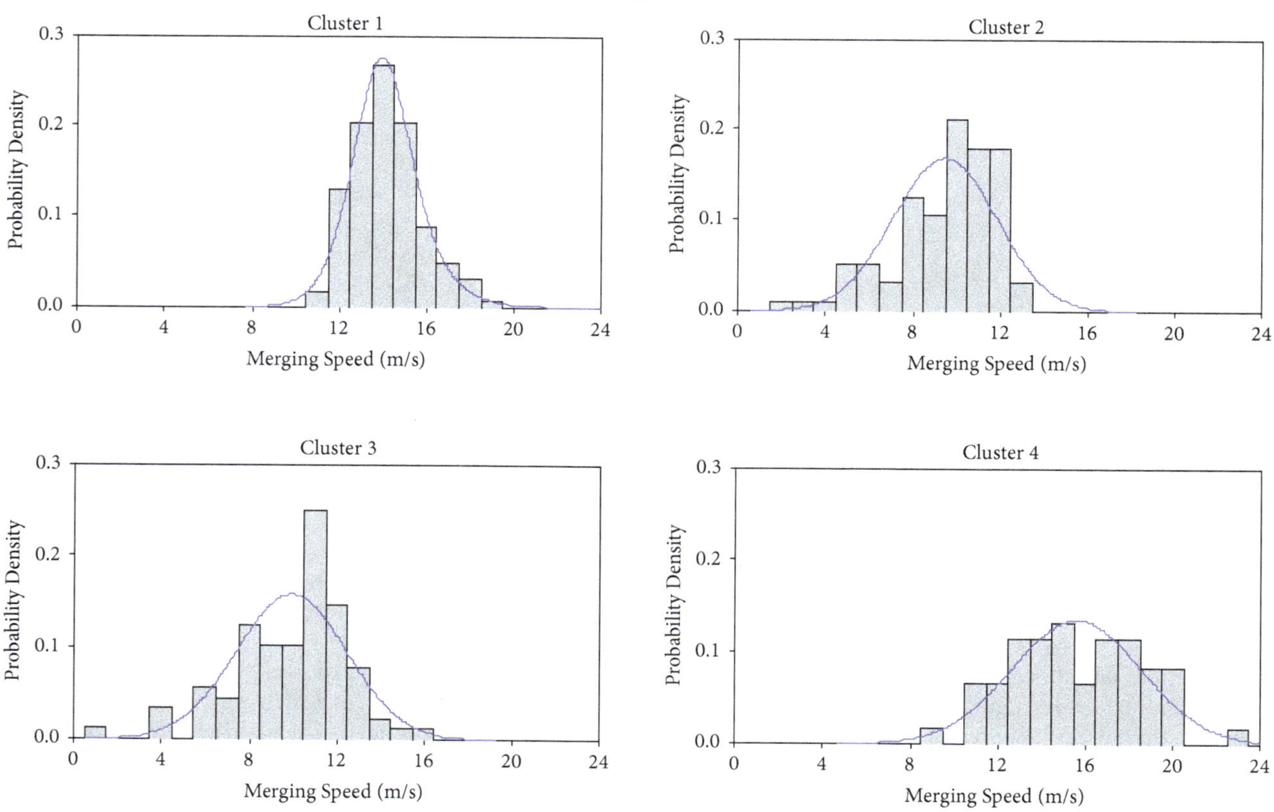

(b)

FIGURE 4: Continued.

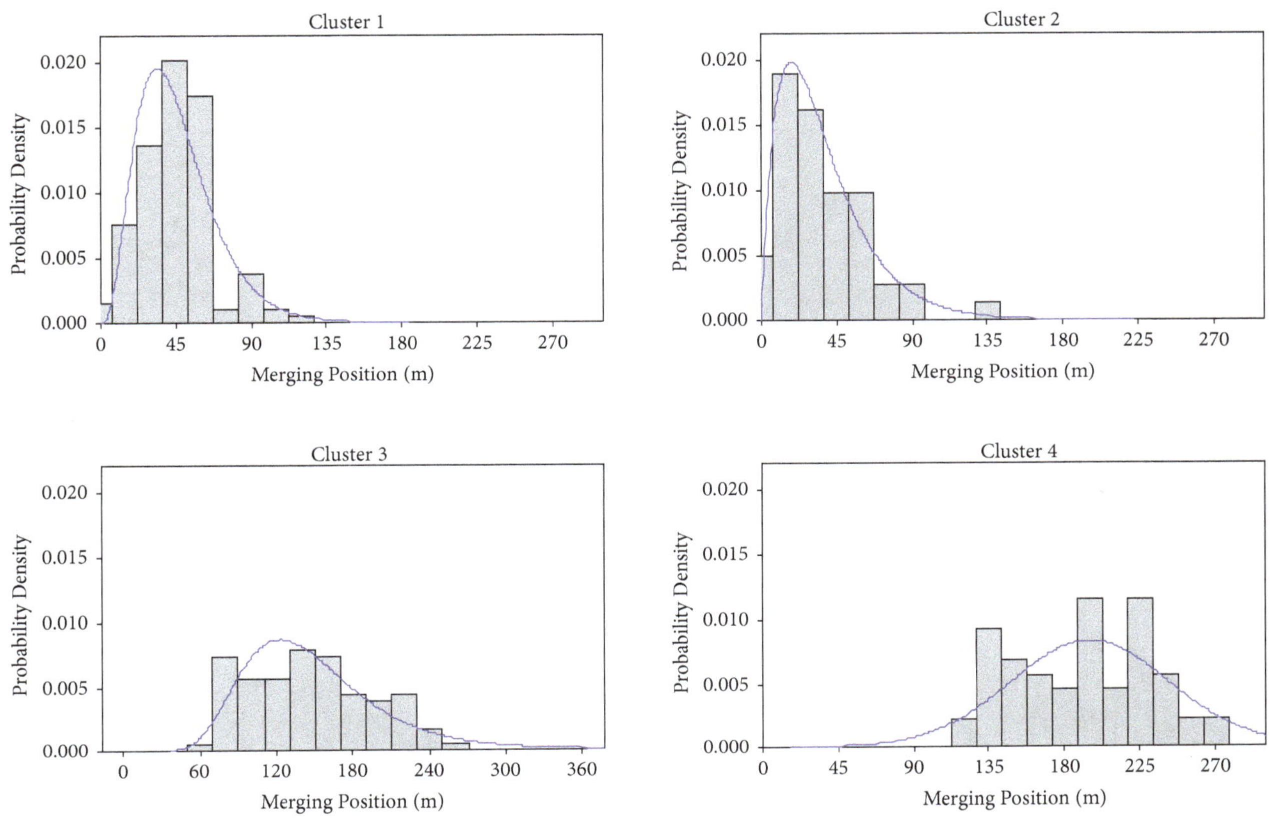

(c)

FIGURE 4: Best fitted distributions for (a) initial speed, (b) merging speed, and (c) merging position for different clusters.

TABLE 3: Distribution fitting results of merging maneuver parameters for different clusters.

Merge Maneuver	Cluster	Best Fitted Distribution	Parameters		Goodness of Fit, K-S test	
			Location/Shape	Scale	K-S Static	p value
Initial Speed	1	Gamma	86.27	5.93	0.0755	0.463
	2	Normal	10.481	2.0346	0.108	0.202
	3	Normal	13.43	2.069	0.0647	0.831
	4	Normal	16.77	1.76	0.102	0.532
Merging Speed	1	Log-Logistic	3.66	0.474	0.0372	0.993
	2	Normal	3.18	0. 445	0.124	0.0996
	3	Normal	9.95	2.51	0.122	0.134
	4	Normal	15.61	2.94	0.109	0.438
Merging Position	1	Gamma	4.56	0.010	0.0671	0.614
	2	Gamma	2.04	0.055	0.05398	0.931
	3	Lognormal	4. 94	0.349	0.0660	0.814
	4	Normal	194.89	47.89	0.0613	0.968

To statistically investigate if there are significant differences in the median values of merging durations and gap acceptance between different clusters, two-sample Mann–Whitney U Tests are carried out. The obtained p values are given in Table 5 and the significance level is set as 0.1

From Table 5, one can find that the median values of merging durations of clusters 1 and 2 are significantly different from those of clusters 3 and 4. It indicates that the merging durations of drivers that merge late (cluster 3 and cluster 4) are significantly shorter than those of drivers merge early (cluster 1 and cluster 2). Previous study has claimed that the merging durations are decreasing with the increase of speed [1]. Although the merging speeds of cluster 2 and cluster 4 are much higher than those of cluster 1 and cluster 4, there are no significant differences between clusters 1 and 2 and between clusters 3 and 4. The Early Merging Drivers might

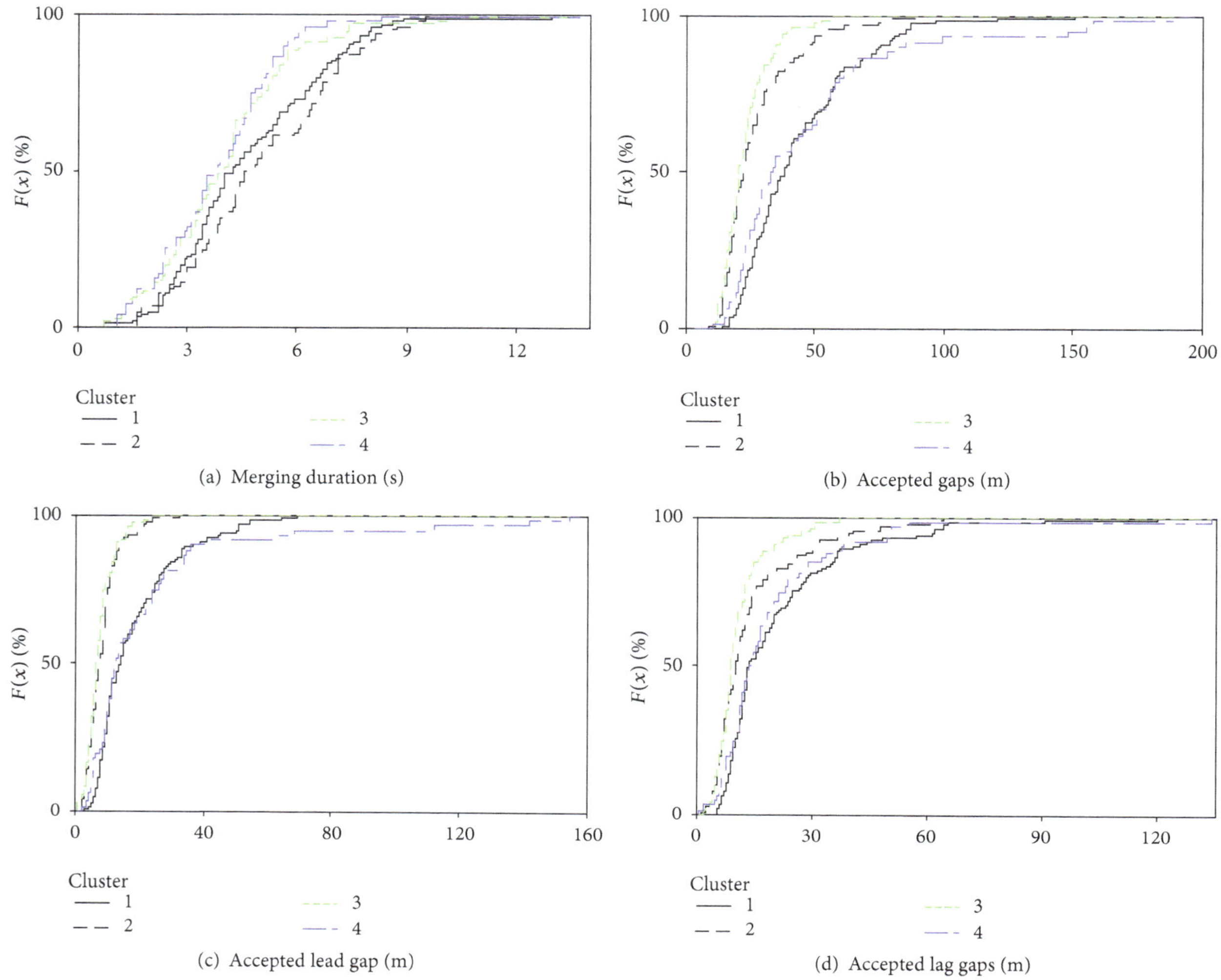

FIGURE 5: Cumulative distributions of (a) merging position (m), (b) merging speed (m/s), and (c) merging duration (s) for different clusters.

TABLE 4: Basic statistics of merging duration and gap acceptance for different clusters.

Parameters	Cluster	Mean	Standard Deviation	Median	Skewness
Merging Duration	1	4.67	2.00	4.2	0.78
	2	5.01	2.02	4.6	0.21
	3	4.00	1.80	3.95	0.81
	4	3.80	1.54	3.75	0.27
Accepted Gaps	1	43.63	22.64	38.15	1.56
	2	26.90	14.56	22.31	1.99
	3	22.85	9.07	20.53	1.42
	4	46.66	36.93	22.88	2.23
Accepted Lead Gaps	1	18.48	13.29	14.12	1.60
	2	8.376	5.09	6.992	1.95
	3	7.463	4.30	6.643	1.22
	4	22.95	29.51	12.63	3.22
Accepted Lag Gaps	1	20.77	17.66	13.25	2.63
	2	14.06	12.38	10.12	2.22
	3	10.70	6.60	8.856	1.87
	4	19.40	19.33	13.74	3.87

TABLE 5: Mann–Whitney U test results between different clusters.

Cluster	Merging Durations	Accepted Gaps	Accepted Lead Gaps	Accepted Lag Gaps
Clusters 1 & 2	0.297	0.000	0.000	0.000
Clusters 1 & 3	0.075	0.000	0.000	0.000
Clusters 1 & 4	0.043	0.399	0.459	0.926
Clusters 2 & 3	0.002	0.450	0.280	0.238
Clusters 2 & 4	0.001	0.000	0.000	0.002
Clusters 3 & 4	0.999	0.000	0.000	0.000

TABLE 6: Mann–Whitney U test results between accepted lead gaps and lag gaps.

Cluster	Accepted Lead Gaps versus Accepted Lag Gaps
Cluster 1	0.3789
Cluster 2	0.000
Cluster 3	0.000
Cluster 4	0.47588

be better prepared for merging before they enter the auxiliary lane, become more patient, and use more time to finish the merging maneuvers, while the long travelling distance might urge the Late Merging Drivers to finish the merging maneuvers as soon as possible even at low speeds.

The hypothesis tests of accepted gaps, accepted lead, and accepted lag gaps have the same results at 90% level. There are no significant differences in the gap acceptance parameters between clusters 1 and 4 and between clusters 2 and 3. However, the accepted gaps and accepted lead and lag gaps of clusters 1 and 4 are significantly larger than those of clusters 2 and 3, indicating that the traffic conditions of clusters 2 and 3 are more congested than those of clusters 1 and 4.

To check if the accepted lag gaps are larger than accepted lead gaps, a Mann–Whitney U Test between accepted lead gaps and lag gaps in different clusters is also carried out. Table 6 shows the obtained p values. One can find that the median values of accepted lag gaps are significantly larger than accepted lead gaps in clusters 2 and 3 and there are no significant differences between the median values of accepted lead gaps and lag gaps in clusters 1 and 4. It means merging drivers are more sensitive to lag gaps than lead gaps under congestion.

The data of merging durations and gap acceptance parameters for different clusters are also used to find the best corresponding fitting distributions using EasyFit 5.4 [48]. 7 distributions (Normal, Log-normal, Student's t, Logistic, Log-Logistic, Gamma, and Weibull) are considered for each sample and the Kolmogorov-Smirnov test is used to select the best fitted distribution. Table 7 shows best fitted distributions and corresponding K-S Test statistics, critical values, and the corresponding p values. The probability distributions of each parameters for different clusters are provided in Figure 6.

One can observe from Table 7 that the best fitted distribution of merging durations for are all Normal, except for the Log-normal distribution for cluster 1. One can also find that the best fitted distributions of gap acceptance parameters are all Log-normal, except for the Gamma distribution for accepted lead gap for cluster 3. There are considerable differences in the parameters and shape of fitted distribution between different classes, indicating the heterogeneity between different classes.

6. Concluding Remarks

To investigate the heterogeneity in merging drivers, a framework that combines two-step cluster analysis, hypothesis, and probability distribution fitting is proposed in this paper. The Tto-step cluster analysis is applied to three merging maneuver parameters (namely, initial speed, merging speed, and merging position) to identify different clusters of merging drivers. Four clusters are automatically and optimally chosen as the best clustering results. Compared with K-means clustering approach, the two-step cluster analysis can not only produce the best clustering results automatically, but also better separate the data.

Mann–Whitney U Tests are carried out and prove that merging maneuvers are indeed different between different clusters. The identified four clusters are called Early Merging Drivers at High Speed, Early Merging Drivers at Low Speed, Late Merging Drivers at Low Speed, and Late Merging Drivers at High Speed. The clustered data are used to find the best corresponding fitting distributions using EasyFit 5.4 and 7 distributions (Normal, Log-normal, Student's t, Logistic, Log-Logistic, Gamma, and Weibull) are considered for each cluster. Kolmogorov-Smirnov test statics are used to select the best fitted distributions.

Merging durations and gap acceptance are analyzed based on the clustering results. It is discovered that merging durations of Late Merging Drivers are significantly shorter than Early Merging Drivers. However, there are no significant differences between Early Merging Drivers at High Speed and Low Speed and between Late Merging Drivers at High Speed and Low Speed. It indicates that drivers are losing patience and getting more aggressive with the increase of time on the auxiliary lane under either congestion or uncongested traffic condition. There are no significant differences in the gap acceptance parameters between clusters 1 and 4 and between clusters 2 and 3. However, the accepted gaps and accepted lead and lag gaps of clusters 1 and 4 are significantly larger than those of clusters 2 and 3, indicating that the traffic conditions of clusters 2 and 3 are more congested than those of clusters 1 and 4. The hypothesis tests between accepted lead gaps and accepted lags show that merging drivers the

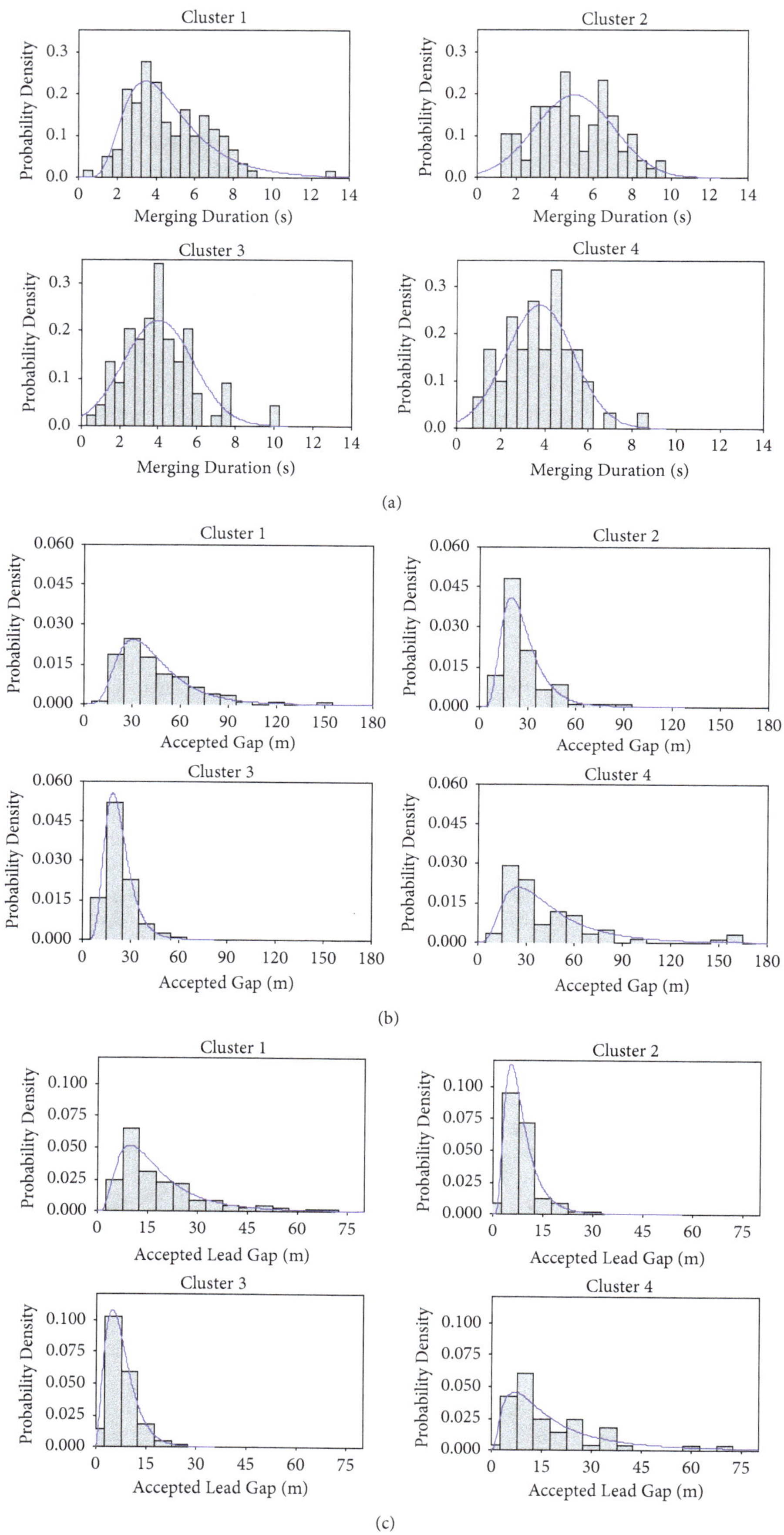

FIGURE 6: Continued.

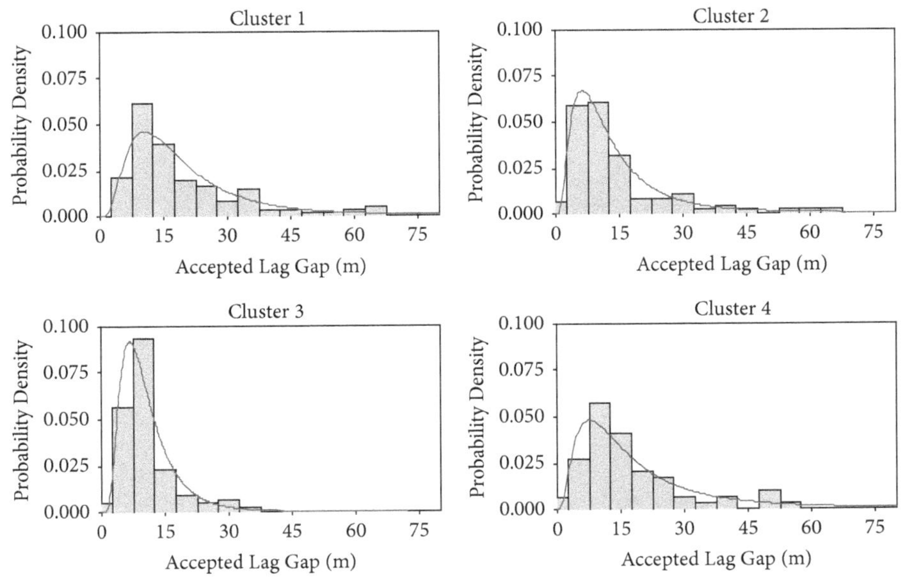

(d)

FIGURE 6: Best fitted distributions for (a) merging duration, (b) accepted gaps, (c) accepted lead gaps, and (d) accepted lag gaps for different clusters.

TABLE 7: Results of the best fitted distributions for merging durations and gap acceptance parameters.

Parameters	Cluster	Best Fitted Distribution	Parameters		Goodness of Fit, K-S test	
			Scale	Shape	K-S Static	p value
Merging Durations	1	Lognormal	4.20	0.852	0.0671	0.612
	2	Normal	5.01	2.02	0.0944	0.344
	3	Normal	4.00	1.80	0.0920	0.421
	4	Normal	3.79	1.54	0.0670	0.934
Accepted Gaps	1	Lognormal	3.66	0.474	0.0676	0.603
	2	Lognormal	3.1823	0. 445	0.0891	0.413
	3	Lognormal	3.06	0.358	0.0618	0.868
	4	Lognormal	3.63	0.625	0.104	0.495
Accepted Lead Gaps	1	Lognormal	2.70	0.644	0.0907	0.248
	2	Lognormal	1.97	0.553	0.0568	0.9021
	3	Gamma	3.01	2.48	0.06678	0.803
	4	Lognormal	2.69	0.873	0.0914	0.664
Accepted Lag Gaps	1	Lognormal	2.79	0.660	0.0978	0.178
	2	Lognormal	2.36	0.728	0.07918	0.564
	3	Lognormal	2.22	0.551	0.0813	0.577
	4	Lognormal	2.66	0.789	0.0920	0.656

accepted lead gaps are significantly smaller than accepted lag gaps under congestion. It indicates that the merging drivers are more sensitive to the lag gap under congestion during merging process. The merging durations and gap acceptance parameters in different clusters are also used to find the best corresponding fitting distributions using EasyFit 5.4.

The proposed method can automatically identify the heterogeneity in merging drivers and the results obtained in this paper can be used to enhance the accuracy of the merge behavior models in microscopic simulation software.

In future research, more data will be collected to investigate the factors that may influence driver heterogeneity, such as the surrounding population and the time of data acquisition. On the other hand, driver heterogeneity will be incorporated into microscopic traffic simulation models to improve the accuracy and authenticity.

Conflicts of Interest

The authors declare that they have no conflicts of interest.

References

[1] Q. Wang, Z. Li, and L. Li, "Investigation of discretionary lane-change characteristics using next-generation simulation data sets," *Journal of Intelligent Transportation Systems*, vol. 18, no. 3, pp. 246–253, 2014.

[2] W. Daamen, S. P. Hoogendoorn, and M. Loot, "Empirical analysis of merging behavior at freeway on-ramp," *Transportation Research Records: Journal of Transportation Research Board*, no. 2188, pp. 108–118, 2010.

[3] Z. Zheng, "Recent developments and research needs in modeling lane changing," *Transportation Research Part B: Methodological*, vol. 60, pp. 16–32, 2014.

[4] M. K. Ardakani, J. Yang, and L. Sun, "Stimulus response driving behavior: An improved general motor vehicle-following model," *Advances in Transportation Studies*, no. 39, pp. 23–36, 2016.

[5] L. Elefteriadou, R. P. Roess, and W. R. McShane, "Probabilistic nature of breakdown at freeway merge junctions," *Transportation Research Record: Journal of the Transportation Research Board*, vol. 1484, pp. 80–89, 1995.

[6] H. Yi and T. E. Mulinazzi, "Urban freeway on-ramp invasive influences on 3 mainline operations," in *Proceedings of the 86th Annual Meeting of the Transportation Research Board*, Washington, DC, USA, 2007.

[7] S. Ossen, S. P. Hoogendoorn, and B. G. H. Gorte, "Interdriver differences in car-following a vehicle trajectory-based study," *Transportation Research Record*, vol. 1965, no. 1, pp. 121–129, 2006.

[8] S. Ossen and S. P. Hoogendoorn, "Car-following behavior analysis from microscopic trajectory data," *Transportation Research Record: Journal of the Transportation Research Board*, vol. 1934, pp. 13–21, 2005.

[9] S. Ossen and S. P. Hoogendoorn, "Heterogeneity in car-following behavior: theory and empirics," *Transportation Research Part C: Emerging Technologies*, vol. 19, no. 2, pp. 182–195, 2011.

[10] J. W. Kim and H. S. Mahmassani, "Correlated parameters in driving behavior models: car-following example and implications for traffic microsimulation," in *Proceedings of the 90th Annual Meeting of the Transportation Research Board*, Washington, DC, 2011.

[11] L. Sun, Y. Pan, and W. Gu, "Data mining using regularized adaptive B-splines regression with penalization for multi-regime traffic stream models," *Journal of Advanced Transportation*, vol. 48, no. 7, pp. 876–890, 2014.

[12] L. Sun, "Spectral and time-frequency analyses of freeway traffic flow," *Journal of Advanced Transportation*, vol. 48, no. 7, pp. 821–857, 2014.

[13] L. Sun, "Stochastic projection-factoring method based on piecewise stationary renewal processes for mid- and long-term traffic flow modeling and forecasting," *Transportation Science*, vol. 50, no. 3, pp. 998–1015, 2016.

[14] Y. Pan and L. Sun, "Characterizing heterogeneity in vehicular traffic speed using two-step cluster analysis," *Journal of Southeast University*, vol. 28, no. 4, pp. 480–484, 2012.

[15] W. Xiong, L. Sun, and J. Zhou, "Spline-based multi-regime traffic stream models," *Journal of Southeast University*, vol. 26, no. 1, pp. 122–125, 2010.

[16] Cambridge Systematics Inc., *NGSIM U.S. 101 Data Analysis Summery Report*, Federal Highway Administration, Washington, DC, USA, 2005.

[17] Q. Yang and H. N. Koutsopoulos, "A microscopic traffic simulator for evaluation of dynamic traffic management systems," *Transportation Research Part C: Emerging Technologies*, vol. 4, no. 3, pp. 113–129, 1996.

[18] K. I. Ahmed, *Modeling Drivers Acceleration and Lane Changing Behavior [Ph.D. thesis]*, Department of Civil and Environmental Engineering, MIT, 1999.

[19] T. Toledo and D. Zohar, "Modeling duration of lane changes," *Transportation Research Record: Journal of the Transportation Research Board*, no. 1999, pp. 71–78, 2007.

[20] T. Toledo, H. N. Koutsopoulos, and M. Ben-Akiva, "Estimation of an integrated driving behavior model," *Transportation Research Part C: Emerging Technologies*, vol. 17, no. 4, pp. 365–380, 2009.

[21] M. A. Ahammed, Y. Hassan, and T. A. Sayed, "Modeling driver behavior and safety on freeway merging areas," *Journal of Transportation Engineering*, vol. 134, no. 9, pp. 370–377, 2008.

[22] R. Herman and G. H. Weiss, "Comments on the highway-crossing problem," *Operation Research*, vol. 9, pp. 828–840, 1961.

[23] D. R. Drew, L. R. Lamotte, J. A. Wattleworth, and J. H. Buhr, "Gap acceptance in the freeway merging process," *Highway Research Record*, vol. 208, 1967.

[24] A. J. Miller, "Nine Estimators of Gap Acceptance Parameters," in *Proceedings of the 5th International Symposium on the Theory of Traffic Flow and Transportation*, 1972.

[25] P. G. Gipps, "A model for the structure of lane-changing decisions," *Transportation Research Part B: Methodological*, vol. 20, no. 5, pp. 403–414, 1986.

[26] P. Hidas, "Modelling lane changing and merging in microscopic traffic simulation," *Transportation Research Part C: Emerging Technologies*, vol. 10, no. 5-6, pp. 351–371, 2002.

[27] J. Wang, "A simulation model for motorway merging behavior," *Transportation and Traffic Theory*, vol. 16, pp. 281–301, 2005.

[28] L. Bloomberg and J. Dale, "Comparison of VISSIM and CORSIM traffic simulation models on a congested network," *Transportation Research Record*, no. 1727, pp. 52–60, 2000.

[29] PTV, User Manual: VISSIM 4.0, 2004.

[30] SIAS., *S-Paramics 2005 Reference Manual*, SIAS Ltd, Edinburgh, UK, 2005.

[31] V. Punzo, D. J. Forminaso, and V. Torrieri, "Nonstationary kalman filter for estimation of accurate and consistent car-following data," in *Proceedings of the 84th Annual Meeting of the Transportation Research Board*, Washington, DC, USA, 2005.

[32] F. Marczak, W. Daamen, and C. Buisson, "Merging behaviour: Empirical comparison between two sites and new theory development," *Transportation Research Part C: Emerging Technologies*, vol. 36, pp. 530–546, 2013.

[33] T. D. Chu, T. Miwa, and T. Morikawa, "An analysis of merging maneuvers at urban expressway merging sections," *Procedia-Social and Behavioral Sciences*, vol. 138, pp. 105–115, 2014.

[34] J. Weng and Q. Meng, "Modeling speed-flow relationship and merging behavior in work zone merging areas," *Transportation Research Part C: Emerging Technologies*, vol. 19, no. 6, pp. 985–996, 2011.

[35] M. Sarvi and M. Kuwahara, "Microsimulation of freeway ramp merging processes under congested traffic conditions," *IEEE Transactions on Intelligent Transportation Systems*, vol. 8, no. 3, pp. 470–479, 2007.

[36] L. Sun, W. Gu, and H. Mahmassani, "Estimation of expected travel time using the method of moment," *Canadian Journal of Civil Engineering*, vol. 38, no. 2, pp. 154–165, 2010.

[37] L. Sun, J. Yang, H. Mahmassani, W. Gu, and B.-J. Kim, "Data mining-based adaptive regression for developing equilibrium speed-density relationships," *Canadian Journal of Civil Engineering*, vol. 37, no. 3, pp. 389–400, 2010.

[38] L. Sun, H. Zhang, R. Gao, W. Gu, B. Xu, and L. Chen, "Gaussian mixture models for clustering and classifying traffic flow in real-time for traffic operation and management," *Journal of Southeast University*, vol. 27, no. 2, pp. 174–179, 2011.

[39] L. Sun and J. Zhou, "Development of multiregime speed-density relationships by cluster analysis," *Transportation Research Record: Journal of the Transportation Research Board*, no. 1934, pp. 64–71, 2005.

[40] C. Hill, L. Elefteriadou, and A. Kondyli, "Exploratory analysis of lane changing on freeways based on driver behavior," *Journal of Transportation Engineering*, vol. 41, article 04014090, no. 4, 2014.

[41] M. J. Norusis, *IBM SPSS Statistics 19 Procedures Companion*, Addison-Wesley, Reading, Mass, USA, 2011.

[42] M. J. Norusis, *SPSS 15.0 Advanced Statistical Procedures Companion*, Prentice Hall, Chicago, Ill, USA, 2007.

[43] V. Alexiadis, J. Colyar, J. Halkias, R. Hranac, and G. McHale, "The next generation simulation program," *Institute of Transportation Engineers (ITE) Journal*, vol. 74, no. 8, pp. 22–26, 2004.

[44] C. Thiemann, M. Treiber, and A. Kesting, "Estimating acceleration and lane-changing dynamics from next generation simulation trajectory data," *Transportation Research Record: Journal of the Transportation Research Board*, vol. 2088, pp. 90–101, 2008.

[45] IBM., "Two-step cluster analysis," https://www.ibm.com/support/knowledgecenter/SSLVMB_20.0.0/com.ibm.spss.statistics.help/alg_twostep.htm, 2016.

[46] W. C. Navidi, *Statistics for Engineers and Scientists*, vol. 2, McGraw-Hill, New York, NY, USA, 2006.

[47] S. Gurupackiam and S. L. Jones, "Empirical study of lane changing in urban streets under varying traffic conditions," *Procedia-Social and Behavioral Sciences*, vol. 16, pp. 259–269, 2011.

[48] K. Schittkowski, "EASY-FIT: A software system for data fitting in dynamical systems," *Structural and Multidisciplinary Optimization*, vol. 23, no. 2, pp. 153–169, 2002.

[49] E. Balal, R. L. Cheu, T. Gyan-Sarkodie, and J. Miramontes, "Analysis of discretionary lane changing parameters on freeways," *International Journal of Transportation Science & Technology*, vol. 3, no. 3, pp. 277–296, 2014.

6

A Simheuristic Method for the Reversible Lanes Allocation and Scheduling Problem at Smart Container Terminal Gate

Wenyuan Wang,[1,2] Ying Jiang,[2] Yun Peng,[1] Yong Zhou,[3] and Qi Tian[1]

[1]*State Key Laboratory of Coastal and Offshore Engineering, Dalian University of Technology, Dalian 116024, China*
[2]*Mobilities and Urban Policy Lab, Graduate School for International Development and Cooperation, Hiroshima University, Hiroshima, Japan*
[3]*College of Transport & Communications, Shanghai Maritime University, Shanghai 201306, China*

Correspondence should be addressed to Yun Peng; yun¯peng@yahoo.com

Academic Editor: Ludovic Leclercq

Under the constraints of limited spaces and imbalanced traffic volumes (for both in and out directions) of container gates, reversible lane layouts become an economical and practical way to improve the service level of container terminal systems and make the maximum use of the current terminal resources. Together with a consideration of minimized total costs (both construction and operating) of terminal gate system, this paper first developed an optimization model to decide the number and scheduling rules of the reversible lanes at a terminal gate. A metaheuristic algorithm was built to solve the optimal model. Meanwhile, to reflect the randomness and dynamics property of the terminal gate system in practice, parameters that cannot be calculated from conventional analytic methods are obtained through a simulation model. Finally, a hub container terminal in the northeast of China was employed to verify the effectiveness of the proposed method and provide a theoretical foundation for the construction and management of terminal gate systems.

1. Introduction

With the continuous increase of containers' throughput, terminal gates have become the bottleneck of restricting the development of container terminals. Container trucks frequently queue up and wait for service out of the terminal gates. Poor performance of the terminal gate will not only restrict the logistics efficiency of whole terminal system but also cause serious air pollution problems due to the long waiting time at gates. Thus, it is urgent and important to improve the service level of the congested terminal gate. Generally, congestion issues at container terminal gates can be alleviated by increasing the number of service lanes. However, due to the restrictions of available space and huge construction cost, it is impractical to expand the gates for all container terminals. Then, smart transformation of existing terminal gates from the perspective of operational management has become a more practical and economical way to solve the current problems. Therefore, it naturally raises a question: how should the service lanes at smart container terminal gate be allocated and scheduled so that the total costs of construction and operation of the gate system are minimized?

In order to solve the congestion problem at container terminal gates, efforts that have been made mainly focus on two aspects: container trucks' management and terminal gates' operation. Then, for the management of container trucks, Namboothiri and Erera [1] studied the management of a fleet of trucks providing container pickup and delivery service (drayage) to a terminal with an appointment-based access control system; Chen et al. [2] proposed an analytical point-wise stationary approximation model to analyze time-dependent truck queuing processes with stochastic service time distributions at gates and yards of a terminal; Chen et al. [3] proposed a method called vessel-dependent time windows to control truck arrivals, involving partitioning truck entries into groups and assigning different time windows to the groups. Later, Chen et al. [4] developed a biobjective model to minimize both truck waiting times and truck arrival pattern change, so that the emissions from idling truck engines at marine container terminals can be reduced; Yang et al. [5] presented an integrated planning model and a sequential planning model to coordinate the major terminal planning activities and developed a heuristic-based genetic algorithm to solve the models; Phan and Kim [6] addressed a negotiation process for smoothing truck arrivals in peak hours among multiple trucking companies and a terminal; Azab and Eltawil [7] developed a discrete event simulation model to study the effect of various truck arrival patterns on length of stay of trucks in container terminals; Ambrosino and Peirano [8] solved a mixed integer linear programming model based on the network flow theory to determine the number of appointments offered by each time window to trucks in the shortest time as possible; Chen and Jiang [9] proposed a solution of managing truck arrivals with time windows based on the truck-vessel service relationship, where trucks delivering containers for the same vessel share one common time window.

On the other hand, from the perspective of terminal gate operation and management, Guan and Liu [10] applied a multiserver queuing model to analyze marine terminal gate congestion and quantify truck waiting time cost in a case study of the port of New York/New Jersey. Focusing on the business bottlenecks of terminal operations, Hu et al. [11] tried to improve operation efficiency from the perspectives of orderly balance and seamless connection in different operational processes at the entrance gate of container terminals; Tsai et al. [12] proposed an image-processing algorithm to automatically extract truck processing time data (by using the low-frame-rate images); then experimental tests were employed to demonstrate the robustness of the proposed algorithm for dealing with the unique technical challenges at a terminal gate; Fleming et al. [13] used agent-based simulation to model the terminal gate system with two queuing strategies, that is, pooled queue and nonpooled queues, to evaluate the performance of pooling trucks into a single queue at a gate; Ambrosino and Caballini [14] addressed the minimization of truck service times at container terminals while respecting a certain level of congestion; Kadir et al. [15] presented an application of Radio Frequency Identification technology, and e-seal in a terminal gate was used to reduce the processing clearance time.

Several studies have also been conducted by emphasizing on the gate scale optimizing from the perspective of terminal planning. For example, Chu (2014) developed a planning-level tool to assess the effectiveness of gate layouts and determine the optimal layout for marine container terminals. The tool can also be used to determine the average truck queuing time for a given gate configuration and to determine how many service gates and queuing lanes are needed to achieve a desired level of service for a given truck arrival rate and truck service rate. Zeng et al. [16] conducted a study by focusing on the queue length and average waiting time of the railway container terminal gate system and the optimal number of service lanes during different time period. Keceli [17] simulated a multipurpose terminal to estimate the gate traffic and determine the necessary gate infrastructure.

In summary, existing literatures are mainly focusing on the layout design and scale optimization of a container terminal gate. Dealing with the congestion problem at container terminal gates, countermeasures usually come from two aspects: (1) Expand the scale of container terminal gates and increase the number of lanes. This method can alleviate the congestion at terminal gates directly and decrease waiting times and fuel costs for container trucks as well. However, it only works under the condition of enough space being available nearby the terminal gate. Besides, the cost on new construction and operation will significantly increase. (2) Adopt effective scheduling methods, for example, ordering arrival and departure time for container trucks, extending operating time for terminal gates, and changing queuing discipline for container trucks, as well as setting reversible lanes (RLs). Then, compared with other scheduling methods, setting RLs had less influences on works of container truck management by transport enterprises and manual operation at container terminals. Although the RL needs higher construction and operation costs than the fixed lane (FL), setting a certain number of RLs with reasonable scheduling rules without increasing the total lane numbers can effectively alleviate the congestion at container terminal gates. Moreover, the imbalanced queuing issue of FLs in arrival and departure directions could also be solved and will further contribute to the reduction of variable cost to container trucks (waiting time and fuel consumption). To the best of the authors' knowledge, this is the first study to explore the dynamic container terminal gate management based on a reversible lane (RL) idea.

The concept of RL is generally used in urban traffic design, tunnels, and bridges [18]. Depending on certain conditions, the traffic in RLs may travel by either direction. Traffic capacity of the corresponding direction could be amplified temporally. For example (Figure 1), in the real container terminal, lanes 1, 2, 3, 4, 5, 6, and 7 are used for entering port, while lanes 7, 8, 9, and 10 are set for leaving port; that is, lane 7 is a RL. Then, by adopting RLs concept, a new terminal gate management method was established with consideration of unbalanced traffic volumes of arriving and departure from a container. It is expected that setting of RLs at a container terminal gate could significantly contribute to the improvement of the terminal gate management system.

In order to further investigate the necessity of RLs setting at a container terminal gate, the corresponding optimal number, and scheduling rule of RLs, a series of scheduling rules of RLs under the special concept of terminal gates are proposed in this study. Then, aiming at minimizing the total costs of the container terminal gate system, which include construction and operating costs, an optimization model of the container terminal gate system was established. Additionally, a metaheuristic algorithm was proposed to solve the model and find the optimal number of the container lanes under the condition of fixed scale gates. Finally, the necessity of setting RLs can be verified through comparison of the total cost between two different terminal gate systems, regardless of whether RLs were equipped or not.

The rest of this paper is organized as follows. Section 2 introduces methods to alleviate congestions at a container

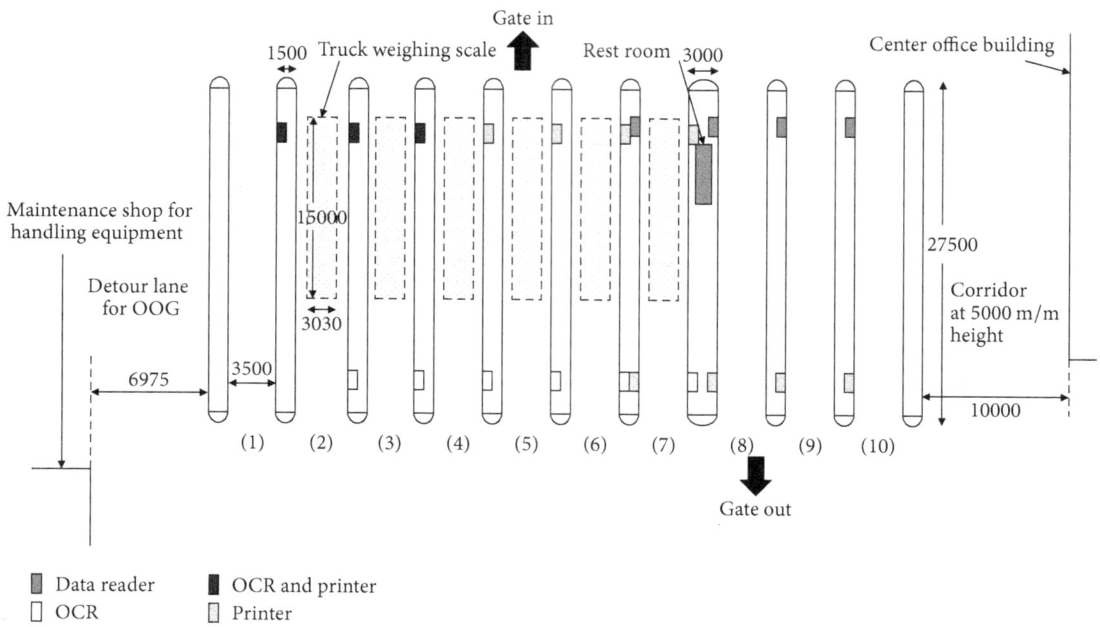

FIGURE 1: Lanes setting and scheduling for gate in a real container terminal.

terminal gate, select imbalanced queuing coefficients, and set scheduling rule of RLs. Section 3 establishes an optimization model of the container terminal gate system to seek the best number of RLs and scheduling rules. Since some parameters in the optimization model cannot be quantified through classical numerical methods, in Section 4, a simulation model is built to determine those parameters. Based on results of the simulation model, a simulated annealing (SA) metaheuristic algorithm is proposed to solve the optimization model. Section 5 introduces the case study and sensitivity analysis is carried out, followed by conclusions and future discussions in Section 6.

2. Problem Description

To alleviate congestions and minimize total costs of the container terminal gate system, the objective of this study is to figure out how to determine the optimal number of RLs and develop reasonable scheduling rules for RLs from the perspectives of both planning and management level. To further explain the operating principles of RLs at a container terminal gate, in Section 2.1, the conception of "imbalanced queuing coefficients" is firstly introduced to figure out conditions when RLs should change their traffic direction. Then, detailed discussion of scheduling rules of RLs is given in Section 2.2.

2.1. Imbalanced Queuing Coefficients. Direction variation of RLs depends on the queuing condition of a container terminal gate. Concretely speaking, if the queue length in direction of the RL is much shorter than that of the opposite direction, the traffic direction of RLs should be changed. In this way, the number of lanes in arrival and departure directions can be dynamically adjusted. Therefore, the thorough capacity of the container terminal gate in two directions can be well balanced. And the congestion problem caused by imbalanced serving number of container trucks at the container terminal gate in two directions can also be solved, which will ultimately improve the service level of the container terminal gate.

In order to quantitatively measure the queuing states at a container terminal gate, imbalanced queuing coefficient τ is introduced and defined, which refers to the ratio of the average queue length in arrival direction divided by that in departure direction. The RL should change the traffic direction only if τ reaches a certain value, defined as the critical value of imbalanced queuing coefficient τ^*. Additionally, the critical value is extended to a critical interval to avoid frequent direction changing problem in actual operation. Then, traffic direction of RLs can be managed as follows:

(i) Change to departure direction when $\tau < 1/\tau^*$

(ii) Change to arrival direction when $\tau > \tau^*$

(iii) No change when $1/\tau^* \leqslant \tau \leqslant \tau^*$

It is noticeable that the queue length counted at a container terminal gate includes not only container trucks in waiting queue but also those being served. This is because during off-peak period, a container truck might be directly served without any waiting queues. In this case, if only waiting queue of container trucks is counted, then the direction of the RL should be kept in departure direction ($\tau = 0$). This is not necessary, since traffic volume in both directions is less during off-peak period.

2.2. Scheduling Rules of RLs. Whether RLs should change their traffic direction or not is based on the rules defined

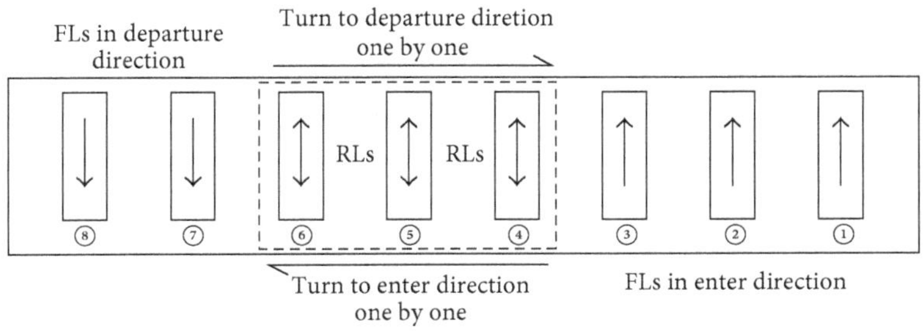

FIGURE 2: Changing order of RLs.

in Section 2.1. If several RLs are equipped, the judgment for each RL is made separately, starting from the nearest lane. Take situation in Figure 2 as an example; all three RLs (numbers ④–⑥) are served at departure direction at first. When $\tau > \tau^*$, which means that the number of lanes in arrival (enter) direction is not enough, one or more of the RLs should be changed. In this case, traffic direction of lane number ④, which is most close to the enter direction, will be changed first. If the congestion in enter direction is still serious ($\tau > \tau^*$), the traffic direction of lane number ⑤ will then be changed in sequence. Similarly, when changing RLs to departure direction, the changing starts from lane number ⑥.

3. Optimization Model

3.1. Model Assumptions. Model assumptions are as follows:

(1) First-come-first-served rule for container trucks at the container terminal gate without any reservation is followed.

(2) Service time for all kinds of container trucks follows the same probability distribution at the container terminal gate, and the influence of special containers is not taken into account.

(3) Service time of all container trucks in the container yard is quite random, which follows a certain specific probability distribution according to investigation of port actual operation.

(4) Only lanes for container trucks are considered.

(5) There is no breakdown or maintenance for lanes at the container terminal gate.

3.2. Symbol Description

3.2.1. Parameter Definition. The parameters of the optimization model are shown in Table 1.

3.2.2. Decision Variables. The decision variables are as follows:

n_{in}, n_{out}: number of FLs in arrival and departure direction

n_{va}: number of RLs

τ^*: critical value of imbalanced queuing coefficient

3.3. Model Formulation. Questions of optimization number of RLs and scheduling rules at the container terminal gate can be described as follows: under the condition of fixed throughput of a container terminal, how should the number of FLs in arrival and departure directions and the number of RLs with reasonable scheduling rules be determined so that the total cost under construction and operating stages are minimized?

The optimization model is built as follows:

$$\min \quad C_T = \left\{ C_{\text{in}} n_{\text{in}} + C_{\text{out}} n_{\text{out}} + C_{\text{va}} n_{\text{va}} + \sum_{i=1}^{N} (l_i) \left(C_w + C_f \right) \right\} T \tag{1}$$

$$C_{\text{in}} = C_{d\text{-in}} + C_{l\text{-in}} + C_{e\text{-in}} \tag{2}$$

$$C_{\text{out}} = C_{d\text{-out}} + C_{l\text{-out}} + C_{e\text{-out}} \tag{3}$$

$$C_{\text{va}} = C_{d\text{-va}} + C_{l\text{-va}} + C_{e\text{-va}} \tag{4}$$

$$\text{s.t.} \quad n_{\text{in}} + n_{\text{out}} + n_{\text{va}} = N \tag{5}$$

$$n_{\text{in}} \geq n_{\text{out}} \tag{6}$$

$$l_i = f\left(n_{\text{in}}, n_{\text{out}}, n_{\text{va}}, \tau^*\right), \quad i = 1, 2, \ldots, N \tag{7}$$

TABLE 1: Parameters of the optimization model.

Symbol	Parameters	Unit
i	Serial number of lanes at the container terminal gate	—
C_T	Total costs of the container terminal gate system	Chinese Yuan (CNY)
$C_{\text{in}}, C_{\text{out}}$	Construction and operating costs per FL in arrival and departure direction at the container terminal gate per hour, including depreciable cost, labor cost, and electric power cost	CNY/lane/hour
$C_{d\text{-in}}, C_{d\text{-out}}$	Depreciable cost per FL in arrival and departure direction at the container terminal gate per hour	
$C_{l\text{-in}}, C_{l\text{-out}}$	Labor cost per FL in arrival and departure direction at the container terminal gate per hour	
$C_{e\text{-in}}, C_{e\text{-out}}$	Electric power cost per FL in arrival and departure direction at the container terminal gate per hour	
C_{va}	Construction and operating costs per RL at the container terminal gate per hour	
$C_{d\text{-va}}$	Depreciable cost per RL at the container terminal gate per hour	
$C_{l\text{-va}}$	Labor cost per RL at the container terminal gate per hour	
$C_{e\text{-va}}$	Electric power cost per RL at the container terminal gate per hour	
C_w	Average waiting time cost per container truck per hour	
C_f	Average fuel cost per container truck per hour	
N	Total number of lanes at the container terminal gate	—
l_i	Average queue length of the ith lane	Vehicle
T	Calculation period	Hour

$$\begin{aligned} &n_{\text{in}}, n_{\text{out}} \in Z_+, \\ &n_{\text{va}} \in Z, \\ &\tau^* > 1, \end{aligned} \tag{8}$$

where objective function (1) minimizes the total cost of the container terminal gate in both construction and operating stages, including construction and operating cost of FLs in arrival direction (the first part) and departure direction (the second part), construction and operating cost for RLs (the third part), and waiting time and fuel costs for all container trucks at the container terminal gate (the fourth part). For a given container terminal, some costs are constant, including construction and operating costs for each lane per time unit (C_{in}, C_{out}, and C_{va}), waiting time cost (C_w), and fuel cost (C_f) for container trucks. The calculation period is also a constant and is determined based on the research condition. Equation (5) restricts the total number of lanes at the container terminal gate to be a constant. Equation (6) restricts the number of FLs, which means that lanes in arrival direction are not fewer than those in departure direction. Equation (7) is the relationship between average queue length for the ith lane and four decision variables (n_{in}, n_{out}, n_{va}, and τ^*). Here, since RLs are changing dynamically, it is impossible to quantify the decision variables through analytical expressions; a simulation model is established to solve this problem, which will be discussed in Section 4. Equation (8) shows value ranges of decision variables.

4. Simulation-Based Metaheuristic Solution Algorithm

Due to challenges inherited in the stage of model formulation, where optimal number of RLs at the container terminal gate and corresponding scheduling rules are targeted, our model cannot be solved directly by any commercial solver (e.g., CPLEX and LINDO). Moreover, some functions proceeded in the model, which are not able to be transformed into any closed-form formulas, need to be described as a black box by simulation method as well. Therefore, to obtain the optimal solution efficiently, an SA metaheuristic algorithm method is designed in this study.

4.1. General Framework. Solution of the model goes into two parts: (1) four decision variables of n_{in}, n_{out}, n_{va}, and τ^* at planning stage and (2) the function l_i with four decision variables at operation stage. Then, a two-loop iteration framework is proposed. The outer loop determines values of four decision variables, n_{in}, n_{out}, n_{va}, and τ^*, while the inner loop determines the value of the function l_i under the given values of decision variables. The SA metaheuristic method is employed to change values of n_{in}, n_{out}, n_{va}, and τ^* in the outer loop iterations. For the inner loop, a simulation model is used to obtain value of l_i. Finally, the optimal solution can be obtained by the SA metaheuristic method within a reasonable time period.

Solution procedures of the SA metaheuristic method (Lu, 2014) are describes as follows.

Step 1. Obtain an initial solution; let $T = T_0$ (the initial temperature in SA).

Step 1.1. Generate an initial setting for decision variables n_{in}, n_{out}, n_{va}, and τ^* (denoted by $u^{(0)}$).

Step 1.2. Based on $u^{(0)}$, use the simulation model to obtain value of l_i (denoted by $v^{(0)}$).

Step 1.3. Calculate the objective value of the solution ($u^{(0)}$ and $v^{(0)}$), denoted by $F^{(0)}$.

Step 2. Repeat the following steps until one of stopping conditions becomes true.

Step 2.1. Generate R neighbors of $u^{(0)}$, that is, $u^{(n)}$, $n \in \{1, 2, \ldots, R\}$.

Step 2.2. For $n = 1$ to R, repeat the following steps.

Step 2.2.1. Based on $u^{(n)}$, use the simulation model to obtain the value of l_i.

Step 2.2.2. Calculate the objective value, denoted by $F^{(n)}$.

Step 2.2.3. Let $\delta = F^{(n)} - F^{(0)}$.

Step 2.2.4. If $\delta < 0$, set $u^{(0)} = u^{(n)}$.

Step 2.2.5. If $\delta \geq 0$, generate a random number $\chi \in (0, 1)$; if $\chi < e^{-\delta/T}$, $u^{(0)} = u^{(n)}$.

Step 2.3. Set $T = r \times T$.

The stopping criteria include the following: (a) fitness value of solution is equal to zero; (b) temperature is smaller than a given threshold value; (c) the best value of fitness cannot be improved within a given number of external loops.

Key issues for implementing the above SA procedure are as follows: (1) initial setting $u^{(0)}$ for decision variables n_{in}, n_{out}, n_{va}, and τ^*; (2) solution obtaining $v^{(0)}$ for l_i by a simulation model; (3) solution evaluating of the objective functions; and (4) neighborhood defining of solution. The first, third, and fourth issues are further addressed in Sections 4.1.1 to 4.1.3, while the second issue is addressed in Section 4.2.

4.1.1. Generating an Initial Solution. The initial setting for four decision variables (n_{in}, n_{out}, n_{va}, and τ^*) is the starting point of the proposed metaheuristic method and is critical for the efficiency of solution process. According to previous experience, when generating an initial solution for the decision variables, throughput of the terminal is divided into two parts: import throughput and export throughput. Next, the numbers of FLs in departure direction n_{out} and arrival direction n_{in} are calculated based on the import and export throughput. Moreover, since the construction and operating costs of a RL are higher than those of a FL, the initial number of the RLs n_{va} is set as zero. Finally, the initial critical value of imbalanced queuing coefficient $\tau^{*(0)}$ is determined. In this way, a set of initial solutions for four decision variables can be obtained; that is, $u^{(0)} = (n_{\text{in}}^{(0)}, n_{\text{out}}^{(0)}, 0, \tau^{*(0)})$.

4.1.2. Evaluating Solutions. After the previous steps, initial solution for the decision variables $u^{(0)}$ is obtained. Then, average queue length of the ith lane l_i can be obtained based on simulation results. After that, the value of $u^{(0)}$ and l_i can be substituted into Expression (1) to calculate the value of the objective function.

4.1.3. Neighborhood of a Solution in SA. During each iteration, the SA heuristic method explores solution space by moving from current solution u to another solution in its neighborhood $N(u)$. The neighborhood of solution u needs to meet all the following requirements: (1) Number of elements of lanes at the container port gate (n_{in}, n_{out}, and n_{va}) changes randomly by plus one or minus one or remains stable. (2) Total number of lanes at the container port gate (N) remains stable. (3) The number of FLs in arrival direction (n_{in}) should not be fewer than that in departure direction (n_{out}). (4) There is at least one FL in each direction (n_{in} and n_{out}), respectively. (5) The number of RLs (n_{va}) should not be less than zero. (6) The critical value of imbalanced queuing coefficient (τ^*) changes randomly by plus 0.1 or minus 0.1 or remains stable. (7) The critical value of imbalanced queuing coefficient (τ^*) should be larger than 1. (8) At least one of four decision variables (n_{in}, n_{out}, n_{va}, and τ^*) changes.

4.2. Simulation Model. In the SA method, since the average queue length of the ith lane l_i cannot be calculated by numerical method, a simulation model for container port gate system is established by using Rockwell Arena software 10.0 to obtain the value of l_i corresponding to each set of decision variables.

4.2.1. Logic Model. According to the operating procedures of the container trucks arriving at the terminal with RLs, the logic model of the container terminal gate system can be determined (shown in Figure 3).

In the logic model, the key operation point of the container terminal gate system is the direction changing of RLs. Logic of detailed lane changings is shown in Figure 4.

In case of container trucks departing from terminals, similar logical framework is defined for direction setting of RLs. The only difference is that, in case of changing the traffic direction of a RL into departure direction, the judgement criteria of $\tau < 1/\tau^*$ should be met and it must be ensured that all the container trucks in arrival direction have already passed the gate.

4.2.2. Simulation Model Establishment. Once the logic model is determined, the Rockwell Arena software is used to

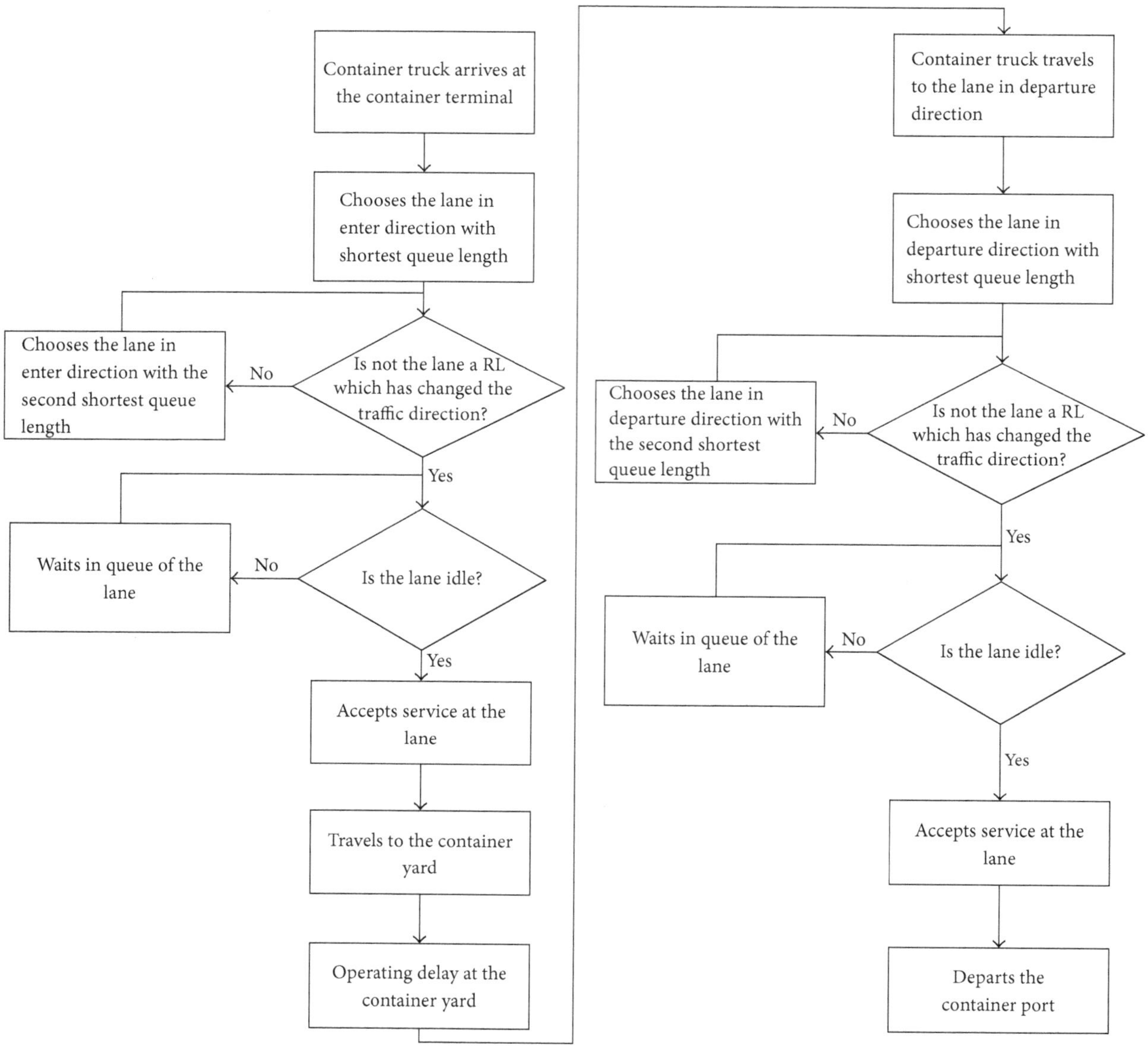

FIGURE 3: Logic model of the container terminal gate system with RLs.

simulate the model of container terminal gate system. The simulation model consists of four submodels, that is, *create container truck*, *RL scheduling*, *entering port*, and *departing port*. Interface of the simulation model is shown in Figure 5.

(1) Submodel of Create Container Truck. Create container truck submodel describes the creation and initialization processes of container truck entities. Firstly, a *Create* module is used to create a container truck entity. Then, by loading from data files of each container truck, the arrival time interval and waiting time are obtained and assigned to each container truck by two *Read/Write* modules separately. Next, a *Delay* module is used to defer the container truck entity based on the real arrival time interval, so that every entity can arrive at a proper time. Finally, a *Separate* module is adopted to send the entity back to *Read Arrival Time* module through the *Original exit point* and read the next arrival time interval from data files. On the other hand, a copy of the container terminal gate entity is generated by the same module and sent to the next submodule at the *Duplicate exit point* by combining the *Route* and *Station* modules.

(2) RLs Scheduling Submodel. This submodel schedules direction changing of RLs. Firstly, a *Decide first container truck* module is used to judge whether the current entity is the first arrival of the system. For the initial state, traffic direction of all RLs is arrival direction, which is used to make sure that numbers of lanes in enter direction are more than those in

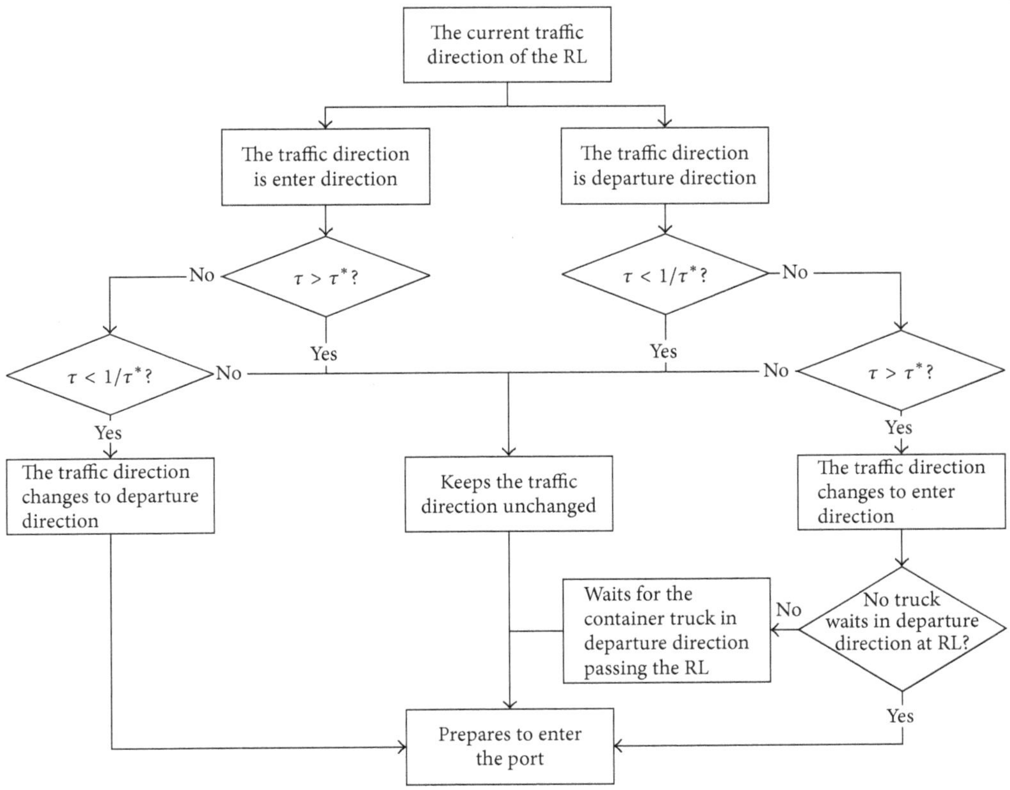

FIGURE 4: Logic of the changing traffic direction of RLs when container truck enters the terminal.

departure direction. An *Assign* module is used to set this initial stage as mode 1.

In case of other following container truck entities, *Decide mode_1* module is firstly used to judge the traffic direction of each RL. For arrival direction, models of *Decide variable in* and *Decide variable change to 2* will be used to judge whether the criteria of changing traffic direction have been met or not. If not, keep the traffic direction of the RL staying in mode 1; otherwise, an *Assign* module is used to convert the traffic direction to mode 2. Similarly, the process of traffic direction changing from departing to arrival direction is accomplished by a series of *Decide* modules in the submodel.

It is noticeable that even if the criteria of changing traffic direction to enter direction have been achieved but there are still some container trucks queueing in departure direction at the RL, then the traffic direction of the RL will keep in departure direction to prevent further congestions. Only when all the container trucks in departure direction have passed the RL will the traffic direction of RL change and serve in arrival direction. This process is finished by a series of *Decide* modules. Finally, four *Assign* modules are used to set the values of current modes, especially the traffic direction of the RL. Then the entities of container trucks will be exported to the next submodel.

(3) Entering Port Submodel. Entering port submodel accounts for two processes: container trucks accepting service at lanes in arrival direction and waiting at the terminal. Two *Choose gate* submodules and two *Decide gate* modules are firstly used to describe the selecting processes of lanes in mode 1 (arrival direction) and mode 2 (departure direction) of the RL. Then, multiple *Process* modules are used to describe service process of container trucks at lanes in arrival direction. Finally, the *Delay time* module is adopted to represent the process of container truck staying in the container terminal and preparing to depart the system.

(4) Departing Port Submodel. Similar to *RLs scheduling* submodel, this submodel starts with container trucks arriving at lanes in departure direction and ends up with the trucks leaving. At first, a *Decide* module is used to decide the traffic direction of each RL. If the mode is mode 1, that is, arrival direction, the container trucks will choose other lanes in departure direction for service. However, if the traffic direction of the RL should be changed to departure direction, that is, mode 2, several *Decide* modules will be used to further judge whether the queue length in arrival direction of the RL is zero. If so, the RL can be used as a lane in departing direction. The container trucks entities will be served at the terminal gate by a combination of several *Choose gate* submodels, *Decide* modules, and *Process* modules. At the end of this submodel, a *Dispose* model is adopted to send truck entities out of the system.

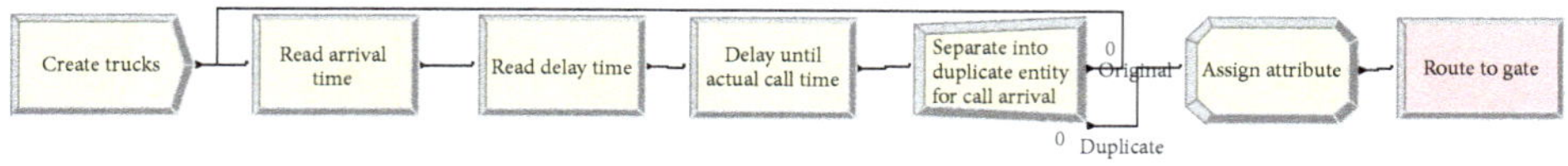

(a) *Create container truck* submodel

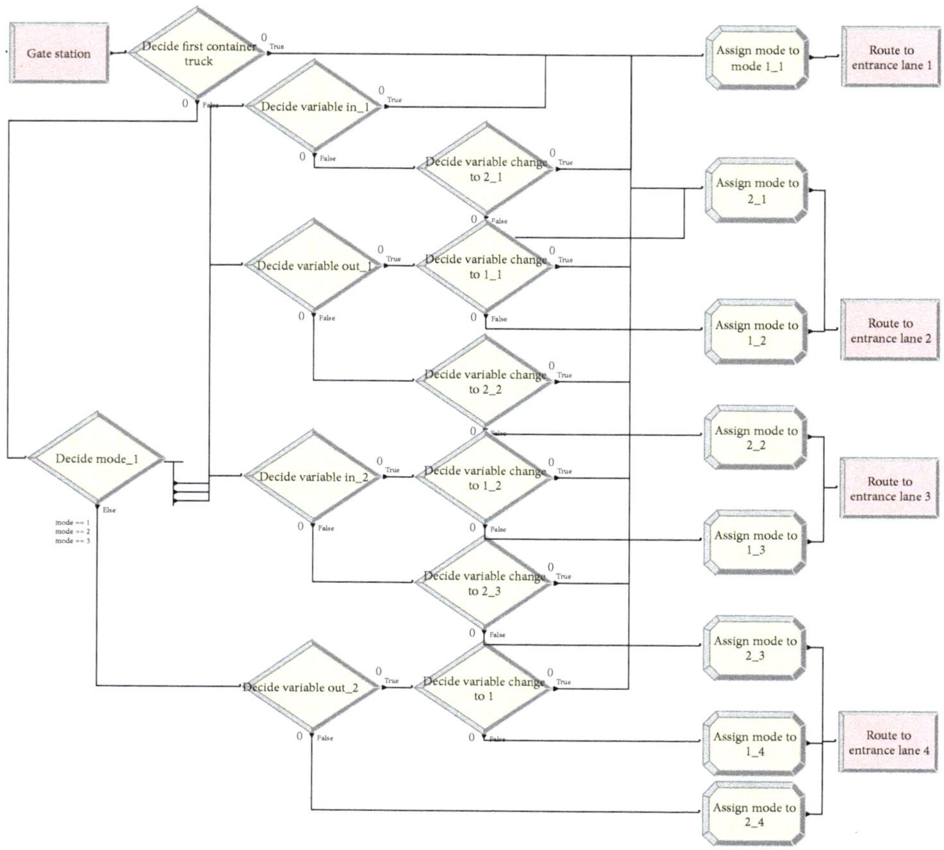

(b) *RLs scheduling* submodel

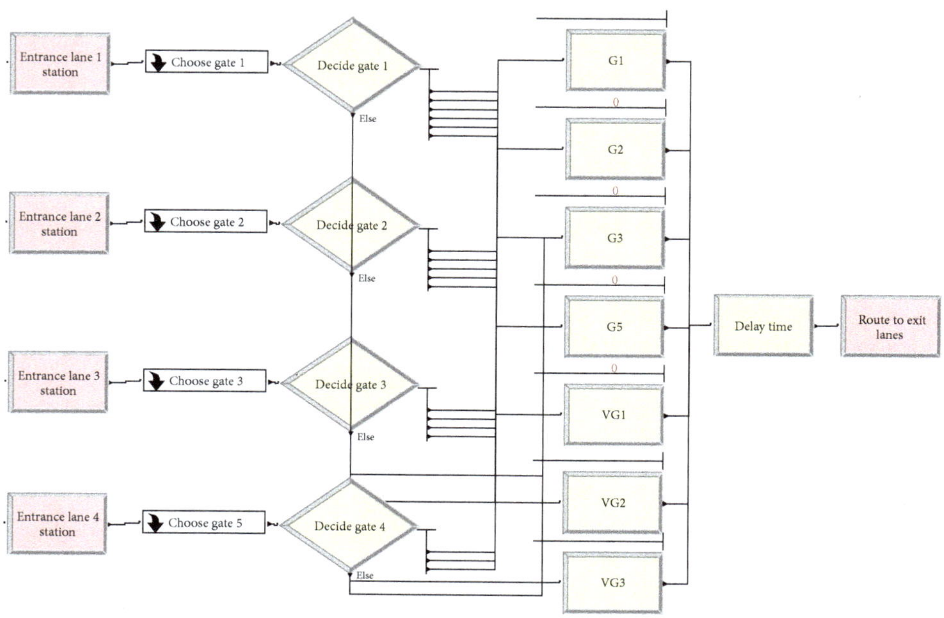

(c) *Entering port* submodel

FIGURE 5: Continued.

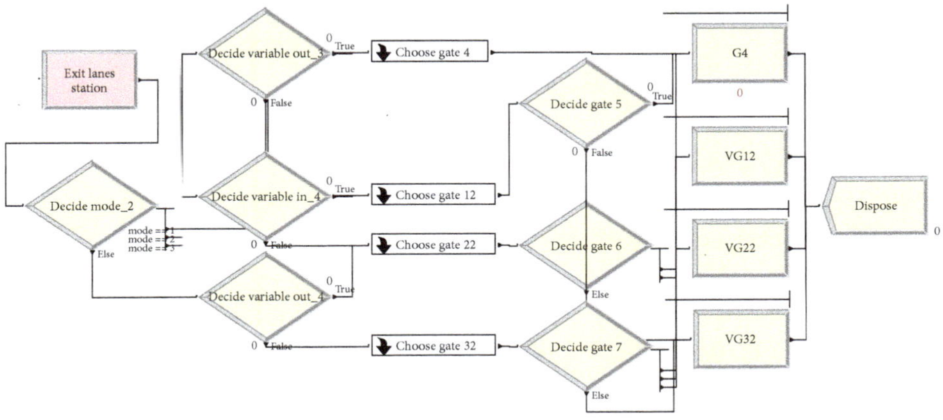

(d) *Departing port* submodel

FIGURE 5: Interface of the simulation model of the container terminal gate system.

TABLE 2: Cost coefficients of lanes at container gates.

Parameters	Symbols	Units	Values
Depreciable costs coefficients of FLs in enter direction	$C_{d\text{-in}}$	CNY/hour	8.00
Depreciable costs coefficients of FLs in departure direction	$C_{d\text{-out}}$	CNY/hour	7.00
Depreciable costs coefficients of RLs	$C_{d\text{-va}}$	CNY/hour	15.00
Labor costs coefficients of lanes	$C_{l\text{-in}}, C_{l\text{-out}}, C_{l\text{-va}}$	CNY/hour	20.00
Electric power cost coefficients of lanes	$C_{e\text{-in}}, C_{e\text{-out}}, C_{e\text{-va}}$	CNY/hour	2.00
Waiting time cost coefficient of container trucks	C_w	CNY/vehicle/hour	25.00
Fuel cost coefficient of container trucks	C_f	CNY/vehicle/hour	7.50

4.2.3. Model Parameters. Input parameters of the simulation model include the following:

(1) Container terminals: scale of container terminals, modes of the berth arrangement, and road network

(2) Containers: container throughput, number and proportion of different types of containers, and storage period of the containers at container yard

(3) Container trucks: running speed and arrival patterns of container trucks

(4) Container terminal gates: position of the terminal gate, the number and service efficiency of different types of lanes, and critical values of imbalanced queuing coefficients

Output parameters of the simulation model are the average queue length of container trucks at each lane during the calculation period which is corresponding to different combinations of lane arrangement scheme and the critical value of imbalanced queuing coefficients.

5. Case Study

In order to verify the effectiveness and practicability of the proposed method, a hub container terminal in northeast of China is used as a case in this paper to study the necessity of setting RLs, the optimal number of RLs, and the reasonable scheduling rules. For the SA metaheuristic method, the initial temperature $T = 10{,}000$, temperature length $R = 20$, and cooling rate $r = 0.6$. Stopping criteria are $T < 0.01$ or no further improvement of the fitness's best value after five external loops.

5.1. Input Parameters

5.1.1. Background of the Container Terminal. As the second largest container transportation hub in China, there are 14 container berths, which can berth fifth- and sixth-generation container ships. The designed capacity of the terminal is 9.1 million TEU. In 2010, the port berthed 5300 liner ships and finished 5.262 million TEU throughput of containers, in which the import and export throughput of containers reached 2.003 million TEU and 3.259 million TEU, respectively. According to the expected throughput of containers, the operators of the port planned to build 8 FLs totally. However, considering the imbalanced traffic volumes in two directions, they wonder whether it is necessary to build some RLs at the container gate.

Cost coefficients are summarized in Table 2.

5.1.2. Initial Decision Variables. According to ratio of the import and export throughput of the terminal, it can be determined that the initial number of FLs in arrival direction $n_{\text{in}}^{(0)}$ is 5, and the number of FLs in departure direction $n_{\text{out}}^{(0)}$ is 3. Moreover, the initial number of RLs $n_{\text{va}}^{(0)}$ is set as 0, and initial critical value of imbalanced queuing coefficient $\tau^{*(0)}$ is

Table 3: The summation of the average queue length of all lanes and the total costs of the container terminal gate system corresponding to different solutions.

Serial number of the decision variable	Value of the decision variable	Summation of the average queue length of all lanes (veh/h)	Total costs of the container terminal gate system C_T (Yuan)
$u^{(1)}$	$(5, 3, 0, 1.4)$	1.4731	2,495,521
$u^{(2)}$	$(5, 3, 0, 1.6)$	1.4731	2,495,521
$u^{(3)}$	$(6, 2, 0, 1.5)$	31.7383	11,120,770
$u^{(4)}$	$(6, 2, 0, 1.4)$	31.7383	11,120,770
$u^{(5)}$	$(6, 2, 0, 1.6)$	31.7383	11,120,770
$u^{(6)}$	$(4, 4, 0, 1.5)$	0.7472	2,280,096
$u^{(7)}$	$(4, 4, 0, 1.4)$	0.7472	2,280,096
$u^{(8)}$	$(4, 4, 0, 1.6)$	0.7472	2,280,096
$u^{(9)}$	$(4, 3, 1, 1.5)$	0.3448	2,235,606
$u^{(10)}$	$(4, 3, 1, 1.4)$	0.3440	2,235,366
$u^{(11)}$	$(4, 3, 1, 1.6)$	0.3429	2,235,070

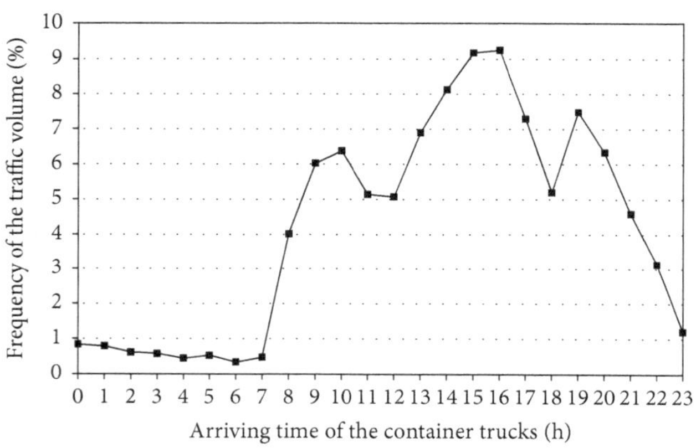

Figure 6: Frequency of traffic volume per hour.

set as 1.5. Therefore, the initial decision variables $u^{(0)}$ are (5, 3, 0, and 1.5).

5.1.3. Arrival Patterns of Container Trucks. According to the historical data of container truck arrival in 2010, frequency of traffic volume per hour can be obtained (shown in Figure 6), which work as the inputs for the simulation model. According to Figure 5, peak periods in day time mostly take place around 15:00 and 16:00, whose frequency is around 9.2%. Meanwhile, peak periods at night usually take place around 19:00 with frequency of 7.5%.

5.1.4. Serving Time at Container Terminal Gates. By employing the nonparametric test, comparison results of serving time are shown in Figure 7. For lanes in arrival direction (a), two-tailed probability (0.095) is larger than 0.05, indicating that the test sample has no significant difference from the theoretical distribution whose significant level α is 0.05. Therefore, the serving time at lanes in arrival direction follows normal distribution with mean value of 0.971 and standard deviation of 0.32. Similarly, testing result of two-tailed probability of serving time at lanes in departure direction (b) is 0.2, which is also larger than 0.05. Therefore it fits to a normal distribution with mean value of 0.796 and standard deviation of 0.34.

In summary, the serving time at the container terminal gate fits to normal distribution in this paper, and the average serving time at lanes in arrival direction is longer than that in departure direction.

5.2. Results Analysis. Once the input parameters are determined, the average queue length of each lane corresponding to the initial solution ($u^{(0)} = (5, 3, 0, 1.5)$) can be obtained in a calculation period. Summation of all the average queue lengths is 1.4731. Thus, total cost of the container terminal gate system, corresponding to initial solution by Expression (1), can be obtained as 2,495,521 Yuan.

Based on the neighborhood analysis of solutions in SA, there are 11 satisfactory solutions in the neighbor of $u^{(0)}$: $u^{(1)} = (5, 3, 0, 1.4)$, $u^{(2)} = (5, 3, 0, 1.6)$, $u^{(3)} = (6, 2, 0, 1.5)$, $u^{(4)} = (6, 2, 0, 1.4)$, $u^{(5)} = (6, 2, 0, 1.6)$, $u^{(6)} = (4, 4, 0, 1.5)$, $u^{(7)} = (4, 4, 0, 1.4)$, $u^{(8)} = (4, 4, 0, 1.6)$, $u^{(9)} = (4, 3, 1, 1.5)$, $u^{(10)} = (4, 3, 1, 1.4)$, and $u^{(11)} = (4, 3, 1, 1.6)$. Next, summation of the average queue lengths of all lanes in calculation period of each solution could be obtained from the simulation model. Then, the total costs of the container terminal gate system corresponding to each solution can be calculated through Expression (1).

As shown in Table 3, among the 11 solutions, $u^{(11)} = (4, 3, 1, 1.6)$ is the optimal solution with lowest total cost of 2,235,070 Yuan.

Following procedures of the SA metaheuristic method, it is finally obtained that the total cost of the container terminal gate system (C_T) reaches the minimum value when decision variable u is $(4, 3, 1, 1.6)$, that is, setting one RL, and the critical value of imbalanced queuing coefficient is 1.6. Compared with the initial solution (set 5 FLs in arrival direction and 3 FLs in departure direction), the optimal

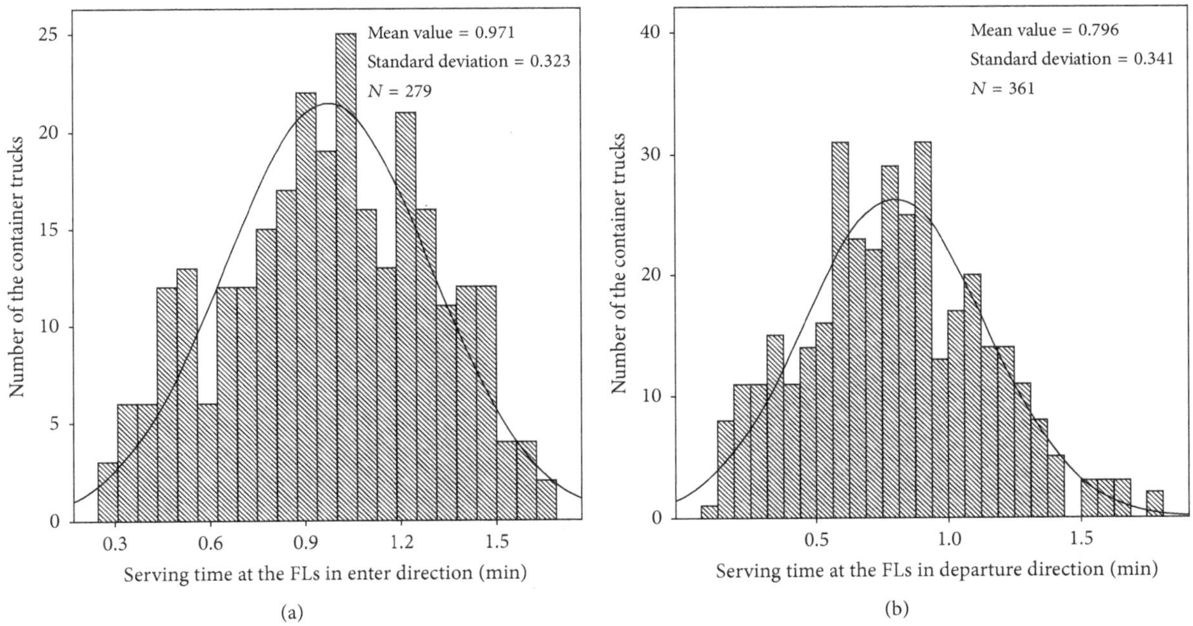

FIGURE 7: Testing result of serving time at the FL.

TABLE 4: Selection of the cost coefficients.

Group number	Cost coefficients which are changed	Change to n times of the control group
(1)	$C_{l\text{-in}}, C_{l\text{-out}}, C_{l\text{-va}}, C_w$	0.5
(2)	$C_{l\text{-in}}, C_{l\text{-out}}, C_{l\text{-va}}, C_w$	1.5
(3)	$C_{l\text{-in}}, C_{l\text{-out}}, C_{l\text{-va}}, C_w$	2
(4)	$C_{d\text{-in}}, C_{d\text{-out}}, C_{d\text{-va}}$	0.5
(5)	$C_{d\text{-in}}, C_{d\text{-out}}, C_{d\text{-va}}$	1.5
(6)	$C_{d\text{-in}}, C_{d\text{-out}}, C_{d\text{-va}}$	2

solution can decrease the summation of all the average queue lengths and total costs by 76.7% and 10.4%, respectively.

5.3. Sensitivity Analysis for the Cost Coefficients. In order to verify the robustness of the proposed metaheuristic method, the sensitivity of cost coefficients is analyzed in this section. Six groups of cost coefficients (shown in Table 3) are selected to investigate the influence caused by variation of local labor and land costs on the proposed method. Then, the variation of optimal total costs along with cost coefficients changing is also analyzed.

In Table 4, groups numbers (1)~(3) change the labor costs, including the labor costs of workers at the container terminal gate and the labor costs of container drivers based on the control group (the optimal solution of the case study). Meanwhile, groups numbers (4)~(6) change the construction costs, which is represented as the depreciable costs of all lanes at the gate. The SA metaheuristic method proposed is used to analyze the total costs of the container terminal gate system and its optimal decision variables corresponding to the six groups. Finally, variation of the total costs compared with the control group can be obtained, as shown in Table 5.

Comparing groups numbers (1)~(3) with the control group, it is found that the variation of total costs of the container terminal gate system relies on linear growth of labor costs. The total costs will be increased by 33.0% as the labor costs rise by 50%. Similarly, comparing groups numbers (4)~(6) with the control group, it is clear that the variation of total costs basically relies on linear growth of the depreciable costs. With slight disturbance, the total costs will rise by about 13.3% when the depreciable costs increase by 50%.

It is noticeable that, for group (1) (labor costs change to half time of the control group) and group (6) (depreciable costs become twice of the control group), the ratio of the depreciable costs to the total costs rises. In this case, the cost of setting a RL increases and the optimal decision variable u changes from (4, 3, 1) to (4, 4, 0), which means that replacing a RL with a FL in departure direction is better. In this way, the depreciable costs will be decreased, although the waiting time costs and fuel costs increase. Then, the total costs of the container port gate system will be reduced.

6. Conclusions and Discussions

This paper proves the necessity of setting RLs, which can efficiently alleviate the congestion at the container terminal gate and improve the service level of the terminal gate. However, since changing the traffic direction of the RL

Table 5: Total costs and their variation corresponding to each group.

Group number	Optimal total costs (Yuan)	Variation of total costs compared with the control group (%)	Optimal decision group u
Control group	2,235,070	0	(4, 3, 1)
(1)	1,496,720	−33.0	(4, 4, 0)
(2)	2,973,421	33.0	(4, 3, 1)
(3)	3,711,771	66.0	(4, 3, 1)
(4)	1,937,230	−13.3	(4, 3, 1)
(5)	2,532,910	13.3	(4, 3, 1)
(6)	2,805,696	25.5	(4, 4, 0)

will take time, to some extent, the service efficiency will definitely decrease. Moreover, the cost of RLs is larger than cost of FLs. Therefore, simply increasing the number of RLs does not mean a consequential total cost decrease, and the optimal number of the RLs needs to be determined cautiously and comprehensively. In this paper, aiming at minimizing the total costs, including the construction costs and operating costs of the whole container terminal gate system, an optimal model is first constructed to determine the optimal number of the RLs and reasonable scheduling rules at a terminal gate. Then, an SA metaheuristic method is proposed to solve the optimal model, and a simulation model is established to obtain parameters that cannot be determined by analytical methods. Finally, the proposed methodology is verified by quantitative conclusions obtained from a specific case study. Specifically, three conclusions are shown as follows:

(1) For the container terminal in case study, it is optimal to set one RL at the terminal gate and the optimal critical value of imbalanced queuing coefficient is 1.6.

(2) Compared with the design plan without RLs, the optimal plan can decrease the average queue length by 76.7%, and the total costs of the container terminal gate system can also be reduced by 10.4%.

(3) The variation of the total costs is more sensitive to the labor costs than the land costs.

The proposed method can provide a theoretical foundation for the container terminal planning and construction. The conception of the RL can also serve as a new idea for the smart container terminal gate system construction and management. However, there are still some limitations in this paper. For example, without simulating the operating process of container trucks in detail, this paper only considered the service time of the container trucks in container yards. As a next-step study, the handling process of container trucks in the yard will be further refined to reflect the actual situation.

Conflicts of Interest

The authors declare that they have no conflicts of interest.

Acknowledgments

The authors would like to acknowledge the National Natural Science Foundation of China (Grants no. 51779037 and no. 51709037) for supporting this research.

References

[1] R. Namboothiri and A. L. Erera, "Planning local container drayage operations given a port access appointment system," *Transportation Research Part E: Logistics and Transportation Review*, vol. 44, no. 2, pp. 185–202, 2008.

[2] X. M. Chen, X. S. Zhou, and G. F. List, "Using time-varying tolls to optimize truck arrivals at ports," *Transportation Research Part E: Logistics and Transportation Review*, vol. 47, no. 6, pp. 965–982, 2011.

[3] G. Chen, K. Govindan, and M. M. Golias, "Reducing truck emissions at container terminals in a low carbon economy: Proposal of a queueing-based bi-objective model for optimizing truck arrival pattern," *Transportation Research Part E: Logistics and Transportation Review*, vol. 55, no. SI, pp. 3–22, 2013.

[4] G. Chen, K. Govindan, and Z. Yang, "Managing truck arrivals with time windows to alleviate gate congestion at container terminals," *International Journal of Production Economics*, vol. 141, no. 1, pp. 179–188, 2013.

[5] Z.-Z. Yang, G. Chen, and D.-P. Song, "Integrating truck arrival management into tactical operation planning at container terminals," *Polish Maritime Research*, vol. 20, pp. 32–46, 2013.

[6] M.-H. Phan and K. H. Kim, "Negotiating truck arrival times among trucking companies and a container terminal," *Transportation Research Part E: Logistics and Transportation Review*, vol. 75, pp. 132–144, 2015.

[7] A. E. Azab and A. B. Eltawil, "A simulation based study of the effect of truck arrival patterns on truck turn time in container terminals," in *Proceedings of the 30th European Conference on Modelling and Simulation, ECMS 2016*, pp. 80–86, Regensburg, Germany, June 2016.

[8] D. Ambrosino and L. Peirano, "Truck arrival management at maritime container terminals," in *Proceedings of the 30th European Conference on Modelling and Simulation, ECMS 2016*, pp. 114–120, Regensburg, Germany, June 2016.

[9] G. Chen and L. P. Jiang, "Managing customer arrivals with time windows: a case of truck arrivals at a congested container

terminal," *Annals of Operations Research*, vol. 244, no. 2, pp. 349–365, 2016.

[10] C. Q. Guan and R. F. Liu, "Container terminal gate appointment system optimization," *Maritime Economics & Logistics*, vol. 11, no. 4, pp. 378–398, 2009.

[11] L. Hu, X. N. Shi, S. Voss, and W. G. Zhang, "Application of RFID Technology at the Entrance Gate of Container Terminals," in *2nd International Conference on Computational Logistics (ICCL 2011)*, vol. 6971 of *Lecture Notes in Computer Science*, pp. 209–220, Springer, Hamburg, Germany, 2011.

[12] Y. Tsai, T. Dusanter, Y. Wu, C. Ai, and Q. Catucoli, "Automatic truck processing time extraction at marine container terminal gates using low-frame-rate images," *Journal of Intelligent Transportation Systems: Technology, Planning, and Operations*, vol. 16, no. 4, pp. 211–225, 2012.

[13] M. Fleming, N. Huynh, and Y. C. Xie, "Agent-based simulation tool for evaluating pooled queue performance at marine container terminals," *Transportation Research Record*, no. 2330, pp. 103–112, 2013.

[14] D. Ambrosino and C. Caballini, "Congestion and truck service time minimization in a container terminal," in *Proceedings of the International Conference on Maritime-Port Technology and Development, MTEC 2014*, pp. 1–10, Trondheim, Norway, October 2014.

[15] E. A. Kadir, S. L. Rosa, and H. Gunawan, "Application of RFID Technology and E-Seal in Container Terminal Process," in *proceedings of the 4th International Conference on Information and Communication Technology (ICoICT)*, Bandung, Indonesia, 2016.

[16] M. Zeng, W. M. Cheng, and P. Guo, "A transient queuing model for analyzing and optimizing gate congestion of railway container terminals," *Mathematical Problems in Engineering*, vol. 2014, Article ID 914706, pp. 1–13, 2014.

[17] Y. Keceli, "A simulation model for gate operations in multi-purpose cargo terminals," *Maritime Policy & Management*, vol. 43, no. 8, pp. 945–958, 2016.

[18] D. Liu, X. S. Cao, and X. Miao, "Research on tidal traffic based variable lane isolation belt setting," *Hebei Journal of Industrial Science and Technology*, vol. 31, no. 05, pp. 14–19, 2014.

7

Predicting Operating Speeds at Urban Multilane Roundabouts in Abu Dhabi, United Arab Emirates

Mohammad Almoarawi and Essam Dabbour

Department of Civil Engineering, Abu Dhabi University, P.O. Box 59911, Abu Dhabi, UAE

Correspondence should be addressed to Essam Dabbour; edabbour@hotmail.com

Academic Editor: Richard S. Tay

There are more than 460 multilane roundabouts located in Abu Dhabi, the capital city of the United Arab Emirates, and its surrounding areas. Most of those roundabouts have three entry/circulatory/exit lanes with large radii, which resulted in the majority of drivers exceeding the speed limits for those roundabouts. Those excessive operating speeds, along with lack of drivers' awareness of the proper rules of driving at roundabouts, have resulted in increased collision frequencies at Abu Dhabi roundabouts. In this paper, operating speeds were measured at 12 roundabouts in Abu Dhabi and those collected speed observations were used to calibrate regression models to predict the 85th percentile operating speeds at roundabouts in Abu Dhabi. Predicting operating speed at a roundabout, during its design stage, is necessary to ensure that the expected operating speed and capacity will meet the design expectations. Three models were calibrated to predict the entry, circulating, and exit speeds, respectively. The calibrated models were validated with data not used in calibration and they were found to be stable and robust. The findings of this research study will help engineers when designing new roundabouts in Abu Dhabi or other cities with similar characteristics. This research study also provides a methodological framework for other researchers when conducting similar speed studies for roundabouts in other cities or metropolitan areas around the world.

1. Introduction

Roundabouts are very common in the United Arab Emirates with more than 460 roundabouts located in its capital city Abu Dhabi and its surrounding suburban areas. Most of those roundabouts have three entry/circulatory/exit lanes with large radii and large operating speeds.

Previous research studies found a correlation between operating speeds at roundabouts and the frequency and severity of collisions at those roundabouts [1–4]. Therefore, predicting the potential operating speed at a roundabout, during its design stage, is vital to ensure that the expected operating speed will meet the design expectations and the safety target for that roundabout. Furthermore, the capacity of a roundabout is highly correlated with critical time headways and follow-up times selected by drivers [5–9], which are related to the operating speed at the roundabout [2, 10–14]. Therefore, predicting the operating speed will also help in estimating the potential capacity of the roundabout being designed and ensure that the capacity meets the design requirements and the expected traffic volume. Predicting the operating speed will also help in estimating emissions [15–17]. Finally, precise prediction of the operating speed is also needed for different software applications related to traffic simulation.

A speed prediction model for US roundabouts [2, 18] is based on the assumption that the operating speed at any segment of the roundabout is correlated with the radius of the vehicle path at that segment. This approach is similar to an earlier prediction model for operating speeds at Australian roundabouts [19]. Džambas et al. [20] found similar correlation between vehicle speeds at turbo roundabouts and the radius of vehicle path. Another approach for speed prediction models was developed for Italian roundabouts [21] where the predicted operating speed was found to be correlated with the diameter of the inscribed angle, the width of the circulatory roadway, and the width of the entry lane. Pilko et al. [22] measured operating speeds at four single-lane roundabouts in the City of Zagreb, the capital city of Croatia, where they found correlation between entry speed and entry radius. Similar correlations were also observed between circulating speed and circulating radius and between exit speed and exit radius.

This paper presents regression models that predict the operating speeds at three-lane roundabouts in the city of Abu Dhabi, the capital city of the United Arab Emirates, based on the traffic hourly volumes and the geometric characteristics of the roundabouts. Radar guns were used to measure speeds of all vehicles, with different paths, at 12 three-lane roundabouts in the city Abu Dhabi and its surrounding suburban areas. Hourly traffic volumes and the geometric characteristics related to those roundabouts were all obtained from Abu Dhabi Department of Municipal Affairs and Transport. The following section provides details on data collection, followed by another section that provides details on how the developed models were calibrated and validated. Conclusions and recommendations for future research are also provided.

2. Data Collection

Operating speeds of individual vehicles were measured at 12 three-lane roundabouts in Abu Dhabi and its surrounding suburban areas. The speeds were measured using Bushnell Velocity speed guns [23], which are capable of measuring vehicle speeds from 16 km/h to 322 km/h at a range up to 457 m. All the roundabouts in the sample had three entry/circulatory/exit lanes, which represent the majority of roundabouts in Abu Dhabi and its surrounding suburban areas. Figure 1 shows satellite images of the roundabouts included in the sample. It must be noted that all the roundabouts in the sample were designed according to the UAE geometric design guide, which is mainly derived from the US design guide [18]. There is currently a newer design guide for Abu Dhabi urban streets, which is Abu Dhabi Urban Street Design Manual [24]. That manual started to be implemented in 2011. However, all the roundabouts included in the sample were designed according to the older design guide that was derived from US design guide [18].

Hourly traffic volumes, proportions of heavy vehicles (including buses and heavy trucks), and the geometric characteristics of the roundabouts were all obtained from Abu Dhabi Department of Municipal Affairs and Transport. The hourly traffic volumes and proportions of heavy vehicles obtained from Abu Dhabi Department of Municipal Affairs and Transport were all based on real-time sensor data that were collected within a month before the experiments; and therefore, they were believed to be up-to-date data at the time of the experiments. There were four roundabouts in the sample where traffic volumes and the proportions of heavy vehicles were not available from the Department of Municipal Affairs and Transport. For those four roundabouts, manual counting was used to obtain hourly traffic volumes and the proportions of heavy vehicles. The hourly traffic volumes used for this study included the morning traffic volume, V_m, the afternoon traffic volume, V_a, and the evening traffic volume, V_e. The proportions of heavy vehicles were observed during the morning, PHV_m, the afternoon, PHV_a, and the evening, PHV_e. Table 1 shows the traffic and geometric characteristics of the roundabouts included in this study. Figure 2 shows the typical locations of the entry radius, R_1, the radius of the central island, R_2, the exit radius, R_3, and the diameter of the inscribed circle, D_I.

For every possible vehicle path within a roundabout (turning right, turning left, going through, or making a U-turn), speeds were typically measured at the following three areas shown in Figure 2:

(a) The entry speed along the entry path arc, with the radius of the vehicle path, is mainly determined by the entry radius R_1.

(b) The circulating speed around the inscribed circle, with the radius of the vehicle path, is mainly determined by the radius of the central island, R_2.

(c) The exit speed along the exit path arc, with the radius of the vehicle path, is mainly determined by the exit radius R_3.

To ensure the highest level of precision in measuring the speeds, the speed guns were placed as close as possible at collision course with the target vehicles to minimize the angle of incidence and therefore eliminate the cosine effect [25]. This is illustrated in Figure 2, which shows that the speed guns were placed at point "A" to measure entry speeds, point "B" to measure circulating speeds, and point "C" to measure exit speeds. To avoid sampling bias, the speed of only the first vehicle was measured in cases when vehicles were traveling in platoons.

For every roundabout, vehicle speeds were measured during four different weekdays. Since weekend in Abu Dhabi is typically Friday and Saturday, speeds were typically measured on Mondays, Tuesdays, and Wednesdays. For each day, vehicle speeds were measured during one hour in the morning (between 7:30 AM and 9:30 AM), one hour in the afternoon (between 2:30 PM and 4:30 PM), and one hour in the evening (between 6:30 PM and 8:30 PM). Due to the weather and travel habits in Abu Dhabi, the hours during which speeds were measured are within the morning, afternoon, and evening peak-hour periods as per the data provided by Abu Dhabi Department of Municipal Affairs and Transport.

There were a total of 38,764 speed observations obtained for all the sites. For every site, the 85th percentile entry/circulating/exit speeds were calculated from those speed observations during every observation hour. The 85th percentile speed is the speed at which 85% of drivers drive at or below that speed [26]. The 85th percentile speed is the most frequently used descriptive statistics for the operating speed associated with a particular location or geometric feature [27]. Since vehicle speeds were measured in each site for three hours on four different weekdays, there were 12 different values observed for the 85th percentile entry/circulating/exit speeds for each site. Those values are shown in Table 2, which shows the 85th percentile speed during the morning hour, V_{85m}, during the afternoon hour, V_{85a}, and during the evening hour, V_{85e}. Based on this, for each of the entry, circulating, and exit speeds, there

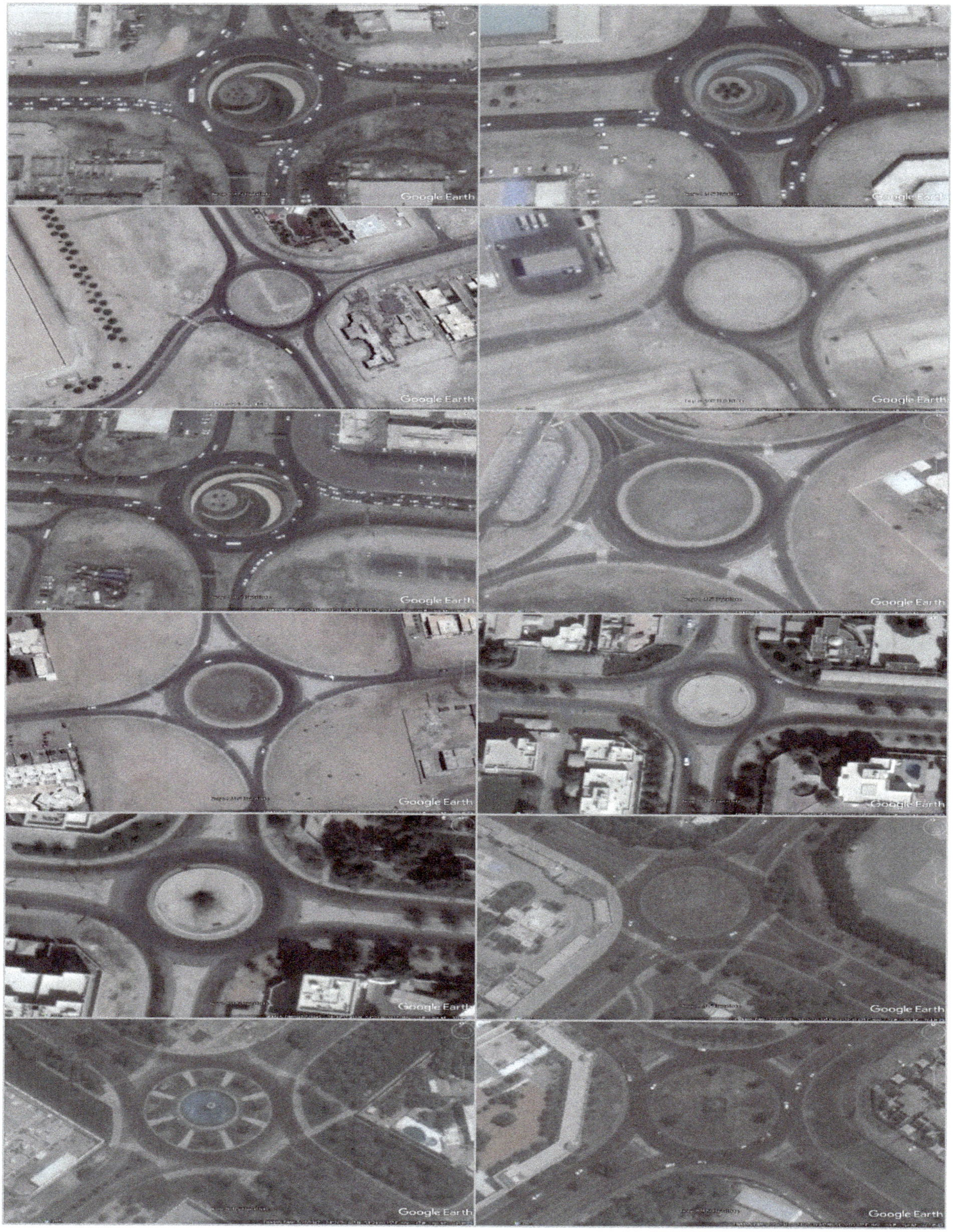

FIGURE 1: Satellite images of the sites selected for the study (map data ©2018 Google).

Table 1: Traffic and geometric characteristics.

Site	Area	Latitude	Longitude	R_1 (m)	R_2 (m)	R_3 (m)	D_I (m)	V_m (vph)	V_a (vph)	V_e (vph)	PHV_m	PHV_a	PHV_e
1	Mussafah	24°21′33.18″N	54°30′06.30″E	35.65	30.55	45.25	85.23	1935	915	1365	0.165	0.148	0.067
2	Mussafah	24°22′05.53″N	54°30′04.39″E	36.85	31.05	48.25	87.25	1875	785	1025	0.173	0.152	0.059
3	Khalifa City	24°25′09.18″N	54°32′48.81″E	32.15	30.65	41.2	80.85	1125	780	625	0.048	0.036	0.022
4	Khalifa City	24°24′20.58″N	54°36′19.10″E	33.2	30.25	40.65	80.15	1025	660	640	0.037	0.028	0.019
5	Mussafah	24°21′35.36″N	54°30′59.77″E	32.9	30.85	44.6	89.91	1825	1140	985	0.159	0.155	0.046
6	MBZ City	24°21′41.30″N	54°33′18.42″E	30.55	28.75	38.45	83.75	685	305	365	0.032	0.025	0.014
7	MBZ City	24°21′48.97″N	54°32′43.02″E	29.65	28.60	34.8	82.25	825	480	720	0.028	0.019	0.010
8	Al Maqta	24°24′26.51″N	54°29′44.23″E	23.55	14.55	29.65	51.55	980	510	680	0.045	0.038	0.024
9	Al Maqta	24°24′37.81″N	54°29′48.30″E	24.15	15.40	31.25	53.50	1050	735	610	0.051	0.040	0.022
10	Abu Dhabi	24°26′50.42″N	54°22′10.07″E	27.45	20.85	35.4	82.85	1260	670	725	0.012	0.008	0.006
11	Abu Dhabi	24°26′26.66″N	54°21′40.15″E	30.25	29.95	37.9	81.25	1540	740	810	0.018	0.011	0.008
12	Abu Dhabi	24°27′18.04″N	54°22′03.07″E	35.45	31.35	44.3	84.50	1650	665	775	0.021	0.016	0.009
		Mean		30.98	26.90	39.31	78.59	1315	699	777	0.066	0.056	0.026
		Standard deviation		4.11	5.97	5.55	11.96	414.7	204.3	242.6	0.059	0.056	0.020

Table 2: Observed 85th percentile operating speeds.

Site	Day 1			Day 2			Day 3			Day 4			Mean	Std. Dev.	Speed Limit
	V_{85m}	V_{85a}	V_{85e}	V_{85m}	V_{85a}	V_{85e}	V_{85m}	V_{85a}	V_{85e}	V_{85m}	V_{85a}	V_{85e}			
1 (Entry)	26.7	32.4	29.7	25.4	33.1	30.6	28.8	34.5	31.2	27.4	33.7	32.9	30.5	2.8	40
1 (Circ.)	30.4	36.1	32.8	32.5	38.9	34.7	35.1	40.2	38.6	34.4	41.5	37.6	36.1	3.2	40
1 (Exit)	32.4	37.1	33.6	33.5	41.2	36.5	38.2	44.2	41.0	36.9	40.1	34.5	37.4	3.5	40
2 (Entry)	28.7	33.1	30.4	29.6	34.8	33.2	32.2	35.8	33.6	36.2	39.5	37.4	33.7	3.1	40
2 (Circ.)	33.2	36.8	35.2	35.1	38.6	37.4	34.7	39.2	37.4	30.2	33.2	31.8	35.2	2.7	40
2 (Exit)	37.9	41.2	40.5	38.9	36.1	35.7	36.8	42.5	39.7	34.7	36.5	33.9	37.9	2.6	40
3 (Entry)	40.6	43.2	44.8	41.5	45.6	46.7	41.2	44.7	48.3	40.1	42.5	46.7	43.8	2.6	50
3 (Circ.)	42.3	44.1	45.2	40.8	44.3	43.8	40.1	42.7	46.0	39.7	41.4	44.8	42.9	2.0	50
3 (Exit)	45.2	47.5	48.1	46.8	48.5	44.3	43.2	44.4	48.9	41.1	40.1	42.8	45.1	2.8	50
4 (Entry)	44.7	48.3	50.1	43.9	46.2	44.5	47.2	46.1	50.3	48.3	50.7	51.6	47.7	2.5	50
4 (Circ.)	46.0	49.3	51.2	44.5	45.6	48.7	49.1	45.5	51.1	48.3	52.5	54.6	48.9	3.0	50
4 (Exit)	48.3	52.1	53.4	48.7	46.3	50.7	53.4	48.9	52.7	49.3	54.6	55.8	51.2	2.8	50
5 (Entry)	28.4	33.4	31.8	29.6	34.8	33.5	33.0	34.7	32.6	37.1	30.4	31.9	32.6	2.3	40
5 (Circ.)	32.4	34.5	33.6	30.1	36.7	39.5	34.8	35.0	34.6	38.7	31.2	33.6	34.6	2.6	40
5 (Exit)	34.5	36.8	38.7	31.5	36.9	41.2	37.0	36.4	35.9	40.1	33.5	39.7	36.9	2.7	40
6 (Entry)	39.8	44.5	43.6	41.8	48.6	45.7	42.3	44.5	48.7	45.0	47.8	43.2	44.6	2.6	50
6 (Circ.)	42.6	48.7	44.6	47.6	47.6	46.9	44.3	47.8	50.2	47.8	49.6	51.3	47.4	2.4	50
6 (Exit)	44.7	51.8	46.3	52.7	53.2	51.4	45.3	48.2	51.4	48.9	52.3	53.7	50.0	3.1	50
7 (Entry)	42.1	48.9	44.6	50.2	51.3	50.8	44.6	47.9	50.7	49.3	50.7	51.2	48.5	3.0	40
7 (Circ.)	44.6	51.2	47.9	53.0	53.4	51.7	48.9	49.0	53.6	51.4	53.2	54.8	51.1	2.8	40
7 (Exit)	46.9	53.8	49.1	55.3	56.8	53.7	51.2	49.8	55.6	52.3	55.4	57.8	53.1	3.2	40
8 (Entry)	39.7	37.4	38.1	36.5	34.1	33.8	32.8	33.9	34.7	30.8	32.5	33.6	34.8	2.5	40
8 (Circ.)	37.1	34.2	33.3	36.1	30.2	29.8	28.7	26.9	27.8	28.6	30.7	27.4	30.9	3.3	40
8 (Exit)	38.3	35.4	31.9	37.5	31.6	32.1	29.4	27.8	28.6	29.0	33.1	28.5	31.9	3.4	40
9 (Entry)	36.1	31.5	33.8	35.6	34.7	30.8	29.7	28.4	27.3	26.8	27.0	28.6	30.9	3.3	40
9 (Circ.)	35.1	34.6	33.7	33.6	32.7	30.1	31.4	28.4	27.6	26.8	27.8	26.3	30.7	3.1	40
9 (Exit)	34.8	35.6	34.1	33.2	33.1	28.9	29.7	28.7	29.0	31.0	30.8	27.4	31.4	2.6	40
10 (Entry)	36.1	37.4	38.0	35.3	33.8	34.7	33.1	34.8	36.0	30.1	31.2	33.2	34.5	2.2	40
10 (Circ.)	37.0	39.1	40.1	37.4	35.0	36.7	34.1	33.8	39.8	30.8	32.4	33.9	35.8	2.9	40
10 (Exit)	40.1	41.2	43.6	39.8	38.6	37.8	39.0	34.7	42.5	33.6	37.4	36.9	38.8	2.8	40
11 (Entry)	32.1	29.7	28.6	28.4	27.0	26.9	30.4	31.2	28.6	30.7	33.1	31.8	29.9	1.9	40
11 (Circ.)	34.8	35.6	38.1	40.2	39.8	41.5	38.6	42.3	41.8	36.9	42.7	41.6	39.5	2.6	40
11 (Exit)	39.6	42.3	44.7	46.9	47.2	44.3	40.8	45.6	43.6	39.6	45.7	44.0	43.7	2.5	40
12 (Entry)	38.4	37.6	41.0	45.6	44.2	40.8	41.2	44.3	42.1	38.6	37.6	39.0	40.9	2.6	40
12 (Circ.)	39.1	40.2	43.6	44.2	43.8	41.3	42.9	41.7	43.0	39.6	38.5	41.0	41.6	1.9	40
12 (Exit)	44.2	45.3	46.8	47.2	45.0	43.6	47.2	44.3	45.6	47.6	41.2	43.1	45.1	1.9	40

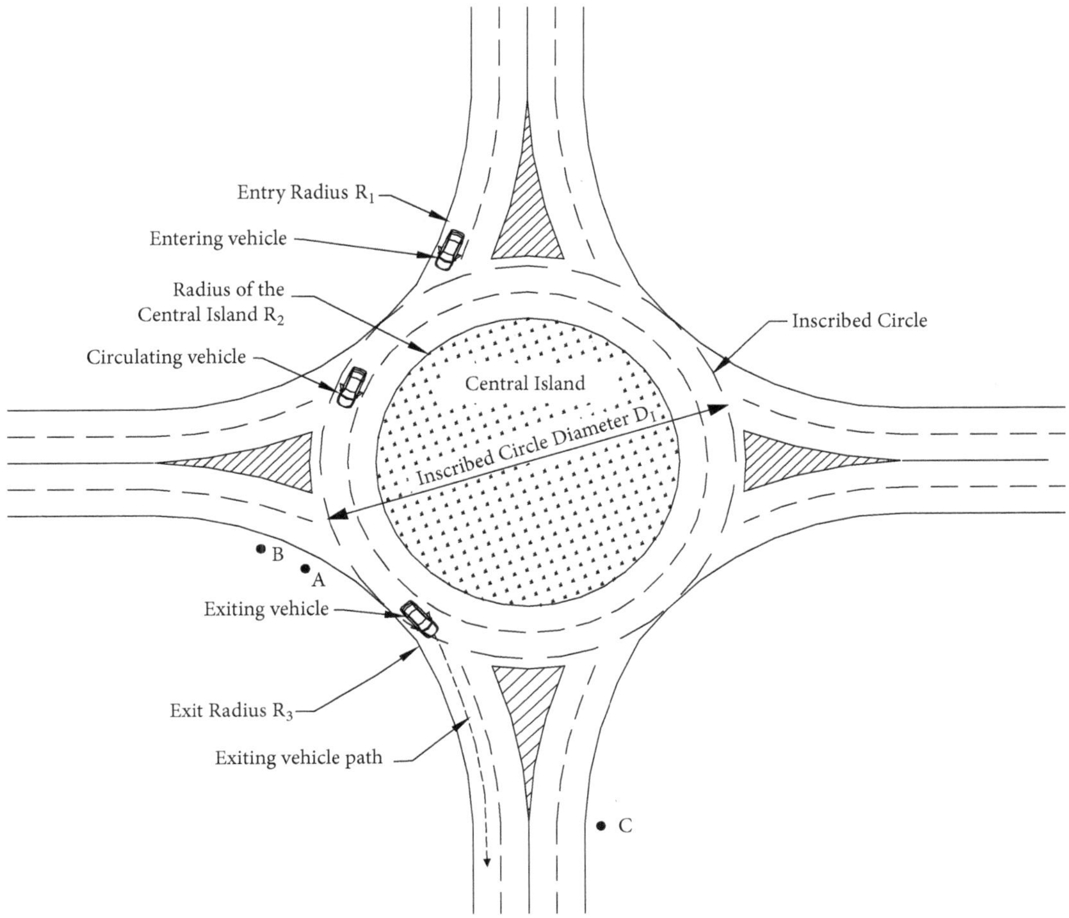

FIGURE 2: Example of speed measuring locations at a multilane roundabout.

were a total of 144 observed values of the 85th percentile speeds for all the 12 sites. From those 144 observations, 108 observations were used to calibrate the prediction models for entry/circulating/exit speeds, and the remaining 36 observations were used to validate those calibrated models.

For each roundabout in the sample, the controlling speed limit is also shown in Table 2. The controlling speed limit is the lowest speed limit posted on the four roads intersecting at the given roundabout. As shown in Table 2, the measured mean speeds exceed the posted speed limits in 4 roundabouts, representing 33.33% of the sample size. The mean values of the entry/circulating/exit speeds in all the sites are 37.7 km/h, 39.6 km/h, and 41.9 km/h, respectively. Those values are higher than those found in literature for roundabouts in other countries [28]. In a previous research study to measure the 85th percentile speeds at 98 roundabout approaches in the United States [29], it was found that the mean 85th percentile entering speed was 34.9 km/h (with a standard deviation of 3.9 km/h). In another study to measure the 85th percentile speeds at 7 roundabouts in Italy [21], the mean 85th percentile circulating speed was found to be 34.9 km/h (with a standard deviation of 6.3 km/h). Kim and Choi [1] measured the 85th percentile operating speeds at 14 roundabouts in South Korea where they found that the mean entry, circulating, and exit speeds, along with their respective standard deviations, were (26.3 ± 5.2) km/h, (23.2 ± 4.0) km/h, and (27.0 ± 6.3) km/h, respectively. Rubio-Martín et al. [30] presented a case study for a single-lane roundabout in Spain where the mean values and standard deviations of entry, circulating, and exit speeds were estimated to be (35.1 ± 4.3) km/h, (33.3 ± 4.6) km/h, and (38.4 ± 7.6) km/h, respectively. Pilko et al. [22] measured operating speeds at four single-lane roundabouts in the City of Zagreb, the capital city of Croatia, where they found those operating speeds to be typically below 33.9 km/h. Several other studies found that vehicle speeds at turbo roundabouts are typically below 40 km/h [20, 31, 32]. Based on these studies, the measured operating speeds at Abu Dhabi roundabouts are typically higher than those in other parts of the world. Those excessive speeds seem to be mainly attributed to the use of large radii used in Abu Dhabi roundabouts. As an example, Table 1 shows that the mean value of the inscribed circle diameter in Abu Dhabi is (78.59 ± 11.96) m. By

TABLE 3: Developed speed prediction models.

Parameter	Coefficient	t-stat
Predicting entry speed:		
Intercept	35.622	9.806
Entry radius[(a)]: $(R_1)^{0.8}$	1.754	8.510
Hourly traffic volume: $(V)^{0.5}$	−0.595	−9.506
Proportion of heavy vehicles: $(\text{PHV})^{0.2}$	−14.728	−3.156
Coefficient of determination (R^2)	0.675	-
Significance of f statistic	0.000	-
Standard error	3.245	-
Predicting circulating speed:		
Intercept	36.971	17.514
Radius of the central island[(a)]: $(R_2)^{0.8}$	1.885	14.521
Hourly traffic volume: $(V)^{0.5}$	−0.456	−8.446
Proportion of heavy vehicles: $(\text{PHV})^{0.2}$	−19.531	−6.124
Coefficient of determination (R^2)	0.787	-
Significance of f statistic	0.000	-
Standard error	2.734	-
Predicting exit speed:		
Intercept	35.729	8.314
Exit radius[(a)]: $(R_3)^{0.8}$	1.914	7.443
Hourly traffic volume: $(V)^{0.5}$	−0.378	−5.218
Proportion of heavy vehicles: $(\text{PHV})^{0.2}$	−36.616	−5.847
Coefficient of determination (R^2)	0.646	-
Significance of f statistic	0.000	-
Standard error	3.611	-

[(a)] See Figure 2 for the typical locations of the entry radius, the radius of the central island, and the exit radius.

comparison, the mean values of the inscribed circle diameter in South Korea and Serbia were (37.88 ± 14.08) m and (36.20 ± 1.47) m, respectively [1, 33]. More strict design standards may be needed to be implemented for roundabouts in Abu Dhabi to force drivers to reduce their speeds.

3. Speed Prediction Models

Multiple linear regression was used to calibrate the prediction models [34], which takes the form:

$$y = \beta_0 + \beta_1 x_1 + \beta_2 x_2 + \cdots + \beta_k x_k + \varepsilon \quad (1)$$

For the purpose of this research, the dependent variable, y, in the above equation is the estimated 85th percentile entry, circulating, or exit speed. The independent variables $(x_1, x_2 \cdots x_k)$ are the hourly traffic volumes, the percentages of heavy vehicles, and the different geometric characteristics of the roundabouts, as shown in Table 1. The regression coefficients $(\beta_0, \beta_1, \beta_2 \cdots \beta_k)$ are calculated using the least-square method. The error term, ε, accounts for random variation in the dependent variables, y, not fully explained by the independent variables.

Several combinations of independent variables were attempted to develop the regression models. The final models that best fit the data are shown in Table 3 where V is the hourly traffic volume (vehicles per hour) and PHV is the proportion of heavy vehicles. Different power combinations were also attempted for the selected independent variables to maximize the coefficients of determination, R^2, and minimize the validation errors. After several attempts, it was decided to use the power values shown in Table 3 that include a power value of 0.8 for the geometric radii, 0.5 for the hourly traffic volume, and 0.2 for the proportion of heavy vehicles. In all the three developed models, the coefficients of all independent variables are significantly different from zero at 0.05 significance level. The residual plots of the independent variables exhibited random dispersion around the horizontal axis in all models, indicating a constant variance of errors. The models shown in Table 3 have logical explanations for the effect of different independent variables on the estimated speed. The positive sign associated with the curve radius coefficient suggests that the speed increases with the increase of the horizontal curve radius, which is consistent with most vehicle dynamics models found in literature and design guides [2, 18, 19, 21, 35]. The negative sign associated with the hourly traffic volume coefficient suggests that the speed decreases with the increase of the traffic volume due to the decrease in vehicle spacing which is also consistent with most traffic flow models [11, 36, 37]. Finally, the negative sign associated with the proportion of heavy vehicles suggests that the speed decreases with the increase of the proportion of heavy vehicles since heavy vehicles usually move at lower speeds when negotiating roundabouts [2, 22, 38]. The values associated with the coefficients of determination (R^2) shown in Table 3 are mainly comparable with those found in similar prior research studies [1, 22, 39, 40].

As was aforementioned, there were 36 speed observations reserved to validate the developed models. The validation statistics for all the models are shown in Table 4. The statistics shown include the sum of errors, the sum of squared errors, the mean squared errors, and the root-mean squared errors. As shown in Table 4, the sums of errors are small indicating that the errors are fluctuating around a zero value. Furthermore, the root-mean squared errors for all models are small with values close to the standard errors associated with model calibration as shown in Table 3. Based on these statistical measures, it could be concluded that the developed models are stable and robust when being validated with data not used in calibration.

As was discussed in the previous section, the observed operating speeds at Abu Dhabi roundabouts are mainly larger than those observed at roundabouts in several other countries including South Korea [1], Spain [30], Croatia [22], Italy [21], and the United States [2, 29]. The authors of this study are not aware of any other speed prediction models where traffic volumes or the proportions of heavy vehicles are used as predictors. However, it is expected that the operating speeds predicted by the models developed in this study will be typically higher than those predicted by other models. This is due to the fact that the models developed in this study are calibrated using data where higher operating speeds are observed.

4. Conclusions

In this paper, regression models were developed to predict the 85th percentile operating speeds at multilane roundabouts in

TABLE 4: Validation statistics of the speed prediction models.

Site	Observed speed			Predicted speed			Error			Squared error		
	Entry	*Circ.*	*Exit*	*Entry*	*Circ.*	*Exit*	*Entry*	*Circ.*	*Exit*	*Entry*	*Circ.*	*Exit*
	27.4	34.4	36.9	29.8	32.4	34.0	−2.4	2.0	2.9	5.6	4.2	8.6
1	37.3	40.5	40.1	38.2	38.9	39.7	−0.9	1.6	0.4	0.8	2.5	0.2
	32.9	37.6	38.5	35.7	37.8	40.8	−2.8	−0.2	−2.3	7.6	0.0	5.5
	32.6	32.0	34.7	30.9	32.9	36.1	1.7	−0.9	−1.4	2.9	0.8	2.0
2	39.5	39.2	40.5	40.3	40.2	42.6	−0.8	−1.0	−2.1	0.6	1.1	4.2
	37.4	38.1	42.3	39.6	40.7	45.4	−2.2	−2.6	−3.1	5.0	6.9	9.4
	37.1	39.7	41.1	35.8	40.2	40.6	1.3	−0.5	0.5	1.7	0.2	0.3
3	37.9	44.1	41.0	39.6	43.3	43.8	−1.7	0.8	−2.8	2.9	0.6	8.0
	43.7	44.8	48.2	42.1	45.6	46.7	1.6	−0.8	1.5	2.7	0.6	2.3
	38.3	42.3	40.9	37.9	41.1	41.8	0.4	1.2	−0.9	0.2	1.4	0.8
4	43.3	45.2	45.6	42.0	44.5	45.2	1.3	0.7	0.4	1.6	0.4	0.2
	41.6	45.6	45.8	42.8	45.4	46.7	−1.2	0.2	−0.9	1.5	0.0	0.8
	27.8	33.7	33.1	28.7	33.3	34.2	−0.9	0.4	−1.1	0.8	0.2	1.2
5	34.0	36.2	35.3	34.1	37.4	37.7	−0.1	−1.2	−2.4	0.0	1.5	5.7
	39.1	39.6	39.7	37.7	41.4	44.0	1.4	−1.8	−4.3	2.0	3.2	18.7
	43.0	42.8	41.9	39.7	42.9	42.9	3.3	−0.1	−1.0	10.9	0.0	1.0
6	47.8	46.9	48.1	45.2	47.4	47.1	2.6	−0.5	1.0	6.6	0.2	1.0
	43.2	48.3	50.7	45.0	47.6	48.4	−1.8	0.7	2.3	3.3	0.5	5.4
	39.1	41.4	42.3	37.7	41.9	39.7	1.4	−0.5	2.6	1.9	0.2	6.7
7	40.7	43.9	44.5	42.3	45.7	43.6	−1.6	−1.8	0.9	2.6	3.3	0.8
	41.1	44.8	47.1	40.2	44.5	43.8	0.9	0.3	3.3	0.8	0.1	11.2
	30.8	28.6	31.9	31.0	28.2	33.0	−0.2	0.4	−1.1	0.1	0.1	1.2
8	35.2	32.7	36.1	36.5	32.6	37.0	−1.3	0.1	−0.9	1.7	0.0	0.7
	36.3	29.4	38.5	35.1	31.9	37.3	1.2	−2.5	1.2	1.5	6.1	1.4
	28.6	28.6	31.9	30.6	28.2	33.3	−2.0	0.4	−1.4	4.1	0.1	2.1
9	33.7	28.7	38.0	34.2	31.1	36.3	−0.5	−2.4	1.7	0.2	6.0	2.9
	38.6	36.1	37.4	36.5	33.4	39.4	2.1	2.7	−2.0	4.5	7.3	3.9
	33.1	33.8	43.1	33.2	34.1	40.4	−0.1	−0.3	2.7	0.0	0.1	7.3
10	41.1	42.3	47.3	39.4	39.1	45.2	1.7	3.2	2.1	2.8	10.0	4.4
	38.2	39.3	46.9	39.1	39.1	45.6	−0.9	0.2	1.3	0.9	0.0	1.7
	31.7	39.6	39.6	32.5	38.9	39.6	−0.8	0.7	0.0	0.7	0.4	0.0
11	42.1	46.1	47.5	40.3	45.2	45.7	1.8	0.9	1.8	3.3	0.7	3.4
	38.1	46.1	44.9	39.9	45.2	46.1	−1.8	0.9	−1.2	3.3	0.9	1.4
	36.8	39.6	46.7	35.1	39.1	43.2	1.7	0.5	3.5	2.9	0.3	12.3
12	42.6	48.5	51.2	44.3	46.3	49.7	−1.7	2.2	1.5	2.9	4.7	2.3
	40.3	45.0	53.1	43.8	46.3	50.7	−3.5	−1.3	2.4	12.1	1.8	6.0
Sum of errors							−4.8	1.4	5.4			
Sum of squared errors										102.7	66.6	144.8
Mean squared error										10.1	8.2	12.0
Root-mean squared error										3.2	2.9	3.5

Abu Dhabi, United Arab Emirates. Three models were developed to predict the entry, circulating, and exit speeds. The speeds are predicted based on the hourly traffic volumes, the proportions of heavy vehicles, and the geometric characteristics of the roundabouts. The models were validated with data not used in calibration and they were found to be stable and robust. The developed models will help roundabout designers to adjust their designs to ensure that the expected operating speeds will meet the design requirements. The developed models will also help in predicting the potential capacities of multilane roundabouts that are being designed to ensure that the capacities meet the design requirements and the expected traffic volumes. Finally, the developed models will also help developers of traffic simulation applications by providing realistic estimates for operating speeds that may be implemented in those applications. Based on the observations collected in this research, it was found that most roundabouts were overly designed, which resulted in operating speeds that exceed those found in other countries around the world. It must be noted that the developed models are based on the unique characteristics and attitudes of drivers in the United Arab Emirates. Further research may be needed to develop other models for other countries where driver characteristics and attitudes are different from those found in the United Arab Emirates.

Conflicts of Interest

The authors declare that there are no conflicts of interest regarding the publication of this paper.

Acknowledgments

This research was financially supported by a Science and Engineering Research Grant provided by the Emirates Foundation. The authors of this paper would also like to thank the staff at Abu Dhabi Department of Municipal Affairs and Transport for providing part of the data used in this research.

References

[1] S. Kim and J. Choi, "Safety analysis of roundabout designs based on geometric and speed characteristics," *KSCE Journal of Civil Engineering*, vol. 17, no. 6, pp. 1446–1454, 2013.

[2] National Cooperative Highway Research Program (NCHRP), "Roundabouts: an informational guide, Second Edition," NCHRP Report 672, Transportation Research Board (TRB), National Research Council, Washington, DC, USA, 2010.

[3] National Cooperative Highway Research Program (NCHRP), "Guidelines for selection of speed reduction treatments at high-speed intersections," NCHRP Report 613, Transportation Research Board, National Research Council, Washington, DC, USA, 2007.

[4] A. Flannery, "Geometric design and safety aspects of roundabouts," *Transportation Research Record*, vol. 1751, pp. 76–81, 2001.

[5] L. Ren, X. Qu, H. Guan, S. Easa, and E. Oh, "Evaluation of roundabout capacity models: An empirical case study," *Journal of Transportation Engineering*, vol. 142, no. 12, Article ID 04016066, 2016.

[6] X. Qu, L. Ren, S. Wang, and E. Oh, "Estimation of entry capacity for single-lane modern roundabouts: case study in Queensland, Australia," *Journal of Transportation Engineering*, vol. 140, no. 7, Article ID 05014002, 2014.

[7] A. Vasconcelos, A. Seco, and A. Silva, "Comparison of procedures to estimate critical headways at roundabouts," *Promet – Traffic & Transportation*, vol. 25, no. 1, pp. 43–53, 2013.

[8] A. Polus, S. S. Lazar, and M. Livneh, "Critical gap as a function of waiting time in determining roundabout capacity," *Journal of Transportation Engineering*, vol. 129, no. 5, pp. 504–509, 2003.

[9] R. M. Kimber, "The traffic capacity of roundabouts," TRRL Report LR942, Transport and Road Research Laboratory, Crowthorne, UK, 1980.

[10] H. Hatami and I. Aghayan, "Traffic efficiency evaluation of elliptical roundabout compared with modern and turbo roundabouts considering traffic signal control," *Promet – Traffic & Transportation*, vol. 29, no. 1, pp. 1–11, 2017.

[11] Transportation Research Board, *Highway Capacity Manual*, Transportation Research Board, National Research Council, Washington, DC, USA, 5th edition, 2010.

[12] A. Polus, Y. Shiftan, and S. Shmueli-Lazar, "Evaluation of the Waiting-Time effect on Critical Gaps at Roundabouts by a Logit Model," *European Journal of Transport and Infrastructure Research*, vol. 5, no. 1, pp. 1–12, 2005.

[13] S. Teply, M. I. Abou-Henaidy, and J. D. Hunt, "Gap acceptance behavior – aggregate and logit perspectives: part 1," *Traffic Engineering and Control*, vol. 38, no. 9, pp. 474–482, 1997a.

[14] S. Teply, M. I. Abou-Henaidy, and J. D. Hunt, "Gap acceptance behavior -aggregate and logit perspectives: Part 2," *Traffic Engineering and Control*, vol. 38, no. 10, pp. 540–544, 1997b.

[15] G. Song, L. Yu, and Y. Wu, "Development of speed correction factors based on speed-specific distributions of vehicle specific power for urban restricted-access roadways," *Journal of Transportation Engineering*, vol. 142, no. 3, Article ID 04016001, 2016.

[16] H. C. Frey, N. M. Rouphail, and H. Zhai, "Speed- and facility-specific emission estimates for on-road light-duty vehicles on the basis of real-world speed profiles," *Transportation Research Record*, vol. 1987, pp. 128–137, 2006.

[17] N. Davis, J. Lents, M. Osses, N. Nikkila, and M. Barth, "Development and application of an international vehicle emissions model," *Transportation Research Record*, vol. 1939, pp. 157–165, 2005.

[18] American Association of State Highway and Transportation Officials, *A Policy on Geometric Design of Highways and Streets*, American Association of State Highway and Transportation Officials (AASHTO), Washington, D.C., USA, 2011.

[19] O. Arndt, *Relationship between Roundabout Geometry And Accident Rates*, Queensland University of Technology, Brisbane, Queensland, Australia, June 1994.

[20] T. Džambas, S. Ahac, and V. Dragčević, "Design of turbo roundabouts based on the rules of vehicle movement geometry," *Journal of Transportation Engineering*, vol. 142, no. 7, Article ID 05016004, 2016.

[21] M. Bassani and E. Sacchi, "Experimental investigation into speed performance and consistency of urban roundabouts: an Italian case study," in *Proceedings of the 3rd International Conference on Roundabouts*, Transportation Research Board, Carmel, Indiana (US), 2011.

[22] H. Pilko, D. Brčić, and N. Šubić, "Study of vehicle speed in the design of roundabouts," *Gradjevinar - Journal of the Croatian Association of Civil Engineers*, vol. 66, no. 5, pp. 407–416, 2014.

[23] Bushnell Inc., *Bushnell Velocity Speed Gun Manual*, 2017, http://bushnell.com/global/customer-service/product-manuals.

[24] Abu Dhabi Urban Planning Council, "Abu Dhabi Urban Street Design Manual (USDM)," 2017, https://www.upc.gov.ae/en/publications/manuals-and-guidelines/abu-dhabi-urban-street-design-manual.

[25] United States Department of Transportation (US DOT), "Speed-measuring device performance specifications: across-the-road radar module," US DOT Report DOT HS 810 845, National Highway Traffic Safety Administration, United States Department of Transportation, Washington, DC, USA, 2007.

[26] K. Fitzpatrick, P. Carlson, M. Brewer, M. Wooldridge, and S. Miaou, "Design speed, Operating Speed, and Posted Speed Practices," NCHRP Report 504, National Cooperative Highway Research Program, Transportation research Board, Washington, D.C., USA, 2003.

[27] K. Fitzpatrick, J. D. Blaschke, C. B. Shamburger, R. A. Krammes, and D. B. Fambro, "Compatibility of Design Speed, Operating Speed, and Posted Speed," Final Report FHWA/TX-95/1465-2F, Texas Department of Transportation, College Station, TX, USA, 1995.

[28] E. Dabbour, M. Al Awadhi, M. Aljarah, M. Mansoura, and M. Haider, "Evaluating safety effectiveness of roundabouts in Abu Dhabi," *IATSS Research*, 2017, https://doi.org/10.1016/j.iatssr.2018.04.003.

[29] M. Angelastro, "The influence of driver sight distance on crash rates and driver speed at modern roundabouts in the United States," *ITE Journal (Institute of Transportation Engineers)*, vol. 80, no. 7, pp. 28–39, 2010, Washington DC.

[30] J. L. Rubio-Martín, R. Jurado-Piña, and J. M. Pardillo-Mayora, "Heuristic procedure for the optimization of speed consistency in the geometric design of single-lane roundabouts," *Canadian Journal of Civil Engineering*, vol. 42, no. 1, pp. 13–21, 2015.

[31] F. Corriere and M. Guerrieri, "Performance Analysis of Basic Turbo-Roundabouts in Urban Context," *Procedia - Social and Behavioral Sciences*, vol. 53, pp. 622–632, 2012.

[32] CROW, "(Dutch Technology Platform for Transport, Infrastructure and Public Space)," CROW Report 257, Netherlands, 2008, Turborotondes [Turbo roundabouts].

[33] M. Vujanić, M. Savićević, B. Antić, and D. Pešić, "Safety effectiveness of converting conventional intersections to roundabouts: Case study in the city of Niš," *Promet – Traffic & Transportation*, vol. 28, no. 5, pp. 529–537, 2016.

[34] A. Rencher and G. Schaalje, *Linear Models in Statistics*, John Wiley & Sons, Inc., Hoboken, NJ, USA, 2nd edition, 2003.

[35] F. Mannering and S. Washburn, *Principles of Highway Engineering and Traffic Analysis*, John Wiley & Son, New York, NY, USA, 5th edition, 2013.

[36] A. May, *Fundamentals of Traffic Flow*, Prentice - Hall Inc., Englewood Cliff, NJ, USA, 2nd edition, 1990.

[37] Transportation Research Board, "Traffic flow theory: A monograph," Special Report 165, Transportation Research Board, National Research Council, Washington, DC, USA, 1975.

[38] C. Lee, "Developing passenger-car equivalents for heavy vehicles in entry flow at roundabouts," *Journal of Transportation Engineering*, vol. 141, no. 8, Article ID 04015013, 2015.

[39] Y. Chen, B. Persaud, E. Sacchi, and M. Bassani, "Investigation of models for relating roundabout safety to predicted speed," *Accident Analysis & Prevention*, vol. 50, pp. 196–203, 2013.

[40] P. Perco, A. Marchionna, and N. Falconetti, "Prediction of the operating speed profile approaching and departing intersections," *Journal of Transportation Engineering*, vol. 138, no. 12, pp. 1476–1483, 2012.

8

A Model for the Evaluation of Brazilian Road Transport: A Sustainable Perspective

Mariana Gonçalves de Carvalho Wolff and Marco Antonio Farah Caldas

Department of Industrial Engineering, Universidade Federal Fluminense, Niterói, RJ, Brazil

Correspondence should be addressed to Mariana Gonçalves de Carvalho Wolff; carvalho.mariana@ymail.com

Academic Editor: Zhi-Chun Li

Freight transportation in Brazil is heavily based on the road mode bringing heavy dependence of this mode to the country and some consequences such as high operating costs and high air emissions. For these reasons, it was considered the necessity of evaluating new investments made in this mode from a sustainable perspective, taking into account not only economic but also social and environmental concepts. Towards this purpose, the study develops an evaluation model of road transport infrastructure, the TIM, which will help decision making in situations of choice between different alternatives of road projects. The model includes ten variables that count from the initial investment in the road to air emissions of carbon dioxide. All are monetarized individually and then consolidated into a single equation. The result of applying the model is innovative when considering the three spheres of sustainability to evaluate future road transport projects and rank them.

1. Introduction

The transport infrastructure of a country boosts its socioeconomic development by enabling a better flow of production in terms of volume and time. Depending on the volume, the origin, and destination of the commodity and the type of consumer, the choice must occur regarding a specific mode of transport. The Brazilian transportation network for freight is based on road (61.1%), followed by railroad (20.7%), waterway (13.6%), pipeline (4.2%), and air (4%) [1]. The consequences of this imbalance are negative for the country in several aspects, such as the higher cost of road transport over railroad. The higher atmospheric emissions of carbon dioxide (CO_2) of those type (highways) compared with others, and so on, will be deepened in the present study.

Regarding Brazilian states the analysis shows the great influence of some types of products in the transport mode choice. In the state of Minas Gerais, for example, there is an option for railroad transportation oriented to iron ore, since there is such availability of the commodity in the region and the cargo has low added value and can be transported in large volumes. Meanwhile, the state of São Paulo strongly uses the waterway mode in the transportation of bulk products, automobiles, and others, through the strong participation of the Port of Santos. However, not all states are benefited from a more diversified matrix relying heavily on road transport alone.

The five modes of cargo transport are differentiated based on the characteristics: possible volume of transport; initial cost of installing a new route; cost of operating an existing transport infrastructure; security that the environment provides to the workers involved and users; the possibility of having a door-to-door service; availability of transportation of any type of cargo; and finally, the social impact generated to the population with the use of this mode of transportation.

The road transport mode, for example, is advisable for door-to-door transportation, especially in the case of fragmented products deliveries to the final customer. Railways are mainly used towards large volumes of low value-added products. Ships are also used for large volumes, which can carry products in container or in bulk. In this case the range is a little more limited, having the product to be in some port or near waterways within the country. In some situations these two modes are combined, resulting in multimodal transportation. The cargo is transported by rail to the port, where it will continue by ship to the next destination, whether in the country or not.

More specifically there is transportation via pipelines, which allows a continuous flow of fluids and gases for some commodities that have these characteristics. It allows a large volume shift and that is from point to point, previously determined. Finally, air transport, with high cost, less time, and greater security, is appropriate for commodities that have high added value. This also requires a complementary mode for delivery and withdrawal of the product at airports.

In contrast to our explanation above, the circulation of goods in practice does not exactly follow the priorities presented because it depends on the existing infrastructure in each region. As we have already seen, in Brazil, road transport is predominant and it is by this means that most of the country's cargo flows. For this reason this mode of transport will be the focus of this study.

The simplest way to analyze the impacts of an infrastructure project is from models that consider direct impacts, mainly represented by the reduction in total travel time, number of accidents, and transportation costs [2–7]. More complex models that take into account indirect impacts, such as the generation of indirect jobs and the increase of municipal GDP, are few and very directed to a specific project [8].

Another issue is the environmental impact, still weakly considered in this type of analysis. According to [9], sustainability in transportation is a recent concept derived from sustainable development and it is based on measurements through various transport indicators and also in proposals to change old habits, such as encouragement to mass transportation.

For the already explained reasons, this article provides an evaluation of the transport infrastructure projects in the road mode in Brazil, allowing comparisons between different alternatives and justifying the choice of one of them through technical arguments. The model seeks flexibility and can be adapted according to the local reality, since the regional conditions influence and show ramifications in various social, economic, and environmental aspects. The model calculates a monetary value for the analyzed alternatives in any time frame.

The article is composed of six sections, including the introduction. They are divided as follows: Section 2 presents the theoretical foundations, surveying international references for the structuring; in Section 3 the methodology used for the construction of the model is described; Section 4 shows the development of the Transport Infrastructure Model (TIM) and the equation that defines it; Section 5 brings the results obtained, and finally, the sixth section concludes this article and suggests future studies that can enrich it.

2. Literature Review

According to [10], economic development, in addition to growth, measured by the increase in Gross Domestic Product (GDP) in the long term, reduces poverty rates and improves health, education, housing, and transportation conditions. One of the indicators used to assess impacts on the population is the Human Development Index (HDI), which calculates the relationship between quality of life and economic development.

For this reason the tripod of sustainability has become the subject of great questions in several countries and different areas, including the transportation of cargo and passengers. The social and environmental criteria are gaining relevance in this scenario, as well as the economic ones. The economic benefits obtained from transportation are usually observed in long-term periods. In this way the quantification of these is made from the knowledge of costs to direct users and their externalities, taking into account the socioeconomic and environmental context [11].

2.1. Relationship between the Influence of Infrastructure and Regional Sustainable Development. Studies indicate the relationship between transport infrastructures and the regional development that accompanies it under social, economic, and environmental aspects. The better the design of a route and the more interconnections it has, the better the consequences for that region. In Brazil some researchers show this relationship and seek to prove it through significant variables [10, 12–15].

The railroads present their role in the economic development of Brazil. Reference [10] evaluates the behavior of the state GDP and the operating income of the company during the year, being the last one obtained through the amount of cargo transported in the period. On the other hand [15], trying to understand the influence of the North-South Railroad analyzes the impacts on the variables: product, income, employment, and tax collection in the municipalities involved. The projections are calculated from the variables of GDP per capita, service tax of any kind (ISSQN), formal employment, and average wage of the worker. The deflation index used is the General Price Index.

Identifying the parts involved in an infrastructure project is a differential in the [14] study on the subway deployment process. The actors are divided between private and public, the first of them being focused on their survival. In addition, the study identifies the economic development and real estate valuation for the region that receives subway stations as benefits to the population, as well as the reduction of traffic in the adjacent road system and urban development centered on public transport.

The proposition of evaluation of transportation projects in Brazil [13] uses a more qualitative multicriteria method to relate the increase of infrastructure efficiency in the country aligned with sustainable development. Thirty stakeholders evaluated under the criteria: intermodal connection, job creation, social welfare, environmental impact, recovery of environmental liabilities, costs and benefits, and project execution period. On the other hand, [12] adopts a more qualitative and economic methodology that uses VPL, TIR, and B/C indicators (benefits and costs). That study proposes that only financial profit is enough to make a project viable.

Based on the American reality, the variables considered valid for the case study are the number of transportation modals in the region; quality of transport (average speed, safety, traffic jam, and physical conditions); sound, air, and water pollution; impacts on the community life and surroundings; access to education and job creation; impacts on economic activities and in the estate of the region; and income distribution [16]. Some US states have developed their own models to set priority on investments in infrastructure and, according to [17], consider direct and indirect benefits under the three pillars of sustainability. The direct ones are the economic ones and the indirect ones are time of the trip, creation of jobs, and security.

Another study that considers cost-benefit analysis (CBA) was conducted by Barfod and Leleur [6] at the Danish Ministry of Transportation. The VPL calculation is given by the difference between the benefits and costs of the project. The primary variables are construction cost, disturbance in the construction period, maintenance costs, work completion costs, travel time reduction, driver costs, and taxes (road toll). The secondary ones are noise pollution, number of accidents, local pollution, and global pollution.

Analyzing the potential impact that each project can generate in the population [18] consolidates the study through sixteen criteria. Traffic, travel time, distance, average speed, local inhabitants, age group, births and deaths, migrations and marriages, family income, GDP, CO_2 emissions and land occupation rate, distance between housing and work, and number of vehicles in the region.

However, according to [19] the analysis is more focused on highway safety. Thus, its main concerns are divided into three groups: security, economy, and social importance. The criteria used are reduction of accidents, reduction of fatal accidents, investment, project continuity period, daily traffic average, and percentage of growth.

2.2. Models for Evaluating Transport Infrastructure Projects. Infrastructure project evaluation models are mostly acknowledged through quantitative methods in order to compare two or more alternatives, or even to justify investments from long-term viability. However, a model may also have references in qualitative methods when it is not possible to give it real values, or if it is to consider the opinion of those involved.

For qualitative variables [20] use the Likert scale to analyze the energetic efficiency of 29 provinces in China through Data Envelopment Analysis (DEA) model. The inputs of the model are capital invested in assets, employment, and energy consumption; the outputs are GDP, CO_2 emissions, and air quality level.

The use of ACB is widely used for engineering, health, and public policy applications, among others. These analyze range from natural hazard mitigation strategies to the development of electric vehicles [21, 22]. For transport infrastructure this analysis is also used since it considers the trade-off between the costs and benefits of the highways, helping in the decision making to invest in an alternative.

Reference [23] developed the Eco-Mobility (EM) that considers, besides the ACB, the risk analysis (Monte Carlo simulation) and the multicriteria complementary analysis of decision. The first two work as inputs for the third. Eco-Mobility evaluates alternatives from the point of view of sustainability. The criteria adopted are socioeconomic strength, improved mobility for drivers and users of public transport, visual impact on cities, economic impact on the region, impact on logistics flexibility, and contribution to the European green corridor. Another similar model is [24] that considers travel time, traffic volume, construction time, and traffic demand variation in the work.

The MicroBenCost model, according to [2], developed in the 1990s by the Texas Transportation Institute and widely used to this day, is one of the ways to evaluate highway improvement projects from the ACB, following the increased transport capacity and improved track safety. The following factors are analyzed as benefits: travel time, vehicle operating cost, and number of accidents.

Developed and used by the World Bank, the HDM-4 is a model that proposes an evaluation of the performance with regard to the following aspects: traffic volume, type of pavement, maintenance standards, and regional characteristics. In addition, the benefits identified for the consumer are reductions in operating costs, travel time, and number of accidents; traffic jam; social benefits and environmental impacts (air pollution and energy efficiency) [3].

From the HDM-4 [7] they also constructed a multicriteria analysis model to optimize investments in highways. The criteria are the economic, social, and environmental impacts that present different weights. The quantified cost-benefits for each of the global impacts are measured from these weights. The following are evaluated: atmospheric emissions, energy efficiency, number of accidents, comfort to users, traffic jam, accessibility to schools, hospitals, and commerce. Another model is *Computational General Equilibrium* (CGE), which contributes to estimating the economic impacts from the travel time inputs and truck operating costs [17].

The study developed by Cascetta et al. [25] is based on three processes for decision making in transport projects: a rational cognitive approach, a 5-step model to identify stakeholders engagement, and a quantitative analysis. The process begins in the analysis of the scenario and formulation of the alternatives until the choice for a region to implement it.

In the European Union, the choice of transport projects is also based on the ACB and considers as influence factors investments in the road, operation and maintenance costs, road toll revenue, travel time, vehicle operating cost, accidents, noise, air pollution, and climate change. The data are monetarized [5]. To support the European model, a manual on external transport costs has been developed [26]. This presents the state of the art in terms of categories: traffic jam, accidents, noise and atmospheric pollution, climate change, other environmental impacts, and the deterioration of railway and road infrastructure. The study shows that there are two approaches to data collection: top-down or bottom-up, where the former is more generic and has lower costs and the second is based on actual data.

Reference [11] uses the AHP (Analytic Hierarchy Process) method to choose criteria that focus on infrastructure projects in Brazil. The model is based on the opinion of thirty-three professionals and evaluated transport costs, financial feasibility of the project, intermodality, transport supply in the region, HDI, environmental impacts, and atmospheric emissions.

Table 1 summarizes the main variables considered by each author in their models. Travel time, for example, is the main one, present in 61% of the models. Environmental impacts are in second place, addressing issues such as air pollution, atmospheric emissions, air quality, and climate change. Economic impacts are shown in fifth position.

3. Research Methodology

Based on the same proposal made by Zheng et al. [27] for the development of an index that evaluates sustainability in transportation, this study follows four of the five proposed steps to evaluate transport infrastructures: (A) identify the positive and negative impacts of a highway; (B) select appropriate variables; (C) monetarize the data to make them comparable; (D) aggregate the considered components.

From the adjustment, the steps that compose the methodology of this study are described. Initially the literature survey is composed by the study of the main models of transport infrastructure evaluation and analysis of the variables used. The second part is the selection of the variables that will compose the TIM model, monetarization, and development.

4. Transport Infrastructure Model

The proposal of the developed model is an evaluation of projects that, according to [28], aims to present a positive relationship between the costs involved in an infrastructure and the benefits generated, comparing road transport projects. Not only technical aspects of the projects, but also social ones will be considered, achieving an efficient model with reduced chances of error and a transparent process. The analysis will be based on the sustainability tripod considering social, environmental, and economic issues.

The scenarios evaluated are dynamic and the result obtained today can be different from the results during the construction of the project and at the end of it, making the analysis a little more complex, as evidenced by Bagloee and Asadi [24]. For this reason the projects are comparable in the base scenario. A highway may be preferable to the other in the initial conditions adopted but not at the end of its operation.

The great advantage of the model is the transferability to other cities or regions in Brazil. It happens because variables consider both national and local data, such as GDP and the number of formal jobs.

4.1. Quantitative Method. The model of this study proposes an evaluation of the road infrastructures of cargo transportation in Brazil that uses as base only quantitative variables. Some of these are presented in monetary terms and others are not, requiring an additional process of transformation. This measure is intended to consolidate the data and make it comparable.

According to [5] investments in new infrastructures can be of eight types, since the increasing demand of the route until the complete improvement of the inoperability of the transport networks. In a simplified procedure, the model considered only three types: (a) improvement, when the initial mileage of the highway is maintained; (b) expansion, when there is increase of the initial mileage and, (c) creation, when there is no initial highway infrastructure in place. Positive and negative impacts are generated for the involved parties, presented in vehicle per kilometer, for example, [16].

4.2. Stakeholders of the Model. It is critical to check who the stakeholders are in this process and what their particular interests are, as done by the National Cooperative Highway Research Program (NCHRP) [29]. Stakeholders are all those who, directly or indirectly, are affected by the development of a project or can contribute positively or negatively in it. Some interests are common to all and others are group-specific, such as the increase in municipal tax collection.

The 5 stages of the tool proposed by Bjugn and Casati [30] for stakeholders analysis are identification, assignment of values, prioritization, engagement plan of the participants, and monitoring. The developed model presents an adaptation, preserving the same logic. Initially, those involved are identified and then the main interests of each group are listed. The third step is to verify the influence exercised by the parts in the infrastructure project and finally to classify according to the degree of this influence. The classification is qualitative, based on the concepts of high, medium, and low. The last step is monitoring, which occurs through constant reevaluation in the course of the project.

The monitoring routine should occur at each critical phase of the project. Six milestones were identified for this routine: initial evaluation to execute the project, beginning of the project implementation, construction, project completion, and beginning of operation and periodically at each predefined period. Review periods should be between 2 and 5 years, at which time those involved may exhibit changes of interest or change the power of influence they perform, depending on the size of the project and the time frame of the project. For monitoring, all stakeholders should be reviewed in its entirety, by surveying the compliance and demand of the interests of the parts and field research with those directly involved.

Mapping all the stakeholders allows an analysis of the economy of the municipalities involved and the transformations that occurred during the period of construction of the road and after its completion. From this information it is possible to understand what the positive and negative impacts are, perceived in short and long terms, and how they modify the lives of those involved.

For [15] the impacted municipalities were divided into three groups: municipalities where goods embark/disembark, transit municipalities, and municipalities located in the area of influence. In the present model only two types are considered: those directly involved and those indirectly as well, since, in this way, it is possible to analyze all that receive some type of impact and to quantify them. This division is important because each one will have interests and will exercise distinct charges on the infrastructure.

The main gains of a model that considers the stakeholders in the process are the concern with all involved and the search

Table 1: Overview of the models (variables × authors).

Variables	Barfod and Salling 2015 [23]	Bagloee and Asadi 2015 [24]	Odoki et al. 2015 [7]	Rabello Quadros and Nassi 2015 [11]	Barfod and Leleur 2015 [6]	Campos 2015 [15]	Sartori et al. 2014 [5]	Nogués and González-González 2014 [18]	Sage et al. 2013 [4]	Kansas & North Carolina DOT 2013	Michigan DOT 2013	Indiana DOT 2013	Montana DOT 2013	MPPP tool 2013	Wang et al. 2013 [17]	Dewan et al. 2013	Yu and Liu 2012 [19]	Da Silva and Netto 2010 [13]	Guenther & Farkavcová 2010	Sehn 2009 [12]	Litman 2002 [16]	Kerali 2000 [3]	McFarland et al. 1993 [2]	*Total*
Travel time	×	×	×		×		×	×	×				×	×	×	×					×	×	×	*14*
Environmental impact (air pollution)			×	×	×		×	×	×						×	×		×	×		×	×	×	*13*
Security			×		×		×		×					×			×				×	×	×	*9*
Operational costs			×		×		×		×					×	×							×	×	*8*
Economic impacts	×		×						×	×	×				×					×	×			*8*
Employment						×				×	×	×						×			×			*6*
Transportation cost				×									×			×					×	×		*5*
Household income						×		×		×	×										×			*5*
Construction and maintenance cost					×		×										×	×					×	*5*
Social impacts			×		×																×	×		*4*
Traffic volume		×					×	×									×							*4*
Quality																×					×	×		*3*
Energy efficiency			×																×			×		*3*
Noise pollution					×		×														×			*3*

Source. Author.

for the balance between them; that is to say, neither part can be harmed compared with another. In addition, equality between the three pillars of sustainability is still foreseen, ensuring that both economic, social, and environmental aspects must all be considered with the same degree of importance.

4.3. Variables. A solid study must consider as many variables as possible which adopt the three spheres intended. On the other hand, one must evaluate which are really relevant in order to avoid many variables, making it impossible to consolidate the equation. The division of variables follows the concept of sustainability for projects developments. The first deals with the economic impact that can be caused to stakeholders. The social variable refers to the way in which the population is directly affected and, finally, the environmental variable is related to the environment in which the projects are being developed.

The variables will have all the same relevance in the final equation, thus showing the importance given to each of the three groups through a horizontal model. The methods for estimating the variables are presented based on best market practices. The top-down approach uses the average of the national data, in a generic way.

4.3.1. Traffic Volume. The prediction of the demand is the variable of greater impact of the model and according to [4] it will mean the main direction of the project, since all the others will be influenced by it. For this reason the prediction must be made from a drive record, considering the matrices of origin-destination (OD matrices). This data will be generated for the first year of track operation and the following years will be projected from the accumulation of the annual growth rate.

The matrix is a quantitative technique widely used in transport engineering and also has application in diverse studies [31–33]. The tool allows knowing in depth the prediction of a route utilisation, facilitating both its dimensioning and diagnosis of the impacts as to highway deterioration and traffic jams, among others. The original matrix tool will not be drawn because it requires a lot of development time.

A table identifies the volumes transported and details the origin and destination of each cargo, that is, where they come from and where they are going to and at what time they cross the highway in question. Finally, the volume and the percentage of each OD absorbed by the highway are dimensioned. The expected result is the sum of the volume expected to travel in the alternative, that is to say, the multiplication of the percentage of each of these loads by the volumes which will be absorbed by this option. The volume is measured in trucks per year because the other variables are calculated in this unit. The measurement of a medium truck with three axles and 40 tons of capacity is used as reference.

4.3.2. Traffic Volume Growth Rate. To plan the behavior of the coming years, as done by Yu and Liu [19], the growth rate is used. For the authors, this criterion has social and environmental importance in the validation of an infrastructure improvement. Its growth is linked to the average GDP growth expected for the country, industrial and agricultural production, and population size [34]. The authors make a demand prediction from the traffic record in the region. The relation between tax rate and these factors is used because there is a direct proportion between the increase in GDP and the traffic on the highways.

In addition to the indicator used above, because it is a new highway, there is also a heating factor in which drivers usually give preference to the highway gradually. Three behavior scenarios are described: (1) highway emerges as a much better alternative than existing ones, either in road toll exemption or in better pavement conditions or with less traffic jam; (2) highway is a good alternative for transportation in the region and has good traffic and safety conditions; (3) highway has some more favorable conditions than other alternatives (good paving, signaling, and less traffic jam) and some less favorable conditions (road toll taxes or poor security). The scenarios will follow the additional growth rate [35] of 3.5%; 2.5%; and 1.5%, respectively. Thus, the final rate will correspond to the annual forecast of GDP added, in the first three years of operation, to the percentage referring to the highway scenario.

Another point which should be observed is that the tax rate will not grow infinitely because there is a saturation point, determined by the track capacity. This measure can vary according to several factors: number and width of the traffic lane, percentage of heavy vehicles, existence of roadside, bus stops, and number of entrances and exits, among others [36]. A visual indicator for saturation is traffic congestion on the highway, showing that its capacity has already been reached, especially at peak times.

4.3.3. Reduction in Travel Time. The reduction in the travel time of each vehicle is presented in minutes from the comparison with the available alternatives. When choosing the infrastructure under analysis it is intended to gain time in the course, since in some cases the goal is to improve the traffic flow and possible reductions in travel. The calculation is based on the congestion band, average speed, and distance traveled in the alternative in question against the average of the others, as shown in Table 2.

In order to assist in measuring the impact of traffic jams, the comparison between the highways is based on the classification of each one in the congestion bands, varying between 1 and 5 [26]. One lane in band 5 operates above its average vehicular traffic capacity, resulting in frequent traffic jams and a percentage reduction of the actual average speed. Alternative lanes with this feature are likely to transfer a significant portion of this volume, as drivers will give preference to avoid traffic jam and drive on less busy highways.

Index 4, when applicable, may also show a decrease in speed, since at peak times the perceived flux is greater than that supported. With the index being between 1 and 3, there is no impact on speed. Classification 5 represents a relationship between volume and capacity greater than 1, and index 4 occurs when this ratio is between 0.75 and 1.

Table 2: Alternatives comparative (travel time).

Alternatives	Congestion band	Average speed	Average distance	Reduction in time

Source. Authors.

Table 3: Alternatives comparative (operating costs).

Alternatives	Traffic flow	Pavement	Signaling

Source. Authors.

4.3.4. Reduction of Vehicle Operating Costs. The reduction in vehicle operating costs is one of the consequences of a more fluid traffic, since the new route will have less traffic jam, better quality of asphalt, and others. Costs are just variables because we want to understand the difference between alternatives, and fixed costs are inherent to the choice of route. According to the WSDOT model they include fuel and lubricant consumption, tyre consumption, and vehicle maintenance, totaling one value per kilometer [17]. Each vehicle has a manual with specifications for the optimal scenario.

The characteristics of the highway can maintain these values or make them worse. In traffic jams the consumption of fuel is greater and in a route poorly conserved there is greater tyre wear and more demanding of maintenance of the vehicle, for example. For [37] there are two classifications that differentiate the conditions of a highway: better and worse.

Reference [38] considers, in relation to urban transport, twenty-one quantitative and qualitative indicators to characterize urban road transport; among them are connectivity with the transportation network, traffic jam, average speed, track monitoring, pavement, and accident rate. Reference [39] uses as influence in the system traffic flow, pavement conditions, and signaling.

The qualitative method used in this study deals with the evaluation of subjective characteristics with complexity and high cost. The low credibility of this method occurs due to the judgment made by the evaluators, which may tend towards their objectives. On the other hand, by selecting the sample correctly, the opinion of the judges can bring a more up-to-date assessment to the problem in question. For this reason the qualitative method, when combined with the quantitative one, becomes more solid [40].

Table 3 shows the three criteria that should be evaluated: traffic flow, pavement, and signaling. The classification of each one of them will be compared to the alternative in question and follows a scale of five levels: terrible, poor, medium, good, and excellent: two positive rates, two negative ones, and one neutral [41]. Switching from a higher level to a lower level means an increase of 1.5% in the factor. There may be a recorded worsening of up to 6% in each criterion, since the change from the highest to the lowest level can be translated into four levels.

Considering the three conditions of the route a multiplicative factor is generated and has a maximum increase of 18% in the total operational cost. The percentage defined is based on Inmetro's research methodology for vehicular fuel consumption, which indicates that it is within a range of more or less 20% of that declared [42]. Considering that the behavior of the other criteria is equal, the maximum percentage of 18% acts on the total cost that considers the vehicle consumption. The larger this number is the worse the alternative is, which will increase the operating cost. The minimum factor is one, present in new highways with excellent conditions for the traffic.

The reduction will be calculated by comparing the highway with the average of all existing alternatives. This is why the factor is obtained from the difference between highways. The reduction of operating costs should be presented in *reais* per vehicle, multiplying variable cost, difference in multiplicative factor, and average difference of mileage traveled.

4.3.5. GDP. The Human Development Index (HDI) is not considered in this analysis because it is an interval between 0 and 1, nonmonetary, and its influence factors are income, education, and health. For this reason it is preferable to consider only the income through the GDP variable. The other index factors are not portrayed.

GDP is considered an important economic variable because the transportation of cargo is directly associated with the purchasing power of the region, since the transport of goods is larger in the area. The measurement is made based on the data before the development of the highway with the projection to the average of the municipalities involved. The difference between the years of analysis is then used.

4.3.6. ISSQN. The ISSQN tax also composes the model, clarifying the development of the companies activities in the municipalities involved. It is considered relevant because any contracted service originated in the municipality in question must pay it, being a relevant metric for the growth of economic activities in this locality.

Likewise as the previous criterion, the ISSQN is measured in the base year and in the year of the survey and then its difference is given. The tax is also projected according to GDP growth in the municipalities involved. The warmer the market, the higher the volume of taxes paid.

4.3.7. Fatal Accidents Reduction. Other data of great relevance for all involved in the project is reducing the number of fatal crashes, improving overall safety. This variable deals with the reduction when comparing the project in question with the other alternatives [6]. In order to measure it, safety aspects are raised, such as risks of collision with another vehicle or with road barriers, number of lanes, pavement conditions, landslides, and signaling, among others.

Table 4 consolidates the main factors based on highway improvements (Sinha & Hu 1985) APUD [43]. Assuming that the road analyzed will adopt all these maintenance tasks over time, the reduction factor totals 1.91, which means that practically only 2 accidents can be avoided for every one

TABLE 4: Accident reduction factors for highways.

Activity	Accident reduction factor or f_{AC}*
Increase in number of lanes (4 to 6)	0,95
Lane and shoulder widening	0,30
Resurfacing	0,33
Pavement maintenance	0,08
Shoulder maintenance	0,25

*Accidents per million vehicles per mile; source [43].

million vehicles per mile, or 1.19 accidents per kilometer. The evaluation of the indices should occur in the comparison between the option analyzed and the average of the other alternatives, being assigned the value of the difference between them. In cases where both highways are new and there is no accident history, the reduction is zero, considering that both will have good traffic conditions.

Thus, the reduction in the number of fatal accidents per year will be multiplied by the projected demand. In (1) the reduction factor (Δf_{AC}) must be multiplied by the difference mileage traveled and divided by 1 million, translated into accidents per vehicle.

$$AC = \frac{\Delta f_{AC} * \Delta \text{km}}{10^6}. \quad (1)$$

4.3.8. Number of Formal Jobs. The number of formal jobs assists in the quantification of social aspects. The inclusion of a new highway or the improvement of an existing one contributes to the installation of new industries and service companies in the region, increasing employability, since there are more opportunities and the dispute for employees tends to increase. For comparison, these data are used before the development of the highway and in the year in question, from the available database. The variable follows the percentage projection of the average GDP growth of the municipalities involved.

4.3.9. Reduction of CO_2 Emissions. The importance of quantifying CO_2 emissions is steadily increasing and takes a lot of space in current literature. Reference [44] analyzes the conflict between the time of travel and the emissions generated in road transport because not always the least travel time corresponds to lower atmospheric emissions.

For the calculation of greenhouse gas emissions, [45] is based on the three main gases (CO_2, N_2O, and CH_4) and compares what was emitted by the main airline companies in relation to the occupation rate of the aircraft. In addition, it considers different emission factors according to the distance traveled because the take-offs and landings have a larger amount of emissions. For road transport there are no great advantages in this segregation because the beginning and end of a travel do not mean a much larger volume.

On the other hand, [26] suggests a cost of air pollution in euros per vehicle-kilometer traveled for several types of vehicles, from cars to heavy trucks. The second division is by region of emission, varying from urban, suburban, countryside, and highway. The more the urbanization, the greater the cost, since the global concentration is higher and the dissipation factor is lower.

In this paper only the reduction of atmospheric emissions of CO_2 is considered. Other gases are also emitted by this mode of transport [46] but are not analyzed because they have an unimpressive percentage. As for the emission of CO_2 the main reason to consider it is because the road transportation mode is a large emitter. Some factors exert an influence: better traffic conditions, less traffic jams, increased collective transport, substitution for less emission modes (rail or waterway), substitution for less polluting fuels, and increase of fuel tariffs, among others [47, 48].

For the purposes of calculation the following are considered: the individual emission index of trucks per kilometer, the average mileage traveled (Δkm relative to the average difference of the alternatives), and the cost of the gas emitted in *reais* per ton. Adopting the premise that loaded trucks have a higher emission rate than the empty ones [49], 50% of the vehicles are given as loaded and 50% are not, since in a complete trip the vehicle leaves the destination full and returns empty.

The index adopted for the emission of the medium-sized diesel trucks loaded (f_C) is 2.75 kg/l, which represents approximately 0.77 kg/km [50]. The study presented by Bartholomeu and Caixeta Filho [37] shows a consumption varying from 2.37 kg/l to 2.07 kg/l, reducing up to 25% of the initial prediction, according to road and truck type conditions. The index of empty vehicles (f_V) is one-third full as the combined total weight of the structure revolves around this percentage. The reduction of these emissions has a direct impact on the air quality of the region and on the health of the population involved.

The importance of the fleet's age to calculate the emissions cannot be denied [46], since the older the vehicle, the higher the CO_2 rates. In this study this characteristic will not be taken into account because it is not possible to predict what fleet will be running. In addition, for the purposes of comparison between highways, the age is the same.

The cost of carbon dioxide (C_{CO2}) emitted is the same for all emission sources, whether it is air transport, industrial activity, or energy sector [51, 52]. Factors that change this cost refer to the country of origin of the emission, the place that receives it, and the situation of the carbon credit market. The value of the ton is dynamic and its variation can be broad depending mainly on political issues and environmental agreements to mitigate them, since they are negotiated between companies. The *cap and trade* variation is large and should be evaluated at the time of the study on BOVESPA. Equation (2) summarizes this cost.

$$CO2 = (0{,}5 * f_C + 0{,}5 * f_V) * \Delta km * C_{CO2}. \quad (2)$$

4.3.10. Variables Consolidation. Of the variables mentioned, three are responsible for the cost of the highway to users: reduction of travel time, vehicle operating costs, and the number of accidents. If there is no regular maintenance on the roads, the tendency is for this number to increase [3].

In the application of the model not always all the variables are available and, therefore, in cases of comparison between projects, they should be considered. All of them present a positive expected signal for the equation, adding up to the evaluation value of an infrastructure project. The positive sign means that the larger the variable, the better for the project. The variables that predict a reduction of value, if this reduction is not effective, their signal must be inverted. The same happens when the variable foresees a growth and in practice there is reduction of its value, as the increase of the number of jobs.

4.4. Monetarization of Variables. The conversion of the variables into a single measure is important to enable a comparison between the routes. Reference [17] transforms all: value of time traveled, operational cost, and the cost of atmospheric emissions. Monetization must occur when the variable is measured in a separate unit. The technique is used for a variety of issues, including environmental issues such as the emission of greenhouse gases in air transport and the impact of forest fires [45, 53]. Travel time, number of fatal accidents, and number of jobs are monetarized.

The calculation of the time value, according to [6], considers some factors such as reason for travel, total distance, mode of transport, urban traffic conditions, trip time, driver's wage, and traffic jam, among others. For the sake of simplicity, this study will consider the trip to work of a driver and a helper of a freight company. The salary of the employees of the region serves as basis for the calculation of the minute worked.

Then the employee's cost is multiplied by the time saved by choosing the infrastructure under analysis. Reduced travel time, in minutes, is based on the average of available options.

In order to transform the reduction in the number of fatal accidents in the route into monetary values, the reduction factor, measured by the comparison between the alternative and the average of the others, is multiplied by the statistical value of life or *value of statistical life* (VSL) [26]. It considers both direct and indirect costs and it is based on the social cost attributed to life in the region analyzed. The value is not constant and varies according to some factors, reflecting risk characteristics and the local population.

Increasing the number of people employed impacts positively on the local economy. To measure it social indicators are designed. Through the difference in the number of employed people, it is multiplied by the current average per capita GDP of the municipalities involved, reaching a monetary increase to the region. GDP is known from the real values for the year of operation. It then multiplies by the growth rate accumulated in the years of analysis.

4.5. Tim Model. In order to facilitate the analysis of a road transport infrastructure project, the model will have as output the number of vehicles that travel in the route in relation to the other variables. The final value will be in *reais* per vehicle, making the comparison of the initial investment fair, since a more expensive project can carry greater traffic. From this relation an alternative will be preferable when compared to the other. The analysis is done individually with the purpose of identifying the monetary return, being able to generate a ranking from the consolidation of several options.

Some municipalities are directly affected by the transportation infrastructure and others indirectly affected. This identifies the radius of influence of the highway. The analysis takes place for the year of future projection of the impacts and the identification of what was generated by the project. The variables participate in the final equation considering only the municipalities directly affected.

Equation (3) consolidates the Transportation Infrastructure Model (TIM) by assessing the reality of the traffic at the proposed time. The consolidation of sustainable benefits is divided into three blocks, as proposed by Odoki et al. [7]. The first deals with economic aspects, the second deals with social aspects, and the last deals with environmental aspects. The final equation subtracts the initial investments and multiplies the three blocks by demand and accumulated growth rate until the period.

The equation has as reference the time and all the data are predicted for the project at the moment. What is important is that all infrastructures in comparison deal with the same time horizon, seeking to assist in the decision making process when making the choice for an option.

The distance to be traveled on the highway is considered in operating costs and atmospheric emissions, which are values influenced by mileage. Different weights will not be assigned to the variables because all the criteria are in the same degree of importance for the final equation.

$$\mathrm{TIM}_{AT} = \frac{-I + (D * R) * [(T + C + \mathrm{PIB}/(D * R) + \mathrm{ISS}/(D * R)) + (\mathrm{AC} + \mathrm{EMP}/(D * R)) + (\mathrm{CO2})]}{(D * R)}, \quad (3)$$

where

A is project under analysis,

T is project analysis year A,

I is investment in new infrastructure,

D is demand forecast,

R is annual cumulative growth rate of traffic up to the year being evaluated,

T is reduction of travel time cost,

C is reduction of operating costs,

GDP is GDP growth,

ISS is ISSQN growth,

AC is reduction of accidents,

EMP is increase in the number of jobs in the municipalities involved,

CO2 is reduction of carbon dioxide (CO_2) emissions.

The developed model was applied in a study case of Arco Metropolitano do Rio de Janeiro. This road project was developed in 2007 and inaugurated in 2014 in the state of Rio de Janeiro, Brazil. The main objective was to create an alternative route to flow the production to the Itaguaí Port.

The model is innovative from the point of view of evaluation of road transport infrastructure projects because it considers the three spheres of sustainability in a single analysis and allows the ranking of projects that are competing. In this way, it is possible to make decisions regarding the priority investment for the benefit of the country, enabling a broader improvement. The final value found is intended to be comparative and will have significance when compared to other values presented by other judged projects. The preferred alternative will yield better results and the bigger the option the better.

5. Conclusions and Discussion

This paper presents a quantitative model for evaluating road transport infrastructure projects with the objective of being transparent and helping decision making when it comes to the choice between investments. The project under analysis may be new, expanded or improved, depending on the difference between the initial and the final mileage. As already predicted in the study of [37] a trip on a highway with better conditions of maintenance presents greater environmental benefits than the same trip made in precarious conditions.

The model is synthesized by the equation involving ten variables, separating them into the three sustainable spheres: social, economic, and environmental. It is possible to consolidate a comparative table between the alternatives and generate the ranking of benefits in monetary value per truck projected for the future highway, which may be in the thirtieth year of operation and when compared to other projects may mean its feasibility or not. The more positive the final number, the more feasible the alternative.

Regarding the variables raised in the referential framework, no model adopts all the same variables and the four most relevant ones being considered in this study are travel time, atmospheric emissions, number of accidents, and operating costs. Two variables are present in only one reference model: annual traffic growth rate and the municipal tax, ISSQN.

The TIM model contributes to the development of research in this segment because it is innovative in considering the three spheres of sustainability together to evaluate future road transport projects and assist in decision making by the competent administration. The final value obtained allows ranking alternatives, whether they are competing or not, justifying the choice. The higher the value, the more the priority of the highway for all stakeholders.

The data obtained in this model also allow a secondary analysis, referring to the percentage of reduction that can be perceived in the global value of the products freight costs. This calculation is done with reference to the travel time variables and the vehicle operating costs since the freight is calculated as a function of these. In some cases, the influence can also occur due to the safety of the track, for example.

For the development of future work it is recommended to include qualitative variables in the environmental and social area, ensuring a greater participation of these two spheres. Air quality, for example, may be an option to measure the environmental impact of a new transport infrastructure. In addition, the perception and feelings of the local residents regarding the enterprise can be captured through the social impact variable.

In order to improve the tool developed it is suggested to evaluate scenarios for expected demand considering optimistic and pessimistic projections. In this way it is possible to reevaluate whether the project remains advisable throughout the period of operation.

Conflicts of Interest

The authors declare that there are no conflicts of interest regarding the publication of this paper.

References

[1] CNT, "Boletim Estatístico," 2016a, http://www.cnt.org.br/Boletim/boletim-estatistico-cnt.

[2] W. F. McFarland et al., *MicroBENCOST User's Manual (Draft)*, Texas Transportation Institute, 1993.

[3] H. G. R. Kerali, *Overview of HDM-4*, The Highwa, Washington, 2000.

[4] J. Sage, K. Casavant, A. Goodchild et al., *Development of a Freight Benefit/Cost Methodology for Project Planning*, 2013.

[5] D. Sartori, G. Catalano, M. Genco et al., *Guide to Cost-Benefit Analysis of Investment Projects*, Brussels, 2014, http://ec.europa.eu/regional_policy/sources/docgener/studies/pdf/cba_guide.pdf.

[6] M. B. Barfod and S. Leleur, *Socio-economic analysis in the transport sector*, 2015, http://orbit.dtu.dk/files/104276045/DTU_Transport_Compendium_Part_1_CBA_1.pdf.

[7] J. B. Odoki, A. Di Graziano, and R. Akena, "A multi-criteria methodology for optimising road investments," *Proceedings of the Institution of Civil Engineers: Transport*, vol. 168, no. 1, pp. 34–47, 2015.

[8] M. A. F. Caldas and M. A. Jaller, "Evaluating Regional Multi-Modal Freight Transport Projects Considering Operational and Socio - Economic Benefits," *Transportation Research Board Annual Meeting*, 2017.

[9] J. Zhou, "Sustainable transportation in the US: A review of proposals, policies, and programs since 2000," *Frontiers of Architectural Research*, vol. 1, no. 2, pp. 150–165, 2012.

[10] L. C. de S. Monteiro, *Contribuição à análise do desempenho econômico das ferrovias concedidas no Brasil*, Universidade Federal do Rio de Janeiro, 2015.

[11] S. G. Rabello Quadros and C. D. Nassi, "An evaluation on the criteria to prioritize transportation infrastructure investments in Brazil," *Transport Policy*, vol. 40, pp. 8–16, 2015.

[12] D. Sehn, *Avaliação econômica de projetos de infraestrutura de transportes: Uma metodologia aplicada à tomada de decisão governamental*, Universidade Federal de Santa Catarina UFSC, 2009.

[13] R. B. Da Silva and M. A. C. Netto, "Uma estrutura de apoio à decisão para orientar a escolha de projetos prioritários para a infraestrutura de transporte do Brasil," in *XLII SBPO, Sociedade Brasileira de Pesquisa Operacional*, Bento Gonçalves, Brasil, 2010.

[14] R. Lisboa, *Atores sociais no planejamento de metrôs no Brasil*, Universidade Federal do Rio de Janeiro, 2014.

[15] F. Campos, "A influência da Ferrovia Norte-Sul no desenvolvimento regional do território goiano," *Universidade Federal de Goiás*, 2015.

[16] T. Litman, "Evaluating transportation equity," *World Transport Policy & Practice*, vol. 8, no. 2, pp. 50–65, 2002.

[17] Z. Wang, J. Sage, A. Goodchild, E. Jessup, K. Casavant, and R. L. Knutson, "A Framework for Determining Highway Truck-Freight Benefits and Economic Impacts," *Journal of the Transportation Research Forum*, vol. 52, no. 2, pp. 27–43, 2013.

[18] S. Nogués and E. González-González, "Multi-criteria impacts assessment for ranking highway projects in Northwest Spain," *Transportation Research Part A: Policy and Practice*, vol. 65, pp. 80–91, 2014.

[19] J. Yu and Y. Liu, "Prioritizing highway safety improvement projects: A multi-criteria model and case study with SafetyAnalyst," *Safety Science*, vol. 50, no. 4, pp. 1085–1092, 2012.

[20] Y. Chen, W. D. Cook, J. Du, H. Hu, and J. Zhu, "Bounded and discrete data and Likert scales in data envelopment analysis: application to regional energy efficiency in China," *Annals of Operations Research*, vol. 255, no. 1-2, pp. 1–20, 2015.

[21] J. Massiani, "Cost-Benefit Analysis of policies for the development of electric vehicles in Germany: Methods and results," *Transport Policy*, vol. 38, pp. 19–26, 2015.

[22] O. Špačková and D. Straub, "Cost-Benefit analysis for optimization of risk protection under budget constraints," *Risk Analysis*, vol. 35, no. 5, pp. 941–959, 2015.

[23] M. B. Barfod and K. B. Salling, "A new composite decision support framework for strategic and sustainable transport appraisals," *Transportation Research Part A: Policy and Practice*, vol. 72, pp. 1–15, 2015.

[24] S. A. Bagloee and M. Asadi, "Prioritizing road extension projects with interdependent benefits under time constraint," *Transportation Research Part A: Policy and Practice*, vol. 75, pp. 196–216, 2015.

[25] E. Cascetta, A. Cartenì, F. Pagliara, and M. Montanino, "A new look at planning and designing transportation systems: A decision-making model based on cognitive rationality, stakeholder engagement and quantitative methods," *Transport Policy*, vol. 38, pp. 27–39, 2015.

[26] European Commission, *Update of the Handbook on External Costs of Transport*, London, UK, 2014.

[27] J. Zheng, N. W. Garrick, C. Atkinson-Palombo, C. McCahill, W. Marshall et al., "Guidelines on developing performance metrics for evaluating transportation sustainability," *Research in Transportation Business & Management*, vol. 7, pp. 4–13, 2013.

[28] A. Clemente and E. Fernandes, *Projetos empresariais e públicos*, Atlas, São Paulo, 2002.

[29] National Cooperative Highway Research Program (NCHRP), *Research on the Relationship Between Economic Development and Transportation Investment*, Washington, 1998.

[30] R. Bjugn and B. Casati, "Stakeholder analysis: A useful tool for biobank planning," *Biopreservation and Biobanking*, vol. 10, no. 3, pp. 239–244, 2012.

[31] O. Z. Tamin and L. G. Willumsen, "Transport demand model estimation from traffic counts," *Transportation*, vol. 16, no. 1, pp. 3–26, 1989.

[32] M. Shahin, "Updating future origin-destination sub-matrices based upon traffic counts," *Traffic, Engineering and Control*, vol. 53, no. 5, pp. 174–181, 2012.

[33] G. Zou and R. Kulkarni, "Simulation-based adaptive calibration and optimization of intelligent transportation systems for highway congestion management," *Simulation*, vol. 90, no. 12, pp. 1360–1374, 2014.

[34] R. Chrobok, O. Kaumann, J. Wahle, and M. Schreckenberg, "Different methods of traffic forecast based on real data," *European Journal of Operational Research*, vol. 155, no. 3, pp. 558–568, 2004.

[35] F. William, I. R. Urquijo, H. Krieg, M. Hughes, L. Mikalsen, and Ó. R. Gutiérrez, "Cost and Environmental Evaluation of Flexible Strategies for a Highway Construction Project under Traffic Growth Uncertainty," *Journal of Infrastructure Systems*, vol. 21, no. 3, p. 05014006, 2015.

[36] TRB, *Highway Capacity Manual*, Washington, 2010.

[37] D. B. Bartholomeu and J. V. Caixeta Filho, "Quantification of the environmental impacts of road conditions in Brazil," *Ecological Economics*, vol. 68, no. 6, pp. 1778–1786, 2009.

[38] Q. Hu, Z. Zhou, and X. Sun, "A study on urban road traffic safety based on matter element analysis," *Computational Intelligence and Neuroscience*, vol. 2014, Article ID 458483, 2014.

[39] CNT, *Pesquisa CNT de Rodovias*, 2016b.

[40] N. R. Sanders and L. P. Ritzman, "Integrating judgmental and quantitative forecasts: methodologies for pooling marketing and operations information," *International Journal of Operations & Production Management. Emerald*, vol. 24, no. 5, pp. 514–529.

[41] L. R. B. Martins, L. S. Pereira, L. M. Almenida, H. R. M. Hora, and H. G. Costa, "Estudo sobre escala mais adequada em questionários: um experimento com o modelo de Kano," *Vértices*, vol. 13, no. 1, pp. 75–103, 2011.

[42] INMETRO, *Metodologia Para Divulgação De Dados De Consumo Veicular*, 2009, http://www.inmetro.gov.br/consumidor/pbe/Metodologia_Consumo_Veicular.pdf.

[43] E. Melachrinoudis and G. Kozanidis, "A mixed integer knapsack model for allocating funds to highway safety improvements," *Transportation Research Part A: Policy and Practice*, vol. 36, no. 9, pp. 789–803, 2002.

[44] H. M. A. Aziz and S. V. Ukkusuri, "Exploring the trade-off between greenhouse gas emissions and travel time in daily travel decisions: Route and departure time choices," *Transportation Research Part D: Transport and Environment*, vol. 32, pp. 334–353, 2014.

[45] S. G. Arul, "Methodologies to monetize the variations in load factor and GHG emissions per passenger-mile of airlines," *Transportation Research Part D: Transport and Environment*, vol. 32, pp. 411–420, 2014.

[46] O. C. Ferreira, C. A. F. A. Silva, O. Y. M. Guidicini, F. Eidelman, R. L. Macedo, and L. Deppe, *II Inventário brasileiro de emissões atmosféricas antrópicas de gases do efeito estufa*, Brasília, 2010.

[47] B. Guo, Y. Geng, B. Franke, H. Hao, Y. Liu, and A. Chiu, "Uncovering China's transport CO_2 emission patterns at the regional level," *Energy Policy*, vol. 74, pp. 134–146, 2014.

[48] B. Lin and O. E. Omoju, "Does private investment in the transport sector mitigate the environmental impact of urbanisation? Evidence from Asia," *Journal of Cleaner Production*, vol. 153, pp. 331–341, 2017.

[49] D. B. Bartholomeu, T. G. Péra, and J. V. Caixeta-Filho, "Logística sustentável: avaliação de estratégias de redução das emissões de CO2 no transporte rodoviário de cargas," *Journal of Transport Literature*, vol. 10, no. 3, pp. 15–19, 2016.

[50] IPCC, *Revised 1996 IPCC Guidelines for National Greenhouse Gas Inventories: Reference Manual*, 1996.

[51] R. R. Martín-Cejas, "Ramsey pricing including CO2 emission cost: An application to Spanish airports," *Journal of Air Transport Management*, vol. 16, no. 1, pp. 45–47, 2010.

[52] K. Van den Bergh and E. Delarue, "Quantifying CO2 abatement costs in the power sector," *Energy Policy*, vol. 80, pp. 88–97, 2015.

[53] D. Mills, R. Jones, K. Carney et al., "Quantifying and monetizing potential climate change policy impacts on terrestrial ecosystem carbon storage and wildfires in the United States," *Climatic Change*, vol. 131, no. 1, pp. 179–181, 2015.

9

Signal Timing Optimization for Transit Priority at Near-Saturated Intersections

Shijie Shu, Jing Zhao, and Yin Han

Department of Traffic Engineering, University of Shanghai for Science and Technology, Shanghai 200093, China

Correspondence should be addressed to Jing Zhao; jing-zhao-traffic@163.com

Academic Editor: Alain Lambert

Transit signal priority is a useful way to improve transit operations in urban networks. Most of the existing studies have been conducted in conditions with low saturation to avoid the detrimental effects of vehicles without priority. However, from the public transit point of view, it is more meaningful to assign transit signal priority when the degree of the saturation intersections is high. This study proposes a signal control model for transit signal priority to minimize the overall delay at near-saturated intersection. The delay increment is calculated in three scenarios for buses and private vehicles according to the dissipation time of the vehicular queue. A set of constrains are set up to avoid queue overflows and to ensure the rationalization of the signal timing. The proposed control model is tested based on a case study and numerical experiments. The results show that the proposed model can reduce the total person delay at near-saturated intersections. The length of priority time, degree of saturation, and number of lanes are the three main influencing factors. More than 6% reductions in person delay can be obtained for undersaturated intersections when the priority time is less than 5 s. Moreover, even when the intersection saturation is 0.95, the bus signal priority can be applied if only the priority time is less than 5 s.

1. Introduction

As cities become more and more congested, land for road construction is limited. Therefore, congestion needs to be alleviated by changing the model of travel [1]. The improvement of public transit service can elicit important social benefits. Transit signal priority and exclusive bus lanes are two common methods used to improve the public transit service [2]. Setting exclusive bus lanes is a cost-effective approach for providing a high-quality transit service, enhancing the reliability of bus services, and increasing the speed of buses [3–6]. Although continuous running ways yield great benefits to transit operations, transit vehicles will be delayed at intersections by traffic signals [7], especially buses that arrive at the end of the green light period. Therefore, to further improve the running efficiency of buses at intersections, the exclusive bus lanes need to be perfected by assigning transit signal priority.

Transit signal priority is a typical strategy used to reduce the delay of public transit, and it has been used extensively to improve transit operations in urban networks. Existing transit signal priority strategies can be divided into three major categories: passive priority strategies, active priority strategies, and real-time priority strategies [8].

Passive priority strategies are developed on the basis of historical data and do not require any detection system. Skabardonis introduced transit priority measures on urban networks [9]. It has been proven that the passive priority strategies are effective in simple network configurations and elicit predictable dwell times and high bus frequencies. However, it was also pointed out that the passive priority strategies were less adaptable when vehicular flow was low. Estrada proposed a passive signal priority simulation model which restricted the maximum car passenger delay increment [10]. The model was applied to a series of trial networks, and the results showed that the model can reduce the bus travel times by 8.5%, while concurrently increasing the car delay by 5%. Correspondingly, the detection system developed numerous active priority strategies which could provide more effective priority to buses.

Active priority strategies include phase extension, phase advance, phase insertion, and phase rotation [11]. These strategies respond to traffic variations in real-time and are therefore more effective than passive priority strategies. Ngan et al. examined the impacts of a set of traffic parameters on the effectiveness of a transit signal priority application [12]. The results show that transit signal priority application would be more effective when the bus volume is moderate to heavy. Jacobson and Sheffi presented an analytical model for the bus preemption analysis [13]. The results showed that the benefits of bus preemption were small when the cross-traffic flow was much lower than the preemption direction. Furth and Muller put forward a method of operational control which can keep buses on schedule [14]. This method has been implemented in the Netherlands. The results showed that the proposed method can improve the bus schedule. In addition, conditional priority elicited minimal delay increments compared to the case where no priority was assigned, while absolute priority caused large delay increments. Zeeshan et al. proposed an analytical model for estimating the delay impacts of several transit priority strategies [15]. Active priority strategies often have detrimental effects on the vehicles that have no priority, so real-time priority strategies were developed which focused on the total vehicular delay or the total passenger delay at signal intersections.

Real-time priority strategies optimize signal timings based on performance criteria, such as passenger delays, vehicular delays, and combinations of vehicular delays. Ling and Shalaby put forward an algorithm which is based on reinforcement learning and the control policy is learned by the algorithm [16]. The best duration of each phase is determined by the reinforce learning. The experimental results show that the proposed algorithm can reduce the transit headway deviation and disrupt cross street traffic smaller. Christofa and Skabardonis presented a real-time signal control system for signal priority on conflicting transit routes that also minimized the negative effects on vehicular traffic [17]. To solve the issue of multiple priority requests, Head proposed a model which could analyze complex controller behaviors [18]. The analyzed results showed that the proposed model for serving priority requests resulted in fewer delays than a first-come, first-served policy. From the perspective of intersection groups, Ma presented an optimization model which could generate the optimal combination of priority strategies for intersection groups [19]. Christofa et al. presented a real-time signal control system for assigning signal priority on conflicting transit routes that also minimized the passenger delay on arterials [20]. Ma et al. proposed a priority signal model for multiple bus requests [21]. The optimal priority serving sequence is generated by the proposed model, so the bus can maximize the utilization of available green times. With the development of communication technologies, optimization methods were proposed in a connected vehicle environment. Hu presented an optimization method for an intelligent transit signal priority logic that enables bus signal cooperation among consecutive signals [22]. The results showed that the intelligent TSP logic can efficiently reduce the bus delay compared to the conventional TSP. Li et al. proposed an adaptive transit signal priority optimization model that optimizes green splits for three consecutive cycles to minimize the vehicle delay [23]. The numerical example results show that the bus delay and the traffic delay along the bus movement direction were all reduced. He and Head presented a request-based, mixed-integer, linear program, which can coordinate multiple priority requests from pedestrians and different modes of vehicles [24]. Lin presented a control model that minimized bus passenger waiting time at downstream bus stops but did not increase the delay for vehicular passengers [25]. The test results show that the proposed model can reduce the total intersection delay and the bus passenger waiting time. Altun and Furth found that the optimal performance that used conditional priority under an aggressive schedule can lead to obvious reductions in running time and headway irregularity [26]. A novel method was also proposed by Guler and Gayah that provided priority to buses at signalized intersections using single-lane approaches [27]. The method used addition signals to stop cars on the opposing travel lane so that the bus could use the travel lane in the opposite direction. The results showed that the presignal strategy could reduce the bus and the total passenger delays. The genetic algorithm was used to optimize four basic signal timing parameters and transit priority settings. The results showed that the genetic algorithm elicited a good performance in the optimization of the transit priority setting [28]. Ma et al. assumed that the bus speed is available to adjust in real-time [29]. A set of integrated operational rules which integrated operation of signal timings and bus speed were developed to provide priority to bus at isolated intersections. The experimental analyses show that the proposed integrated operational rules perform better than the priority strategies which do not adjust the bus speed.

Most of the existing transit priority studies have been conducted in low-saturation conditions. It was argued that the intersection under near-saturation conditions is difficult to adopt transit signal priority, owing to the detrimental effects on the no-priority vehicles, which will lead to long vehicular queue dissipation times and long queue lengths [30]. Accordingly, serious traffic jam will be easily caused by assigning transit signal priority. However, from the public transit point of view, it is more meaningful to adopt transit signal priority when the intersection saturation is high. It can improve the level of service of public transit systems which would encourage more travelers to choose the transit mode for their travels.

To remedy this deficiency, this study aims to create a signal control model for transit priority at intersections with high degree of saturation. A passenger-delay-based optimization method is used, in which both the performance of buses and private vehicles are considered. The rest of this study is organized as follows. In Section 2, a green extension model is presented for signal priority on buses that minimizes the total passenger delay at intersections with increased saturations. In Section 3, a case is presented, whereby the signal control model is tested in a near-saturated intersection in Shanghai, China. In Section 4, the sensitivity analysis of the proposed model is presented. In Section 5, conclusions are outlined.

2. Methodology

2.1. Basic Idea. To reduce bus delays, especially when these arrive at the end of the green traffic light periods, the green

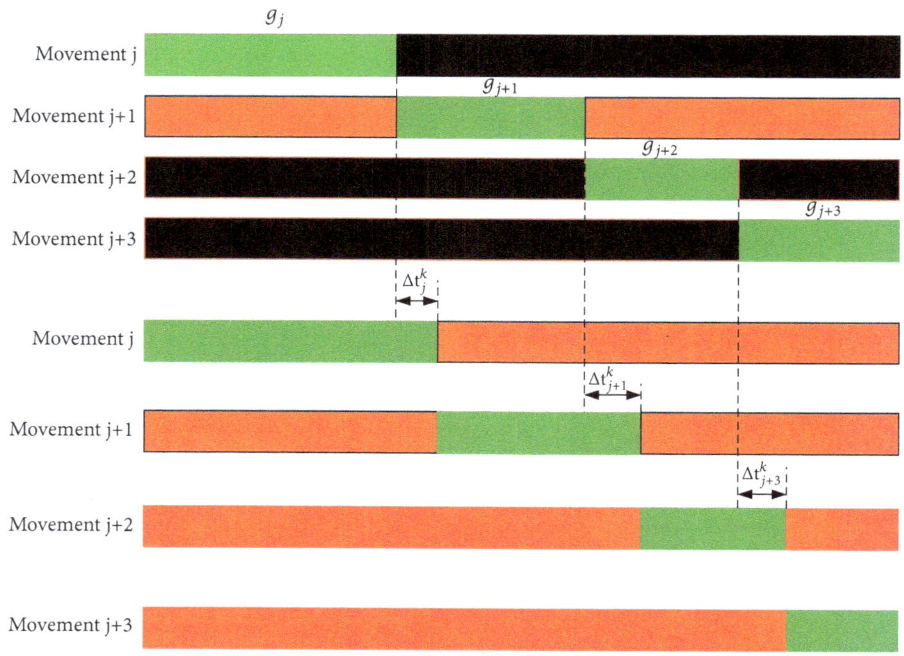

FIGURE 1: Illustration of the green extension strategy.

extension strategy is a common priority strategy that has used in numerous optimization models. Owing to the increase in the time period the green light lasts for buses, the corresponding periods for which the green light is on for other conflicting phases should be shortened. Therefore, the green light period for other phases should be resplit. Correspondingly, three commonly used ways to achieve this include the distribution of all the increased red-light times to a certain phase [31, 32], distribution of the increased red-light period to subsequent phases according to the phase splits [33], and the resplitting of the green light period according to the degree of saturation. However, these red-light time distribution strategies may not lead to optimal signal timing results. This study proposed a signal timing optimization model for transit priority at intersections with high degree of saturation to minimize the total intersection passenger delay. As shown in Figure 1, the green light period of movement j, which has the same direction as that of the bus, is extended to $g_j + \Delta t_j^k$. The delay of movement $j + 1$ will increase owing to the green extension Δt_j^k. To decrease the delay of movement $j + 1$, the green light period of movement $j + 1$ will be extended to $g_{j+1} - \Delta t_j^k + \Delta t_{j+1}^k$, and the same strategy will be adopted in the following movements.

The proposed model aims to minimize the total passenger delay at the intersection by changing the lengths of the green light periods for each movement j. Transit vehicles are assumed to travel on exclusive transit lanes. The dual-ring, eight-phase structure is considered herein, because it is universally applied in the traffic control industry [34]. The advantage of the dual-ring concept is that it allows the first (usually left-turn) phases in Rings 1 and 2 to terminate independently after their respective demands have been satisfied, which can enhance the flexibility of the signal timing [34]. A four-legged intersection with eight movements is shown in Figure 2(a), and each ring in the controller contains four phases, as depicted in Figure 2(b).

The model will be established in three steps: (1) calculation of the benefits (decreased delay) for private vehicles with the same movements as those for buses, and negative effects (increased delay) for other private vehicles when the green light extension strategy is adopted, (2) establishment of a minimized optimization model that considers the total delay at the intersection, and (3) use of enumeration methods to solve the model. For an easier understanding, a summary of the model notations is presented in Table 1.

2.2. Passenger Delay Increment Calculation. In this study, the total passenger delay increment was used as the indicator and comprised two parts: (1) the passenger delay increment of private vehicles at the intersection and (2) the passenger delay increment of buses at the intersection. The total passenger delay increment was defined as the delay in instances when the bus priority strategy was adopted at intersections minus the delay in instances when no bus priority was allocated at intersections. Therefore, the value of the total passenger delay increment should be negative when the transit signal priority strategy was adopted. Since the proposed model was expected to be used at near-saturated intersections, the application of the transit signal priority may lead to the oversaturation of other traffic movements. For a fair comparison, the analysis included all the cycles until the queue lengths of all the movements were restored to their original conditions. According to the vehicular queue dissipation time, three scenarios are considered herein: (1) scenario 1, the queue length of movement j can be restored to zero before the end of the green light period

Table 1: Symbol definitions.

Type	Symbol	Definition
Input parameters	j	Set of movements, $j = 1$ for movement 1
	q_j	Arrival rate for movement j (veh/s)
	s_j	Saturation flow for movement j (veh/s)
	C	Cycle length (s)
	g_j	Length of green time for movement j (s)
	n	Amount of movement j
	P_a	Average passenger occupancy of private vehicles (per/veh)
	P_b	Passenger occupancy of buses (per/veh)
	k	Set of scenarios, $k = 1$ for scenario 1, $k = 2$ for scenario 2, and $k = 3$ for scenario 3
	d_v	Space headway for queuing vehicles (m)
	L_j	Queue length limitation for movement j (m)
	g_p^{min}	Minimum green time for pedestrian crossing the street (s)
Intermediate variables	$N_{1,j}^k, N_{2,j}^k$	Number of vehicles before stop line when the green light period is over in scenario k
	$h_{A,j}^k, h_{B,j}^k, h_{C,j}^k, h_{D,j}^k, h_{E,j}^k, h_{F,j}^k$	Cumulative number of vehicles of movement j at points A, B, C, D, E and F in scenario k
	$t_{1,j}^k, t_{2,j}^k, t_{3,j}^k$	Time of queue dissipation of movement j in scenario k (s)
	$l_{max,j}^k$	Largest queue length of movement j in scenario k (m)
Decision variables	$S_{1,j}^k, S_{2,j}^k, S_{3,j}^k, S_{4,j}^k$	Increased delay of movement j caused by the green light extension in scenario k (s)
	Δt_j^k	Length of green extension of movement j in scenario k (s)
	$\Delta D_{a,j}^k$	Passenger delay increment of private vehicles on the movement j in scenario k (s)
	ΔD_a	Passenger delay increment of private vehicles at the intersection (s)
	ΔD_b	Passenger delay increment of buses at the intersection (s)

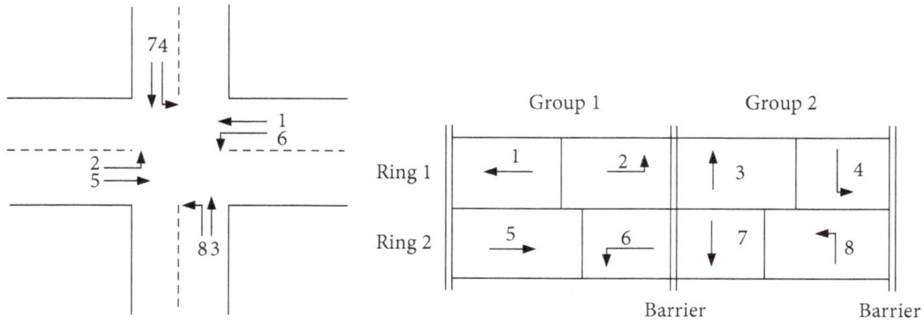

(a) Intersection layout showing the associated phases

(b) Dual-ring phase structure

FIGURE 2: Phase plan.

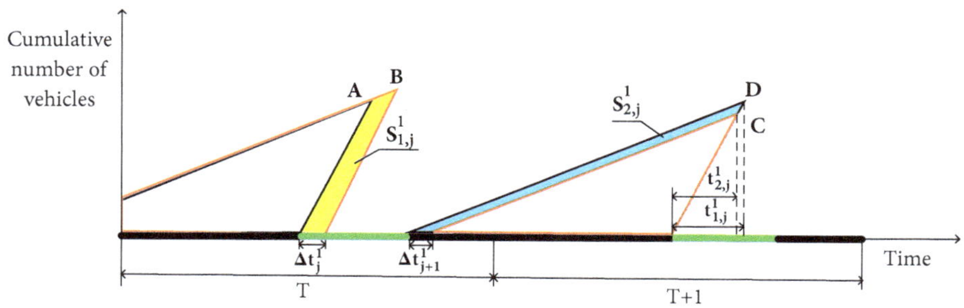

FIGURE 3: Calculation of vehicular delay for movement j in scenario 1.

that can serve j in the first signal cycle, as shown in Figure 3; (2) scenario 2, the queue length of movement j can be restored to zero before the end of the green light period that can serve j in the second signal cycle, as shown in Figure 4; (3) scenario 3, the queue length of movement j can be restored to zero before the end of the green light period that can serve j in the third signal cycle, as shown in Figure 5. When more than three cycles are required for the queue length of movement j to be restored to zero, the total passenger delay increment will be greater than zero. These cases are beyond the scope of the current analyses and are not discussed.

2.2.1. Passenger Delay Increment of Private Vehicles

(1) Scenario 1. In scenario 1, the queue length of movement j can be restored to zero before the end of the green light period that can serve j in the first cycle. Figure 3 shows the cumulative number of vehicles present at an intersection for cycle T and cycle T + 1. Without loss of generality, the beginning of the phase that adopts the bus signal priority is set to be the beginning of the signal cycle. The area of the yellow part is the delay increment caused by the extension of the green light, $S^1_{1,j}$. The area of blue part is the delay decrement caused by the extension of the green light, $S^1_{2,j}$. Then, the passenger delay increment of private vehicles on the movement j in scenario 1 can be calculated by the following:

$$\Delta D^1_{a,j} = P_a \cdot \left(S^1_{1,j} - S^1_{2,j}\right) \quad (1)$$

Equation (1) is subject to the constraint:

$$q_j \cdot \left(C + \Delta t^1_{j+1}\right) - \left(g_j - \Delta t^1_j + \Delta t^1_{j+1}\right) \cdot s_j \leq 0 \quad (2)$$

Variables in (1), $S^1_{1,j}$ and $S^1_{2,j}$, can be calculated by (3) and (4), respectively. Intermediate variables, $h^1_{A,j}$, $h^1_{B,j}$ $h^1_{C,j}$, and $h^1_{D,j}$, are the cumulative number of vehicles of movement j at points A–D, which can be calculated by (5)-(8), respectively. Intermediate variables, $t^1_{1,j}$ and $t^1_{2,j}$, are the time of queue dissipation of movement j, which can be calculated by (9) and (10), respectively. Intermediate variable, $l^1_{max,j}$, is the largest queue length of movement j in scenario 1, which can be calculated by (11).

$$S^1_{1,j} = \frac{1}{2}\left(C - g_j + \Delta t^1_j\right) h^1_{B,j} - \frac{1}{2}\left(C - g_j\right) h^1_{A,j} \quad (3)$$

$$S^1_{2,j} = \frac{1}{2}\left(C - g_j\right) h^1_{D,j} - \frac{1}{2}\left(C - g_j - \Delta t^1_{j+1}\right) h^1_{C,j} \quad (4)$$

$$h^1_{A,j} = \frac{s_j q_j \left(C - g_j\right)}{s_j - q_j} \quad (5)$$

$$h^1_{B,j} = \frac{s_j q_j \left(C - g_j + \Delta t^1_j\right)}{s_j - q_j} \quad (6)$$

$$h^1_{C,j} = \frac{s_j q_j \left(C - g_j - \Delta t^1_{j+1}\right)}{s_j - q_j} \quad (7)$$

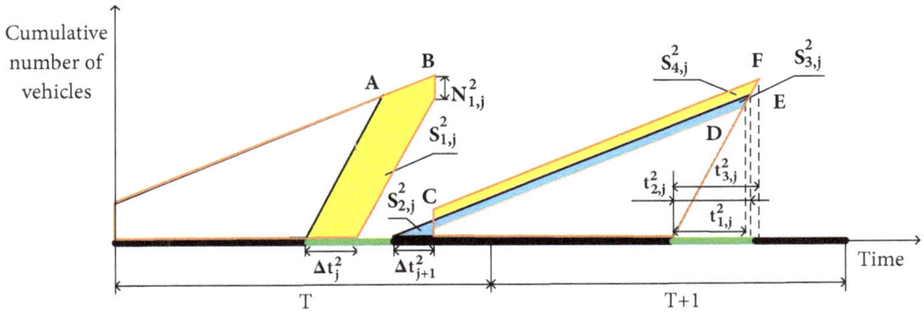

FIGURE 4: Calculation of vehicular delay for movement j in scenario 2.

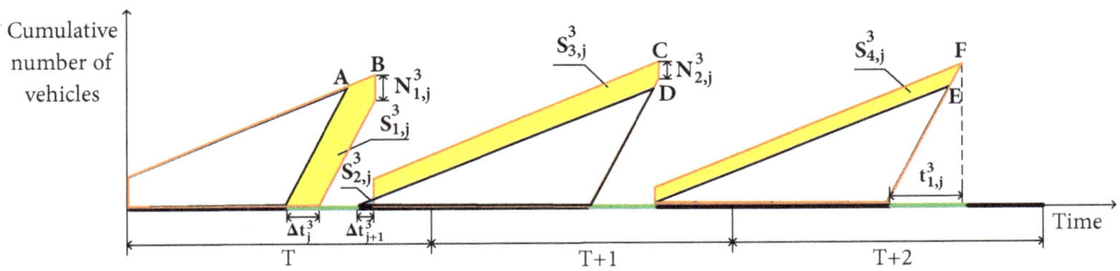

FIGURE 5: Calculation of vehicular delay for movement j in scenario 3.

$$h^1_{D,j} = h^1_{A,j} \tag{8}$$

$$t^1_{1,j} = \frac{q_j \cdot (C - g_j)}{s_j - q_j} \tag{9}$$

$$t^1_{2,j} = \frac{q_j \cdot (C - g_j - \Delta t^1_{j+1})}{s_j - q_j} \tag{10}$$

$$l^1_{max,j} = h^1_{B,j} d_v = \frac{s_j d_v q_j (C - g_j + \Delta t^1_j)}{s_j - q_j} \tag{11}$$

(2) Scenario 2. In scenario 2, the queue length of movement j can be restored to zero before the end of the green light period that can serve j in the second cycle. Figure 4 shows the cumulative number of vehicles present at an intersection for cycle T and cycle T + 1. The areas of the yellow parts, $S^2_{1,j}$ and $S^2_{4,j}$, are the delay increments caused by the extensions of the green light periods Δt^2_j and Δt^2_{j+1}, respectively. The areas of the blue parts, $S^2_{2,j}$ and $S^2_{3,j}$, are the delay decrement caused by the extension of the green light Δt^2_{j+1}. Then, the passenger delay increment of private vehicles on the movement j in scenario 2 can be calculated by (12). In (12), the condition $N^2_{1,j} \geq \Delta t^2_{j+1} q_j$ implies that the queue length $N^2_{1,j}$ is larger than or equal to the queue length when the movement j does not adopt the extension of the green light period Δt^2_{j+1}. The condition $N^2_{1,j} < \Delta t^2_{j+1} q_j$ implies that the queue length $N^2_{1,j}$ is less than the queue length in the instance when the movement j does not adopt the extension of the green light period Δt^2_{j+1}.

$$\Delta D^2_{a,j} = \begin{cases} P_a \left(S^2_{1,j} - S^2_{2,j} + S^2_{4,j}\right), & N^2_{1,j} \geq \Delta t^2_{j+1} q_j \\ P_a \left(S^2_{1,j} - S^2_{2,j} - S^2_{3,j}\right), & N^2_{1,j} < \Delta t^2_{j+1} q_j \end{cases} \tag{12}$$

Equation (12) is subject to the restrictions:

$$q_j \left(C + \Delta t^2_{j+1}\right) - s_j \left(g_j - \Delta t^2_j + \Delta t^2_{j+1}\right) \geq 0 \tag{13}$$

$$q_j (2C) - s_j \left(2g_j - \Delta t^2_j + \Delta t^2_{j+1}\right) \leq 0 \tag{14}$$

Variables in (12), $S^2_{1,j}$, $S^2_{2,j}$, $S^2_{3,j}$, and $S^2_{4,j}$, can be calculated by (15)-(18), respectively. The intermediate variable $N^2_{1,j}$ denotes the number of vehicles when the green light period is completed, which can be calculated by (19). The intermediate variables $h^2_{A,j}$, $h^2_{B,j}$, $h^2_{C,j}$, $h^2_{D,j}$, $h^2_{E,j}$, and $h^2_{F,j}$ are the cumulative numbers of vehicles for movement j at points A–F, which can be calculated by (20)-(25), respectively. The intermediate variables $t^2_{1,j}$, $t^2_{2,j}$, and $t^2_{3,j}$ are the queue dissipation time for movement j, which can be calculated by (26)-(28), respectively. The intermediate variable $l^2_{max,j}$ is the largest queue length of movement j, which can be calculated by (29).

$$S^2_{1,j} = \frac{1}{2}\left(C + \Delta t^2_{j+1}\right) h^2_{B,j} - \frac{1}{2}\left(C - g_j\right) h^2_{A,j} - \frac{1}{2}\left(g_j - \Delta t^2_j + \Delta t^2_{j+1}\right)\left(h^2_{B,j} - N^2_{1,j}\right) \tag{15}$$

$$S^2_{2,j} = \frac{1}{2}\Delta t^2_{j+1} \Delta t^2_{j+1} q_j \tag{16}$$

$$S^2_{3,j} = \frac{1}{2}\left(N^2_{1,j} + h^2_{D,j}\right)\left(C - g_j - \Delta t^2_{j+1} + t^2_{j,1}\right) - \frac{1}{2} t^2_{1,j} h^2_{D,j} - \frac{1}{2}\left(\Delta t^2_{j+1} q_j + h^2_{E,j}\right)\left(C - g_j - \Delta t^2_{j+1} + t^2_{2,j}\right) + \frac{1}{2} t^2_{2,j} h^2_{E,j} \tag{17}$$

$$S_{4,j}^2 = \frac{1}{2}\left(N + h_{F,j}^2\right)\left(C - g_j - \Delta t_{j+1}^2 + t_{3,j}^2\right) - \frac{1}{2}t_{3,j}^2 h_{F,j}^2 - \frac{1}{2}\left(\Delta t_{j+1}^2 q_j + h_{E,j}^2\right)\left(C - g_j - \Delta t_{j+1}^2 + t_{2,j}^2\right) + \frac{1}{2}t_{2,j}^2 h_{E,j}^2 \quad (18)$$

$$N_{1,j}^2 = q_j \cdot \left(C + \Delta t_{j+1}^2\right) - s_j\left(g_j - \Delta t_j^2 + \Delta t_{j+1}^2\right) \quad (19)$$

$$h_{A,j}^2 = \frac{q_j\left(C - g_j\right)}{s_j - q_j} s_j \quad (20)$$

$$h_{B,j}^2 = q_j\left(C + \Delta t_{j+1}^2\right) \quad (21)$$

$$h_{C,j}^2 = N_{1,j}^2 \quad (22)$$

$$h_{D,j}^2 = s_j t_{1,j}^2 \quad (23)$$

$$h_{E,j}^2 = s_j t_{2,j}^2 \quad (24)$$

$$h_{F,j}^2 = s_j t_{3,j}^2 \quad (25)$$

$$t_{1,j}^2 = \frac{N_{j,1}^2 + q_j \cdot \left(C - g_j - \Delta t_{j+1}^2\right)}{s_j - q_j} \quad (26)$$

$$t_{2,j}^2 = \frac{q_j\left(C - g_j\right)}{s_j - q_j} \quad (27)$$

$$t_{3,j}^2 = \frac{N_{1,j}^2 + q_j \cdot \left(C - g_j - \Delta t_{j+1}^2\right)}{s_j - q_j} \quad (28)$$

$$l_{max,j}^2 = h_{B,j}^2 d_v = q_j\left(C + \Delta t_{j+1}^2\right) d_v \quad (29)$$

(3) Scenario 3. In scenario 3, the queue length of group j can be restored to zero before the end of the green time period that can serve j in the third cycle. Figure 5 shows the cumulative number of vehicles present at an intersection for cycles T, T + 1, and T + 2. The area of the yellow part, $S_{1,j}^3$, $S_{3,j}^3$, $S_{4,j}^3$, is the delay increments caused by the extensions of the periods when the green light was on. The area of the blue part, $S_{2,j}^3$, is the delay decrement caused by the extension of the green light period. Then, the passenger delay increment of private vehicles on the movement j in scenario 3 can be calculated by (30).

$$\Delta D_{a,j}^3 = P_a\left(S_{1,j}^3 - S_{2,j}^3 + S_{3,j}^3 + S_{4,j}^3\right) \quad (30)$$

Equation (30) is subject to the restrictions:

$$q_j(2C) - \left(2g_j - \Delta t_j^3 + \Delta t_{j+1}^3\right)s_j \geq 0 \quad (31)$$

$$q_j(3C) - \left(3g_j - \Delta t_j^3 + \Delta t_{j+1}^3\right)s_j \leq 0 \quad (32)$$

Variables in (12), $S_{1,j}^3$, $S_{2,j}^3$, $S_{3,j}^3$, and $S_{4,j}^3$, can be calculated by (33)-(36), respectively. Intermediate variables $N_{1,j}^3$ and $N_{2,j}^3$ are the number of vehicles when the time period of the green light is completed, which can be calculated by (37) and (38), respectively. The intermediate variables $h_{A,j}^3$, $h_{B,j}^3$, $h_{C,j}^3$, $h_{D,j}^3$, $h_{E,j}^3$, and $h_{F,j}^3$ are the cumulative number of vehicles of movement j at points A–F, which can be calculated by (39) and (44), respectively. The intermediate variable $t_{1,j}^3$ is the time of queue dissipation of movement j, which can be calculated by (45). The intermediate variable $l_{max,j}^3$ is the largest queue length of movement j, which can be calculated by (46).

$$S_{1,j}^3 = \frac{1}{2}\left(C + \Delta t_{j+1}^3\right)h_{B,j}^3 - \frac{1}{2}\left(C - g_j\right)h_{A,j}^3 - \frac{1}{2}\left(g_j - \Delta t_j^3 + \Delta t_{j+1}^3\right)\left(h_{B,j}^3 - N_{1,j}^3\right) \quad (33)$$

$$S_{2,j}^3 = \frac{1}{2}\Delta t_{j+1}^3 \Delta t_{j+1}^3 q_j \quad (34)$$

$$S_{3,j}^3 = \frac{1}{2}\left(N_{1,j}^3 + h_{C,j}^3\right)\left(C - \Delta t_{j+1}^3\right) - \left[\frac{1}{2}\left(C - g_j\right)h_{D,j}^3 - S_2\right] - \frac{1}{2}g_j\left(h_{D,j}^3 - N_{2,j}^3\right) \quad (35)$$

$$S_{4,j}^3 = \frac{1}{2}\left(N_{2,j}^3 + h_{F,j}^3\right)\left(C - g_j + t_{1,j}^3\right) - \frac{1}{2}\left(C - g_j\right)h_{E,j}^3 - \frac{1}{2}t_{1,j}^3 h_{F,j}^3 \quad (36)$$

$$N_{1,j}^3 = q_j\left(C + \Delta t_{j+1}^3\right) - s_j\left(g_j - \Delta t_j^3 + \Delta t_{j+1}^3\right) \quad (37)$$

$$N_{2,j}^3 = N_{1,j}^3 + q_j\left(C - \Delta t_{j+1}^3\right) - s_j g_j \quad (38)$$

$$h_{A,j}^3 = \frac{s_j q_j\left(C - g_j\right)}{s_j - q_j} \quad (39)$$

$$h_{B,j}^3 = q_j\left(C + \Delta t_{j+1}^3\right) \quad (40)$$

$$h_{C,j}^3 = N_{1,j}^3 + q_j\left(C - \Delta t_{j+1}^3\right) \quad (41)$$

$$h_{D,j}^3 = \frac{s_j q_j\left(C - g_j\right)}{s_j - q_j} \quad (42)$$

$$h_{E,j}^3 = \frac{s_j q_j\left(C - g_j\right)}{s_j - q_j} \quad (43)$$

$$h_{F,j}^3 = \frac{N_{j,2}^3 + q_j \cdot \left(C - g_j\right)}{s_j - q_j} s_j \quad (44)$$

$$t_{1,j}^3 = \frac{N_{j,2}^3 + q_j\left(C - g_j\right)}{s_j - q_j} \quad (45)$$

$$l^3_{max,j} = h^3_{C,j} d_v = \left[N^3_{1,j} + q_j\left(C - \Delta t^3_{j+1}\right)\right] d_v \tag{46}$$

In summary, the passenger delay increment of buses at the intersection on movement j in scenarios 1–3 can be expressed as (47). For the scenario 1, 2, and 3, (2), (13)-(14), and (31)-(32) should be satisfied, respectively. The passenger delay increment of private vehicles, ΔD_a, equals the sum of the passenger delay increment for private vehicles of all movements, as shown in (48).

$$\Delta D_{j,a} = \begin{cases} P_a \cdot \left(S^1_{1,j} - S^1_{2,j}\right), & scenario\ 1 \\ P_a \cdot \left(S^2_{1,j} - S^2_{2,j} + S^2_{3,j} + S^2_{4,j}\right), & scenario\ 2 \\ P_a \cdot \left(S^3_{1,j} - S^3_{2,j} + S^3_{3,j} + S^3_{4,j}\right), & scenario\ 3 \end{cases} \tag{47}$$

$$\Delta D_a = \sum_{j=1}^{n} \Delta D_{j,a} \# \tag{48}$$

2.2.2. Passenger Delay Increment of Buses. The passenger delay increment of buses, ΔD_b , is the negative value of the length of the remaining red light of movement 1, as shown in the following:

$$\Delta D_b = -P_b \cdot \left(C - g_1 - \Delta t^k_1\right) \tag{49}$$

2.2.3. Validation of the Delay Model. The delay increment calculation should be validated through the microscopic simulation package VISSIM. The basic idea of delay increment calculation validation is as follows: as shown in (48), the passenger delay increment of private vehicles, ΔD_a, equals the sum of the passenger delay increment for private vehicles of all movements, so the delay calculation of each movement is calculated by the same equation (47). In other words, the validation of the passenger delay increment calculation is actually to validate (47).

In the following validation process, a test movement is selected, the basic parameters of the test movement and the green light extension time are input, and, after the simulation, the total vehicular delay of the movement is output. Because the correctness of the delay increment calculation under different scenarios needs to be tested, different green light extension times are input, which can result in shorter green times assigned to the vehicle and the queue length needing multiple cycles to return to zero. Finally, the total vehicular delay of the movement under the three scenarios is recorded and compared with the calculated values of the model.

The value of the input parameters are as follows: the cycle length is 100 s, the green time of the test lane is 25 s, the arrival rate and the saturation rate of the test lane are 360 and 1800 veh/h respectively, and the green extension time will be from 1 to 10. If the green extension time is 1,2,3, the queue length of test movement will be restored to zero in first cycle, so the delay increment calculation in scenario 1 will be validated. If the green extension time is 4,5,6, the queue length of test movement will be restored to zero in second cycle, so the delay increment calculation in scenario 2 will be validated. If the green extension time is 7,8,9,10, the queue length of test movement will be restored to zero in third cycle, so the delay increment calculation in scenario 3 will be validated. The simulation time of VISSIM is set to 4000s. The delay increment calculation validation results are presented in Figure 6. As depicted in Figure 6, the delay increment calculation value is close to the simulation value, and the maximum difference does not exceed 16%, the delay increment calculation is proved to be valid.

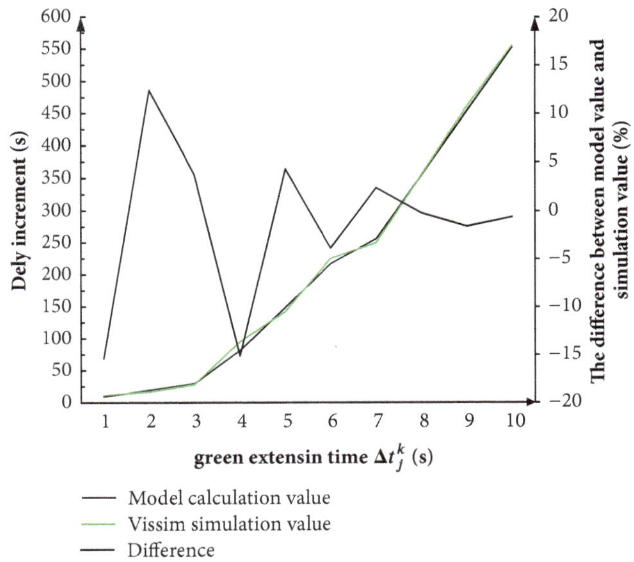

FIGURE 6: Comparison between the model calculation value and the simulation value.

2.3. Optimization Model

2.3.1. Objective. The transit signal priority strategy is to improve the operational efficiency. Minimizing the total passenger delay was selected as the optimization objective, which is equivalent to minimizing the total passenger delay increment caused by the transit signal priority strategy, as shown in (50), in which the passenger delay increment of private vehicles and buses can be calculated by (48) and (49), respectively.

$$\min \Delta D_a + \Delta D_b \tag{50}$$

2.3.2. Constraints. To avoid vehicle queue spillovers, the queue length cannot be greater than the maximum queue length limitation, L_{max}, as shown in the following:

$$l^k_{max,j} \le L_{max}, \quad \forall k = 1, 2, 3 \tag{51}$$

The green light extension for movement $j+1$ is less than or equal to the green light extension for movement j, as shown in (52). For Δt^k_4 and Δt^k_8, they are at the end-periods of the ring and cannot extend the green light period to the next movement. Therefore, they are equal to zero, as shown in (53). Moreover, in the dual-ring, all phases in one group must

TABLE 2: Traffic parameters of the tested intersection.

Intersection approach	Movements	Traffic volume (veh/h)	Saturation flow rate (veh/h)
	Left turn	728	3100
East	Through	1108	5800
	Right turn	588	1550
	Left turn	294	1400
West	Through	1368	8000
	Right turn	696	1550
	Left turn	864	3200
North	Through	540	2000
	Right turn	104	775
	Left turn	124	700
South	Through	524	3400
	Right-turn	92	775

TABLE 3: Number of cycles required for recovery.

Cycles for recovery	M1	M2	M3	M4	M5	M6	M7	M8
Proposed method	1	2	1	1	1	1	1	2
Traditional method A	1	1	2	3	1	1	2	2
Traditional method B	1	1	2	3	1	1	2	1

Note: M = movement.

be terminated before the beginning of any of the phases of the next group. Thus, Δt_2^k should be equal to Δt_6^k, as shown in (54).

$$0 \le \Delta t_{j+1}^k \le \Delta t_j^k, \quad \forall j = 1, 2, 3, 6, 7; \ k = 1, 2, 3 \tag{52}$$

$$\Delta t_4^k = \Delta t_8^k = 0 \tag{53}$$

$$\Delta t_2^k = \Delta t_6^k \tag{54}$$

The green time under the dual-ring phase plan should meet the constraint (55) due to the setting of the barriers shown in Figure 2(b). Moreover, the minimum green time period should be larger than or equal to the minimum green light period needed for the pedestrians to cross the street, as shown in (56).

$$g_j + g_{j+1} + \Delta t_{j+1}^k = g_{j+4} + g_{j+5} + \Delta t_{j+5}^k, \quad \forall j = 1, 3. \ k = 1, 2, 3 \tag{55}$$

$$g_j - \Delta t_j^k + \Delta t_{j+1}^k \ge g_p^{min}, \quad \forall k = 1, 2, 3 \tag{56}$$

2.3.3. Solution. The proposed model was coded in MATLAB and tested on an Intel i5, 2.5 GHz processor and 4.0 GB RAM, running Windows, and was solved using an enumeration method. The program running interface is presented in Figure 7.

3. Case Study

The proposed model is applied to a real-world intersection, namely, the intersection of the Huashan and Middle Yan'an roads located in Shanghai, China. The intersection's layout is presented in Figure 8(a). As shown in this figure, the main through movement is via Yan'an road. The exclusive bus lane is set on Middle Yan'an road. The traffic volumes during the morning peak hours (8 to 9 a.m.) were obtained based on a field survey. The real data acquired from the Huashan and Middle Yan'an roads are presented in Table 2. The intersection is operated on a fixed four-phase cycle with a cycle length of 190 s, as presented in Figure 8(b). The saturation of the intersection is 85.2%. The average passenger occupancy of private vehicles is 1.5 per/veh; the passenger occupancy of buses is 20 per/veh.

Now, there is an application of the bus priority. The required bus priority time is 18 s. In order to better reflect the advantages of the proposed model, the basic timing plan will be adjusted, the existing timing plan will be optimized according to actual traffic flow. In this way, the problem of unreliable model results due to the irrationality of the basic signal timing can be solved. The adjusted signal timing is presented in Figure 9. The proposed model is used to optimize the adjusted signal timing. The running time in MATLAB is 1.72 s. The optimization results include the extension time for the green light period $\Delta t_1^k = 18s$, $\Delta t_2^k = 8s$, $\Delta t_4^k = 2s$, $\Delta t_6^k = 8s$, $\Delta t_7^k = 5s$. The optimized signal timing is presented in Figure 10.

The performance of the proposed model is compared with two traditional methods, namely, traditional methods A and B as shown in Tables 3–5. The traditional method A is the method which distributes red-light time to the following phase according to the phase splits. The traditional method B is the method which resplits the green light time period according to the degree of saturation.

As Table 3 shows, the queue length of movement j will be restored to zero in two cycles. As Table 4 shows, the

TABLE 4: Largest queue length.

Queue length (m)	M1	M2	M3	M4	M5	M6	M7	M8
Proposed method	77.2	81.7	126.0	68.4	62.5	94.5	118.5	66.8
Traditional method A	77.2	80.5	130.5	74.1	62.5	94.5	122.7	66.8
Traditional method B	77.2	80.8	130.2	72.2	62.5	94.5	122.5	65.4

Note: M = movement.

TABLE 5: Comparisons of three methods about delay variation.

Schemes	Index	Total passenger delay (s)	Bus passenger delay (s)	Auto passenger delay (s)
No transit signal priority	Mean	33333.6	2640.0	30693.6
Proposed method	Mean	31849.9	0	31849.9
	variation	-4.5%	-100%	3.8%
Traditional method A	Mean	34291.3	0	34291.3
	variation	2.9%	-100%	11.7%
Traditional method B	Mean	33498.3	0	33498.3
	variation	0.5%	-100%	9.1%

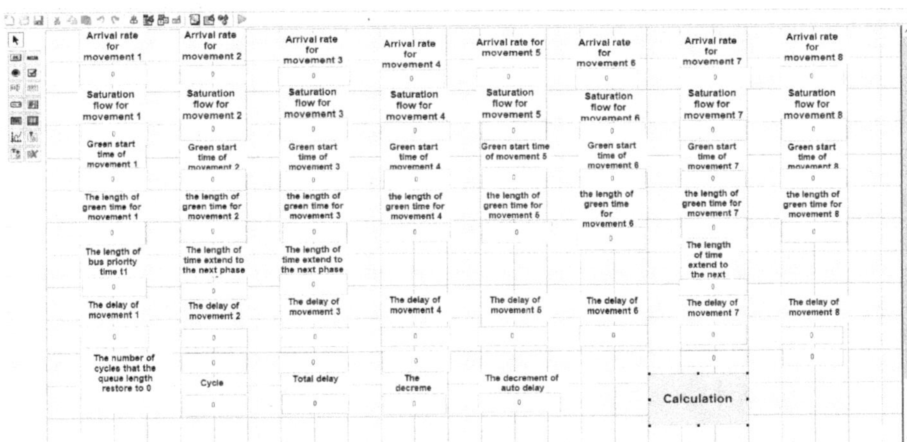

FIGURE 7: Program running interface.

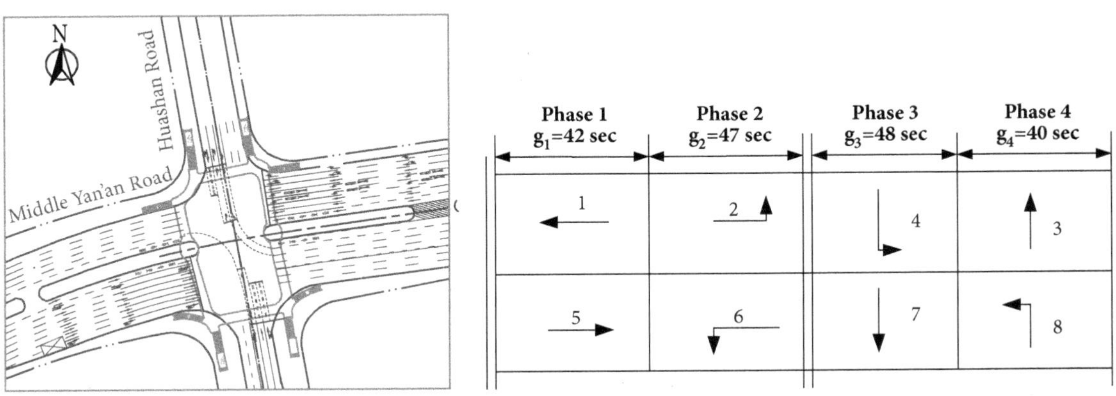

(a) Original layout

(b) Original signal timing

FIGURE 8: Original layout and signal timing of the tested intersection.

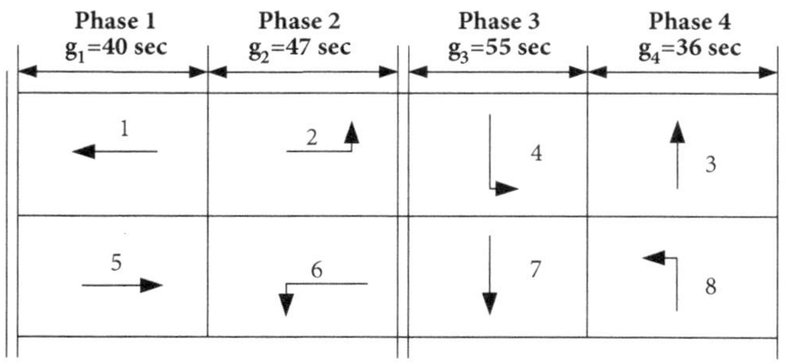

FIGURE 9: Adjusted signal timing.

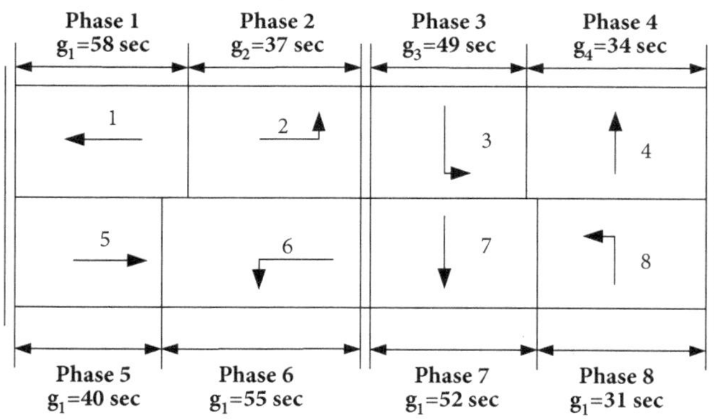

FIGURE 10: Optimized signal timing with transit signal priority.

largest queue length of movement j is less than 150 m, which is less than the road length. Thus, the proposed method is suitable for this intersection. Moreover, the proposed method outperforms the traditional methods in terms of the recovery time and queue length. All the movements can be recovered within 2 cycles for the proposed method, while it requires 3 cycles for recovery. The average and longest queue length for the proposed method is shortened by 1.7% and 3.4%, respectively, with the comparison of the traditional methods.

Table 5 compares the three methods from the perspective of the delay variation in three signal cycles. As Table 5 shows, the proposed method can reduce the total passenger delay and the percentage of delay decrement by 4.5%. Conversely, the traditional method will increase the total passenger delay. In addition, the percentage of the auto-delay increment is more than 9.1% after the optimization of the traditional method, which is larger than 3.8%.

4. Sensitivity Analysis

A sensitivity analysis was conducted in order to analyze the influence of the intersection's geometry and traffic conditions on the efficiency of the proposed model. The length of the priority time for buses, the intersection saturation, and the number of lanes were selected as the three key parameters for the sensitivity analysis. The value of the input parameters are as follows: in Figures 11, 12, and 14, the cycle length is 80 s, there are four phases and the time of each phase is 20 s, the arrive rate of each movement is the same, which is from 315 to 472 veh/h depending on the saturation, the saturation rate of each movement is 1800 veh/h, the average passenger occupancy of private vehicles is 1.5 per/veh, the passenger occupancy of buses is 20 per/veh, and the number of lanes is three. In Figure 13, the arrival rate of each movement is 382 veh/h, and other parameters are the same as above.

In order to avoid the irrationality of the basic signal timing results to unreliable model results, the signal timing and the intersection conditions are set as above described, the signal timing has been set to be optimal. Figure 11 shows the relationship between the delay increment and the three key parameters. Figure 11 is composed of 9 subfigures. The ordinates and abscissa of Figure 11 are priority time and number of lanes, respectively. The ordinates and abscissa of each subfigure are saturation and delay increment, respectively. The nine subfigures of Figure 9 are arranged such that the priority time gradually increases from the bottom to the top, and the number of lanes gradually increases from the left to the right.

From the perspective of each of these subfigures, the vehicular delay increases as the saturation increases; however, the delay of the bus will not change. Additionally, comparison of these subfigures from the bottom to the top indicates that the vehicular delay increases as the priority increases. In contrast, the bus delay decrement becomes smaller. Comparison of the data in the subfigures from the left to the right indicates that the vehicular delay increases as the number of lanes increase, while the bus delay does not change. The reason for which the vehicular delay increases as the number of lanes increases is attributed to the fact that as the number of lanes increases, the number of vehicles will increase if the saturation of the intersection remains constant.

Figure 12 further shows the combination effects of the saturation of intersection and the priority time. As indicated, the total delay will increase as the priority time and the saturation increase. The delay increment increases rapidly and becomes very large as the saturation of intersection becomes larger than 0.9. Similarly, the delay increment will increase rapidly and will become very large as the priority time becomes larger than 10 s.

As Figure 12 shows, the delay increment is sensitive to the priority time. Thus, a better optimization model is needed which is less sensitive to the priority time. As Figure 13 shows, the two traditional methods get the same results, the reason is that the basic signal timing is optimal, the splits and degree of saturation of each phase is the same. Figure 13 compares the proposed model and the traditional model, based on which several inferences can be drawn: (1) the proposed model is less sensitive than the traditional model to the priority time; (2) the difference of the effect of optimization is not obvious between the proposed model and traditional method when the priority time is less than 5 s, and (3) when the priority time is larger than 5 s, the difference of the effects of the optimization between the two methods will increase, as the priority time will increase. When the priority time becomes larger than 9 s, the proposed model is 15% better than the traditional models A and B.

Figure 14 presents a table listing the recommended values for bus signal priorities, which can provide suggestions for practical applications. As Figure 14 shows, the percentages in this figure are the percentage reductions of the total

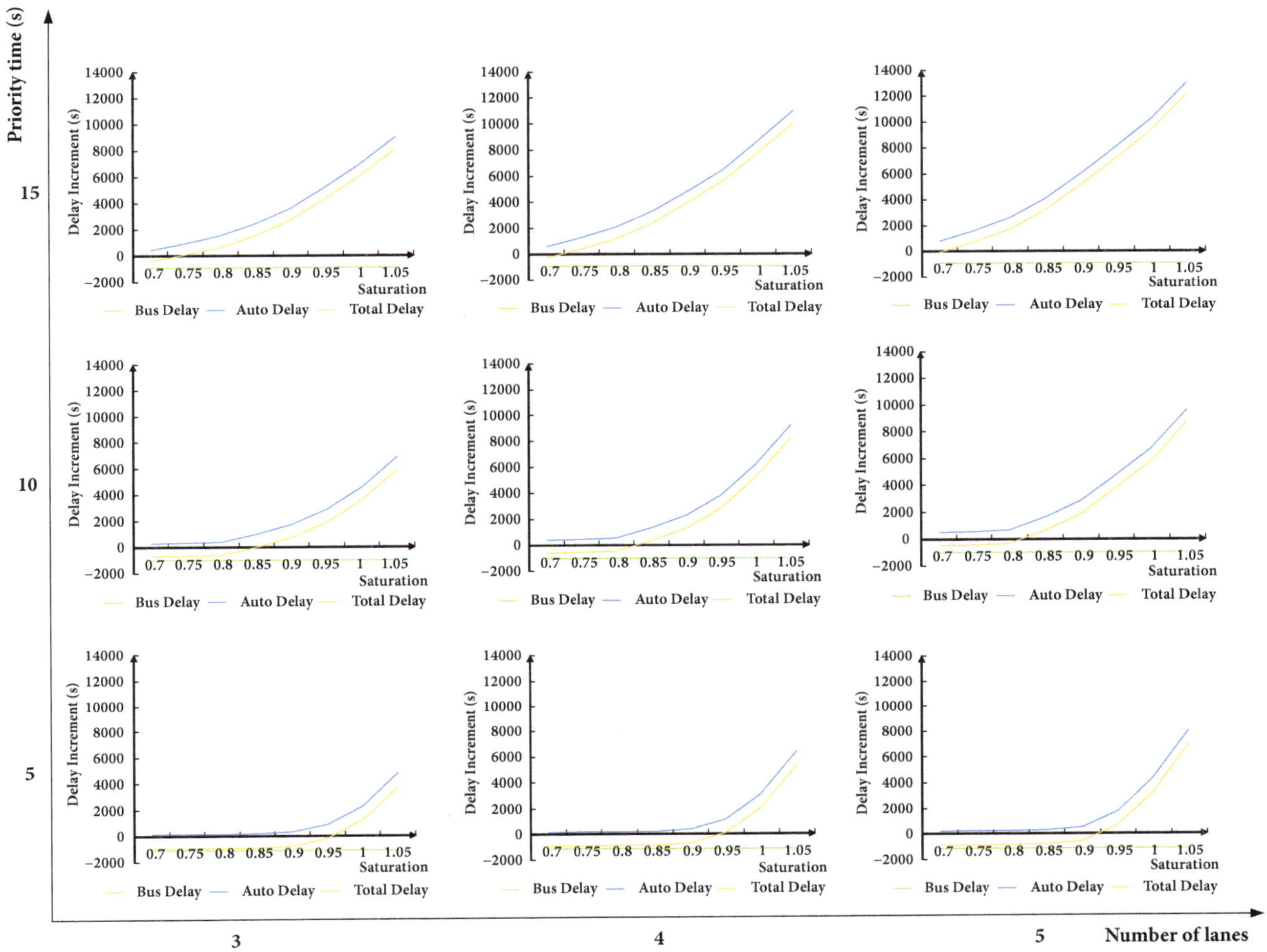

FIGURE 11: Sensitivity plots of the total delay increment to the priority time and the number of lanes.

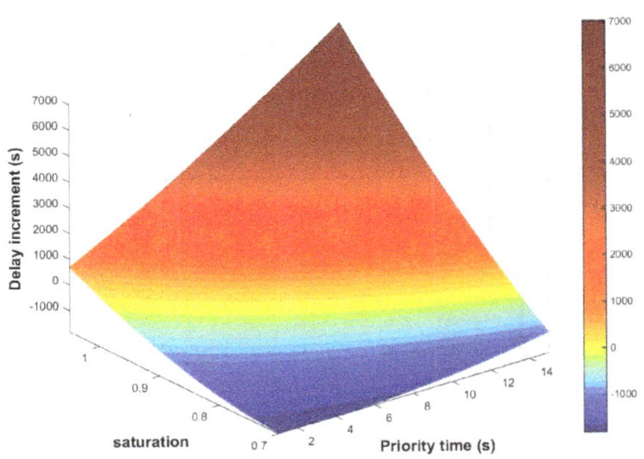

FIGURE 12: Sensitivity of the total delay increment as a function of the priority time and saturation.

passenger delays at the intersection when bus signal priorities are assigned. Figure 14 contains three parts, namely, the green part that is suitable for application of the bus signal priority strategy, the red part that is unsuitable for application of the bus signal priority strategy, and the yellow part that indicates that the percentage reduction of the total delay is larger than zero but less than 10%. In the yellow highlighted part, the effect of the bus signal priority strategy is minor.

5. Conclusions

This paper presented a signal control model for assigning bus signal priorities that also minimized the total passenger delay increment at intersections with the near-saturation. The proposed model was tested with simulations at a near-saturated intersection in Shanghai, China. Several meaningful conclusions can be drawn from the sensitivity analyses:

(1) Bus signal priority can be applied in near-saturated intersections, whereby the length of priority time, the saturation of the intersection, and the number of lanes were the three main factors that influenced whether the intersections should have adopted the bus signal priority strategy.

(2) The method on resplitting the green light time period affected the benefits of the bus signal priority considerably.

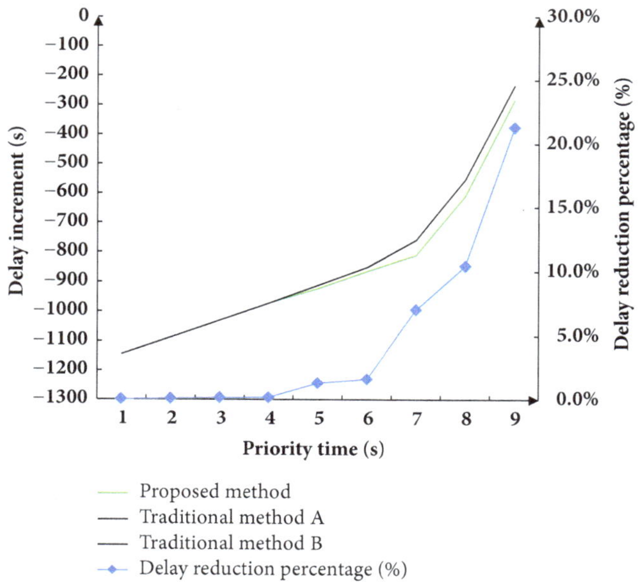

FIGURE 13: Comparison of the proposed model to the traditional model.

Saturation \ Priority time (s)	1	2	3	4	5	6	7	8	9	10	11	12	13	14	15
1.05	53%	65%	77%	89%	102%	28%	127%	140%	153%	166%	179%	193%	207%	220%	235%
1.0	-22%	-10%	2%	14%	26%	39%	52%	64%	77%	90%	104%	117%	131%	145%	159%
0.95	-33%	-30%	-25%	-16%	-6%	6%	19%	32%	45%	58%	71%	84%	98%	112%	126%
0.9	-35%	-33%	-32%	-30%	-25%	-17%	-8%	1%	10%	23%	37%	50%	64%	78%	92%
0.85	-37%	-35%	-33%	-32%	-30%	-28%	-27%	-18%	-10%	-1%	9%	17%	28%	41%	55%
0.8	-38%	-37%	-35%	-34%	-32%	-30%	-28%	-27%	-24%	-21%	-12%	-3%	6%	16%	26%
0.75	-40%	-39%	-37%	-36%	-34%	-32%	-31%	-29%	-27%	-25%	-23%	-21%	-15%	-6%	3%
0.7	-43%	-41%	-40%	-38%	-36%	-35%	-33%	-31%	-29%	-28%	-26%	-24%	-22%	-20%	-18%

FIGURE 14: Table with recommended values for bus signal priorities.

Simple distribution of the added red-light time to the following phases according to the phase splits or resplitting the green light time period according to the degree of saturation was shown to be unsuitable.

(3) A table with recommended bus priority values was presented according to the saturation of the intersection and the length of the priority time. More than 6% reductions in person delay can be obtained for undersaturated intersections when the priority time was less than 5 s. Moreover, even when the intersection saturation was 0.95, the bus signal priority could be applied if only the priority time was less than 5 s.

The signal control model proposed in this study can be applied in near-saturated intersection. However, the issue on the multiple bus requests was not discussed and constitutes a research topic for our ongoing and future research endeavors.

Conflicts of Interest

The authors declare that they have no conflicts of interest.

Acknowledgments

The research is supported by the National Natural Science Foundation of China under Grant no. 51608324.

References

[1] J. Shi, Z. Wu, and J. Jin, "Reform Beijing to a public transit oriented city - From the view of transportation equity," *Journal of Advanced Transportation*, vol. 45, no. 2, pp. 96–106, 2011.

[2] K. M. N. Habib, L. Kattan, and T. Islam, "Model of personal attitudes towards transit service quality," *Journal of Advanced Transportation*, vol. 45, no. 4, pp. 271–285, 2011.

[3] M. Eichler and C. F. Daganzo, "Bus lanes with intermittent priority: strategy formulae and an evaluation," *Transportation Research Part B: Methodological*, vol. 40, no. 9, pp. 731–744, 2006.

[4] W. J. Ma, K. L. Head, and Y. H. Feng, "Integrated optimization of transit priority operation at isolated intersections: a person-capacity-based approach," *Transportation Research Part C: Emerging Technologies*, vol. 40, pp. 49–62, 2014.

[5] M. Mesbah, M. Sarvi, and G. Currie, "New methodology for optimizing transit priority at the network level," *Transportation Research Record*, vol. 2089, no. 12, pp. 93–100, 2008.

[6] S. I. Guler and M. J. Cassidy, "Strategies for sharing bottleneck capacity among buses and cars," *Transportation Research Part B: Methodological*, vol. 46, no. 10, pp. 1334–1345, 2012.

[7] A. H. Alomari, H. Al-Deek, A. Sandt, J. H. Rogers, and O.

Hussain, "Regional evaluation of bus rapid transit with and without transit signal priority," *Transportation Research Record*, vol. 2554, no. 6, pp. 46–59, 2016.

[8] K. N. Balke, C. L. Dudek, and T. Urbanik II, "Development and evaluation of intelligent bus priority concept," *Transportation Research Record*, vol. 1727, no. 1543, pp. 12–19, 2000.

[9] A. Skabardonis, "Control strategies for transit priority," *Transportation Research Record*, vol. 1727, no. 1584, pp. 20–26, 2000.

[10] M. Estrada, C. Trapote, M. Roca-Riu, and F. Robusté, "Improving bus travel times with passive traffic signal coordination," *Transportation Research Record*, vol. 2111, no. 9, pp. 68–75, 2009.

[11] J. Lee, A. Shalaby, J. Greenough, M. Bowie, and S. Hung, "Advanced transit signal priority control with online microsimulation-based transit prediction model," *Transportation Research Record*, vol. 1925, no. 1, pp. 185–194, 2005.

[12] V. Ngan, T. Sayed, and A. Abdelfatah, "Impacts of various parameters on transit signal priority effectiveness," *Journal of Public Transportation*, vol. 7, no. 3, pp. 71–93, 2004.

[13] J. Jacobson and Y. Sheffi, "Analytical model of traffic delays under bus signal preemption: theory and application," *Transportation Research Part B: Methodological*, vol. 15, no. 2, pp. 127–138, 1981.

[14] P. G. Furth and T. H. J. Muller, "Conditional bus priority at signalized intersections: better service with less traffic disruption," *Transportation Research Record*, vol. 1731, no. 1, pp. 23–30, 2000.

[15] Z. R. Abdy and B. R. Hellinga, "Analytical method for estimating the impact of transit signal priority on vehicle delay," *Journal of Transportation Engineering*, vol. 137, no. 8, pp. 589–600, 2011.

[16] K. Ling and A. Shalaby, "Automated transit headway control via adaptive signal priority," *Journal of Advanced Transportation*, vol. 38, no. 1, pp. 45–67, 2003.

[17] E. Christofa and A. Skabardonis, "Traffic signal optimization with application of transit signal priority to an isolated intersection," *Transportation Research Record*, vol. 2259, no. 1, pp. 192–201, 2011.

[18] L. Head, D. Gettman, and W. Zhiping, "Decision model for priority control of traffic signals," *Transportation Research Record*, vol. 1978, no. 1, pp. 169–177, 2006.

[19] W. Ma, X. Yang, and Y. Liu, "Development and evaluation of a coordinated and conditional bus priority approach," *Transportation Research Record*, vol. 2145, no. 6, pp. 49–58, 2010.

[20] E. Christofa, K. Ampountolas, and A. Skabardonis, "Arterial traffic signal optimization: a person-based approach," *Transportation Research Part C: Emerging Technologies*, vol. 66, pp. 27–47, 2016.

[21] W. Ma, Y. Liu, and X. Yang, "A dynamic programming approach for optimal signal priority control upon multiple high-frequency bus requests," *Journal of Intelligent Transportation Systems: Technology, Planning, and Operations*, vol. 17, no. 4, pp. 282–293, 2013.

[22] J. Hu, B. B. Park, and Y.-J. Lee, "Coordinated transit signal priority supporting transit progression under connected vehicle technology," *Transportation Research Part C: Emerging Technologies*, vol. 55, pp. 393–408, 2015.

[23] M. Li, Y. Yin, W.-B. Zhang, K. Zhou, and H. Nakamura, "Modeling and implementation of adaptive transit signal priority on actuated control systems," *Computer-Aided Civil and Infrastructure Engineering*, vol. 26, no. 4, pp. 270–284, 2011.

[24] Q. He, K. L. Head, and J. Ding, "Multi-modal traffic signal control with priority, signal actuation and coordination," *Transportation Research Part C: Emerging Technologies*, vol. 46, pp. 65–82, 2014.

[25] Y. J. Lin, X. F. Yang, G.-L. Chang, and N. Zou, "Transit priority strategies for multiple routes under headway-based operations," *Transportation Research Record*, vol. 2356, no. 5, pp. 34–43, 2013.

[26] S. Z. Altun and P. G. Furth, "Scheduling buses to take advantage of transit signal priority," *Transportation Research Record*, vol. 2111, no. 7, pp. 50–59, 2009.

[27] S. I. Guler, V. V. Gayah, and M. Menendez, "Bus priority at signalized intersections with single-lane approaches: A novel pre-signal strategy," *Transportation Research Part C: Emerging Technologies*, vol. 63, pp. 51–70, 2016.

[28] J. Stevanovic, A. Stevanovic, P. T. Martin, and T. Bauer, "Stochastic optimization of traffic control and transit priority settings in VISSIM," *Transportation Research Part C: Emerging Technologies*, vol. 16, no. 3, pp. 332–349, 2008.

[29] W. Ma, Y. Liu, and B. Han, "A rule-based model for integrated operation of bus priority signal timings and traveling speed," *Journal of Advanced Transportation*, vol. 47, no. 3, pp. 369–383, 2013.

[30] F. Dion and B. Hellinga, "A rule-based real-time traffic responsive signal control system with transit priority: application to an isolated intersection," *Transportation Research Part B: Methodological*, vol. 36, no. 4, pp. 325–343, 2002.

[31] M. Janos and P. G. Furth, "Bus priority with highly interruptible traffic signal control: simulation of San Juan's Avenida Ponce de Leon," *Transportation Research Record*, vol. 1811, no. 1, pp. 157–165, 2002.

[32] J. Hu, B. Park, and A. E. Parkany, "Transit signal priority with connected vehicle technology," *Transportation Research Record*, vol. 2418, no. 3, pp. 20–29, 2014.

[33] J. Li, W. Wang, H. J. van Zuylen, N. N. Sze, X. W. Chen, and H. Wang, "Predictive strategy for transit signal priority at fixed-time signalized intersections case study in Nanjing, China," *Transportation Research Record*, vol. 2311, no. 12, pp. 124–131, 2012.

[34] Transportation Research Board, "Highway capacity manual 2010. Washington D C: National Research Council," 2010.

10

Evaluation of the Rain Effects on Gap Acceptance Behavior at Roundabouts by a Logit Model

Dongmin Lee,[1] Sooncheon Hwang,[2] Eunhan Ka,[3] and Chungwon Lee[4]

[1]*Associate Professor, Department of Transportation Engineering, University of Seoul, 163 Seoulsiripdae-ro, Dongdaemun-gu, Seoul 02504, Republic of Korea*
[2]*Graduate Student, Department of Transportation Engineering, The University of Seoul, Korea Seoulsiripdaero 163, Dongdaemun-gu, Seoul 02504, Republic of Korea*
[3]*Researcher, Institute of Construction and Environmental Engineering, Seoul National University, 1 Gwanak-ro, Gwanak-gu, Seoul 08826, Republic of Korea*
[4]*Associate Professor, Department of Civil and Environmental Engineering, Seoul National University, Korea 1 Gwanak-ro, Gwanak-gu, Seoul 151-744, Republic of Korea*

Correspondence should be addressed to Chungwon Lee; chungwon@snu.ac.kr

Academic Editor: Raffaele Mauro

A roundabout is generally known as an efficient, safe, and environmentally friendly intersection. Since 2010, the Korea government has taken the lead in constructing roundabouts as part of a special project. During that time, many ideas have been put forward to improve the safety, operation, and design of such roundabouts. In terms of improvements, it is particularly important to understand roundabout gap acceptance behavior. As such, we investigated gap acceptance behaviors at four roundabouts based on field observation during both good weather and rainy conditions. Based on the observed data, roundabout critical gaps were estimated, and a logit model for gap acceptance using various roundabout variables was developed to investigate gap acceptance maneuvering at roundabouts. A total of 2,421 data events for gap acceptance were collected from the field observation. Out of these events, 64.6% of drivers (1,564 drivers) accepted the given gaps and 35.4% of drivers (857 drivers) rejected them. The values for critical gaps were estimated using several different estimation methods and ranged from 3.3 to 4.7 seconds. The model was developed using four variables including gap size, type of circulating vehicle, traffic volume at the circulating lane, and weather conditions. The developed model shows that a longer gap results in a 3.669 times higher probability of entering roundabouts when the gap is sufficiently great for acceptance than when the gap is smaller. The effects of other variables, such as circulating vehicle types, circulating traffic volume, and weather conditions, are relatively lower than that of a gap size. Rain conditions influenced gap acceptance maneuvering around a roundabout. Drivers need about a 10 percent longer gap to accept entry into roundabouts during rainy conditions, and gap acceptance probabilities are 10 to 20 percent lower for the same given gap time during rainy conditions compared to good weather conditions.

1. Introduction

A roundabout is generally defined as a form of a circular intersection in which traffic travels counterclockwise around a central island and in which entering traffic must yield to circulating traffic [1]. Roundabouts are generally considered an efficient intersection system to reduce intersection travel time, a safe intersection design to decrease the number of accidents and damage from accidents, and environmentally friendly. As such, many countries have incorporated roundabouts in their road designs, including Korea. Since 2010, roundabouts have been constructed in Korea as part of a special government-led project. After 87 roundabouts were constructed in 2010, about 20 to 100 roundabouts are constructed every year. There have been improvements in safety and efficiency benefits due to such construction. However, there have also been many issues and ideas put forward to improve those roundabouts in terms of safety, operation, and

TABLE 1: Mean crash reductions of conversion to roundabouts.

Country	Mean Reductions (%)	
	All Crashes	Injury Crashes
Australia	41-61%	45-87%
Frances	-	57-78%
Germany	36%	-
Netherlands	47%	-
United Kingdom	-	25-39%
United States	35%	76%

Source: [1].

design. Of particular importance for design considerations is an understanding of roundabout gap acceptance behavior.

Although many studies regarding safety and efficiency effects have investigated roundabouts, there were few empirical studies investigating driver behavior at roundabouts. In roundabouts, traffic flows move based on a yield mechanism of the right-of-way principle that circulating traffic has a right-of-way compared to entering traffic. Entering traffic have to yield to circulating traffic inside roundabouts. Due to this roundabout operation principle, the availability of entering into roundabouts, which is a critical element that influences capacity, is determined by considering the gap size of circulating traffic inside the roundabouts. As such, gap acceptance is very significant in roundabouts. In the US Highway Capacity Manual (HCM), the most important elements are the critical gap in circulating traffic and follow-up time of entering traffic [2].

Therefore, our study investigated gap acceptance behaviors at roundabouts based on field observation. The purpose of this study is to estimate the critical gap for roundabouts by using various estimation methods. We also develop a logit model for gap acceptance using various influencing factors at roundabouts to investigate gap acceptance maneuvers. The gap acceptance maneuver at roundabouts is generally affected by weather conditions including rain and snow. Thus, we developed a logit model for gap acceptance at roundabouts that included weather variables.

2. Literature Review

2.1. Roundabouts. The first roundabout was a circle-shaped intersection built in 1905 at Columbus Circle in New York City [1]. This circle shape intersection was called a traffic circle and was designed to enable high-speed entry with priority and slow circulation that yielded to entering traffic. Before traffic volumes significantly increased, traffic circles worked well. However, as traffic volumes increased, traffic circles started to have safety and efficiency problems. Rectifying these problems led to the development of the modern roundabout in the 1960s. The United Kingdom imposed a new rule that required entering traffic to yield and give way to circulating traffic. This rule prohibited vehicles from entering the intersection as incoming traffic until there were sufficient gaps in the circulating traffic. This approach improved the safety and efficiency of circular intersections. Due to the advantages of roundabouts, many countries adopted modern roundabouts as a common circular intersection and have developed extensive design guides and methods to evaluate operational performance. Many studies have evaluated the effects of roundabout adaptation, typically accident reduction effects. In the United States, conversion to a roundabout results in a reduction of 35% and 76% in total and injury crashes, respectively. These accident reduction effects are similar to results from studies in other countries, as shown in Table 1.

Meanwhile, there have been recently many studies to overcome disadvantages of standard roundabouts in particular actual circumstance. And they suggested to implicate some alternative types of roundabouts such as a Turbo-roundabout, an Assembly roundabout, a traffic signal controlled roundabout, a Dog-bone roundabout, and other new types of roundabouts [3]. Out of these alternative types of roundabouts, a Turbo-roundabout was developed and installed at the end of the 1990s in The Netherlands. The Turbo-roundabout is a type of roundabouts to improve two-lane roundabouts in safety and capacity through eliminating the necessity of weaving and conflicts by the divided curbs [4–6].

2.2. Critical Gap and Follow-Up Time in Roundabouts. The capacity of a roundabout is determined by three traffic flows around the roundabouts: entering traffic, circulating traffic, and exiting traffic. The capacity of a roundabout decreases as the circulating traffic flow increases. Maneuvering is similar to the effect of a right-turning stream [2]. In the process of estimating capacity, there are two main measures: critical gap in circulating traffic and follow-up time for entry traffic. The critical gap from circulating traffic is the minimum time gap that an entering driver would accept to merge into the circulating lane of the roundabout. A driver rejects any gaps less than the critical gap and accepts any gap that is greater. In US studies, critical gap and follow-up time were 4.5 to 5.1 seconds and 3.2 to 3.4 seconds, respectively, in single roundabouts [7]. In Korea, a field experiment was also conducted to estimate critical gap and follow-up time in roundabouts and found them, estimated using Wu's method, to be 5.6 seconds and 2.3 seconds, respectively [8]. The critical gap and the follow-up time measured from the experiment were relatively greater and smaller than the results in other countries.

2.3. Previous Studies of Gap Acceptance Behavior. Due to the importance of gap acceptance maneuvers in unsignalized intersections and roundabouts, methods have been developed to estimate critical gap, including the Raff method, the Ashworth method, the Troubeck method, and the Wu method. The Raff method is a critical gap estimation method using two cumulative distributions of accepted and rejected gaps [9], and the Ashworth method is based on the hypothesis that average critical gap may be evaluated from the mean of accepted gaps [10]. Meanwhile, the Troubeck method estimates critical gap estimation using a maximum likelihood estimation [11] and is applied with an assumption that the distribution of critical gaps follows a log-normal distribution and calculates the probability that the critical gap would be between the largest rejected and accepted gaps. The Wu method was developed recently and estimates critical gaps based on an equilibrium of a probability concept [12].

These critical gap estimation methods assume, however, that drivers' behavior remains consistent and all drivers are homogeneous in a deterministic approach. As is well known, a deterministic gap acceptance estimation can estimate only the mean critical gap acceptance from the population, and careful drivers may be overrepresented as a result of information loss [13]. To overcome these limitations, a discrete choice model is used to estimate gap acceptance as probability and only the mean value of gap acceptance based on the probability function for success in accepting a gap. Originally in 1981, Daganzo developed a discrete choice model to estimate gap acceptance based on this probability function. This discrete choice model produces the probability function for success in accepting a gap [14]. This discrete choice model to estimate gap acceptance can be formularized based on the distribution of the mean and variance of the gap acceptance function parameter [15]. Mahmassani and Sheffi used a probit model to investigate gap acceptance behavior. They found that the critical gap decreases as long as drivers are waiting for an acceptable gap [16]. The effects of waiting time on gap acceptance were explained by Polus and his research team [17]. A logit model is one of the currently more popular models to explain gap acceptance behavior [18–25].

Weather conditions generally influence driver behaviors such as for speeding, braking maneuvers, gap acceptance behaviors, and other driving behaviors. A FHWA study showed that inclement weather influences such microscopic traffic behavior [26]. This study investigated changes in driver gap acceptance behavior on intersections during rainy weather conditions. They found that there is a more careful driving behavior for left-turn gap acceptance in the rain, and drivers need greater critical gaps. Zohdy and colleagues found that a larger value for the critical gap for a left turn is needed as the intensity of rain increases [18].

2.4. Studies of Gap Acceptance Maneuver in Roundabouts. Investigation of gap acceptance behavior in roundabouts has been conducted thru two approaches. Most of studies estimated critical gap and follow-up time, which are the two key parameters for gap acceptance. Other studies developed roundabout gap acceptance behavior models and there are many different values for critical and follow-up time for roundabouts, which are summarized in Table 2. Xu and Tian investigated gap acceptance characteristics at roundabouts in California, US [27], and the critical gaps were estimated in the range of 4.5 to 5.3 seconds at eight single roundabouts. They found that circulating flow rate and speed are two major influencing factors for critical gap and follow-up time.

Polus and his colleagues developed a disaggregate logit model to evaluate the waiting time effect on critical gaps at a roundabout [17] and included only two variables: waiting time and gap size. In their study, critical gaps at roundabouts were calculated in the range of 2 to 5 seconds and the waiting time may affect gaps in those ranges. However, gaps lower than 2 seconds were not likely to be accepted because the risk was considered too high.

2.5. Effects of Roundabout Construction in Korea. In Korea, roundabout construction has been led by federal agencies, typically the Presidential Council on National Competitiveness, the Ministry of Land, Transport and Maritime Affairs, the Ministry of Public Administration and Security, National Police Agency, and some research institutes. Currently, there are more than 500 roundabouts that have been constructed with federal government subsidies and with only local government funding. The roundabouts have been considered successful, and the construction of roundabouts continues to accelerate. Half of the roundabouts were constructed in urban areas and the others in rural areas. In most of these roundabouts, the number of crashes and the damage from the crashes were significantly reduced, and intersection traffic operational performance improved. Most residents were also satisfied with the conversion from signalized and unsignalized intersections to roundabouts. Overall, there has been an observed reduction of 46.1% in total crashes and 66.7% in fatality as shown in Table 3.

3. Methodology

3.1. Data Collection. Field observations for this study were conducted in Seoul, South Korea, at four typical roundabouts designed using the Korean roundabout design guide [15]. Preliminary observations were conducted at 19 roundabouts, and then four roundabouts were selected for study sites after a final consideration of final design layouts and traffic volume. The field observations were conducted via video recording on both sunny and rainy days. Rainy days selected for data collection were days with more than 5 mm of rainfall. Videos were recorded at time periods from 7:00 A.M. to 9:00 A.M. for morning peak time, from 12:00 P.M. to 2:00 P.M. for afternoon nonpeak time, and from 6:00 P.M. to 8:00 P.M. during evening peak time. Basic characteristics of the four intersections are summarized in Table 4. Data for gap acceptance behaviors were extracted from the recorded video images using video editing equipment to measure more detailed values of gaps.

TABLE 2: Critical and follow-up headways values in previous studies.

Study name	Country	Critical Gap (Sec.)	Follow-up headway (Sec.)
Abdullah Ahmad	India	2.5 ~ 3.0	-
Abrams et al.	US	2.20	-
Brilon	Germany	4.07 ~ 4.45	2.89 ~ 2.99
Cheng Jie	China	4.1 ~ 5.4	-
Dahl & Lee	Canada	3.90 ~ 5.30	2.10 ~ 4.20
Feng Xu	US	4.5 ~ 5.3	-
Fortuijn	The Netherlands	3.16 ~ 3.28	2.10
Gazzarri et al.	Italy	3.54 ~ 4.10	2.52 ~ 2.76
Guo Ruijun	China	2.7	-
HCM (2010)	US	4.1	-
Iraj Bargego	Iran	2.52 ~ 4.03	-
KHCM (2013)	South Korea	3.21	-
Kim, T	South Korea	2.42 ~ 2.7	-
Liang Ren	Australia	4.6 ~ 4.8	-
Mensah S. el al.	US	2.50 ~ 2.60	-
NCHRP (3-65)	US	4.2 ~ 5.9	-
Nicolosi et al.	Italy	3.19 ~ 3.99	3.15 ~ 2.11
Park, S	South Korea	3.7 ~ 4.1	-
Qu X et al.	Australia	-	-
Rodegerdts et al.	US	3.90 ~ 5.90	2.60 ~ 4.30
Shweta Rao	India	1.36 ~ 2.52	-
Vasconcelos et al.	Portugal	3.23 ~ 4.50	-
Vasconcelos et al.	Portugal	3.37 ~ 4.28	2.08 ~ 2.20
Wu	Germany	4.12	2.88
Xu & Tian	US	4.50 ~ 5.30	2.30 ~ 2.80
Zheng et al.	US	3.80 ~ 5.50	2.30 ~ 3.80

Sources: [27–37].

TABLE 3: Accident reduction effects of roundabout in Korea.

Before/After Roundabouts	Average Annual Crashes				
	Total	Fatality	Severe Injury	Injury	Minor Injury
Before	571	15	258	274	24
After	308	5	118	171	14
%	-46.1%	-66.7%	-54.3%	-37.6%	-41.7%

Source: Korean Roundabout Research Center (http://www.roundabout.or.kr/).

Figure 1 shows the maneuvering of gap acceptance at roundabouts according to the sequence of arriving vehicles at the entry lane and circulating vehicles in the circulatory lane. As shown in Figure 1, the difference in the times when circulating vehicles are passing the gap acceptance decision line can be explained as a gap. This decision line is an imaginary line that drivers at the entry lane use to decide whether to enter or not a circulatory lane while looking at oncoming circulatory vehicle traffic. The gap acceptance maneuver is preceded by the decision to accept or decline the circulating vehicle gap as adequate for entry.

This maneuver is illustrated in Figure 1. When entering vehicle #1 arrives at the yield line ① in the middle of the gap of the conflicting vehicles (#1 and #2), the entering vehicle using gap (a) is rejected because the gap (a) is insufficient to safely enter the roundabout as shown in Figure 1(a). In comparison, the gap between circulating vehicles #2 and #3 is sufficiently large that entering vehicle #1 accepts the gap and leaves the yield line on the entry lane to enter the roundabout. At this time, entering vehicle #2 is waiting behind vehicle #1 before vehicle #1 accepts the gap, and then it might leave immediately after the first vehicle starts to enter a roundabout using

Table 4: Data collection sites.

#	Intersection Name	Location	# of circulation lanes	# of legs	Diameter of inscribed circle (m)	Volume (v/h/lane)	
						On circulation	On entrance
#1	Jongro Roubdabout	Seoul	1	4	25	315	135
#2	Mansu-dong Roundabout	Incheon	1	3	22	195	75
#3	Dongducheon Roundabout	Dongducheon	1	4	28	425	210
#4	Pyongtaek Roundabout	Pyongtaek	1	4	28	370	180

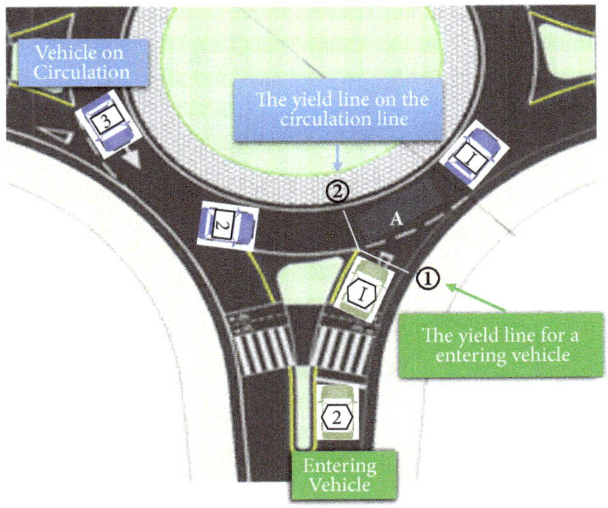

(a) An illustration of sequential gap acceptance maneuvers

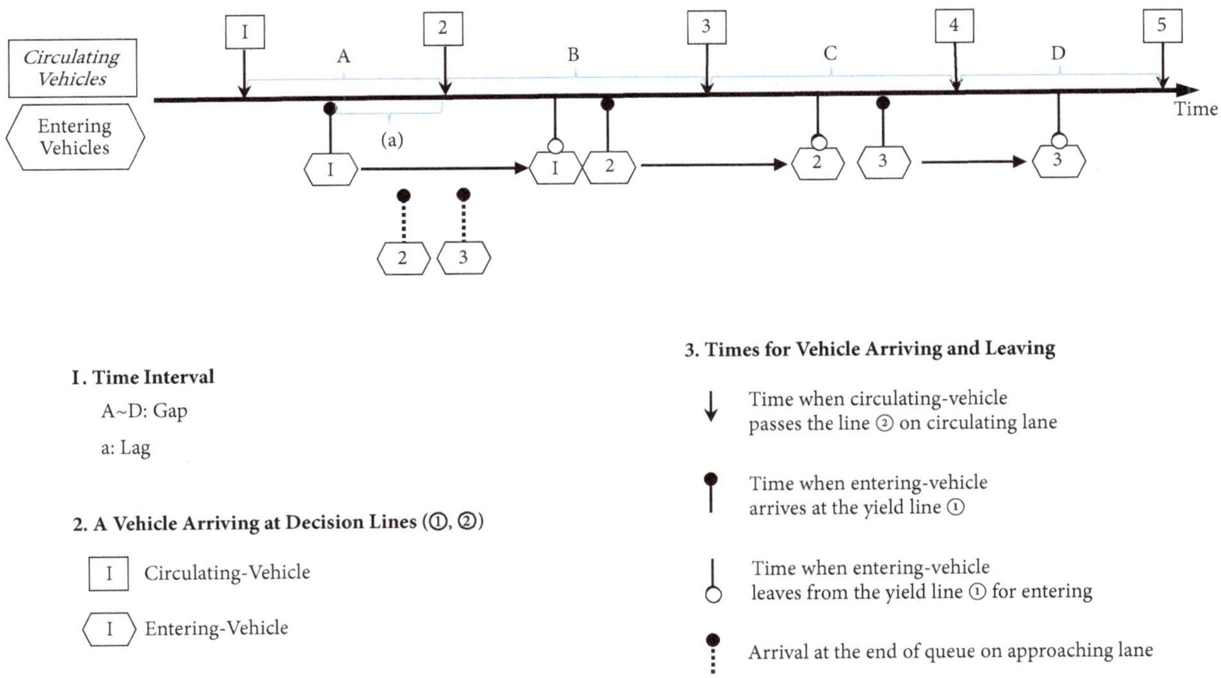

(b) The concept of gaps and lags based on time sequence

FIGURE 1: The concept of gap acceptance data extraction from time series maneuvers of roundabouts.

the gap between circulating vehicles #2 and #3. Figure 1(b) illustrates the concept of measured gaps and lags based on a time sequence.

3.2. Model Estimation. A gap acceptance maneuver is a binary problem having only two events: success or failure. A driver waiting at the entrance to a roundabout faces a series of binary choices. For each gap at the circulatory lane, the driver has to decide whether to accept it and enter the roundabout or reject it and wait for the next gap. This process continues until a sufficiently large-enough gap occurs, and the driver accepts the gap to enter the roundabout. Therefore, the probability of gap acceptance occurrence under given conditions has to be estimated as dependent variables, which can change due to driving environments instead of the event itself. Since this assists in analysis of the gap acceptance decision-making process and for simplicity and wide applicability, a logit model was selected to model estimation of acceptance probability. This study assumes that all gap acceptance decisions are made independently. For every moment, the driver has a utility from accepting or rejecting a confronting gap. The utility from accepting a gap results from avoiding having to wait

Table 5: Data summary and variable descriptive statistics.

Category		Normal	Rainy	Total
# of data		1,681	740	2,421
Gap (sec)	Mean	5.12	5.25	5.16
	S.D	2.81	2.97	2.86
	Acceptance (%)	1,135 (67.5%)	429 (58.0%)	1,564 (64.6%)
	Reject (%)	546 (32.5%)	311 (42.0%)	857 (35.4%)
Waiting Time (sec)	Mean	4.06	6.73	4.88
	S.D	5.59	8.99	6.92
Follow-up time (sec)	Mean	17.30	15.78	16.83
	S.D	18.62	15.90	17.84
# vehicles observed on Entrance	Mean	265.1	323.5	282.9
	S.D	84.50	102.05	94.13
	Passenger Car (%)	1,498 (89.1%)	678 (91.6%)	2,176 (89.9%)
	Heavy Vehicle (%)	183 (10.9%)	62 (8.4%)	245 (10.1%)
# vehicles observed on Circulation	Mean	374.5	502.6	413.6
	S.D	143.01	152.45	157.41
	Passenger Car (%)	1,466 (87.2%)	670 (90.5%)	2,136 (88.2%)
	Heavy Vehicle (%)	215 (12.8%)	70 (9.5%)	285 (11.8%)
Weather (%)		1,681 (69.4 %)	740 (30.6%)	-

longer to enter the roundabout, whereas the utility from rejecting a gap is the added safety from not accepting a short and dangerous gap. The traffic situation variables that affect these utilities and the decision whether to accept or reject a gap includes traffic volume, waiting time, types of circulating and entering vehicles, and other attributes.

Probability that a driver n accepts a gap:

$$P_n(\text{accept}) = P_n(U_{an} \geq U_{rn}) = \frac{1}{1 + e^{-(V_{an}-V_{rn})}} \quad (1)$$

$$U_{an} = V_{an} + \varepsilon_{an} \quad (2)$$

$$U_{rn} = V_{rn} + \varepsilon_{rn} \quad (3)$$

where U_{an} and U_{rn} are utility function of accept and reject, respectively; V_{an} and V_{rn} are the systematic components of the utility of accept and reject, respectively, and are assumed linear in their parameters; ε_{an} and ε_{rn} are the independent and identically Gumbel distributed random components of the utility of accept and reject, respectively.

3.3. Model Validation. Because the logit model in this study produces two outcomes, either "accept" or "reject" for each gap, two different errors type can occur: Type 1 and Type 2. A Type I error is committed if the model predicts that a driver rejects the gap when the driver actually accepts it. Meanwhile, a Type II error is committed if the model predicts that a driver accepts the gap when the driver actually rejects it. The validation for a logit model is generally conducted using the receiver operating characteristics (ROC) curve to test for these two types of errors. ROC curves can show model performance in predicting gap acceptance decisions [15, 38]. The ROC graph is a 2D graph having two axes that include a False Positive Rate (FPR) and True Positive Rate (TPR). If the ROC graph deducted from a model is closer to the point (FDR = 0, TPR = 1), it can be concluded that the developed model is more accurate and is validated. Statistical significance by a numerical value in the ROC graph method can be evaluated thru a C-statistic value. This C-statistic value means the area under the curve (AUC). If the C-statistic value is closer to 1, it means the ROC curve is close to the point (FPR = 0. TPR = 1), and means that it is a better model.

4. Results

4.1. Data Collection Results. The 2,421 events for gap acceptance were collected from the field observation. Out the total, 64.6% of drivers (1,564 drivers) accepted the given gaps and 35.4% of drivers (857 drivers) rejected it as shown in Table 5. The percentage of events for accepted gaps in rainy conditions was relatively smaller than that in good weather conditions at 58.0% and 67.5%, respectively. Data for 1,681 vehicles were observed under good weather conditions, with 740 vehicles during rainy conditions. About 90% of cars were classified as a passenger car under good and rain weather conditions. The traffic flow was not congested for the observation, and the average volumes for entering roundabouts and for circulating were 283 vehicles/hour and 414 vehicles/hour, respectively.

To test statistical significance of the difference of gap acceptance between good weather and rain weather conditions, K-S tests were conducted as can be seen in Table 6. This K-S test results show that there is statistical difference of gap distributions related weather conditions for both the accepted gap and rejected gap. From these results, it was found that the data analyzed in this study can explain sufficiently the weather's influence on gap acceptance behavior for both gap acceptance and rejection on roundabouts.

TABLE 6: K-S test results for effects of weather conditions (Rain).

Statistics	K-S test for accepted gaps	K-S test for rejected gaps
Z	2.091	1.911
P	< 0.001	0.001

TABLE 7: Critical gaps using various methods.

Location/Time		Raff (Sec.)		Wu (Sec.)	
		Normal	Rainy	Normal	Rainy
Total		3.75	4.25	3.84	4.23
Location	RB #1	4.05	4.05	4.14	4.11
	RB #2	3.80	3.75	3.76	3.59
	RB #3	3.70	4.05	3.80	4.12
	RB #4	3.85	4.65	3.33	4.69
Time	Morning Peak	3.55	-	3.77	-
	PM Non-peak	4.00	-	4.03	-
	Evening Peak	3.75	4.30	3.82	4.23

TABLE 8: A Logit model development.

Variables	β	S.E,	P	Exp (β)
X1 (Gap, Seconds)	1.300	0.059	< 0.001	3.669
X2 (Type of Circulating Vehicles, 0: Passenger Car, 1: Heavy Vehicle)	-0.437	0.217	0.045	0.646
X3 (Traffic Volume at the Circulating lane)	-0.003	0.000	< 0.001	0.997
X4 (Weather Condition, 0: Good weather, 1: Rainy Condition)	-0.815	0.170	< 0.001	0.443
Constant	-3.158	0.274	< 0.001	0.043
Number of Observations		2,181		
Log-likelihood Function		-635.530		
Restricted Log-likelihood Function		-1417.961		

4.2. Critical Gap Estimation. This study estimated the values of critical gaps using two different estimation methods, the Raff method and Wu method, as shown in Table 7. Critical gap estimation was conducted under normal weather and rainy conditions, as well as for specific data collection times, such as morning peak, P.M. nonpeak, and evening peak time. Critical gaps under rainy conditions were higher than during good weather. As such, rain conditions influenced gap acceptance maneuvering in a roundabout, and drivers needed about an 8 to 13 percent longer gap to accept entry into a roundabout under rainy conditions. These critical gap estimation results are similar to the critical gap times from previous studies in many different countries (Table 2).

4.3. Development of a Logit Model for Gap Acceptance at Roundabouts. A logit model with various independent variables investigates the effects of the factors on gap acceptance at roundabouts. The model parameter coefficients were estimated using the maximum likelihood method. Before developing the model, the dataset was separated into two groups, one for model development and the other for model validation through a random selection process. Out of the total data, about 90 percent of data (2,181 data) were used for the model development and other data (240 data) were used for the model validation.

In order to develop the gap acceptance model using a logit model formula, eight variables were originally reviewed, and the variables that are significantly correlated with gap acceptance at a 95% significance level were selected. The model was finally developed using four variables including a gap size, a type of circulating vehicle, traffic volume at the circulating lane, and weather conditions, and the model development results are shown in Table 8. The results were as expected, where the utility from accepting a gap increases as the gap increases due to a decrease of the risk in entering roundabouts. The developed model shows that a longer gap results in a 3.669 times higher probability of entering roundabouts when the gap is sufficiently great for acceptance than

TABLE 9: Validation results using the ROC graph method.

C-Statistic	S.D.	P	95% Confidence Interval	
			Min	Max
0.936	0.016	< 0.001	0.904	0.968

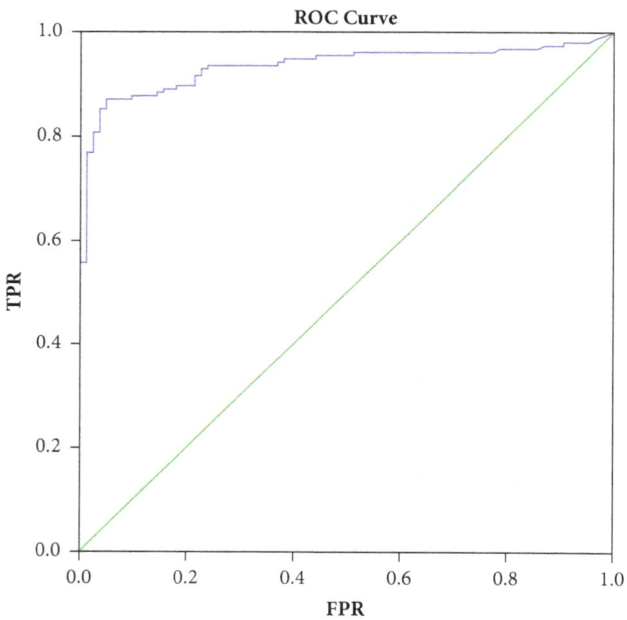

(a)

FIGURE 2: Deducted ROC graph from the model.

when the gap is smaller. When heavy vehicles are circulated inside roundabouts, drivers have a 0.646 times more difficult time accepting gaps to enter the roundabout compared to when there were only passenger cars inside the roundabouts. When the circulating traffic volume is lower, the probability of gap acceptance increases. Lastly, drivers need a longer gap time to accept when it is raining due to a more conservative driving behavior. In the developed model, there was a 0.443 times lower probability of entering roundabouts when it is raining compared to good weather conditions. Through this model development, the limitation of critical gap estimation that the estimation assumes homogenous driving behaviors for all drivers which is unlikely in real situations can be overcome.

This research used 240 data events out of a total 2,421 data events to validate the developed model. The deducted ROC graph is closer to the point (FPR = 0. TPR = 1) as shown in Figure 2 and Table 9. The C-statistic value is 0.936 and validates that the results from the developed model are statistically significant and accurate.

Figure 3 shows the estimated probabilities and elasticity of accepting a gap as a function of the gap size for good weather and rain conditions. Figure 3(a) shows the effects of rain conditions on drivers' gap acceptance probabilities. Drivers need about a 10 percent longer gap to accept entry into roundabouts under rainy conditions, and their probabilities of gap acceptance are 10 to 20 percent lower with the same given gap time under rainy conditions than good weather conditions. In the case of three-second gaps, there was a 20% lower probability in a rainy condition compared to good weather conditions. The elasticity of the logit model represents the responsiveness of an individual's choice probability in regard to a change in the value of some attribute [17]. As Figure 3(b) explains, when the given gaps are less than five seconds, drivers' gap acceptance probabilities in rain conditions did not change more than the change of gap time in good weather conditions because drivers are not willing to enter roundabouts due to riskier conditions during rain. Meanwhile, when the given gaps are greater than five seconds, drivers' acceptance probabilities in rain conditions changed more with a change in gap time.

5. Conclusions

A roundabout is generally defined as a form of a circular intersection in which traffic travels counterclockwise around a central island and in which entering traffic must yield to circulating traffic. This roundabout is generally known as an efficient, safe, and environmentally friendly intersection. In Korea, roundabouts have been constructed as a government-led special project since 2010, and many issues and improvements for roundabouts have been suggested in terms of safety, operation, and design. Of particular importance is understanding gap acceptance behavior in roundabouts.

Gap acceptance behaviors at a roundabout were investigated based on field observations at four roundabouts during both good weather and rainy conditions. Based on the observed data, roundabout critical gaps were estimated, and a logit model for gap acceptance was developed using various variables to investigate gap acceptance maneuvering at roundabouts. Data for 2,421 gap acceptance events were collected from the field observation. Out these events, 64.6% of drivers (1,564 drivers) accepted the given gaps and 35.4% of drivers (857 drivers) rejected them. Critical gap values were estimated using the different estimation methods and ranged from 3.3 to 4.7. These critical gap estimation results are similar to the critical gap times from previous studies in many different countries (Table 3). The model was developed using four variables: gap size, type of circulating vehicle, traffic volume at the circulating lane, and weather conditions. Model validation was conducted using 240 events out of the 2,421 events, which were not used in the model development. For this model validation, the receiver operating characteristics (ROC) curve was tested for two types of errors that occur in binary choice models. This validation process found that the results from the developed model are statistically significant and accurate. This gap acceptance probability model enables overcoming weakness of critical gap estimation that the critical gap estimation assumes homogenous driving behaviors for all drivers unlikely in real situations.

The developed model showed that a longer gap results in a 3.669 times higher probability of entering roundabouts when the gap is sufficiently great for acceptance than when the gap is smaller. The effects of other variables, such as circulating vehicle types, circulating traffic volume, and weather

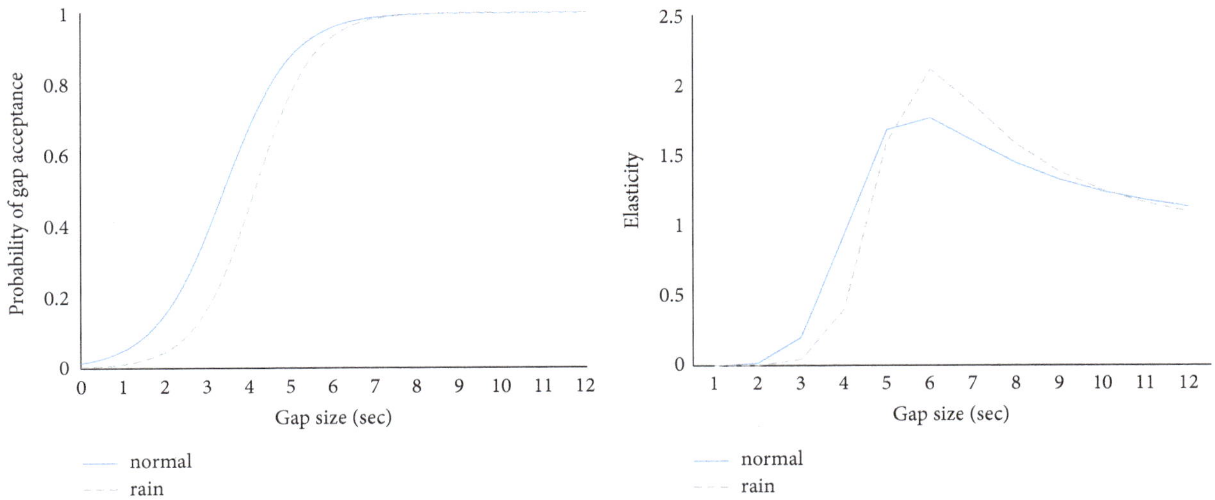

(a) Comparison of gap acceptance probabilities between good weather and rain conditions

(b) Elastics of gap acceptance between good weather and rain conditions

Figure 3: Evaluation of the rain condition effects on gap acceptance behavior.

conditions, are relatively lower than that of a gap size. When heavy vehicles are circulated inside roundabouts, drivers had difficulties reflected in a 0.646 times higher value for accepting gaps for entering roundabouts compared to when there were only passenger cars inside the roundabouts. When there is less circulating traffic, the probability of gap acceptance increases. Lastly, drivers need a longer gap acceptance time when it is raining due to a more conservative driving pattern. The developed model gave a 0.443 times lower probability of entering roundabouts when it is raining compared to good weather conditions. Thus, rain conditions influenced gap acceptance maneuvering into a roundabout. Drivers need about a 10 percent longer gap to accept entering into roundabouts under rainy conditions, and the probabilities of gap acceptance are 10 to 20 percent lower given the same given gap time under rainy conditions than good weather conditions.

This study investigated gap acceptance behaviors at roundabouts by analyzing and estimating critical gaps and the developed a gap acceptance logit model for good weather and rainy conditions. Waiting time was not considered as a statistically significant influence factor based on the analysis results in this study, even though previous studies indicated that waiting time affected gap acceptance behaviors.

There are some limitations to this study, and more field data should be collected as roundabout driving behavior and attitudes mature in Korean. There are still many problems when driving thru roundabouts in Korea due to violations of the right-of-way rule. For more sophisticated weather effect analysis, other types of severe weather conditions such as snow and heavy fog should be considered. Lastly, precipitation may also be an interesting influence factor for gap acceptance behavior analysis.

Conflicts of Interest

The authors declare that they have no conflicts of interest.

Acknowledgments

This research was partially supported by a Grant [no. 17TLRP-B085437-04] from Transportation & Logistics Research Program funded by Ministry of Land, Infrastructure and Transport of Korean government. This research was partially supported by Institute of Construction and Environmental Engineering at Seoul National University. The authors wish to express their gratitude for the support.

References

[1] Transportation Research Board, *NCHRP Report 672, Roundabouts: An Informational Guide*, 2nd edition, 2010.

[2] Transportation Research Board, *Highway Capacity Manual*, National Research Council, Washington, DC, USA, 2010.

[3] T. Tollazzi, *Alternative Types of Roundabouts: An Informational Guide*, Springer International Publishing, 2015.

[4] W. Brilon, L. Bondzio, and F. Weiser, "Experiences with turbo roundabouts in Germany," in *Proceedings of the National Roundabout Conference*, pp. 546–580, Kansas City, MO, USA, 2008.

[5] M. Guerrieri, R. Mauro et al., "Analysis of kinematic parameters and driver behaviour at turbo-roundabouts," *Journal of transportation engineering, Part A. Systems*, vol. 144, no. 6, 2018.

[6] E. Macioszek, "Analysis of significance of differences between psychotechnical parameters for drivers at the entries to one-lane and turbo roundabouts in Poland," in *Proceedings of the International Conference Transport System Theory and Practice*, vol. 505 of *Intelligent Transport Systems and Travel Behaviour*, Springer International Publishing, Katowice, Poland, 2017.

[7] L. Rodegerdts, M. Blogg et al., "Roundabout in the United States," Tech. Rep. NCHRP Report 572, Transportation Research Board of the National Academics, Washington, DC, USA, 2007.

[8] D. Lee and J. You, "A study on appropriate traffic volume calculation for revitalizing roundabout installation," *Journal of Korean Society of Transportation*, vol. 31, no. 6, pp. 43–52, 2013.

[9] M. S. Raff and J. W. Hart, "A Volume Warrant for Urban Stop Signs," The Eno Foundation for Highway Traffic Control, 1950.

[10] R. Ashworth, "The capacity of priority-type intersections with a non-uniform distribution of critical acceptance gaps," *Transportation Research*, vol. 3, no. 2, pp. 273–278, 1969.

[11] R. J. Troutbeck, "Evaluating the performance of a roundabout," *Australian Road Research Board, ARRB Group Limited*, SR no. 45, 1989.

[12] N. Wu, "Equilibrium of probabilities for estimating distribution function of critical gaps at unsignalized intersections," *Transportation Research Record, Journal of the Transportation Research Board*, no. 2286, pp. 49–55, 2012.

[13] A. J. Miller, "Nine estimators of gap acceptance parameters," in *Proceedings of the International Symposium on the Theory of Traffic Flow and Transportation*, G. Newell, Ed., American Elsevier Publishing, Berkeley, Calif, USA, 1972.

[14] C. F. Daganzo, "Estimation of gap acceptance parameters within and across the population from direct roadside observation," *Transportation Research Part B: Methodological*, vol. 15, no. 1, pp. 1–15, 1981.

[15] X. Chen, Y. Qi, and G. Liu, "Empirical study of gap-acceptance behavior of right-turn-on-red drivers on dual right-turn lanes," *Journal of Transportation Engineering*, vol. 139, no. 2, pp. 173–180, 2013.

[16] H. Mahmassani and Y. Sheffi, "Using gap sequences to estimate gap acceptance functions," *Transportation Research Part B: Methodological*, vol. 3, no. 15, pp. 143–148, 1981.

[17] A. Polus, Y. Shiftan, and S. Shmueli-Lazar, "Evaluation of the waiting-time effect on critical gaps at roundabouts by a logit model," *European Journal of Transport and Infrastructure Research*, vol. 5, no. 1, pp. 1–12, 2005.

[18] I. Zohdy, S. Sadek, and H. A. Rakha, "Empirical analysis of effects of wait time and rain intensity on driver left-turn gap acceptance behavior," *Transportation Research Board: Transportation Research Record*, no. 2173, pp. 1–10, 2010.

[19] T. H. Maze, "A probabilistic model of gap acceptance behavior," *Transportation Research Board: Transportation Research Record*, no. 795, pp. 8–13, 1981.

[20] S. M. Madanat, M. J. Cassidy, and M.-H. Wang, "Probabilistic delay model at stop controlled intersection," *Journal of Transportation Engineering*, vol. 120, no. 1, pp. 21–36, 1994.

[21] J. L. Gattis and S. T. Low, "Gap acceptance at atypical stop-controlled intersections," *Journal of Transportation Engineering*, vol. 125, no. 3, pp. 201–207, 1999.

[22] P. Liu, X. Wang, J. Lu, and G. Sokolow, "Headway acceptance characteristics of U-turning vehicles at unsignalized intersections," *Transportation Research Board: Transportation Research Record*, no. 2027, pp. 52–57, 2007.

[23] P. C. Devarasetty, Y. Zhang, and K. Fitzpatrick, "Differentiating between left-turn gap and lag acceptance at unsignalized intersections as a function of the site characteristics," *Journal of Transportation Engineering, ASCE*, vol. 138, no. 5, pp. 580–588, 2012.

[24] S. Teply, M. I. Abou-Henidy, and J. D. Hunt, "Gap acceptance behavior-aggregate and logit perspectives: part 1," *Traffic Engineering and Control*, vol. 38, no. 9, pp. 474–482, 1997.

[25] S. Teply, M. I. Abou-Henidy, and J. D. Hunt, "Gap acceptance behavior-aggregate and logit perspectives: part 2," *Traffic Engineering and Control*, vol. 38, no. 10, pp. 540–544, 1997.

[26] H. Rakha, K. Daniel, C. Gustave et al., *Microscopic Analysis of Traffic Flow in Inclement Weather*, FHWA-JPO-09-066, U.S. Department of Transportation, Research and Innovative Technology Administration, 2009.

[27] F. Xu and Z. Z. Tian, "Driver behavior and gap-acceptance characteristics at roundabouts in California," *Transportation Research Record: Journal of the Transportation Research Board*, vol. 2071, no. 1, pp. 117–124, 2008.

[28] T. Kim, M. Park, and B. Park, "A critical gap model for roundabouts in Korea," *Journal of Korean Society of Transportation*, vol. 30, no. 2, pp. 93–100, 2012.

[29] S. Park, D. Kim, and J. Jeong, "Evaluation of multi-legged roundabout using surveyed critical gap acceptance," *The Journal of the Korea Contents Association*, vol. 13, no. 9, pp. 400–409, 2013.

[30] A. Ahmad, R. Rastogi, and S. Chandra, "Estimation of critical gap on a roundabout by minimizing the sum of absolute difference in accepted gap data," *Canadian Journal of Civil Engineering*, vol. 42, no. 12, pp. 1011–1018, 2015.

[31] J. Cheng, X. Yang, W. Deng, and X. Huang, "Driver's critical gap calibration at urban roundabouts: a case study in China," *Tsinghua Science and Technology*, vol. 13, no. 2, pp. 237–242, 2008.

[32] R. Guo and B. Lin, "Traffic operation performances at roundabout weaving sections," *Journal of Transportation Systems Engineering and Information Technology*, vol. 10, no. 3, pp. 29–34, 2010.

[33] I. Bargegol, S. H. Hosseini, and M. J. Samet, "Determining the capacity model of urban roundabouts, considering the drivers' behaviour in accepting and rejecting of gaps," in *Proceedings of the IOP Conference Series: Materials Science and Engineering*, vol. 254, June 2017.

[34] L. Ren, X. Qu, H. Guan et al., "Evaluation of roundabout capacity models: an empirical case study," *American Society of Civil Engineers: Journal of Transportation Engineering*, vol. 142, no. 12, 2016.

[35] S. Rao, Y. Krishna, P. Atmakuri et al., "Calibration of performance of roundabouts based on gap acceptance parameters using simulation for indian scenario," in *Proceedings of the 12th International Conference on Transportation Planning and Implementation Methodologies for Developing Countries (TPMDC '16)*, pp. 19–21, IIT Bombay, Mumbai, India, 2016.

[36] X. Shi and J. Zhu, "Modification of a gap acceptance theory model of roundabouts' capacities," in *Proceedings of the 16th COTA International Conference of Transportation Professionals*, pp. 1655–1662, Shanghai, China, July 2016.

[37] O. Giuffrè, A. Granà, and M. L. Tumminello, "Gap-accepteance parameters for roundabouts: a systematic review," *European Transport Research Review*, vol. 8, no. 2, pp. 1–20, 2016.

[38] T. Fawcett, "An introduction to ROC analysis," *Pattern Recognition Letters*, vol. 27, no. 8, pp. 861–874, 2006.

11

Design of Signal Timing Plan for Urban Signalized Networks including Left Turn Prohibition

Qinrui Tang[1,2] **and Bernhard Friedrich**[2]

[1]*Institute of Transportation Systems, German Aerospace Center, Berlin, Germany*
[2]*Institute of Transportation and Urban Engineering, Technische Universität Braunschweig, Braunschweig, Germany*

Correspondence should be addressed to Qinrui Tang; qinruitang@gmail.com

Academic Editor: Wai Yuen Szeto

Urban road networks may benefit from left turn prohibition at signalized intersections regarding capacity, for particular traffic demand patterns. The objective of this paper is to propose a method for minimizing the total travel time by prohibiting left turns at intersections. With the flows obtained from the stochastic user equilibrium model, we were able to derive the stage generation, stage sequence, cycle length, and the green durations using a stage-based method which can handle the case that stages are sharing movements. The final output is a list of the prohibited left turns in the network and a new signal timing plan for every intersection. The optimal list of prohibited left turns was found using a genetic algorithm, and a combination of several algorithms was employed for the signal timing plan. The results show that left turn prohibition may lead to travel time reduction. Therefore, when designing a signal timing plan, left turn prohibition should be considered on a par with other left turn treatment options.

1. Introduction

Left turns at signalized intersections may cause efficiency problems, because they have comparatively high potential for conflicts with other movements. When permitted left turns are used, the delay of vehicles is determined both by the traffic signal and by the opposing vehicles. As a result, vehicles turning left would usually wait longer than other movements. Protected left turns can be applied at the expense of increasing the intergreen time and decreasing the effective green time, resulting in efficiency problems. Therefore, to improve efficiency, left turn prohibition (LTP) is investigated in urban networks as a remedy for urban congestion problems.

Previous left turn treatment guidelines provide the treatment of permitted/protected left turns at signalized intersections [1–6], while very few refer to left turn prohibition. Although left turns can be avoided by constructing unconventional intersections such as U-turns in the middle of roads [7–12], mid-block left turns [13, 14], continuous flow intersections, jughandle intersections [15], exit lanes for left turn [16], special width intersection [17], and tandem intersection/presignals [18–20], due to space limitation in cities and high construction expense, these unconventional intersections may not be feasible in the urban area. The impact of these unconventional intersections on networks is also not clear because the intersections/locations are usually studied in isolation. Another alternative to left turn prohibition, which does not necessitate infrastructure alteration, would be to redesign the network. By redesigning a network in this way, flow redistribution can be obtained using traffic assignment models [21–24]. However, when the turning restriction problem is formulated as a pure network design problem, signal settings are ignored. Consequently, this is making it unsuitable for the use in the urban signalized networks. In these environments, it is necessary to design signal timing plan for the LTP problem.

The methods for signal timing plan design are the stage-based method [25–29], the group-based method [30–32], and the lane-based method [33–35]. The stage-based method, whose cycle is composed of stages, determines green durations of each stage by minimizing delays or other objectives. In practice, all existing methods for network signal setting design follow the stage-based method [29]. The most famous signal control method using the stage-based method

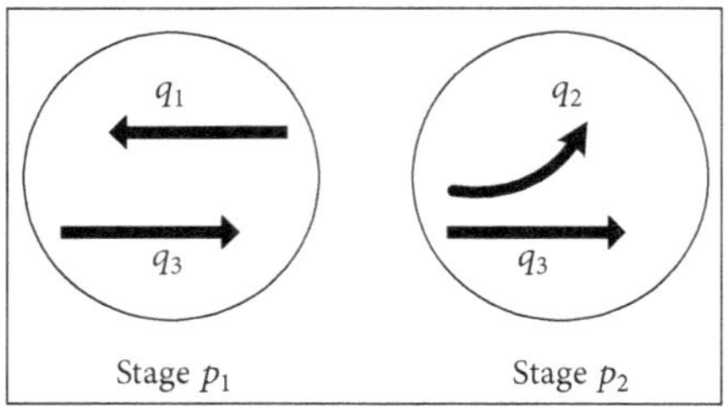

FIGURE 1: An example of two stages sharing movements [38].

is TRANSYT. However, the existing stage-based methods rarely include the stages sharing movements when generating stages and optimizing/calculating signal timing. The group-based method and the lane-based method have a similar formulation. The difference is that the lane-based method can also assign lane markings. The lane-based method was used in turning restriction problems [36] and was integrated with traffic assignment models [37], but both methods have not been practically applied. In consideration of the practical extension, the stage-based method is used in this paper. The research gap filled for the stage-based is to explain the treatment of the case that stages are sharing movements (see Figure 1).

This paper presents a method for selecting left turns with a genetic algorithm for prohibition and designing a fixed signal timing plan including LTP for a time in a day, by minimizing total travel times. After prohibiting left turns, the proposed method adjusts signal settings using the stage-based method. Once the selected left turns are prohibited, all stages are regenerated and then the stage sequence is optimized. Besides, cycle length and green duration are calculated. The intersections in networks are coordinated by the common cycle length, which is the largest cycle length of all intersections. Stochastic user equilibrium (SUE) is integrated for forecasting redistributed traffic flows, and it is used for the evaluation of the LTP. To test the proposed method, one should set the total travel time without LTP as a base case and compare it with the total travel time with LTP.

The following abbreviations are used in this paper: LTP represents left turn prohibition; SUE means stochastic user equilibrium; OD indicates Origin-Destination; TSP stands for traveling salesman problem.

2. Methodology

2.1. Overall. The methodology consists of three parts: selecting the prohibited left turns, signal timing plan design including permitted/protected/prohibited left turns for each intersection, and an SUE used for a realistic assignment of the demand.

The flowchart in Figure 2 presents the approach in a chronological sequence. LTP influences the network link flows. The link flows are estimated using SUE based on a link cost function, which does not consider signal timing. Then for each left turn that is not prohibited it is decided whether it is permitted or protected. Feasible stages are generated, and then the signal sequence is determined by minimizing the total intergreen time. In parallel, lane flows are calculated, and saturation flows are adjusted to determine signal timing. The SUE model is run once again using a cost function based on signal timing. This is performed to evaluate the total travel time with the LTP and the signal timing. Left turns are selected to be prohibited with genetic algorithms. These processes will be repeated until all selected left turns are tested.

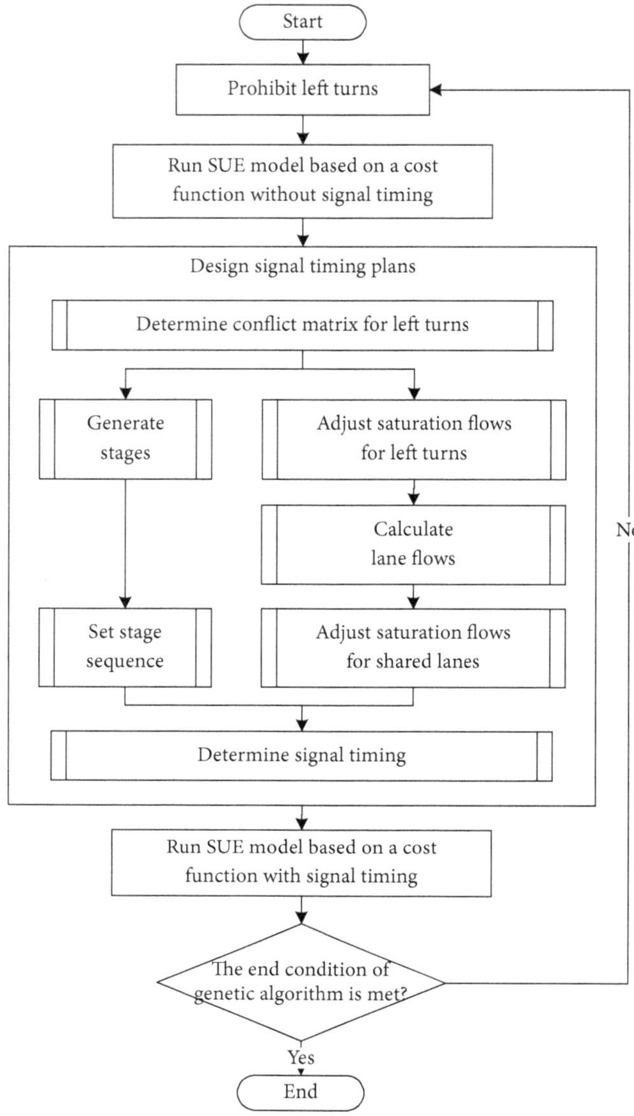

FIGURE 2: Flowchart of signal timing plan design including left turn prohibition.

The following sections explain each processing box and subprocessing step of the flowchart in detail.

2.2. Left Turn Prohibition. Each network includes many left turns. Each left turn has two possible states: it can be either allowed or prohibited. An LTP combination is defined as the decision regarding the state of each left turn in the network. Considering the number of left turns in the network, the total number of LTP combinations can be very large. Therefore, using the algorithm detailed in Section 3, we reduced the number of possible LTP combinations.

The overall objective is to minimize the total travel time in the network as described in (1). This equation sums the total travel time, separating left turn links from other links. The total travel time without LTP is the reference case. Only if the total travel time with LTP is smaller than the reference total travel time can the LTP combination be applied. Eventually, the LTP combination with the minimal total travel time is selected as the solution.

For each selected prohibited left turn, we assumed that the lane would be used for through movement. We also made sure that each selected left turn can be prohibited. Specifically, a check was performed that the prohibition of this turn would still allow network connectivity. Thus each OD path set includes at least one path (see (2)). Additionally, the number of through exit lanes is greater than or equal to the number of through approaching lanes (see (3)). If both of these constraints cannot be met, the left turn is not prohibited.

$$\min \quad TT = \sum_{a \in M/L} q_a t_a(q_a, \eta) + \sum_{a \in L} (1 - x_a) q_a t_a(q_a, \eta) \tag{1}$$

$$\text{subject to} \quad K_{od} \neq \emptyset, \quad \forall o \in O,\ d \in D \tag{2}$$

$$N_{\alpha,z,i} \leq N_{\epsilon,z,i}, \quad \forall z = 1, \ldots, N_Z,\ i = 1, \ldots, N_{A,z}, \tag{3}$$

where TT is total travel time (h); M is set of links; L is set of left turn links; q_a, $\forall a \in M$, is flow of link a (veh/h); $t_a(q_a, \eta)$, $\forall a \in M$, is travel time of link a (h); η is signal settings defined as stage, stage sequence, cycle length, and green durations; x_a, $\forall a \in L$, is indicator of prohibited left turns: $x_a = 1$, if the left turn is prohibited and $x_a = 0$, otherwise; O is origin set; D is destination set; K_{od} is path set between origin o and destination d; z is intersection index; N_Z is number of intersections in the network; i is arm index; $N_{A,z}$ is the number of arms at intersection z; $N_{\alpha,z,i}$ is the number of approaching lanes for through movements in arm i at intersection z; $N_{\epsilon,z,i}$ is the number of exit lanes for through movements in arm i at intersection z.

2.3. Stochastic User Equilibrium Model. A traffic assignment model describes how traffic flows are distributed in the network. In SUE, one of the traffic assignment models, drivers choose routes with minimal perceived travel time. The relevant link flows change according to the result of path choice and influence the travel time. Drivers iteratively react to updated travel time until any change of routes makes the perceived travel time longer.

Link flow calculation and travel time estimation are critical steps in SUE. Link flow is calculated by the product of OD demand and probabilities of paths being chosen. The probabilities are calculated with a simple logit model in which the utility function is path travel time. In this research, we applied the simple logit model with the same configuration as Tang and Friedrich [39].

Link travel times are initialized by applying free flow travel time, and then travel time is differently estimated before and after signal timing plan design. Before signal timing plan is designed, as no signal timing information is available, the link costs are updated by BPR function (see (4)). After signal timing plan is designed, the link travel time is updated by the sum of free flow travel time and link delay (see (5)). Link delay is the function of signal timing. Since during the first few iterations of SUE some of the links can be oversaturated, Akcelik's formula [40] is used. In Akcelik's formula, the delay consists of uniform delay and incremental delay which are represented as the first and the second term of (6), respectively.

$$t_a(q_a) = t_{0,a}\left(1 + \alpha\left(\frac{q_a}{Q_a}\right)^{\beta}\right) \tag{4}$$

$$t_a(q_a, \eta) = t_{0,a} + d_a(q_a, \eta), \tag{5}$$

where $a \in M$ is link index; $t_a(q_a)$ is travel time of link a; $t_{0,a}$ is free flow travel time of link a (h); Q_a is capacity of link a which is assigned as saturation flows of link a (veh/h); α, β are parameters in BPR function; $d_a(q_a, \eta)$ is delay of link a (h).

If link a is on lane k of arm i at intersection z, the link delay $d_a(q_a, \eta)$ is then equal to $d_{z,i,k}(q_{z,i,k}, \eta)$.

$$d_{z,i,k}(q_{z,i,k}, \eta) = \frac{0.5c(1 - g_{z,i,k}/c)^2}{1 - \min(1, x_{z,i,k}) g_{z,i,k}/c} + \min\left(1, \left\lfloor \frac{x_{z,i,k}}{x_{0,z,i,k}} \right\rfloor\right) 900T \left[(x_{z,i,k} - 1) + \sqrt{(x_{z,i,k} - 1)^2 + \frac{12(x_{z,i,k} - x_{0,z,i,k})}{Q_{z,i,k}T}}\right] \tag{6}$$

where k is lane index; $d_{z,i,k}(q_{z,i,k}, \eta)$ is delay of lane k (s); $q_{z,i,k}$ is total traffic flow in lane k (veh/h); c is common cycle length (s); $g_{z,i,k}$ is green duration of lane k (s); $Q_{z,i,k}$ is capacity of lane k (veh/h); $x_{z,i,k} = q_{z,i,k}/Q_{z,i,k}$ is degree of saturation flow of lane k; T is observation time period (h); $x_{0,z,i,k} = 0.67 + s_{z,i,k}g_{z,i,k}/600$ is turning point where the delay dramatically increases; $s_{z,i,k}$ is saturation flow of lane k (veh/s).

2.4. Signal Timing Plan Design. This section first introduces the method of conflict matrix determination according to left turn phasing types. With the conflict matrix, the stages can be generated and their sequences can be optimized. To estimate the delays, one should also calculate the lane flow and adjust the saturation flow. Finally the signal timing is calculated, especially for the stages sharing movements.

2.4.1. Determination of Conflict Matrix for Left Turns. Conflict matrix records conflicts between movements which cannot be in the same signal stage for safety reasons. If left turns are protected, they will not be on the same stage as opposing through movements; if they are permitted, they can be on the same stage as opposing through movements. Thus, types of left turn phasing affect conflict matrix and saturation flow of left turns. As the permitted left turn vehicles are interrupted by opposing through movements, the saturation flows decrease compared with the saturation flows of protected left turns.

To determine whether left turns need to be protected, one can apply recommendations from different guidelines. However, most guidelines focus on safety aspects [1, 4]. In our method, we apply the volume conditions. The volume condition refers to the left turn flows and the product of left turn flow and opposing through flow. Roess et al. [5] suggested that the left turn, whose flow is more than 200 veh/h and product is more than 50000, should be protected. HCM [41] set these conditions more specific, and they are applied in this paper. However, if a left turn has no opposing through movement, it is unnecessary to consider the types of left turn phasing. Any of the following conditions lead to protected left turns:

(i) $q_{\mathrm{LT}} > 240$.

(ii) if $N_{L,\mathrm{TH}} = 1$, $q_{\mathrm{LT}} q_{\mathrm{TH}} > 50000$.

(iii) if $N_{L,\mathrm{TH}} = 2$, $q_{\mathrm{LT}} q_{\mathrm{TH}} > 90000$.

(iv) if $N_{L,\mathrm{TH}} \geq 3$, $q_{\mathrm{LT}} q_{\mathrm{TH}} > 110000$.

Here $N_{L,\mathrm{TH}}$ is number of opposing through lanes; q_{LT} is left turn flow (veh/h); q_{TH} is opposing through flow (veh/h).

Please note that when a left turn is permitted according these conditions, the left turn could be either permitted or protected. In the conflict matrix, if two movements do not conflict, they still could be in the different stages. For the same reason, the permitted left turn and the opposing through movement could be in the different stage, so that the left turn is protected in this case, and when calculating the saturation flow of the left turn, one should regard the left turn as protected left turn. The final left turn phasing type is determined according to the generated stages.

2.4.2. Stage Generation. When generating the stages, one of the conventions is to include all nonconflict movements in the same stage (compatibility). In this regard, a feasible set would be one which contains different stages with nonconflict movements. It is equivalent to the problem of finding all subgraphs in graph theory. The adjacent matrix indicating the compatibility among movements is generated from the conflict matrix. If the value of the conflict matrix is 1, two movements conflict; if the value is 0, the movements are not in conflict. Respectively, if the value in the adjacent matrix is 1, meaning that both movements are compatible, they can be in the same stage; if this value is 0, the two movements are incompatible and therefore cannot be in the same stage, as shown in Figure 3. In some cases, the conflict matrix alone is not sufficient, since the lane configuration may generate additional incompatibility. For example, in Figure 3(a), lane 1 and lane 2 cannot share a stage with either lane 5 or lone 6, because the left turn in lane 2 conflicts with both lanes 5 and 6. However, once the adjacent matrix is generated, the compatibility of all moments can be extracted from there. As can be seen in Figure 3(b), in the conflict matrix, the through movement in arm 1 is not in conflict with the through moment in arm 3. Nevertheless, both of these movements are incompatible, as shown in the respective cell of the adjacent matrix (see Figure 3(c)).

Once we have all possible stages by finding all subgraphs (see Figure 3(d)), we should keep the number of stages as small as possible, to decrease the intergreen time. However, we need to make sure that all movements are included in our plan. To this end, we formulate the stage selection as a 0/1 integer programming problem whose objective is to minimize the number of stages. The decision variables model whether a stage should be selected and the constraints represent the requirement that each movement must be included at least in one stage (completeness). Consequently, the minimum number of stages including all movements is generated. After the stages are generated, the left turn phasing types are fixed. Figure 3(e) illustrates an example of the stage generation. "G" means green time and "R" means red time. "LT" represents left turn, "TH" represents through movement, and "RT" represents right turn.

When the stages are generated, one movement may be shared by multiple stages. For example, in the numerical example of Memoli et al. [29], two stages share the through movement. This is rarely applied in the stage-based method. Sharing a movement by multiple stages influences the stage sequence and green duration of these stages. We further address the issue of stages sharing the same movements in Sections 2.4.3 and 2.4.7.

2.4.3. Stage Sequence Optimization. Generally the most favorable stage sequence is determined by the total necessary intergreen times, which leads to shortest cycle time [3]. In this paper, the stage sequence is optimized by minimizing the total intergreen time. From the adjacent matrix in the "stage generation" (see Figure 4(a)), the intergreen time matrix can be generated. In this matrix, the compatible movements have no intergreen time, whereas the incompatible movements do. The intergreen times between incompatible movements are mainly determined by the geometry of the intersection. The intergreen time between stages is determined based on the intergreen time matrix and the composition of the stages (see Figure 4(b)).

If one movement appears in multiple sequential stages, the signal of the movement can keep green in the intergreen time between stages; i.e., the intergreen time is 0. The movement thereby has extra green time and capacity. However, if these stages are not sequential, the extra green time cannot be gained. Therefore, when determining the stage sequence, one should consider the potential gain of extra green time. In this paper, the stage sequence optimization problem is represented as the traveling salesman problem (TSP), where the sequence of stages is equivalent to the visited cities and the minimum total intergreen time is equivalent to the minimum distance (see Figure 4(e)). The distance between two stages is the sum of intergreen times among the related movements (see Figures 4(c) and 4(d)). In this regard, if the stages sharing the same movements are sequential, the distance between

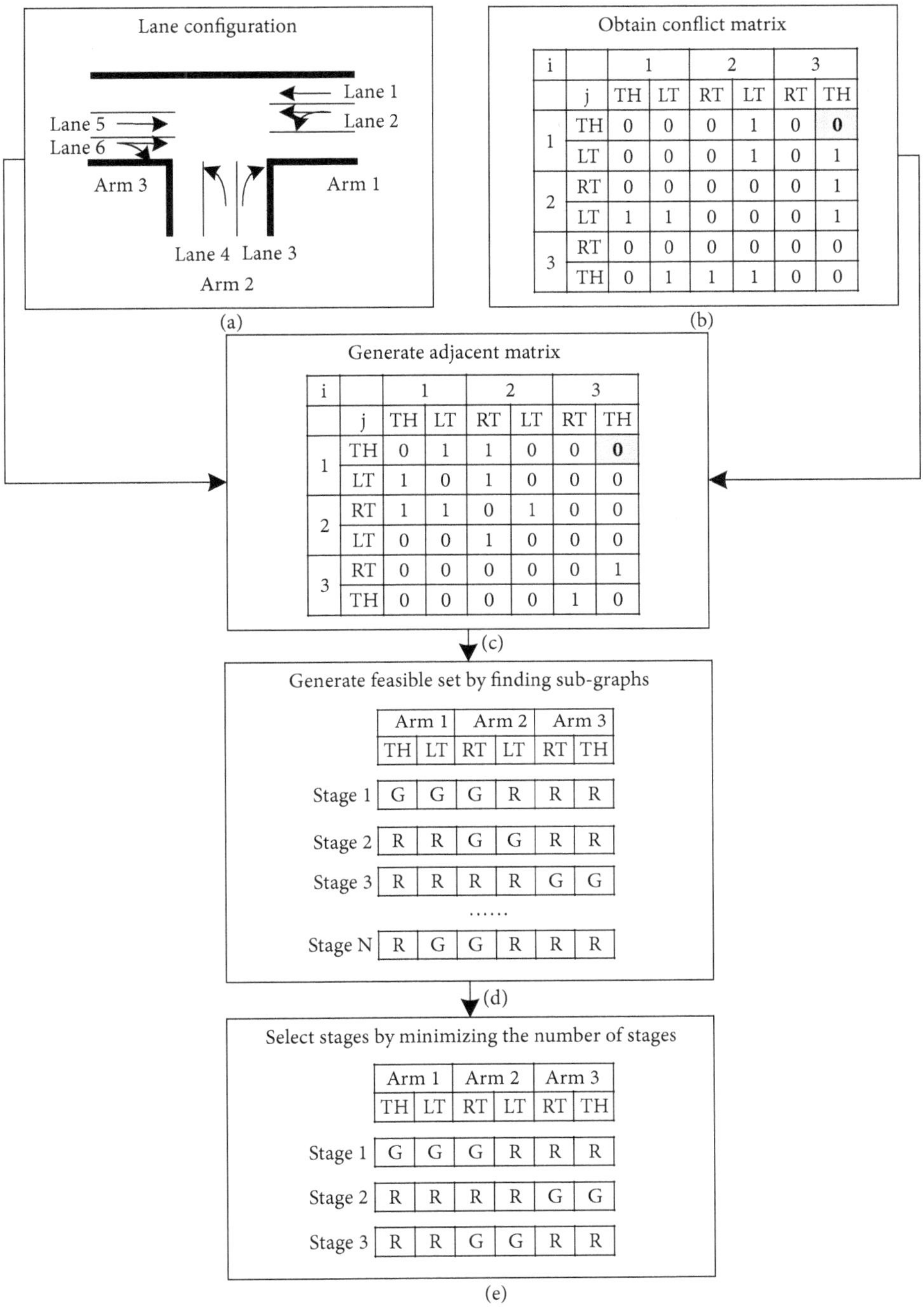

FIGURE 3: An example of stage generation: (a) lane configuration; (b) obtaining conflict matrix; (c) generation of adjacent matrix; (d) generation of feasible set by finding subgraphs; (e) selecting stages by minimizing the number of stages.

each stage is shorter than the distance of nonsequential stages sharing the same movements.

Denote i as the arm index and $N_{A,z}$ is the number of arms at intersection z. Denote movement index j in the turning direction set $J = \{\text{LT}, \text{TH}, \text{RT}\}$, where LT is left turn, and TH is through movement, and RT is right turn. Considering the types of left turn phasing, LT may also be denoted as permLT for permitted left turns and protLT for protected left turns. A movement is then defined as (i, j) which indicates a turning direction j from arm i.

Set formally, for intersection z, if a movement (i, j) is in stage p and a movement (u, v) is in stage p', then the distance between both stages can be expressed as follows:

$$l_{p,p'} = \sum_{i=1}^{N_{A,z}} \sum_{j \in J} \sum_{u=1}^{N_{A,z}} \sum_{v \in J} l_{i,j,u,v}, \tag{7}$$

where $l_{p,p'}$ is distance between stage p and stage p'; i, u denote arm index; j, v denote turning direction index; $l_{i,j,u,v}$ denotes intergreen time between movements (i, j) and (u, v).

Obtain adjacent matrix

i	j	1 RT	1 TH	1 LT	2 RT	2 TH	2 LT	3 RT	3 TH	3 LT	4 RT	4 TH	4 LT
1	RT	0	1	1	0	0	0	0	0	0	1	1	1
	TH	1	0	1	0	0	0	0	0	0	0	0	0
	LT	1	1	0	0	0	0	0	0	0	0	0	0
2	RT	0	0	0	0	1	1	0	0	0	1	0	0
	TH	0	0	0	1	0	1	0	0	0	0	0	0
	LT	0	0	0	1	1	0	0	0	0	0	0	0
3	RT	0	0	0	0	0	0	0	1	1	0	0	0
	TH	0	0	0	0	0	0	1	0	1	0	0	0
	LT	0	0	0	0	0	0	1	1	0	0	0	0
4	RT	1	0	0	1	0	0	0	0	0	0	1	1
	TH	1	0	0	0	0	0	0	0	0	1	0	1
	LT	1	0	0	0	0	0	0	0	0	1	1	0

(a)

Build intergreen time matrix

i	j	1 RT	1 TH	1 LT	2 RT	2 TH	2 LT	3 RT	3 TH	3 LT	4 RT	4 TH	4 LT
1	RT	0	0	0	4	4	4	4	4	4	0	0	0
	TH	0	0	0	4	4	4	4	4	4	4	4	4
	LT	0	0	0	4	4	4	4	4	4	4	4	4
2	RT	4	4	4	0	0	0	4	4	4	0	4	4
	TH	4	4	4	0	0	0	4	4	4	4	4	4
	LT	4	4	4	0	0	0	4	4	4	4	4	4
3	RT	4	4	4	4	4	4	0	0	0	4	4	4
	TH	4	4	4	4	4	4	0	0	0	4	4	4
	LT	4	4	4	4	4	4	0	0	0	4	4	4
4	RT	0	4	4	0	4	4	4	4	4	0	0	0
	TH	0	4	4	4	4	4	4	4	4	0	0	0
	LT	0	4	4	4	4	4	4	4	4	0	0	0

(b)

Obtain generated stages

	Arm 1 RT	Arm 1 TH	Arm 1 LT	Arm 2 RT	Arm 2 TH	Arm 2 LT	Arm 3 RT	Arm 3 TH	Arm 3 LT	Arm 4 RT	Arm 4 TH	Arm 4 LT
Stage 1	G	G	G	R	R	R	R	R	R	R	R	R
Stage 2	R	R	R	G	G	G	R	R	R	R	R	R
Stage 3	G	R	R	R	R	R	R	R	R	G	G	G
Stage 4	R	R	R	R	R	R	G	G	G	R	R	R

(c)

Generate sum of intergreen time matrix

Stage	1	2	3	4
1	0	36	24	36
2	36	0	44	36
3	24	44	0	48
4	36	36	48	0

(d)

Determine stage sequence by solving traveling salesman problem

	Arm 1 RT	Arm 1 TH	Arm 1 LT	Arm 2 RT	Arm 2 TH	Arm 2 LT	Arm 3 RT	Arm 3 TH	Arm 3 LT	Arm 4 RT	Arm 4 TH	Arm 4 LT
Stage 1	G	G	G	R	R	R	R	R	R	R	R	R
Stage 3	G	R	R	R	R	R	R	R	R	G	G	G
Stage 2	R	R	R	G	G	G	R	R	R	R	R	R
Stage 4	R	R	R	R	R	R	G	G	G	R	R	R

(e)

FIGURE 4: An example of stage sequence optimization: (a) obtaining adjacent matrix; (b) building intergreen time matrix; (c) obtaining generated stages; (d) generating sum of intergreen time matrix; (e) determining stage sequence by solving traveling salesman problem.

For example, in Figure 4, the intergreen times between each incompatible movements are 4 s. As stages 1 and 2 do not share any movements, the distance between stages 1 and 2 is 36 s. However, the right turn in arm 1 is shared by both stages 1 and 3, so the distance between stages 1 and 3 sums up to 24 s. The signal for the right turn keeps green, so the intergreen time is gained. By applying TSP, stages 1 and 3 are sequential.

2.4.4. Lane Flow Calculation. As one movement may occupy multiple lanes, drivers have to select one lane for turning preparation. To well adjust signal timing, it is necessary to calculate the number of vehicles turning a direction on each lane, i.e., the assigned flow. Lane flow is the sum of assigned flows of each movement.

Denotations k and $N_{L,i}$ are the lane index and the number of lanes on arm i, respectively. For all intersection $z = 1, \ldots, N_Z$, arm $i = 1, \ldots, N_{A,z}$, and turning direction $j \in J$, demands $F_{i,j}$ from each iteration of SUE with signal timing are equal to the sum of assigned flows of movement (i, j).

$$F_{i,j} = \sum_{k=1}^{N_{L,i}} q_{i,j,k}, \quad \forall i = 1, \ldots, N_{A,z}; \; j \in J, \tag{8}$$

where $q_{i,j,k}$ is the assigned flow of movement (i, j) on lane k.

If a movement (i, j) is not on lane k, the assigned flow $q_{i,j,k}$ should be 0.

$$q_{i,j,k} = 0, \quad \forall i = 1, \ldots, N_{A,z}; \; j \in J; \; k = 1, \ldots, N_{L,i} \tag{9}$$

If movement (i, j) is on multiple lanes, the saturation flow rates of all lanes tend to be identical because drivers enter the lane with the shortest queue, resulting in the even distribution of assigned flows [26].

$$\sum_{j\in J}\frac{q_{i,j,k}}{s_{j,k}}-\sum_{j\in J}\frac{q_{i,j,k+1}}{s_{j,k+1}}=0, \quad \forall i=1,\ldots,N_{A,z};\ j\in J;\ k=1,\ldots,N_{L,i}-1, \tag{10}$$

where $s_{j,k}$ is the saturation flow of movements on exclusive lanes.

By solving linear equations in (8)–(10), the assigned flows are determined. Specifically, if an assigned flow is less than 0, the related movement cannot be assigned to that lane. Instead, the assigned flow is assigned to its adjacent feasible lanes until all assigned flows are nonnegative.

Consequently, the lane flow can be calculated via summing the assigned flow of each movement on the lane:

$$q_{i,k}=\sum_{j\in M}q_{i,j,k}, \quad \forall i=1,\ldots,N_{A,z};\ k=1,\ldots,N_{L,i} \tag{11}$$

2.4.5. Saturation Flow Adjustment. In Figure 2, after "determining types of left turn phasing" and before "calculating lane flows", the saturation flow of permitted/protected left turns can be adjusted. The saturation flow is adjusted once again, this time for shared lanes, after the "calculating lane flows" step.

The saturation flow of permitted left turns per lane is adjusted by estimating the filtered saturation flow based on gap acceptance theory and adjusting the saturation flow of an exclusive lane [40].

$$s_{\text{filtered}}=\frac{q_{\text{TH}}\exp\left(-q_{\text{TH}}l_c\right)}{1-\exp\left(-q_{\text{TH}}l_f\right)}$$
$$s_{\text{permLT},k}=s_{\text{TH}}\frac{s_{\text{filtered}}g_u+N_f}{0.5g_{\text{LT}}}, \tag{12}$$

where s_{filtered} is filtered saturation flow of permitted left turns (veh/s); q_{TH} is opposing through movement flow (veh/s); l_c is critical gap with a value 4.5 s; l_f is follow-up headway with value 2.5 s; $s_{\text{permLT},k}$ is saturation flow of permitted left turns per lane (veh/s); s_{TH} is saturation flow of through movement per lane (veh/s); g_{LT} is green duration of left turns (s); $g_u = \max((s_{\text{TH}}g_{\text{LT}}-q_{\text{TH}}c)/(s_{\text{TH}}-q_{\text{TH}}),0)$ is unsaturated part of the green period for the opposing through movement (s); c is cycle length (s); N_f is the number of vehicles passing during the amber period and is predefined as 1.5 (veh) in Akcelik (1981) and HCM (2000) unless field data are available.

The saturation flow of movements on the shared lane is weighted according to the flow of each movement [26, 40].

$$s_{i,k}=\frac{1}{\sum_{j\in J}r_j/s_{j,k}}, \tag{13}$$

where k is lane index; r_j is ratio of lane flow $q_{i,j,k}$ to the total flow on lane k; s_j is saturation flow of turning direction j (veh/h).

2.4.6. Signal Timing. The formulas of signal timing calculation are originally deduced by Webster [25]. HBS [26] and RiLSA [42] adopt these formulas in the manual. Webster's formulas can only be applied in unsaturated conditions which are regarded as the limitation. However, considering the constraints of cycle length, cycle length can be the maximum value if the degree of saturation is more than 1. Thus, Webster's formulas are feasible and applicable at the saturated intersections. Further, by selecting the largest cycle length of the network as the common cycle length, Webster's formulas can be also used in the network [27].

To calculate cycle length, the flow ratio, which is the flow to the saturation flow, for each stage must be determined. As green times are determined by the lane with maximum flow in a stage, for all lanes in the same stage, the maximum flow ratio of lanes is the flow ratio of the stage (see (14)).

For all i, k related to stage p,

$$b_{z,p}=\max\left(\frac{q_{z,i,k}}{s_{z,i,k}}\right) \tag{14}$$

$$c_z=\frac{1.5\left(\sum_{p=1}^{N_{P,z}}l_p+5\right)}{1-B_z} \tag{15}$$

$$g_{z,p}=\frac{b_{z,p}}{B_z}\left(c_z-\sum_{p=1}^{N_{P,z}}l_p\right) \tag{16}$$

subject to $c_z\in[c_{\min},c_{\max}]$, $g_p\in[g_{\min,p},c_z]$, where $p=1,\ldots,N_{P,z}$ is stage index; l_p is intergreen time of stage p (s); $B_z=\sum_{p=1}^{N_{P,z}}b_p$ is sum of flow ratios at intersection z; $N_{P,z}$ is the number of phases at intersection z; c_z is cycle length of intersection z; $c_{\min}$ is minimum cycle time (s); $c_{\max}$ is maximum cycle time (s); $g_{\min,p}$ is minimum green duration of stage p (s).

Cycle length must be greater than or equal to the minimum cycle length and not larger than the maximum cycle length. Thus, if $B_z>1-1.5(\sum_{p=1}^{N_{P,z}}l_p+5)/c_{\max}$, the cycle length is always maximum.

Due to the constraints of minimum green times, if the green time of stage p is less than the minimum green time, the green time is equal to the minimum green time. These minimum green times are treated as the components of intergreen time. Because the value of the green time is fixed to the minimum, the green times can only be assigned for the rest of stages. The cycle time should be recalculated as the actual sum of lost time increases. The rest of effective green time has to be accommodated according to the stage flows.

$$c_z=\frac{1.5\left(\sum_{p=1}^{N_{P,z}}l_p+\sum_{p=1}^{N_{P',z}}g_{\min,p}+5\right)}{1-B'_z} \tag{17}$$

$$g_{z,p}=\frac{b_{z,p}}{B'}\left(c_z-\sum_{p=1}^{N_{P,z}}l_p-\sum_{p=1}^{N_{P',z}}g_{\min,p}\right) \tag{18}$$

subject to $c_z\in[c_{\min},c_{\max}]$, $g_p\in[g_{\min,p},c_z]$, where $N_{P',z}$ is the number of stages with minimum green time P' and $B'_z=B_z-\sum_{p=1}^{N_{P',z}}b_{z,p}$.

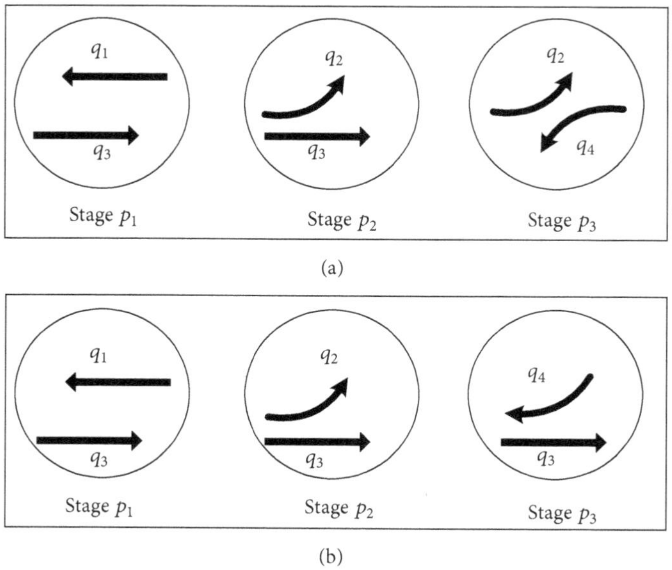

FIGURE 5: Examples of three stages sharing movements.

Cycle lengths are firstly calculated for each intersection, and the largest cycle length is decisive as the common cycle length for the network. Green times are then recalculated with the common cycle length by (16) and (18).

$$c = \max c_z \tag{19}$$

2.4.7. Signal Timing of Stages Sharing Movements. As mentioned, to calculate signal timing, the flows of movements/lanes in each stage should be known. However, if one movement is in two stages, it is hard to separate the movement flows into two stages beforehand and one cannot calculate the flow ratio. This case is neglected by HBS [26] and RiLSA [42] but considered by Pohlmann [38].

The rule to deal with the stages sharing movements is that green times are determined by the lane with the maximum flow. When calculating the green time of stages sharing movements, two cases need to be considered: either the required green time for the movement shared in multiple stages is bigger than the required green time for the other movements included in these stages or this green time is smaller. The final green time will be the larger one of these two cases. For example, in Figure 1, we assume each movement occupies one lane. For the movement with the flow q_3, we cannot know in advance the number of vehicles in stages p_1 and p_2, respectively. Hence, following HBS [26] recommendations, one cannot solve this case. As the largest flow determines the flow ratio in a stage, Pohlmann [38] separately calculated the flow ratio for stages p_1 and p_2 and combined stage $p_1 p_2$. Stages only include the movement not shared by two stages, and the combined stage only include the movement shared by two stages. In Figure 1, $b_{z,p_1} = q_1/s_1$ and $b_{z,p_2} = q_2/s_2$, but $b_{z,p_1 p_2} = q_3/s_3$, and then the flow ratio of stages p_1 and p_2 is determined by $\max(b_{z,p_1} + b_{z,p_2}, b_{z,p_1 p_2})$. With the same logic, the green time of stage p_1 and p_2 is determined by $\max(g_{z,p_1} + g_{z,p_2}, g_{z,p_1 p_2})$.

$$g_{z,p_1 p_2} = \frac{b_{z,p_1 p_2}}{B'_z}\left(c_z - \sum_{p=1}^{N_{P,z}} l_p - \sum_{p=1}^{N_{P',z}} g_{\min,p}\right) \tag{20}$$

If $g_{z,p_1 p_2} > g_{z,p_1} + g_{z,p_2}$,

$$\begin{aligned} g_{z,p_1} &= \frac{b_{z,p_1}}{b_{z,p_1} + b_{z,p_2}} g_{z,p_1 p_2} \in \left[g_{\min,p_1}, c_z\right] \\ g_{z,p_2} &= g_{z,p_1 p_2} - g_{z,p_1}; \end{aligned} \tag{21}$$

else g_{z,p_1} and g_{z,p_2} do not change.

Pohlmann [38] considered only two stages sharing movements. Although this is usually the case, there may also be cases including more than two stages. Generalizing Pohlmann's method [38] to include more than two stages is not straightforward. For example, taking into account three stages sharing movements, the following process should be performed. First, we should consider the fact that not all three stages necessarily share the same movements, and movements may be shared only by two stages, as demonstrated in Figure 5(a). There, stages p_1 and p_2 share the through movement, and stages p_2 and p_3 share the left turn. Therefore, in order to calculate the green time, we should calculate $\max(b_{z,p_1} + b_{z,p_2} + b_{z,p_3}, b_{z,p_1} + b_{z,p_2 p_3}, b_{z,p_1 p_2} + b_{z,p_3})$. A different calculation should be carried out for the case presented in Figure 5(b). There, the through movement is shared by all three stages, and therefore the flow ratio would be $\max(b_{z,p_1} + b_{z,p_2} + b_{z,p_3}, b_{z,p_1 p_2 p_3})$. The green duration is calculated in a similar manner.

3. Algorithms

In "prohibiting left turn" of Figure 2, the left turns are selected by the genetic algorithm with the default configuration of Java

Require: $b_{z,np}, b_{z,cp}, b_{z,sp}, B_z$ ▷ $b_{z,np}, b_{z,cp}, b_{z,sp}$ is the flow ratio of normal stages, combined stages, stages, respectively.
Ensure: c_z
function CALCULATECYCLELENGTH
  $B'_z \leftarrow B_z, \sum_{p=1}^{N_{p',z}} g_{\min,p} \leftarrow 0$
  repeat
    $flag \leftarrow 0$
    if $B_z > 1 - 1.5(\sum_{p=1}^{N_{p,z}} l_p + 5)/c_{\max}$ **then**
      $c_z \leftarrow c_{\max}$
    else
      $c_z \leftarrow Eq.$ (17)
      if then $c_z < c_{\min}$
        $c_z \leftarrow c_{\min}$
      end if
    end if
    for normal stages np **do**
      $g_{z,np} \leftarrow Eq.$ (18)
      if $g_{z,np} < g_{\min,np}$ **then**
        $g_{z,np} \leftarrow g_{\min,np}$
        $\sum_{p=1}^{N_{p',z}} g_{\min,p} \leftarrow g_{z,np}, B'_z \leftarrow B'_z - b_{z,np}$
        $flag \leftarrow 1$
      end if
    end for
    for stages sharing movements cp and their related stages sp **do**
      $g_{z,cp} \leftarrow Eq.$ (18), $g_{z,sp} \leftarrow Eq.$ (18)
      if $g_{z,sp} < g_{\min,sp}$ **then**
        $g_{z,sp} \leftarrow g_{\min,sp}$
      end if
    end for
    for stages sharing movements cp and their related stages sp **do**
      if $g_{z,cp} < \sum g_{z,sp}$ and $g_{z,sp} == g_{\min,sp}$ **then**
        $\sum_{p=1}^{N_{p',z}} g_{\min,p} \leftarrow g_{z,p}, B'_z \leftarrow B'_z - b_{z,p}$
        $flag \leftarrow 1$
      end if
    end for
  until $flag == 0$
  return c_z
end function

ALGORITHM 1: Calculation of cycle length for intersection z.

genetic algorithm package (Jgap) [43] which refers to an elitist ranking selector cloning the top 90% of the user-specified population size with crossover rate 35% and mutation rate 1/12.

SUE is solved with the Method of Successive Average [44]. The link travel times are initialized with the free flow travel times of the links and then the initial link flows are estimated with stochastic loading based on the free flow travel times. In each iteration, the link travel times are updated with the link flows from the previous iteration, and the link flows are updated with the adjustment on the difference between the link flows from stochastic loading and the link flows from the previous iteration. The interaction between the link flows and the link travel times repeats until the convergence criteria are met.

For signal-relevant steps in Figure 2, "generating stages" is solved by finding all subgraphs with depth-first-search algorithm and minimization of the number of stages with routine branch and cut algorithm in IBM Ilog Cplex. The traveling salesman problem in stage sequence is determined by applying backtracking algorithm. The focus of this section is to explain signal timing calculation in the case of combined stages which is extended from [38]. The cycle length for each intersection is firstly calculated in Algorithm 1 and then the largest cycle length is selected as the common cycle length with which the green duration for each stage is calculated in Algorithm 2.

4. Numerical Analysis

4.1. Test Networks. The test network is an artificial network with six signalized intersections, ten origins, and ten destinations (Figure 6). There are 24 left turns. The free flow speed of each link is 50 km/h, and each link has three lanes.

Require: $b_{z,np}, b_{z,cp}, b_{z,sp}, B_z, c$
Ensure: $g_{z,p}$

```
function CalculateGreens
    B'_z ← B_z
    Σ_{p=1}^{N_{p',z}} g_{min,p} ← 0
    repeat
        flag ← 0
        for normal stages p do
            g_{z,np} ← Eq. (18)
            if g_{z,np} < g_{min,np} then
                g_{z,np} ← g_{min,np}
                Σ_{p=1}^{N_{p',z}} g_{min,p} ← g_{z,np}
                B'_z ← B'_z − b_{z,np}
                flag ← 1
            end if
        end for
        for stages sharing movements cp and their related stages sp do
            g_{z,cp} ← Eq. (18)
            g_{z,sp} ← Eq. (18)
            if g_{z,sp} < g_{min,sp} then
                g_{z,sp} ← g_{min,sp}
            end if
        end for
        for stages sharing movements cp and their related stages sp do
            if g_{z,cp} < Σ g_{z,sp} and g_{z,sp} == g_{min,sp} then
                Σ_{p=1}^{N_{p',z}} g_{min,p} ← g_{z,sp}
                B'_z ← B'_z − b_{z,sp}
                flag ← 1
            else
                g_{z,sp} ← (b_{z,sp}/ Σ b_{z,sp}) g_{z,cp} ∈ [g_{min,p_1}, c]
            end if
        end for
    until flag == 0
    for all stages p do
        g_{z,p} ← g_{z,np}, g_{z,sp}
    end for
    return g_{z,p}
end function
```

Algorithm 2: Calculation of green duration for each stage.

The default conflict matrices are extracted from network files of SUMO, which is a microscopic simulation tool developed by the German Aerospace Center (DLR). The network is tested with the OD matrix in Table 1. The values of parameters are in Table 2. Specifically, the intergreen times are assumed to independent of the intersection geometry and the intergreen times between all stages are constant. In SUE, the convergence criterion is that the difference of average link flows between adjacent iterations is less than 1.0×10^{-4}. In the genetic algorithm, the initial population size is 50, and the generation size is 30.

4.2. Results. With the OD matrix and test network, 3 of 24 left turns are prohibited: the left turn in the northern arm at Intersection 2, the left turn in the southern arm at Intersection 3, and the left turn in the western arm at Intersection 6. Before LTP, the total travel time is 564.9 h whereas, after LTP, the minimal total travel time is 435.5 h with reduction of 22.9%. Prohibiting left turns reduces the total travel time. The average degrees of saturation for each intersection before and after LTP are also compared (see Figure 7). Except Intersection 4, the average degrees of saturation of the intersections decrease, especially at the intersections with LTP.

Signal settings are also adjusted before and after LTP. The common cycle lengths before LTP and after LTP are 100 s. The stage generation results and their sequences before and after LTP are shown in Figure 8. The green durations before and after LTP are given in Table 3. After LTP, the number of stages at Intersection 6 becomes smaller which indicates a less intergreen time and longer effective green time.

TABLE 1: Demands between origins and destinations (unit: veh/h).

Origin/Destination	A	B	C	D	E	F	G	H	I	J	Total
A	0	70	90	160	100	60	150	50	180	200	1060
B	200	0	130	60	140	70	90	130	80	110	1010
C	180	90	0	180	90	160	70	80	120	70	1040
D	40	70	140	0	80	120	60	50	150	80	790
E	160	130	70	140	0	90	140	150	50	50	980
F	90	80	110	80	170	0	50	100	120	50	850
G	60	70	200	150	80	80	0	90	140	100	970
H	90	80	40	110	200	70	160	0	80	120	950
I	60	80	100	120	50	100	90	200	0	90	890
J	90	180	40	70	0	120	60	90	100	0	750
Total	970	850	920	1070	910	870	870	940	1020	870	9290

TABLE 2: Values of parameters.

Parameters	Notations	Values
Parameter in BPR function	α	0.15
Parameter in BPR function	β	4
Number of exit lanes	$N_{\epsilon,z}$, $\forall z = 1,\dots,N_Z$	3
Saturation flow of through movements	$s_{\mathrm{TH},k}$, $\forall k = 1,\dots,N_{L,i}$	1900 veh/h
Saturation flow of right turns	$s_{\mathrm{RT},k}$, $\forall k = 1,\dots,N_{L,i}$	1615 veh/h
Saturation flow of protected left turns	$s_{\mathrm{protLT},k}$, $\forall k = 1,\dots,N_{L,i}$	1805 veh/h
Saturation flow of permitted left turns	$s_{\mathrm{permLT},k}$, $\forall k = 1,\dots,N_{L,i}$	by Eq. (12)
Maximum cycle length	$c_{\max}$	60 s
Minimum cycle length	$c_{\min}$	100 s
Minimum green duration	$g_{\min,p}$, $\forall p = 1,\dots,N_{P,z}$	6 s
Intergreen time	I_p, $\forall p = 1,\dots,N_{P,z}$	4 s
Observation time	T	0.25 h

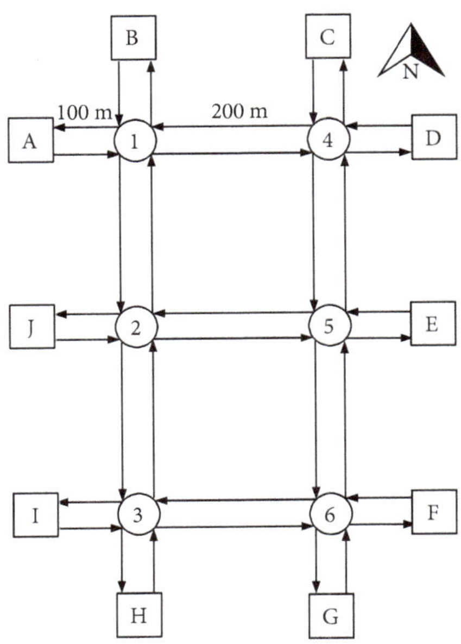

FIGURE 6: Layout of "toy" network with six intersections.

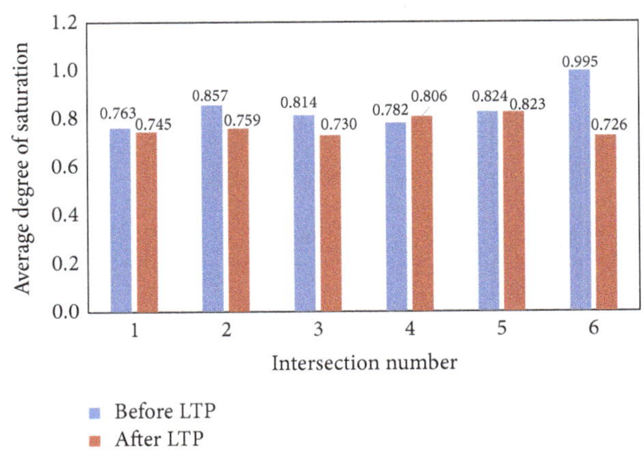

FIGURE 7: Average degree of saturation for each intersection before and after LTP.

4.3. Evaluation of Algorithms. The algorithms of the proposed model are evaluated with increasing scale of networks. Table 4 shows the information of the increasing network scales. The number of origins and the number of destinations are fixed to ten, which is the same as the original network (Network 1). For Network 2 to Network 7, the number of intersections increases one by one, resulting in the increasing number of nodes and links. The experiments have been

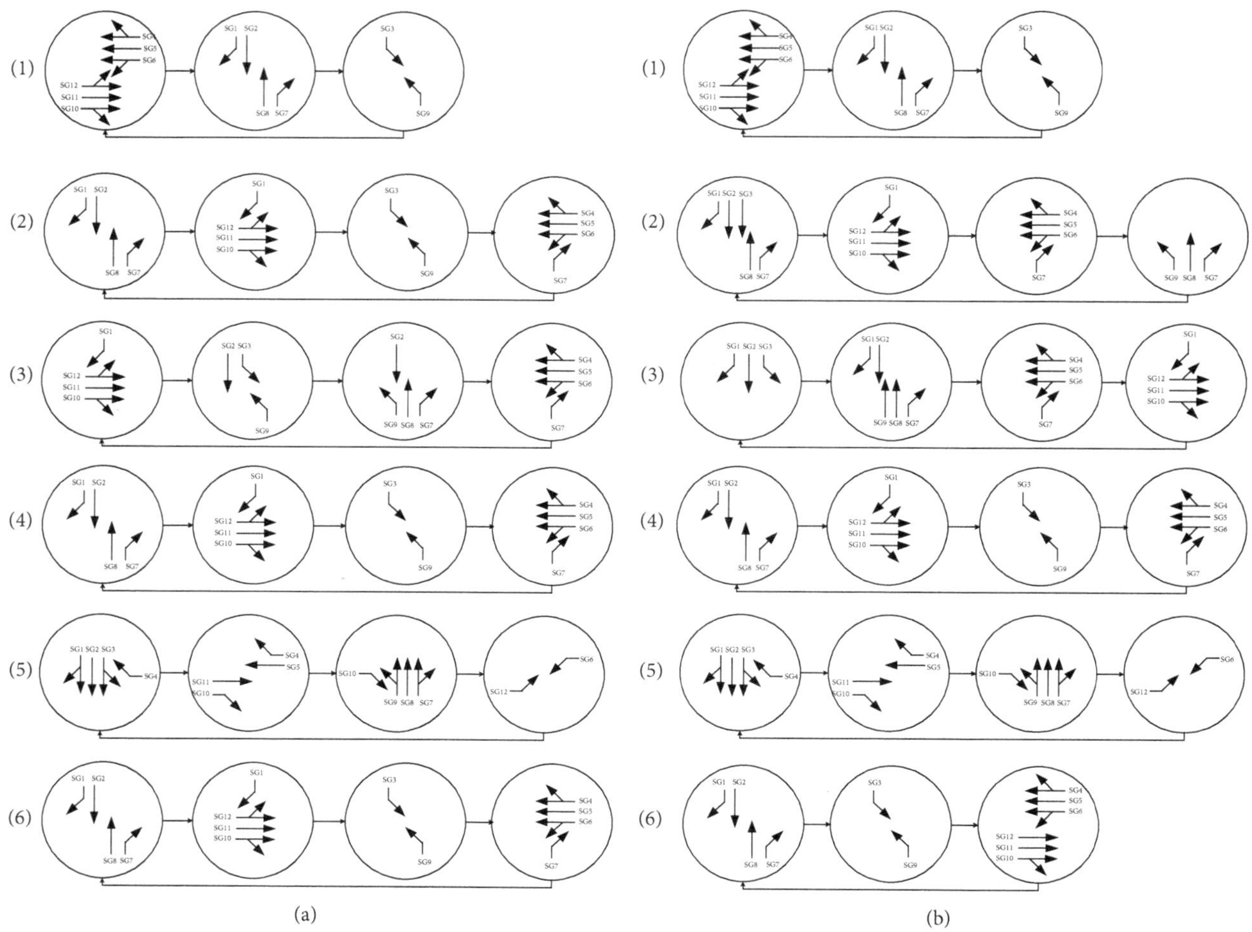

FIGURE 8: (a) Generated stages of test network and their sequence before LTP; (b) generated stages of test network and their sequence after LTP.

TABLE 3: Green durations of stages before and after LTP (s).

(a) Green durations of movements before LTP

Intersection	Stage 1	Stage 2	Stage 3	Stage 4
1	38	27	23	
2	34	22	14	14
3	33	11	24	16
4	27	18	12	27
5	23	16	26	19
6	22	17	22	23

(b) Green durations of movements after LTP

Intersection	Stage 1	Stage 2	Stage 3	Stage 4
1	35	28	25	
2	22	18	22	22
3	15	19	17	33
4	26	21	11	26
5	27	17	20	20
6	30	27	31	

conducted on an IPC with a 2.9 GHz quad-core Intel Core i7 and with 16 GB memory. By applying these networks with the proposed model, the computing time for each network is illustrated in Figure 9. The computing time dramatically increases when the number of intersections is larger than 10. As routing search is important in the SUE algorithm, the network scale significantly influences the computing time. However, as the genetic algorithm may randomly select infeasible left turns, it immediately receives penalty and the total travel time is assigned to a constant large value, which means the SUE model and the design of signal timing plan are skipped. Thus, the computing time varies without an obvious tendency.

4.4. Analysis of Prohibited Left Turns. The flows of prohibited left turn and their types are analyzed. LTP is related to left turn flows according to previous research. Hajbabaie et al. [45] found that low LT volumes and oncoming high volumes led to LTP at some intersections. Pline [4] summarized that LT volume was a critical factor leading to prohibition. In consideration of a protected left turn for small flows wasting

TABLE 4: Network scale information for algorithm evaluation.

Network	Layout	Number of intersections	Number of nodes	Number of links
1		6	68	96
2	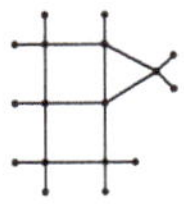	7	76	122
3	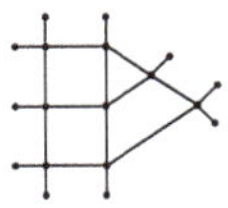	8	84	138
4		9	88	140
5	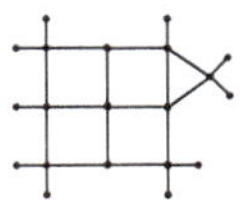	10	96	156
6	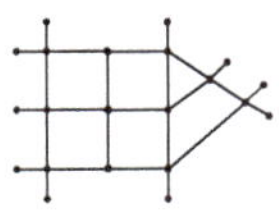	11	104	172
7	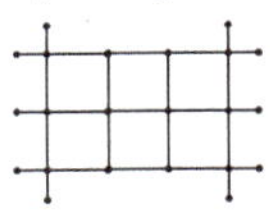	12	108	174

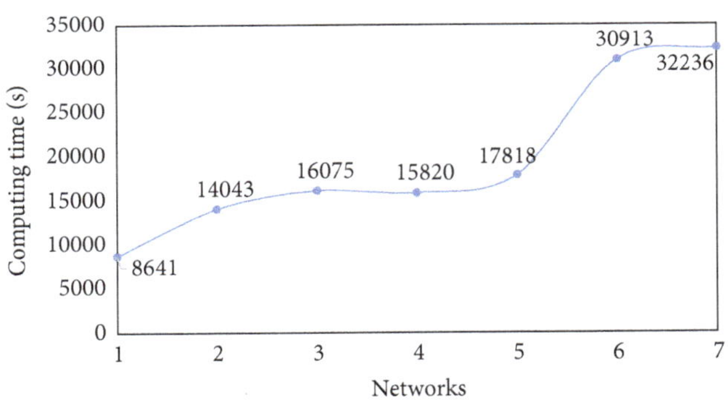

FIGURE 9: Computing time with increasing scale of networks.

capacity, we suppose that LTP is also related to the types of left turn phasing types. Permitted left turn flows justify conflict with the opposing straight vehicles, so whether permitted left turns should be prohibited depends on both left turn flows and opposing flows.

The flows and types of left turns without LTP and the LTP results from running the proposed model are collected via testing eight OD matrices. The OD matrices are generated by the demands in Table 1 plus random numbers in the ranges of $[-40, -30]$, $[-30, -20]$, $[-20, -10]$, $[-10, 0]$, $[0, 10]$, $[10, 20]$, $[20, 30]$, and $[30, 40]$, respectively. Including the original OD matrix, the flows of 214 left turns are collected and 36 of them are prohibited. We summarized the minimum, maximum, and the average value of all permitted left turns and protected left turns. We also calculated the average flow of all prohibited/permitted left turns and prohibited/permitted left turns.

The results are displayed in Figure 10. The green dots represent the average flows of permitted/protected left turns, and the high points of the vertical lines represent their maximum flow, and the low points of the vertical lines represent their minimum flow. The blue dots represent the average flows of prohibited/permitted/protected left turns. As no reduction from LTP can be found with the OD matrix having the random demands in the range of $[-10, 0]$, no left turns are prohibited with this OD matrix. Further, as the demands increase, there are no permitted left turns for the OD matrices having random demands in the ranges of $[20, 30]$ and $[30, 40]$, respectively; and thus for these ranges no data is depicted in Figure 10(a). In Figure 10, all of the flows of the prohibited permitted left turns are smaller than the flows of all permitted left turns and the same as those for the protected left turns.

4.5. Discussion. The goal of this paper is to design signal timing plane including prohibiting left turns. The numerical example shows that prohibiting left turns reduces the total travel time. This study could potentially lead to useful insights regarding congestion management.

The main reason why prohibiting left turns can reduce the total travel time is that LTP reduces the number of conflict points at intersections. Due to the reduction of the conflict points, the number of stages decreases so that the intergreen

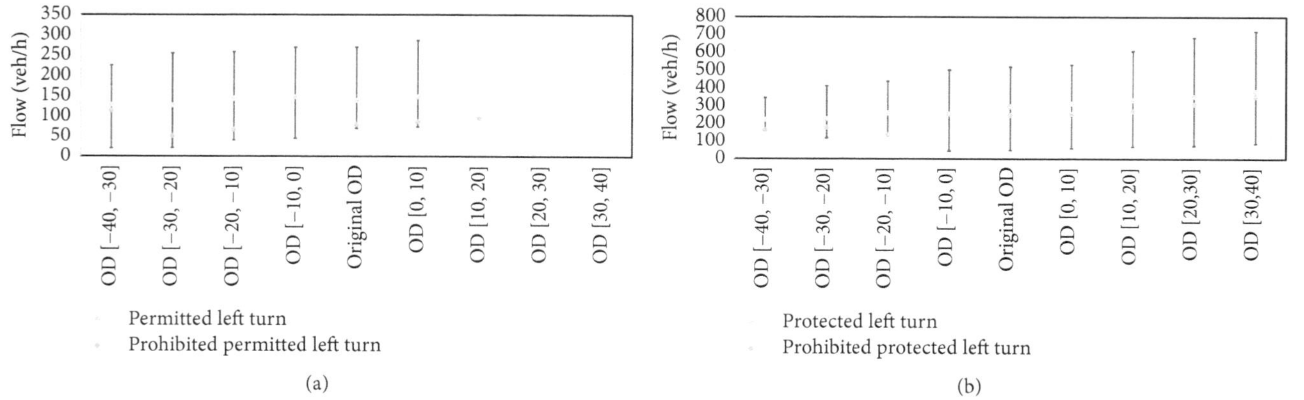

Figure 10: (a) Flow comparison between permitted left turns and prohibited/permitted left turns; (b) flow comparison between protected left turns and prohibited protected left turns.

time between two stages decreases (e.g., Intersection 6). If the intergreen time between two stages is longer, e.g., 6 s, the total travel time reduction is expected to be larger because longer intergreen time indicates a high potential of more effective green time. Moreover, some movements are in multiple stages which also lengthens effective green times for those movements (e.g., the right turn in the northern arm of Intersection 3). Further, the lanes of prohibited left turns are assigned for through movements which also increases the capacity of the through movements (e.g., the through movement in the northern arm of Intersection 2). Therefore, the delays at the intersection go down and the total travel time decreases.

The prohibited left turns may be related to types of left turn phasing. The protected left turns with minor flows should be prohibited, but as the influence of opposing flows is not clear, we cannot conclude that the permitted left turns with minor flows should also be prohibited. Meanwhile, these findings are specific for the network. More generalized findings would require an additional analysis.

The signal timing plan with the proposed method does not aim at reaching global optimum. The design of signal timing consists of "stage generation", "stage sequence optimization", and "signal timing determination". Although "stage generation" and "stage sequence optimization" are conducted with optimization methods, "signal timing determination" is based on the formulas from the relevant manuals. The accumulation of the reasonable steps does not ensure the global optimization. Meanwhile, the interaction between signal timing and traffic flow is conducted in a straightforward way: with given traffic flows obtained from an SUE with a cost based on BRP function, a signal timing plan is designed for each intersection, and then the traffic flows adopt to the signal timing plan. The multiple interactions are not considered as the case in the dynamical scenarios.

5. Conclusions

We propose a method of designing signal timing plan including LTP by minimizing total travel time. The total travel time reduces after LTP. Types of left turn phasing and relevant left turn flows may be related to LTP. This paper provides an idea of congestion management in urban road networks. Prohibiting left turns should actually be considered among other left turn treatments in signal timing plan design. Planned future research includes analyzing the other factors influencing LTP.

Conflicts of Interest

The authors declare that they have no conflicts of interest.

Acknowledgments

This research is supported by the German Research Foundation (DFG) through the Research Training Group SocialCars (GRK 1931). The focus of the SocialCars Research Training Group is on significantly improving the city's future road traffic through cooperative approaches. This work is also carried out within the project OptimUM (Optimization of Urban Traffic Management towards Environment Friendly and Safe Mobility) which is financially supported by the Helmholtz CAS Joint Research Groups (HCJRG) of the Helmholtz Association of German Research Centers. The support is gratefully acknowledged.

References

[1] K. R. Agent, "Guidelines for the Use of Protected/permissive Left-turn Phasing," Technical Report No. UKTRP-85-19, Kentucky Transportation Research Program, College of Engineering, University of Kentucky, 1985.

[2] J. H. Kell and I. J. Fullerton, *Manual of traffic signal design*, Prentice Hall, 1991.

[3] RiLSA., "Guidelines for Traffic Signals RiLSA," Tech. Rep., Cologne, Germany, 1992.

[4] J. L. Pline, "Left-Turn Treatments at Intersections," Number NCHRP Sythesis of Highway Practice 225, Transportation Research Board, 1996.

[5] R. P. Roess, E. S. Prassas, and W. R. McShane, *Traffic Engineering*, Prentice Hall, 2004.

[6] ADOT. ADOT Traffic Engineering Guidelines and Processes: section 612. Technical Report, Arizona Department of Transportation, 2015.

[7] J. Lu, S. Dissanayake, H. Zhou, and X. K. Yang, "Operational evaluation of right turns followed by u-turns as an alternative to direct left turns *the report based on "Methodology to quantify the effects of access management on roadway operation and safety"*, III," Tech. Rep., 2001.

[8] J. G. Bared and E. I. Kaisar, "Median U-turn design as an alternative treatment for left turns at signalized intersections," *ITE Journal (Institute of Transportation Engineers)*, vol. 72, no. 2, pp. 50–54, 2002.

[9] M. Chowdhury, N. Derov, and P. Tan, "Evaluating the effects of prohibiting left turns and the resulting U-turn movement," FHWA/OH-2003/001, University of Dayton,Ohio Department of Transportation, Federal Highway Administration, 2003.

[10] P. Liu, J. J. Lu, H. Zhou, and G. Sokolow, "Operational effects of U-turns as alternatives to direct left-turns," *Journal of Transportation Engineering*, vol. 133, no. 5, pp. 327–334, 2007.

[11] L. Junqiang, Z. Hantao, and Z. Qian, "Research on the impact of U-turn location on operation efficiency at intersection," in *Proceedings of the 2009 International Conference on Measuring Technology and Mechatronics Automation, ICMTMA 2009*, pp. 567–570, April 2009.

[12] J. Zhao, W. Ma, K. L. Head, and Y. Han, "Improving the operational performance of two-quadrant parclo interchanges with median U-turn concept," *Transportmetrica B*, pp. 1–21, 2016.

[13] J. A. Bonneson and P. T. McCoy, "Capacity and operational effects of midblock left-turn lanes," Technical Report NCHRP Rep. 395, Transportation Research Board, Washington, DC, USA, 1981.

[14] M. Chowdhury, N. Derov, P. Tan, and A. Sadek, "Prohibiting left-turn movements at mid-block unsignalized driveways: Simulation analysis," *Journal of Transportation Engineering*, vol. 131, no. 4, pp. 279–285, 2005.

[15] S. Gyawali, *A New Decision Making Approach for Indirect Left Turn Treatments by Utilizing Decision Assistance Curves [Ph.D. thesis]*, 2014.

[16] J. Zhao, M. Yun, H. M. Zhang, and X. Yang, "Driving simulator evaluation of driversâ response to intersections with dynamic use of exit-lanes for left-turn," *Accident Analysis & Prevention*, vol. 81, pp. 107–119, 2015.

[17] J. Zhao, Y. Liu, and T. Wang, "Increasing Signalized Intersection Capacity with Unconventional Use of Special Width Approach Lanes," *Computer-Aided Civil and Infrastructure Engineering*, vol. 31, no. 10, pp. 794–810, 2016.

[18] W. Ma, H. Xie, Y. Liu, L. Head, and Z. Luo, "Coordinated optimization of signal timings for intersection approach with presignals," *Transportation Research Record*, no. 2355, pp. 93–104, 2013.

[19] Y. Xuan, C. F. Daganzo, and M. J. Cassidy, "Increasing the capacity of signalized intersections with separate left turn phases," *Transportation Research Part B: Methodological*, vol. 45, no. 5, pp. 769–781, 2011.

[20] Y. Zhou and H. Zhuang, "The optimization of lane assignment and signal timing at the tandem intersection with pre-signal," *Journal of Advanced Transportation*, vol. 48, no. 4, pp. 362–376, 2014.

[21] J. Long, Z. Gao, H. Zhang, and W. Y. Szeto, "A turning restriction design problem in urban road networks," *European Journal of Operational Research*, vol. 206, no. 3, pp. 569–578, 2010.

[22] X. Guang and L. Wu, "A model of the urban road intersection left-turning restriction," *Journal of Theoretical and Applied Information Technology*, vol. 48, no. 3, pp. 1619–1625, 2013.

[23] J. Long, W. Y. Szeto, and H.-J. Huang, "A bi-objective turning restriction design problem in urban road networks," *European Journal of Operational Research*, vol. 237, no. 2, pp. 426–439, 2014.

[24] L. R. Foulds, D. C. Duarte, H. A. do Nascimento, H. J. Longo, and B. R. Hall, "Turning restriction design in traffic networks with a budget constraint," *Journal of Global Optimization*, vol. 60, no. 2, pp. 351–371, 2014.

[25] F. V. Webster, "Traffic signal settings," Technical Report Road Research Technical Paper, No. 39, Road Research Laboratory, London, UK, 1958.

[26] HBS, *Handbuch für die Bemessung von Straßenverkehrsanlagen (HBS, German Highway Capacity Manual)*, Road and Transport Association, Cologne, Germany, 2001.

[27] K. B. Kesur, "Optimization of mixed cycle length traffic signals," *Journal of Advanced Transportation*, vol. 48, no. 5, pp. 431–442, 2014.

[28] M. Di Gangi, G. E. Cantarella, R. D. Pace, and S. Memoli, "Network traffic control based on a mesoscopic dynamic flow model," *Transportation Research Part C: Emerging Technologies*, vol. 66, pp. 3–26, 2016.

[29] S. Memoli, G. E. Cantarella, S. de Luca, and R. D. Pace, "Network signal setting design with stage sequence optimisation," *Transportation Research Part B: Methodological*, vol. 100, pp. 20–42, 2017.

[30] G. Improta and G. E. Cantarella, "Control system design for an individual signalized junction," *Transportation Research Part B: Methodological*, vol. 18, no. 2, pp. 147–167, 1984.

[31] J. P. Silcock, "Designing signal-controlled junctions for group-based operation," *Transportation Research Part A: Policy and Practice*, vol. 31, no. 2, pp. 157–173, 1997.

[32] X. Ma, J. Jin, and W. Lei, "Multi-criteria analysis of optimal signal plans using microscopic traffic models," *Transportation Research Part D: Transport and Environment*, vol. 32, pp. 1–14, 2014.

[33] C. K. Wong and S. C. Wong, "Lane-based optimization of signal timings for isolated junctions," *Transportation Research Part B: Methodological*, vol. 37, no. 1, pp. 63–84, 2003.

[34] C. K. Wong and B. G. Heydecker, "Optimal allocation of turns to lanes at an isolated signal-controlled junction," *Transportation Research Part B: Methodological*, vol. 45, no. 4, pp. 667–681, 2011.

[35] Y. Liu and C. K. Wong, "Refining Lane-Based Traffic Signal Settings to Satisfy Spatial Lane Length Requirements," *Journal of Advanced Transportation*, vol. 2017, Article ID 8167530, 27 pages, 2017.

[36] J. Zhao, W. Ma, K. L. Head, and X. Yang, "Dynamic turning restriction management for signalized road network," *Transportation Research Record*, vol. 2487, pp. 96–111, 2015.

[37] J. Zhao, Y. Liu, and P. Li, "A network enhancement model with integrated lane reorganization and traffic control strategies," *Journal of Advanced Transportation*, vol. 50, no. 6, pp. 1090–1110, 2016.

[38] T. Pohlmann, *New Approaches for Online Control of Urban Traffic Signal Systems [Ph.D. thesis]*, Fakultaet Architektur, Bauingenieurwesen und Umweltwissenschaften der Technischen Carolo-Wilhelmina zu Braunschweig, 2010.

[39] Q. Tang and B. Friedrich, "Minimization of Travel Time in Signalized Networks by Prohibiting Left Turns," *Transportation Research Procedia*, vol. 14, pp. 3446–3455, 2016.

[40] R. Akcelik, "Traffic signals: capacity and timing analysis," Technical Report Research report ARR No. 123, Australian Road Research Board, 1981.

[41] HCM, *Highway Capacity Manual. Chapter 16: Signalized Intersections*, Transportation Research Board (TRB), Washington, DC, USA, 2000.

[42] RiLSA. Richtlinien für Lichtsignalanlagen (RiLSA, Guidelines for Traffic Signals). Technical Report, Cologne, Germany, 2015.

[43] K. Meffert and N. Rotstan, Jgap: Java genetic algorithm package. 2015, http://jgap.sourceforge.net/.

[44] Y. Sheffi, *Urban transportation networks*, Prentice Hall, 1985.

[45] A. Hajbabaie, J. C. Medina, and R. F. Benekohal, "Effects of ITS-Based Left Turn Policies on Network Performance," in *Proceedings of the 13th International IEEE Conference on Intelligent Transportation Systems, ITSC 2010*, pp. 80–84, September 2010.

12

Lane Change Behavior on Freeways: An Online Survey Using Video Clips

V. L. Knoop,[1] M. Keyvan-Ekbatani,[2] M. de Baat,[1] H. Taale,[1,3] and S. P. Hoogendoorn[1]

[1] *Delft University of Technology, Netherlands*
[2] *University of Canterbury, New Zealand*
[3] *Rijkswaterstaat and Delft University of Technology, Netherlands*

Correspondence should be addressed to V. L. Knoop; v.l.knoop@tudelft.nl

Academic Editor: Jose E. Naranjo

Freeways form an important part of the road network. Yet, driving behavior on freeways, in particular lane changes and the relation with the choice of speed, is not well understood. To overcome this, an online survey has been carried out. Drivers were shown video clips, and after each clip they had to indicate what they would do after the moment the video stopped. A total of 1258 Dutch respondents completed the survey. The results show that most people have a strategy to choose a speed first and stick to that, which is the first strategy. A second, less often chosen, strategy is to choose a desired lane and adapt the speed based on the chosen lane. A third strategy, slightly less frequently chosen, is that drivers have a desired speed, but contrary to the first strategy, they increase this speed when they are in a different lane overtaking another driver. A small fraction have neither a desired speed nor a desired lane. Of the respondents 80% use the right lane if possible, and 80% avoid overtaking at the right. Also 80% give way to merging traffic. The survey was validated by 25 survey respondents also driving an instrumented vehicle. The strategies in this drive were similar to those in the survey. The findings of this work can be implemented in traffic simulation models, e.g., to determine road capacity and constraints in geometric design.

1. Introduction

Freeways constitute a principal part of the transport infrastructure. However, if traffic exceeds capacity, it gets congested. Therefore, capacity is an important traffic variable. It is determined by the average minimum headway between two successive vehicles, and the lane they choose. In practice, the road is often underutilized since not all lanes are operating at capacity in high demand (e.g., Daganzo [1]; Knoop et al. [2]). This unequal lane distribution is a consequence of the lane change behavior of drivers.

Lane changes may cause disruptions and oscillations [3] and consequently influence the capacity of the road by leaving voids and reducing flow [4]. Also, oscillations can result into problems including safety hazards, extra fuel consumption, emissions, delays, and driver discomfort. To mitigate these problems, it would be good if the behavior of drivers was known, and the traffic dynamics could be predicted. This could be useful for road design or measures to improve driver behavior. In this paper, the lane choice of drivers is investigated under European conditions, which means keep-right-unless-overtaking regulations. In a previous study [5] it was shown with a test with 10 drivers that various drivers had different strategies (summarized in Section 2.1). Using a large scale survey the research described in this paper quantifies further which part of the population uses which strategy.

The research objective is to get insights into the distribution of drivers over the different strategies. These insights can be used to better describe traffic operations in microscopic simulation models. To be able to use these models for (ex-ante) assessing road design or traffic management measures, a proper model for lane change and lane choice behavior is crucial, which should be based on the insight of driving behavior. To obtain insights into this behavior, a large scale state preference survey among Dutch drivers was conducted. The research has been submitted and approved by the Ethics Committee of the Delft University of Technology. Preliminary results of the research have been presented earlier [6]. The main contributions of the work presented in the current paper are the use of video clips to introduce possible choices and the large response of this realistic stated preference survey, which gives confidence in the results.

Since the survey gives a stated preference which might differ from a revealed preference, the survey was (face) validated. To that end 25 respondents to the survey also participated in a drive on the road in an instrumented vehicle, recording drivers' actions. For practical reasons, the research started with the drives which were recorded on video. This gave footage of many different traffic situations. Clips thereof were utilized to introduce various questions in the online survey. Typical situations were reconstructed in an approximately ten-second video clip. Respondents were asked how they would cope with the situation which was shown in the video clip. Generally, that could involve a speed and/or a lane adaptation. For the 25 participants in the drive, an interview followed right after the drive. As validation for the stated preference survey, this group's choice in the drive and the outcomes of their interview were compared with their choices in the survey. As stated before, the main goal was to describe the driving behavior of Dutch drivers. It is possible that driving behavior differs for drivers in other countries. To check this, the same survey has been distributed in other countries.

The paper is structured as follows. First, Section 2 discusses the background for driving behavior and survey techniques. The survey setup is then presented in Section 3, followed by the results from the survey in Section 4. The validation of the survey using the instrumented vehicle drives is presented in Section 5. Section 6 shows the validity of the survey for different countries. In the remainder of the paper, the drive in the instrumented vehicle will be called *drive*. The word *respondent* will be used to denote respondents to the survey, whereas an *participant* is a person participating in the drive.

2. Literature Review

In this section, previous research on this topic is discussed from two points of view. First, it is important to know which aspects related to this topic deserve attention. To that end, the literature on lane change behavior on multilane freeways is studied in Section 2.1. Having found the relevant aspects, it is important to study them in the right way, because various techniques are available. Section 2.2 discusses advantages and disadvantages of particular ways of studying driving behavior.

2.1. Lane Change Behavior on Multilane Freeways. In order to quantify the complex features of multilane traffic, understanding lane change behavior on freeways is essential. As addressed by Toledo et al. [7] and Kesting et al. [8], modeling lane changing maneuvers is a key component of microscopic traffic simulation tools. In the vast majority of microscopic simulation models, a desired speed is chosen first, and the lane is being adapted if needed. Several studies have been carried out on developing realistic lane change models. However, despite its importance in mimicking traffic operations on freeways, among the researchers it did not attain as much attention as the car-following models (for a review see Moridpour et al. [9] and Zheng [10]). Rule-based and discrete-choice-based models are the most popular microscopic lane changing algorithms in literature [11]. In rule-based models, considering the heterogeneity among drivers, different gap acceptance conditions have been taken into account. It was argued that oscillations are the result of driver heterogeneous reactions to deceleration waves on freeways [12]. Many lane changing models incorporate the heterogeneity of drivers by different parameter sets. For the car-following behavior, it has been shown that there is variation of strategies among drivers [13]. A recent study [5] confirmed that drivers also have completely different strategies to choose lanes and that speed and lane changing choices are related. This shows that considering different parameter sets may not be a reliable approach to model heterogeneous driving behavior in a microscopic simulation environment, because it does not capture the driver choices realistically. The more detailed discrete-choice-based algorithms apply logit models to simulate driver behavior [14]. This group of microscopic algorithms can describe lane changing behavior in detail, but they often contain a large number of parameters which makes it hard to calibrate and validate the model [8]. The recently developed LMRS model [15] combines the route, speed, and *keep right* incentives for lane changing with the synchronization and relaxation of car-following behavior, but also in this model lane choice strategies are not considered.

As mentioned before, the study described by Keyvan-Ekbatani et al. [5] has led to a categorization of the lane change decision process (i.e., strategies), based on a drive with 10 participants and a series of interviews. Particularly, it has been revealed that people drive on the freeway according to the following four strategies.

Strategy 1 (speed leading). Drivers choose a desired speed for driving on a freeway stretch and try to stick to that speed as much as possible. If required, they change lanes and overtake to continue driving their speed. Also during the overtaking maneuver, they stick to their desired speed.

Strategy 2 (speed leading with speed increase at overtaking). As in the previous strategy, drivers choose a desired speed and try to maintain that speed as much as possible. They may change lane to overtake a slower driver. Different from the speed leading strategy, the drivers increase speed when they are overtaking and do not stick to the original desired speed. (Keyvan-Ekbatani et al. [5] call this strategy "speed leading with overtaking", which might lead to misunderstandings since overtaking is also possible in Strategy 1).

Strategy 3 (lane leading). Drivers choose a specific lane and adapt their speed to the speed of that lane, but within margins (typical accepted range is approximately 40 km/h).

Strategy 4 (traffic leading). Drivers have neither a specific lane nor a desired speed in mind and "go with the flow".

2.2. Studying Lane Change Behavior. Several approaches are available to study driving behavior on freeways. Our work focuses on lane changing, which, in general, is not studied that much. More often, merging, as a special case of lane changing, is studied. Therefore, we look into the methodologies used to study merging behavior, since they are potentially useful for lane changing in general. The following four approaches have been used to study freeway driving behavior: interviews or group discussions (focus groups), an instrumented vehicle, a driving simulator, or a survey. These approaches will all be briefly discussed. First, the focus groups: Kondyli and Elefteriadou [16] performed a study on merging behavior at freeways using a focus group. Participants had to discuss and indicate their thinking process during a merging maneuver from an on-ramp to a freeway. Secondly, the use of an instrumented vehicle: in Van Koningsbruggen and Stranner [17] the driving behavior at merging areas was studied using an instrumented vehicle. In this case the drivers commented in real-time on their behavior. This may lead to a driving behavior which might not reflect reality properly, since, due to the requirement of commenting in real-time, the drivers might make decisions more consciously. Thirdly, the use of driving simulator: other studies apply driving simulators to investigate the driving behavior [18]. Driving simulators are attractive to researchers, since they can be used to create a goal-oriented virtual world for implementing different traffic situations which cannot be tested easily in the real world. However, a potential disadvantage of driving simulators could be that they only provide a representation of reality, not reality itself. Finally, the use of surveys: several studies utilized (static) surveys (also called questionnaires, which we consider the same here) to investigate different aspects of driving behavior, for example, De Winter and Dodou [19] and Davey et al. [20]. The survey used the most is the Driver Behavior Questionnaire (DBQ) [21], which has been used many times, for example, by Bener et al. [22], Bener et al. [23], De Winter and Dodou [19], and Li et al. [24].

Most surveys require people to envisage the situation which might be an extra hurdle. For example, the study by De Craen et al. [25] used pictures of traffic situations to identify novice, unsafe, and overconfident drivers. Although many forms of the DBQ exist, a questionnaire using video clips is not common practice. In some other fields, in particular traffic safety and accident analysis, video clip based traffic surveys have recently been pioneered, for instance, for cyclist safety [26], or for the effect of visibility of work zones on the speed choice of car drivers [27]. But a questionnaire with video clips of particular traffic situations has never, to the best of the authors' knowledge, been used to ask people about their actions concerning speed and lane choice. Applying video clips can help to overcome the shortcoming of surveys that the situation is difficult to envisage. Though, the researchers have to be aware that drivers can have different driving behavior than shown in the video, because they feel they are in a situation they usually do not face.

Since we would like to investigate the distribution of people over the 4-lane change strategies mentioned before, a large response is needed, which is infeasible by test drives or focus groups. Video seemed a good means to that end. We found a way to adapt the videos to the driving style which became visible from the earlier part of the survey. Thus, with these personalized surveys with videos, we maintain the core quality of reaching out to many people (online), while still benefiting from relevant questions. The survey was also validated, by letting several respondents to the survey also drive in an instrumented vehicle to validate their choices. Section 5 will show how the validation was executed.

Table 1: Categorization of respondents into three branches.

Desired speed	Category	Speed in video clip
< 103 km/h	"Slow" driver	~100 km/h
103- 110 km/h	"Average" driver	~107 km/h
> 110 km/h	"Fast" driver	~115 km/h

3. Survey Setup

In this section, we describe the setup of the survey, starting with the design (Section 3.1), followed by the recruitment and characteristics of the respondents (Section 3.2), and finally the analyses in Section 3.3.

3.1. Survey Design. The survey starts with a picture of an empty three-lane freeway with a speed limit of 100 km/h. For this situation, it is asked which lane the respondent will choose and what his speed will be in this situation. Based on the answers, a personalized survey is presented to the respondents, in which the indicated speed of driving in the videos is related to his or her own quoted driving behavior. The categories used for this are given in Table 1.

The survey is set up to find out the driver's strategy to choose a lane using a series of questions containing videos. The videos are reworked to give respondents the idea that the situation is as close as possible to their normal driving behavior. That means that a rear view mirror is added showing the video of a rear-facing camera. Also, a speed limit indication was added to the video clip. The speed shown as current speed is adapted to the desired speed of the respondent. Figure 1 shows a screenshot of the resulting video clip. In all video clips, a view of the traffic situation from the driver's perspective is given, and respondents were asked what they would do if they were in that vehicle.

The main research themes of the survey are to determine the following.

Theme 1. Does a respondent drive according to strategy 1,2, or 3?

Theme 2. Does a respondent drive according to strategy 4?

Theme 3. Does a respondent strictly obey the keep right rule?

Theme 4. Would a respondent overtake another driver at the right side?

Theme 5. Would a respondent make a courtesy lane change to create space for a merging vehicle?

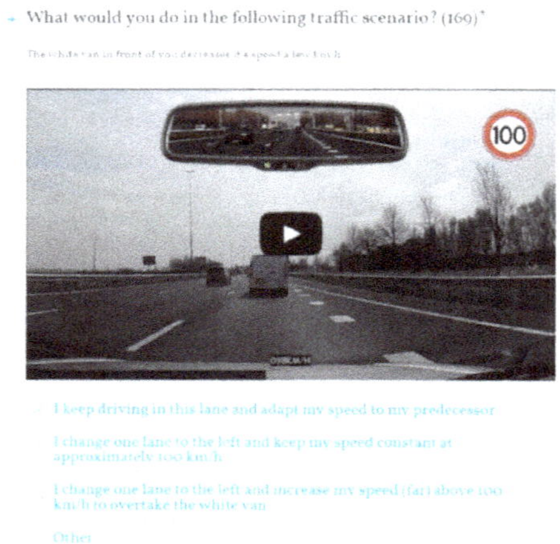

FIGURE 1: Screenshot of a video clip.

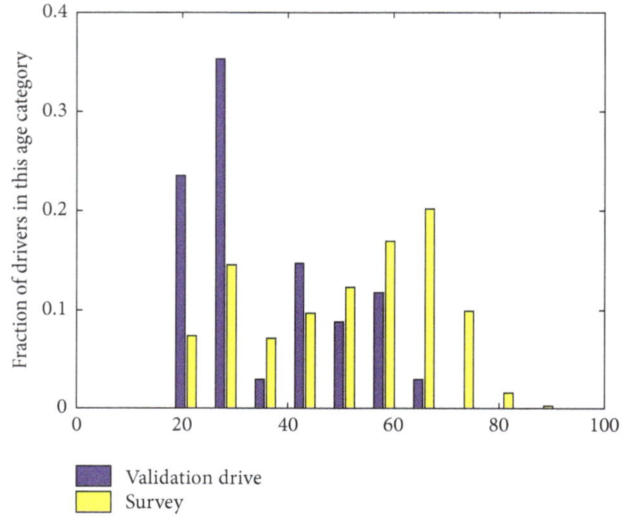

FIGURE 2: Age distribution of the respondents in the survey and participants in the drive.

Based on these themes, 14 questions with corresponding video clips were formulated, which are reported in Table 2. In most cases, an answer would lead to a single strategy, since, apart from the action, also the motivation for the action was asked. In some cases, an answer might exclude some strategies, but not point at a single strategy. Some of the traffic situations in the different questions are similar, which is to test the consistency of the respondents. The videos can be downloaded [28]; note that they are available for reuse. Apart from the driving-strategy related questions, also the basic characteristics of the respondents such as age and yearly mileage were asked. On average, the survey took 17 minutes to complete.

3.2. Respondents. The survey was distributed via personal network of the researchers. More importantly, it has been distributed via communication of the Royal Dutch Touring Club (ANWB). Generally, the social media and e-mail newsletter of this organization are distributed to millions of Dutch motorists. To stimulate people to participate, a gift card of 50 euros was promised as a prize awarded to one randomly chosen respondent. In total, 1,258 respondents completed the survey. As demonstrated in Figure 2, the resulting group is not homogeneously distributed over ages. There is a peak for younger drivers (20-25 years old), possibly due to university involvement, and a peak for more senior drivers (over 60 years old).

3.3. Analyses. The analyses are structured along the 5 themes. For each theme, it is checked which fraction of respondents chooses the particular strategy in the applicable question.

The population of the respondents could cause a bias in the survey, because driving behavior is influenced by personal characteristics, training, and experience; see Fuller [29]. A particular subgroup that shows deviating driving behavior will affect the results if this subgroup is overrepresented in the sample. Hence, we would like to know the dependencies. The two most obvious dependencies have been tested: (1) the relation between desired speed at one hand and age and experience at the other and (2) the relation between desired speed at one hand and frequency of driving and weekly mileage at the other. In Section 4.3 they are presented in pairs, since they are most correlated. Significance is tested with Cramer's V test.

4. Survey Results

This section describes the results. First, it discussed how often respondents choose a particular lane change strategy. Then, Section 4.2 presents results on overtaking at the right, and finally, Section 4.3 presents the desired speed of the respondents.

4.1. Lane Change Strategy Choice. Table 3 shows how the responses of the survey led to a distribution over strategies. Not all questions lead to the same distribution of respondents over the categories, which is in line with the effect seen in the drive. We also see that drivers change strategies for some questions. Nevertheless, we can derive a tendency: Strategy 1 is chosen most often, then strategy 3, then strategy 2, then strategy 4. The lower part of Table 3 shows that strategy 1 is used by almost every respondent and also strategy 3 is used by 74% of the respondents. Strategy 2 is used by just over half of the drivers and strategy 4 is rarely used.

Table 4 shows to which extent respondents are keeping the right lane. Approximately 75% of the respondents would normally go back to the right lane if there is a possibility. Only for question 7, this fraction is considerably lower than 75%. This is likely due to the fact that another vehicle is present in the center lane. Drivers do not prefer to switch lanes frequently.

4.2. Overtaking at the Right and Cooperating to Other Lane Changes. The vast majority of the respondents indicated that they would not overtake at the right; see Table 4. This varied

TABLE 2: Overview of the questions: screenshot (all but the first row cropped due to space limitation), question number, theme of question (in bold), and description.

1: **Right overtaking**. You are driving in a tunnel on a 4-lane freeway on the third lane from the left. A truck is driving up ahead in the rightmost lane and a person car is driving in the second lane from the left while there is space for that driver to keep right using your lane. You are getting closer to that person car.

2: **Strategies 1, 2, and 3**. You are driving on a 3-lane freeway in the right lane. A slightly slower predecessor is driving in front of you and you are approaching that vehicle, while a black car just passed you with high speed.

3: **Strategies 1, 2, and 3**. You are overtaking the white car, which increases its speed a bit, resulting in a very slow overtaking maneuver (this scenario is only shown to the respondents who want to overtake the vehicle with a constant speed).

4: **Strategies 1, 2, and 3**. You are driving on a 2-lane freeway during congestion in the left lane. You are approaching the tail of a stop-and-go wave, while traffic in the right lane drives smoother and with a higher speed.

5: **Right overtaking**. You just entered the freeway via an on-ramp, while in the lane to the left of you a black car is driving with a low speed. You need to proceed on this carriageway to follow your route.

6: **Strategies 1, 2, and 3**. You merge onto the main road from an on-ramp, while to continue your route you need to take the second exit from that point which is 3100 meters up ahead.

7: **Keep right/Strategy 4**. You are driving on a 3-lane freeway in the center lane while having passed a truck in the right lane. There is space to change lanes to the right after you have passed the truck, while a faster driver is approaching you from behind.

8: **Keep right**. You are driving on a 3-lane freeway in the center lane, while passing two trucks in the right lane. After you have passed the trucks there is an empty road ahead of you.

9: **Strategies 1, 2, and 3**. You are driving on a 3-lane freeway in the rightmost lane. A truck is driving in front of you in the same lane, while a faster driver is approaching you from behind on the center lane.

10: **Keep right**. You are driving on a 2-lane freeway in the left lane, while you pass a truck that is driving in the right lane. Up ahead another truck is driving there, while a faster driver is approaching you from behind.

11: **Strategy 4**. You are driving on a 3-lane freeway in the rightmost lane with a speed of 100 km/h. However, all other vehicles around you drive much faster and you are overtaken by them.

12: **Courtesy lane change**. You are driving on a 2-lane freeway in the right lane, while you approach a merging lane on which a vehicle is driving. However, the driver is not using its blinker.

13: **Strategies 1, 2, and 3**. You are driving on a 3-lane freeway in the center lane. To follow your route you have to take the exit that starts at 600 meters from the moment the video stops. On that moment a truck is driving ahead in the right lane.

14: **Keep right**. (Static picture). You are driving on a 3-lane freeway in the rightmost normal lane. This road section is equipped with a rush-hour lane which is currently open for traffic. You will pass the truck that drives on the rush-hour lane.

TABLE 3: Distribution of respondents over the lane change strategies.

		Lane change strategy							Other answers
		1	2	3	4	1/2	3/4	3*	
		Responses per question							
Question	Theme								
2	Strategies 1, 2, and 3	77%	10%	12%	1%	-	-	-	0%
3	Strategies 1, 2, and 3	56%	43%	-	-	-	-	-	1%
4	Strategies 1, 2, and 3	-	-	-	-	27%	73%	-	0%
6	Strategies 1, 2, and 3	-	-	44%	-	28%	-	27%	1%
9	Strategies 1, 2, and 3	85%	11%	3%	-	-	-	-	1%
11	Strategy 4	-	-	-	12%	87%	-	-	1%
13	Strategies 1, 2, and 3	20%	19%	61%	-	-	-	-	1%
		Cumulative use of strategies							
Number of respondents		1211	665	931	11	1220	336	341	
Percentage of respondents		96%	53%	74%	1%	97%	27%	27	

*: Question 6: drivers were instructed to exit in 3200 meters.

TABLE 4: Distribution of respondents to behavior characteristics.

Question	Theme	Changing to rightmost lane?		
		Yes		No
7	Keep right/Strategy 4	57%		42%
8	Keep right	88%		12%
10	Keep right	76%		23%
14	Keep right	73%		25%
		Right overtaking?		
		Yes		No
1	Right overtaking	20%		80%
5	Right overtaking	16%		84%
		Cooperative behavior?		
		Yes		No
		Lane change	Speed adjustment	
12	Courtesy lane change	72%	11%	16%

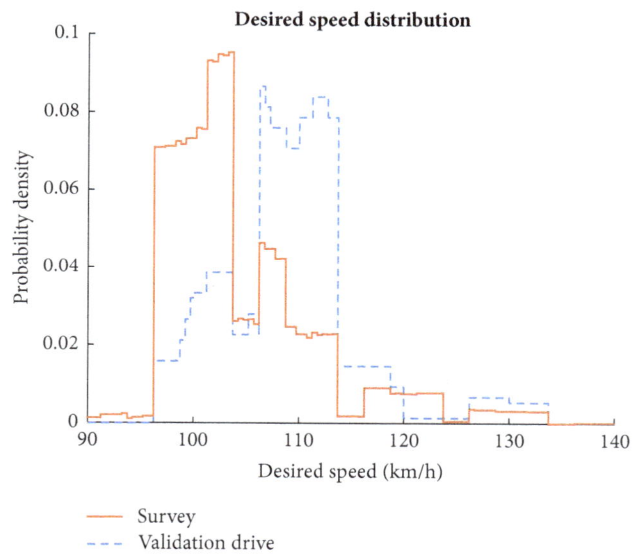

FIGURE 3: Distribution of quoted desired speeds at a road of a 100 km/h speed limit.

for the two different questions on this topic: in question 1, 20% would overtake at the right, and in question 5, 16% would overtake at the right.

Table 4 also shows the cooperative behavior. If another driver indicates the desire to change lane (e.g., the on-ramp lane ends, blinker on), 83% of the drivers will facilitate this lane change. Most respondents will do so by changing one lane to the left. Others (11%) will reduce speed to allow the vehicle to merge in front.

4.3. Desired Speed. Figure 3 shows the quoted desired speeds of the respondents on the 100 km/h section. For illustration purposes in the figure, we added a 3.75 km/h margin (positive and negative). If this was omitted, the figure would have some spikes. The desired speed is on average 104.5 km/h, with a standard deviation of 8.0 km/h. This is hence slightly above the speed limit of 100 km/h.

The correlations with other factors are shown in Figure 4. We find that (1) younger and inexperienced drivers have a higher desired speed than older drivers (statistically significant and strong) and (2) frequent drivers and those with a higher weekly mileage have a higher desired speed than infrequent drivers (statistically significant and weak).

5. Survey Validation

The validation of the survey was done via drives. From the people who filled out the survey, 25 participants also drove an instrumented vehicle. The drive was recorded on video,

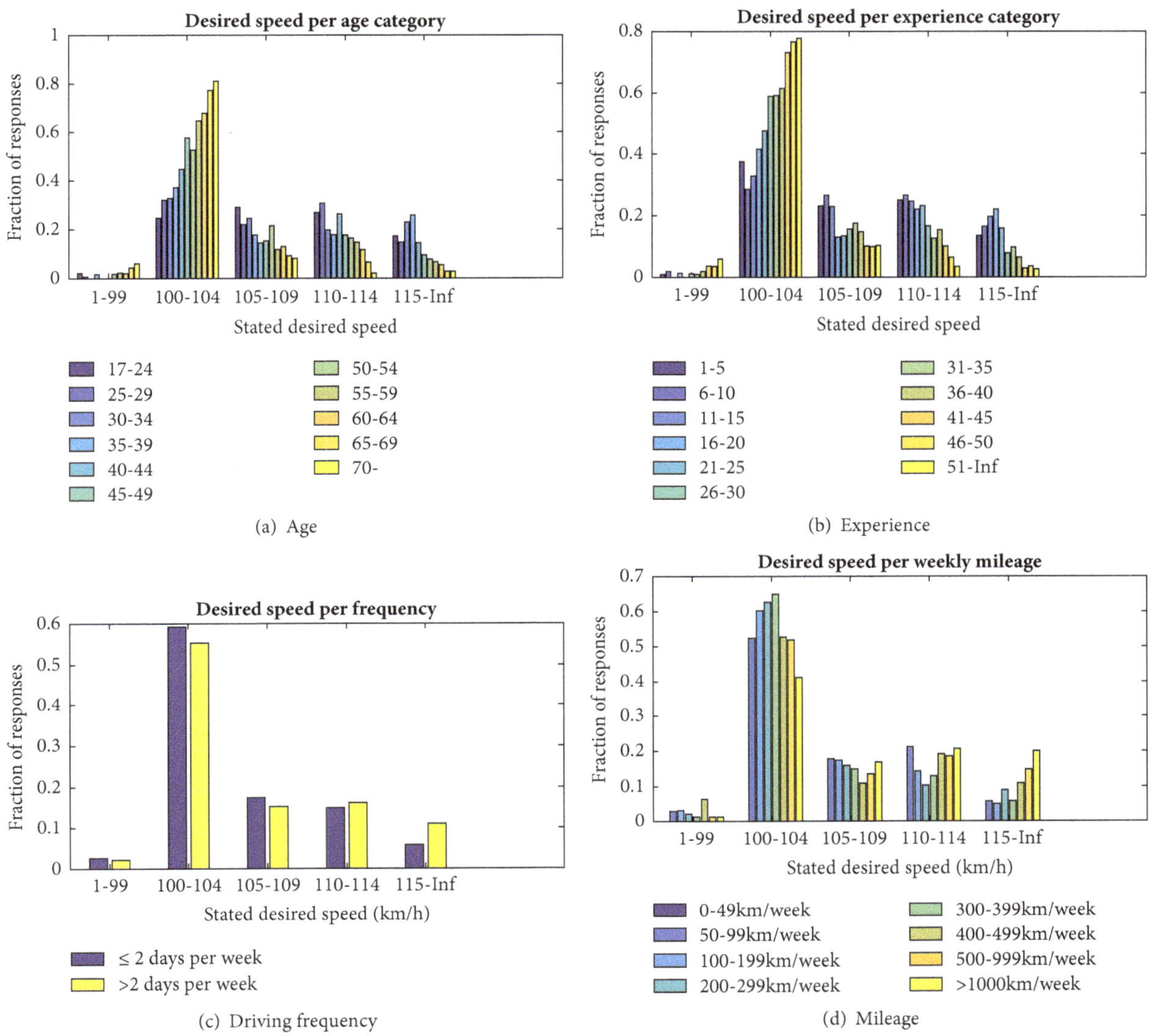

FIGURE 4: Overview of the impacts of various factors to desired speed.

which has been used to discuss the trip with the participants afterwards. Chronologically, the drives took place before the survey. This means that situations from the drives could be used in the survey, and the choice was recorded for the same person—once as participant and once as respondent. With the information from both sources, the validation took place in three stages.

First, the strategy/strategies of a participant was compared with the answers he gave in the survey. For each participant, it can be checked whether the answers given in the survey are in line with the strategy shown during the drive. For several questions it has been analyzed whether the strategy in the survey matches the profile as found during the drive. The answers to 6 questions (i.e., questions 2, 3, 4, 6, 9, and 13) were analyzed and checked whether the chosen strategies match the set of strategies found during the drive. Besides this, it has been investigated whether the respondents who answered that they would follow strategy 4 in question 11 also have strategy 4 in the set of strategies applied during the drive. Unfortunately, strategy 4 has many different appearances and only 1 question explicitly referred to strategy 4. So, even if the results of the drive in the instrumented vehicle and the survey do not match, results from the survey might still hold.

Second, we can identify specific videos to situations that an individual driver encountered and check to which extent the stated action is the same as the revealed action in this situation. This is limited to a maximum of one driver per question in the survey.

Third, the desired speed is compared for the drivers. This can be done for the drivers individually, as well as for the whole population, linked to the drivers characteristics.

The remainder of this section first discusses the setup of the drives and then discusses the validation results.

5.1. Setup of Drives. For the drives and interviews, participants were recruited using the personal and professional networks of the researchers involved. The professional networks included people from the university, the Dutch main road authority (Rijkswaterstaat), and the local gardening center (part-time job of the student). These networks were employed as the base starting point, leading to a wider variety of participants. As a compensation for their contribution, all participants received a 25-euro gift card. In total, a group of 25 people both participated in the drive and took part in the survey. During their trip, three cameras were installed in the vehicle: one pointing forwards, one backwards, and one at the driver.

The drives took place in a period of 7 days, between 1 and 10 February 2016, during which the weather conditions were normal and did not vary much. The 40 km route for the drive includes the freeways A13, A20, and A4 in the east, south, and north direction, respectively. The freeways have different speed limits, various number of lanes, and speed enforcement (by police patrol, permanent and temporary speed cameras, and average speed check). Most participants come from the area and are familiar with the main freeways. A part of the route (the A4 stretch) was opened less than a month before the drives took place, which means most participants were unfamiliar with that road stretch. Ideally, people with different levels of route familiarity would be needed to participate in the drive.

Before the drive, drivers were instructed on the use of the vehicle; this was a normally operated vehicle, a Toyota Prius. The instrumentation of the car was not intrusive, nor limiting the operation of the car in any way. The car was equipped with an automatic gearbox (less common in the Netherlands than a manual gearbox). The drivers were informed about the route, and, in addition to that, if preferred by the participant, a hand-held navigation device was programmed to guide them. The drivers were instructed to drive safely and to drive as they normally would. During the drive, participants were alone in the car.

After the drive, an interview was carried out. This interview was semi-structured and consisted of two parts: one general part related to the driving style in general and another part concerning the particular drive of the participant. During the interviews, the four strategies were discussed with the drivers. They were asked which strategies they have been applying throughout the trip and which they would apply in other conditions. From this interview, their considerations and strategy were derived.

5.2. Validation Results. This section describes the validation results: first the chosen strategies of the participants and then the choice in the identical situation as part of the survey. Finally, the desired speed is compared.

5.2.1. Strategies per Participant. In the survey, all 25 participants are requested to choose 6 times between strategies 1,2, and 3. Accounting for 7 missing or other answers, that gives 6×25-7=143 strategies. Moreover, 4 out of these 25 respondents answered to use strategy 4 in question 11. Hence, in total, we have 143+4=147 chosen strategies, which we can compare with the set of applied strategies found in the drive. For the majority of these chosen strategies (127 choices, or 86%) the given response is in line with one of the strategies derived from the drive, and so the rest (20 choices, or 14%) is not in line. This has been built up as follows: 1 respondent indicated a different strategy in 4 questions; 1 respondent indicated a different strategy in 3 questions and the same strategy in 1 question; 2 respondents indicated a different strategy in 2 questions and the same strategy in 2 other questions; 9 respondents indicated a different strategy in 1 question and the same strategy in 3 questions. This means that 12 respondents indicated all questions in line with the strategies from the drive. With this 86% match between choices in the survey and the choices of the participants in the experimental drive, we can conclude that the survey is well in line with the results shown in the drive and can therefore be used as representative for real driving behavior.

5.2.2. Choice in Identical Situation. For the second validation approach, using exactly the same traffic situation for exactly the same driver, 7 questions could be checked. In 4 out of the 7 cases, the stated preference (in the survey) was exactly the same as the behavior the participant showed during the drive. Given the number of options (approximately 5 per question), this is pretty consistent. However, since the drivers have been shown to have multiple strategies while driving (see Table 3), other approaches of coping with a specific situation did not come as a surprise.

5.2.3. Desired Speed. The desired speeds are given in Figure 3. For the participants, the mean desired speed is 108.0 km/h with a standard deviation of 8 km/h. This is higher than the desired speed of the respondents of the survey (104.5 km/h). This might be due to the age of the participants. Figure 4 shows how the desired speed depends on the age. Section 4.3 showed that the desired speed is lower for higher ages, and the age for the respondents was significantly higher than for the participants in the test drive (see also Figure 2).

6. Lane Changing in Other Countries

To put the results in the perspective, a comparison has been made. The same survey has been distributed to drivers in two other countries: Switzerland and the United States of America (USA). This section describes the setup and results of this comparison.

6.1. Setup of the International Comparison. In Switzerland, traffic regulations are comparable to the Netherlands. The main difference is how these laws are enforced. Switzerland is known for their high traffic fines in case of noncompliance. Also, road design aspects are different. For example, the Swiss guidelines show a shorter length of the deceleration lane (150-170 m) than the Dutch guidelines (250 m).

The rules in the United States are more different. Federal laws do not prohibit overtaking at the right on multilane freeways; however, some states have this prohibition, or a

TABLE 5: Comparison of responses of Dutch and Swiss drivers for scenarios 3, 5, and 13.

Nr	Question	Response	Nationality	
			Dutch	Swiss
3	If you overtake a slower predecessor (passenger car), would you do this according to Strategy 1 or 2?	Strategy 1	56%	73%
		Strategy 2	43%	27%
5	Would you overtake another vehicle via its right side, when there is no possibility to overtake via the left side?	Yes	16%	7%
		No	84 %	93%
13	If you have to take the exit 600 metres up ahead, would you overtake a truck just before the off-ramp?	Yes, with constant speed	20%	12%
		Yes, with speed increase	19%	3%
		No	61%	85%

prohibition of driving left for a prolonged period of time without overtaking. For an overview of the regulations in different states, see [30].

For the comparison, first the desired speeds are tested for differences. The independent sample t-test cannot be used for this since this test assumes that the dependent variable is normally distributed. Therefore, the nonparametric Mann–Whitney test is used to check whether there are differences in the medians of the desired speeds between (1) the Dutch and Swiss drivers, and (2) the Dutch and the American drivers.

Since the goal of the international survey is to identify differences, a smaller sample suffices. To find differences in resulting behavior, some answers of respondents can be grouped. For example, in scenario 10 the four multiple choice answers have been aggregated to two categories, namely, keep right behavior and keep lane behavior. This increases the statistical power of the survey, while at the same time justifying the goal of finding differences in revealed behavior. Using a Chi-square test, statistical significance was tested for confidence level of 95% and a confidence interval of 5%. Important to note is that besides the statistical correlation, also the association is measured. The first indicates how unlikely it is that the result could be obtained by chance; the second one indicates how strong the relation is. To this end, the ϕ value of Cramer's V test is used, with values under 0.1 being considered negligible [31].

The survey has been spread in Switzerland and the USA via the academic network. In total, the survey has been completed by 59 Swiss respondents and 40 respondents from the USA.

6.2. Results: Small Differences. It was found that the desired speed was not significantly different in Switzerland compared to the Netherlands. Mean speeds are close to each other: the Netherlands 104.5 km/h and Switzerland 104.1 km.h. The results of the test show no significant difference in the median values (U=36, z=0.208, p=0.835). In the USA, a higher speed was found: 109.1 km/h as average. Also the median speed was considerably and significantly higher: 110 in the USA versus 100 km/h in the Netherlands (Mann–Whitney test results: U=33, z=4.055, p=0.0005).

Then the lane choice strategy was considered. With the clustered answers, most questions led to statistically significant different fraction of people choosing a lane change strategy. However, all differences found, even though statistically significant, are negligible in size, according to Cramer's V test.

Table 5 presents an overview of the unclustered responses that showed a small size difference for Dutch and Swiss drivers. The results show that Swiss drivers tend to increase their speed less often when they overtake a slower predecessor in comparison with Dutch drivers. Furthermore, less Swiss drivers overtake another vehicle on the right side than Dutch drivers. This is observed in both scenarios that concerned right overtaking behavior, although in only one scenario this difference was found significantly different by the Chi-square test. Furthermore, only 15% of the Swiss respondents would overtake the truck just before the off-ramp, while 39% of the Dutch respondents would do so. This results could be due to the shorter ramps in Switzerland.

For American drivers, more clear differences exist concerning the lane choice. Figure 5 shows the most relevant examples. In short, US drivers are less bound to go to the right and have a higher tendency to overtake at the right. This can be explained by the different legislation in the USA.

All in all, we conclude that, with similar legislation and similar conditions (European driving), similar driving behavior can be expected.

7. Discussion and Conclusions

In this paper, people's driving strategies on the freeway have been studied. A large scale survey has been carried out. In total 1,258 respondents have filled out an online survey containing video clips. The survey has been validated by drives in an instrumented vehicle. For 25 people the answers in the survey were compared with the choice of strategy in a real-world case. Apart from the general chosen strategy, also a few identical situations could be identified, which further validated the survey.

The survey showed that most people choose a speed first and adapt the lane based on it. Less often people would have a strategy with a desired speed, but change their desired speed during overtaking. Also the strategy to choose a lane and adapt the speed accordingly is chosen less often than the strategy with the fixed desired speed. Only a small fraction of the drivers would neither have a desired speed nor a

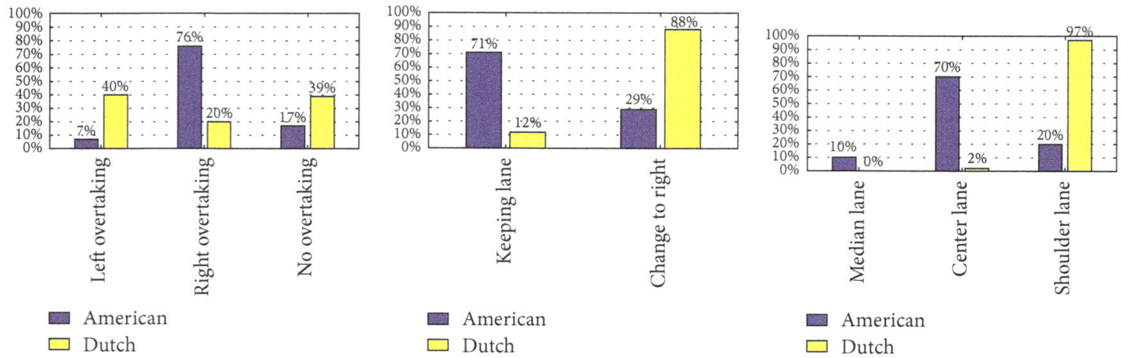

(a) Overtake slower driver that is unnecessarily driving in the second lane from the left in question 1

(b) Keep right on a 3-lane motorway

(c) Stated lane choice

FIGURE 5: Differences in lane choice for US drivers.

desired lane. A large number (75-80%) of the respondents indicated that they would return to the right lane, and over 80% refrained from overtaking at the right.

In over 80% of the cases, drivers were willing to adapt their speed or (more frequently) their lane to facilitate a mandatory lane change of another driver. This courtesy behavior should definitely be incorporated into microscopic simulation models.

In addition, the paper investigated a desired speed distribution for freeways with a 100 km/h speed limit. On average, the desired speed lies slightly above the speed limit, and younger drivers had a significantly higher desired speed. Moreover, drivers, who had higher mileage, had a higher desired speed than those who drive less; this effect is statistically significant, but very small in size.

The drives revealed that people have different strategies throughout their trip. For instance, they sometimes stick to a specific speed and adapt the lane accordingly, and sometimes they stick to a lane and adapt their speed. However, in general it can be concluded that there are various lane change and lane choice strategies. The research steps focus on developing the specific algorithms describing the decision-making in each of the strategies, and the effects thereof. These can then be implemented in microscopic simulation models. Preliminary progress steps have been published [32]. That work can now be expanded using the distributions found here.

The results presented here are valid for Dutch drivers. For Switzerland, with similar driving regulations, similar behavior was found. However, for the USA where the keep-right-unless-overtaking rule does not apply as in Europe, behavior was found to be different. Percentages in each of the categories of behavior are therefore expected to be similar for Europe but will differ significantly for other continents. A similar study could reveal the driving behavior in these other continents. The setup presented in this paper can be used for this goal. Recommendation for future research hence includes performing this study for an international group of respondents and comparing driving strategies of drivers from various countries.

The current study contributes to a more realistic description of driving behavior and would hence improve the applicability of microscopic simulation tools for ex-ante traffic engineering changes. These tools describe how drivers will drive with a particular road layout. The resulting traffic patterns will follow from the microscopic interaction in the simulation. It would be interesting to assess how the effects of heterogeneity in lane change strategies work out for traffic operations. Also for assessing a new road design or new lane change or merging regulations, the insights from this paper could be used to improve microscopic traffic simulation models and to change them into powerful tools, also for these kinds of applications.

Conflicts of Interest

The authors declare that there are no conflicts of interest regarding the publication of this paper.

Acknowledgments

This research is supported by the Netherlands Organization of Scientific Research; the research has been conducted at the ITS Edulab, a cooperation between Rijkswaterstaat and Delft University of Technology.

References

[1] C. F. Daganzo, "A behavioral theory of multi-lane traffic flow. Part I: long homogeneous freeway sections," *Transportation Research Part B: Methodological*, vol. 36, no. 2, pp. 131–158, 2002.

[2] V. L. Knoop, A. Duret, C. Buisson, and B. Van Arem, "Lane distribution of traffic near merging zones influence of variable speed limits," in *Proceedings of IEEE Intelligent Transportation Systems*, 2010.

[3] S. Ahn and M. Cassidy, "Freeway traffic oscillations and vehicle lane-change," in *Proceedigs of the International Symposium of Transportation and Traffic Theory*, R. E. Allsop, M. G. H. Bell,

and B. G. Heydecker, Eds., pp. 691–710, Elsevier, Amsterdam, the Netherlands, 2007.

[4] J. A. Laval and C. F. Daganzo, "Lane-changing in traffic streams," *Transportation Research Part B: Methodological*, vol. 40, no. 3, pp. 251–264, 2006.

[5] M. Keyvan-Ekbatani, V. L. Knoop, and W. Daamen, "Categorization of the lane change decision process on freeways," *Transportation Research Part C: Emerging Technologies*, vol. 69, pp. 515–526, 2016.

[6] V. Knoop, M. Keyvan-Ekbatani, M. De Baat, H. Taale, and S. Hoogendoorn, "Strategy-based driving behaviour on freeways: Findings of test-drive and on-line survey study," in *Proceedings of the 96th Annual Meeting of the Transportation*, 2017.

[7] T. Toledo, C. Choudhury, and M. Ben-Akiva, "Lane-changing model with explicit target lane choice," *Transportation Research Record: Journal of the Transportation Research Board*, pp. 157–165, 1934.

[8] A. Kesting, M. Treiber, and D. Helbing, "General lane-changing model MOBIL for car-following models," *Transportation Research Record: Journal of the Transportation Research Board*, vol. 1999, no. 1, pp. 86–94, 2007.

[9] S. Moridpour, M. Sarvi, and G. Rose, "Lane changing models: a critical review," *Transportation Letters*, vol. 2, no. 3, pp. 157–173, 2010.

[10] Z. Zheng, "Recent developments and research needs in modeling lane changing," *Transportation Research Part B: Methodological*, vol. 60, pp. 16–32, 2014.

[11] T. L. Pan, W. H. K. Lam, A. Sumalee, and R. X. Zhong, "Modeling the impacts of mandatory and discretionary lane-changing maneuvers," *Transportation Research Part C: Emerging Technologies*, vol. 68, pp. 403–424, 2016.

[12] J. A. Laval and L. Leclercq, "A mechanism to describe the formation and propagation of stop-and-go waves in congested freeway traffic," *Philosophical Transactions of the Royal Society A: Mathematical, Physical & Engineering Sciences*, vol. 368, no. 1928, pp. 4519–4541, 2010.

[13] S. Ossen, S. Hoogendoorn, and B. G. H. Gorte, "Interdriver Differences in Car-Following: A Vehicle Trajectory-Based Study," *Transportation Research Record: Journal of the Transportation Research Board*, no. 1965, pp. 121–129, 2006.

[14] K. Ahmed, M. Ben-Akiva, H. Koutsopoulos, and R. Mishalani, "Models of freeway lane changing and gap acceptance behavior," in *Transportation and Traffic Theory*, vol. 13, pp. 501–515, 1996.

[15] W. J. Schakel, V. Knoop, and B. Van Arem, "Integrated lane change model with relaxation and synchronization," *Transportation Research Record*, vol. 2316, pp. 47–57, 2012.

[16] A. Kondyli and L. Elefteriadou, "Driver behavior at freeway-ramp merging areas: Focus group findings," *Transportation Research Record: Journal of the Transportation Research Board*, no. 2124, pp. 157–166, 2009.

[17] P. Van Koningsbruggen and T. Stranner, "EGNOS from an European Service Provider Perspective—Test results and Conclusions," Technical report, ERTICO, GNSS-enabled Services Convergence, 2011.

[18] J. C. F. de Winter, S. de Groot, M. Mulder, P. A. Wieringa, J. Dankelman, and J. A. Mulder, "Relationships between driving simulator performance and driving test results," *Ergonomics*, vol. 52, no. 2, pp. 137–153, 2009.

[19] J. C. F. De Winter and D. Dodou, "The driver behaviour questionnaire as a predictor of accidents: A meta-analysis," *Journal of Safety Research*, vol. 41, no. 6, pp. 463–470, 2010.

[20] J. Davey, D. Wishart, J. Freeman, and B. Watson, "An application of the driver behaviour questionnaire in an Australian organisational fleet setting," *Transportation Research Part F: Traffic Psychology and Behaviour*, vol. 10, no. 1, pp. 11–21, 2007.

[21] J. Reason, A. Manstead, S. Stephen, J. Baxter, and K. Campbell, "Errors and violations on the roads: A real distinction?" *Ergonomics*, vol. 33, no. 10-11, pp. 1315–1332, 1990.

[22] A. Bener, M. G. A. Al Maadid, T. Özkan, D. A. E. Al-Bast, K. N. Diyab, and T. Lajunen, "The impact of four-wheel drive on risky driver behaviours and road traffic accidents," *Transportation Research Part F: Traffic Psychology and Behaviour*, vol. 11, no. 5, pp. 324–333, 2008.

[23] A. Bener, T. Özkan, and T. Lajunen, "The Driver Behaviour Questionnaire in Arab Gulf countries: Qatar and United Arab Emirates," *Accident Analysis & Prevention*, vol. 40, no. 4, pp. 1411-1417, 2008.

[24] J. Li, H. van Zuylen, and E. van der Horst, "The Driver Behaviour Questionnaire: An Investigation Study Applied to Chinese Drivers," in *Computer-based Modelling and Optimization in Transportation*, J. F. de Sousa and R. Rossi, Eds., vol. 262 of *Advances in Intelligent Systems and Computing*, pp. 433–447, Springer International Publishing, Cham, Switzerland, 2014.

[25] S. De Craen, D. A. Twisk, M. P. Hagenzieker, H. Elffers, and K. A. Brookhuis, "The development of a method to measure speed adaptation to traffic complexity: Identifying novice, unsafe, and overconfident drivers," *Accident & Analysis Prevention*, vol. 40, no. 4, pp. 1524–1530, 2008.

[26] E. Lehtonen, V. Havia, A. Kovanen, M. Leminen, and E. Saure, "Evaluating bicyclists' risk perception using video clips: Comparison of frequent and infrequent city cyclists," *Transportation Research Part F: Traffic Psychology and Behaviour*, vol. 41, pp. 195–203, 2016.

[27] R. T. Steinbakk, P. Ulleberg, F. Sagberg, and K. I. Fostervold, "Analysing the influence of visible roadwork activity on drivers' speed choice at work zones using a video-based experiment," *Transportation Research Part F: Traffic Psychology and Behaviour*, vol. 44, pp. 53–62, 2017.

[28] M. De Baat and V. Knoop, "Movies lane changing," available at: http://doi.org/10.4121/uuid:771cdc12-ddd8-4607-8ec2-bab096351824, 2018.

[29] R. Fuller, "Towards a general theory of driver behaviour," *Accident Analysis & Prevention*, vol. 37, no. 3, pp. 461–472, 2005.

[30] M. I. O. T. MIT, "State "keep right" laws," URL: http://www.mit.edu/jfc/right.html, Accessed: 26-07-201.

[31] L. M. Rea and R. A. Parker, *Designing and conducting survey research: A comprehensive guide*, John Wiley & Sons, 2014.

[32] M. Keyvan Ekbatani, V. Grebert, W. Daamen, and V. L. Knoop, "Lane changing and speed interaction on freeways: An analytical microscopic study," in *Proceedings of the 11th Traffic and Granular Flow Conference, TGF15*, pp. 1–8, Springer, Nootdorp, The Netherlands, 2016.

13

The Surrogate Safety Appraisal of the Unconventional Elliptical and Turbo Roundabouts

Giovanni Tesoriere,[1] Tiziana Campisi,[1] Antonino Canale,[1] and Tedi Zgrablić[2]

[1] *Faculty of Engineering and Architecture, University of Enna Kore, Italy*
[2] *Faculty of Engineering and Architecture, University of Maribor, Slovenia*

Correspondence should be addressed to Tiziana Campisi; tiziana.campisi@unikore.it

Guest Editor: Daiva Zilioniene

Double-lane roundabouts have been created in many European countries over the past few centuries and are now characterized by an unsafe geometric development and by a low sustainability capacity or level. In this regard, new double-lane geometries have been implemented to overcome to these critical points. This article shows a comparison of two nonconventional double-lane roundabout schemes defined as elliptical and turbo. Considering this research on the unsafe and congested conditions for each road schemes at grade, the microsimulation approach allows comparing schemes of intersections not yet realized in order to be able to evaluate the critical issues. A symmetric traffic distribution and an identical vehicle mix for both design solutions are considered. The research was conducted considering two different double-lane roundabout-turbo roundabout and the elliptical roundabout. By comparing their geometry and technical elements, this article assumes that turbo roundabout due to its physical separating traffic lanes in the central circulatory carriageway will enable potentially better traffic safety conditions. This article has the following main goal: a comparison of traffic safety using VISSIM microsimulator and SSAM tools. The results can provide to show safety level on investigated scenario considering level of service (LOS) and also the possibility of obtaining time to collision (TTC) and postencroachment time (PET) through the use of surrogate parameters obtained by SSAM tool. In fact, the surrogate safety parameters allows evaluating the possible collision scenarios between them, according to the trajectories of the single vehicles. This assessment is useful in order to be able to evaluate by the local authorities which of the examined schemes can provide greater negativity in the construction and operation phase. Therefore this comparative analysis allows reducing, in the preliminary phase, possible security impacts and also economic ones for the community.

1. Introduction

Today, the different road intersections are increasingly developing. In general, the road intersections in roundabout allow a substantial reduction of the points of conflict between vehicles and vehicles with pedestrians compared to the traditional intersections with precedence or with traffic lights. Proper planning of the roundabout is necessary to reduce the percentage of deaths and injuries on the roads of each countries.

The research was conducted considering two different double-lane roundabouts titled elliptical and turbo. The following paragraphs show a geometric description of the intersections and a detailed analysis of the steps that led to the comparison of the two geometries in safety terms through the use of the VISSIM microsimulator [1] and the dedicated SSAM tool [2].

There was the first phase of selection of the geometries to be analyzed, the traffic scenarios with relative distribution of the vehicle typologies and after the calibration of the model of the microsimulator in order to make the most reliable results. Finally, they were processed in terms of surrogate safety through a dedicated tool.

The comparison of the results shows how to vary the flows and geometries of unconventional satin road intersection you can have more or fewer distributions of certain types of collisions between vehicles along the geometry. In fact, the evaluation and also the comparison of supposed collision are investigated considering surrogate parameters. Surrogate parameters represent a summary assessment in terms of safety, by placing a double-lane rotary comparison on the central circulatory carriageway.

The hypotheses on which this research is based are related to the definition of the surrogate safety of nonconventional roundabout geometries. Specifically, attention was focused on the comparison of two widespread schemes in Europe, in order to be able to investigate in advance which of these, with the same outflow conditions, be better in terms of service level and therefore road safety.

2. Double-Lane Roundabout Geometry

The geometries chosen for the comparison are of a satin type and are referred to as turbo and elliptical roundabout, respectively. Both have in the case examined a double lane in the central circulatory carriageway. At the end of the 90s, in the Netherlands it is following the design and construction of the turbo-type roundabouts. Scheme of stories has evolved over the years by presenting the preselection of the manoeuvres along the arms and therefore they are vehicles of not manoeuvring abrupt or wrong.

The limitation of the manoeuvres was made possible not only by the preselection along the lanes of the arms but also by the presence of curbs or signs of separation between the manoeuvres along the main ring. From here there has been an increase in safety with a reduction in the number of deaths and injuries along these intersecting geometries. Fortuijn was the first to study and design the turbo-type intersections at grade [3].

Unlike the turbo roundabout, the elliptic schemes are not defined by separated lanes of the circulatory ring. Sometimes there are not even the preselection lanes along one or more arms of the intersection.

Several studies have been conducted in recent years on the assessment of the potential of unconventional roundabout in terms of safety and environmental impact, evaluating both the single and double lane on the central ring in different traffic flow conditions like those described by [5, 6].

The evolution of turbo-type rotary geometries has in recent years been carried out in parallel with the development of public and private transport systems in terms of both eco-mobility and traffic decongestion systems with the reduction of the automatic system in favour of collective passengers transport [7].

The elliptical roundabout is a form of a classical double-lane roundabouts with the emphasis that it is not a proper circular shape, but its geometry corresponds to a closed curve-ellipse. This form of double-lane roundabout is characterized by two semiaxes (R and 2R). Usually the higher semiaxis (2R) is located in the main traffic direction while the lower semiaxis is located in the secondary traffic direction. This form of elliptical roundabout is most often applied in areas where, due to spatial conditions, a classical circular roundabout cannot be performed and where there is more traffic at the main traffic direction. Mainly in the world it is not so much represented, but according to [8], elliptical roundabout is widely used in Iran. Figure 1 represents, respectively, the two geometries examined, considering the main and secondary traffic directions. Both road geometries are not signalized; therefore the precedence of vehicles circulating on the ring has been estimated with respect to those that are introduced by the arms in the various directions, respectively.

3. Calibration of Microscopic Traffic Simulator: Methods and Application

Microsimulation approach and microanalytic simulation are linked to a category of analytical tools that perform highly detailed analysis of activities such as highway traffic flowing through road networks. Traffic microsimulation models generally is defined considering a large number of parameters that must be calibrated before the prediction of applied model.

In this paragraph, the details of the traffic microsimulator used and the tool dedicated to the assessment of road safety have been reported. The definitions of the parameters and the calibration of the models followed the processing of the data and the comparison of the two geometries.

3.1. Microsimulation Tool Description. The modelling phase of the compared roundabouts was made by first considering the geometric parameters and the main and secondary traffic directions. It was carried out considering some hypotheses related to the models on which the VISSIM software developed by PTV is based.

The parameters mentioned above mainly include the characteristics of the vehicle, the composition of the traffic flow in terms of percentages of heavy or other light vehicles, the distribution of the speed of desire, the traffic flows, and some behavioural parameters linked to the style of driving and the driver's psychophysical status [9].

The simulated model considers a random generation in the distribution along the vehicle queues in terms of both composition and volume. It also considers that the priority of the fastest moving vehicle may decrease when it reaches and perceives the presence of a slower vehicle in front.

The basic concept of this model is that the driver of a faster moving vehicle starts to decrease own speed as he reaches his individual perception threshold to a slower moving vehicle. Vehicles would be randomly generated as per the given volume and composition.

In order to be able to evaluate surrogate security parameters using the SSAM specific tool, it is first necessary to generate a layout from VISSIM and through specific files in "trj" format possible to trace the trajectories of individual vehicles and therefore to the propensity that they have to enter in fact that the SSAM tool analyzes vehicle interactions to identify events and catalogue all the events found.

For each such event, SSAM also calculates several surrogate safety measures, including the following:

(i) Minimum time to collision (TTC)

(ii) Minimum postencroachment (PET).

The simulation of traffic through the VISSIM software initially considers the definition of traffic flows and their

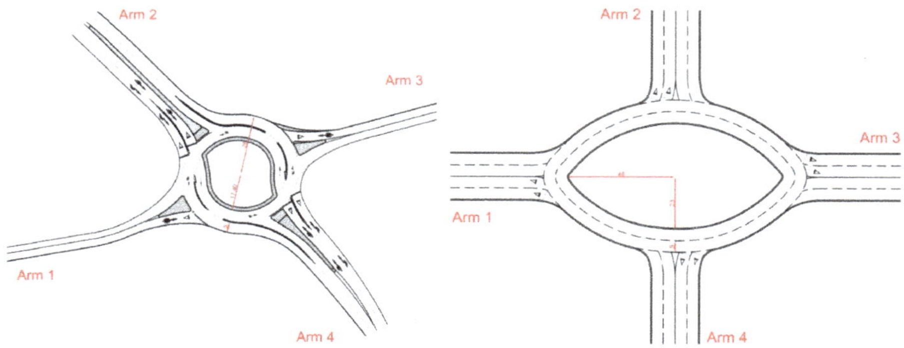

FIGURE 1: Details of turbo (left) and elliptical (right) geometry.

distribution along the intersection arms through the creation of special O/D matrices. In the rotary intersections, the deflection curves created by the motion of the vehicles depend on the geometry of the roundabout itself and in particular on the size of the central island. Another important geometric parameter is the determination of the input radius of the branches on the circulatory ring. They depend on both the capacity and the safety of the examined scheme. The aforesaid parameter is related to the width of entry, the width of the carriageway on the circulatory ring, and the geometry of the central island. These parameters affect the travel speed of the vehicle and therefore the driving comfort and road safety.

Federal Highway Administration (FHWA) Roundabout Guidebook [10] was considered to select roundabout design parameters. After designing all of the roundabouts in VISSIM for both geometries and conditions, traffic loads were assigned. In this study, the actuated traffic control was used in roundabouts. In addition, the right-of-way control was considered for not signalized roundabouts that were given priority to circulating flow. This controlling system allowed roundabouts to perform better than uncontrolled systems. Each roundabout was designed considering double lane on circulatory carriageway.

The following assumptions were considered for this study are linked to urban areas conditions. The analyzed geometries are characterized by a speed value of about 50 km/h on circulatory carriageway while the access speeds from each arm are assumed to be less than 35 km/h. In elliptical roundabouts, large radius is twice the small radius.

The exact disposition of the data and the traffic flows have been calibrated through the origins of origin and are examined in the next paragraphs.

In general, it is possible to resort to traffic microsimulation instruments if, in a preventive manner and for small areas, the possible introduction of different geometrical schemes is to be analyzed and can be compared from both the point of views of safety-related issues (i.e., precedence, percentage of thinking vehicles, and weak road users) in relation to impacts of various kinds such as the environment [11].

This research investigates on two unconventional geometrical layouts (elliptical and turbo roundabouts). They are very recent or just theoretic, and it is often not possible to perform an observational evaluation by collecting crash and operational data.

The microsimulation software tool VISSIM (Version 9) [1] was used to simulate operations at the roundabouts. This software can be used to generate all types of outputs simultaneously (safety, traffic performance, and capacity), allowing a more complete and comprehensive picture of the combined effects in making comparisons of various roundabout design scenarios.

The VISSIM output, in particular files linked to trajectories details, was imported into the Surrogate Safety Assessment Model (SSAM) [2], a software application designed to consider vehicle trajectory data output from microscopic traffic simulation models to derive proximal measures such as conflicts, based on thresholds for either time to collision (TTC) or postencroachment time (PET).

The estimated conflicts were then applied in crash-conflict models for roundabouts developed in other research to compare the roundabout designs on the basis of expected crashes.

The approach of using proximal safety indicators such as conflicts has been suggested as an alternative to the use of crash data, especially where the facility being evaluated does not exist and thus does not have a crash record.

It is possible to consider the frequency of conflicts and the severity measures through the use of intersection conflict index in agreement with Sayed and Zein [12].

This suggests the possibility of using significantly shorter study periods to establish statistically reliable results. The fundamental parameters beyond those previously described are related to the capacity of the analyzed geometry, to the estimated or hypothesized traffic volume and to their relationship. The comparison of the results took place considering the geometries shown in the Table 1.

According to Hatami and Aghayan [8], it is possible to admit that the elliptic roundabouts have now become an unconventional type of roundabout.

The geometry of the aforesaid roundabout leads to a reduction of the positive effects especially with the increase of the radius: in fact, there is a roundabout an elliptic not semaphore, like that object of study, to an increase of the

Table 1: Details of double-lane roundabouts simulated.

Number of lanes	Types of roundabout	Speed limits[km/h]	Traffic light	Central island radii[m]	Total Traffic Volume[veh/h]	Max Capacity
Two lanes	Turbo	50	Absence	Main radius R=56m	3040veh/h	3480veh/h
Two lanes	Elliptical	50	Absence	Main radius R=56m Secondary radius r=23m	2400veh/h	3000veh/h

criticalities both in terms of capacity and of delay. Against the increase of the speed in it involves a positive effect both on the capacity and on the delay.

The Surrogate Safety Assessment Model (SSAM) is a specific tool created by Federal Highway Administration focused on identification, automatically classification of the presence of potential conflicts generated by the trajectories of vehicles circulating on road schemes. The tool also has integrated statistical analysis functions for the frequency and severity of conflicts that can help in the design of traffic safety structures. The following paragraphs describe the calibration steps of the model and report the relative results of the unidentified roundabout, providing the results with the help of the VISSIM software and the SSAM tool.

3.2. Surrogate Safety Tool SSAM. The name of SSAM tool is the acronym of Surrogate Safety Assessment Model. It has many parameters to configure that allow the user to visualize the analysis and customize the analysis results. Some of these data would be best imported from the simulation application(s) itself (e.g., "import traffic-stream data") using a common file format related to trajectories generated by microsimulator software.

Generally the data required to adequately analyze the traffic situation includes the following:

(i) Network geometry (i.e., number of lanes in each direction, turning pockets, and driveways)

(ii) Traffic-stream data (volume of traffic in each direction, change in the volumes over the simulation period)

(iii) Definition of signalized or not signalized intersection with specific priority of each arms or circulatory carriageway

(iv) Driver behaviour data (i.e., aggressiveness distribution and gap-acceptance criteria).

The evaluation of surrogate parameters allows an evaluation of the safety of the schemes analyzed, providing a greater propensity of the intersection to generate different types of conflict points.

3.3. Calibration Process to Microsimulation Tool. Calibration is the adjustment process to set model parameters and to improve the model's ability to reproduce local driver behaviour and traffic performance characteristics.

It is impossible for the researcher to have knowledge of the number of repetitions to be calibrated in advance, but it is possible to use an estimate of how many times it is necessary to calibrate the parameter and/or models investigated to obtain a statistically valid result. The required minimum number of model repetitions is computed using the following equation:

Table 2: Minimum number of repetitions needed to obtain the desired confidence interval.

Desired Range (CI/S)	Desired Confidence	Minimum Repetitions
0,5	99%	130
0,5	95%	83
0,5	90%	64
1,0	99%	36
1,0	95%	23
1,0	90%	18
1,5	99%	18
1,5	95%	12
1,5	90%	9
2,0	99%	12
2,0	95%	8
2,0	90%	6

$$CI_{1-\alpha\%} = 2 * \frac{t_{(1-\alpha/2),N-1} s}{\sqrt{N}} \tag{1}$$

where

CI(1-α)% = (1-α)% is confidence interval for the true mean, where alpha equals the probability of the true mean not lying within the confidence interval

t(1-α/2), N-1 is Student's t-statistic for the probability of a two-sided error summing to alpha with N-1 degrees of freedom, where N equals the number of repetitions

s is standard deviation of the model results.

Table 2 shows the link among above equation, the minimum number of repetitions for various desired confidence intervals, and desired degrees of confidence according to [2].

3.4. Calibration Targets. The model calibration process provides the best possible match between model performance estimates and real measurements.

However, there is a limit to the amount of time and effort that anyone can make to eliminate the error in the model, since the high repetition of the process often does not result in greater accuracy of the data.

Table 3 shows the criteria and calibration measures considering specific vehicle flow intervals adopted by DOT [13].

Table 3: Wisconsin DOT freeway model calibration criteria.

Criteria and Measures	Traffic Flow range	Calibration Acceptance Targets
	Individual Link Flows	
Within 15%	700veh/h<Flow<2700veh/h	>85%of cases
Within 100veh/h	Flow<700veh/h	>85%of cases
Within 400veh/h	Flow>2700veh/h	>85%of cases
Sum of All Link Flows		**Within 5%of sum of all link counts**
GEH statistic<5 for Individual link Flow		**>85%of cases**
GEH statistic<5 for all link Flow		GEH<4 for sum of all link counts
Travel times, Model Versus Observed		
Journay Time, Network Within 15%(or 1min if higher)		>85%of cases
Visual Adults		
Individual Link Speeds-visually Acceptable Speed-Flow relationship		To analyst's satisfaction
Bottlenecks		
Visually Acceptable Queuing		To analyst's satisfaction

Table 4: GEH value obtained in accordance with [4].

Roundabout schemes with 2 lanes on circulatory carriageway					
	Double lane	Turbo	Flower	Target	Elliptical
GEH on right lane	87,5	87,5	87,5	87,5	87,5
GEH on left lane	100	87,5	87,5	87,5	87,5

Table 5: O/D matrix selected to simulation.

O/D	Arml	Arm 2	Arm 3	Arm 4
Arml	0	0.15	0.7	0.15
Arm2	0.15	0	0.15	0.7
Arm3	0.7	0.15	0	0.15
Arm4	0.15	0.7	0.15	0

The calibration of the software model used is based on the model defined Wiedemann 74 used by PTV VISSIM (version 9) to simulate in a reliable and realistic way the traffic of vehicles inside unconventional roundabouts.

The car-following model was considered to be satisfactorily calibrated when the two curves overlap.

To further explore the model's validity, the Geoffrey E. Havers (GEH) statistic index (1,2) was used as the criterion for accepting the template.

The formula for the "GEHi Statistic" is

$$\text{GEHi} = \frac{\sqrt{2\,(M-C)^2}}{M+C} \tag{2}$$

And

$$\text{GEH} = \frac{\sum_{i=1}^{n} \text{GEHi}}{n} \tag{3}$$

GEH index is a global indicator widely used for the validation of traffic simulation models, especially when only aggregate values such as traffic flow counts in time-based detection stations and input capacity are available [14].

The GEH index value is the average over "n"; it was considered equal to 9, simulations of the index for each simulation i, as shown in Table 4.

3.5. Traffic Flow Conditions. Microsimulation approach is helpful to simulate different hypothesis of traffic operations changing schemes and different traffic conditions. The authors investigated the safety results of the two analyzed road schemes considering the 80% of the maximum value of traffic flow, corresponding to each roundabout capacity.

The simulation of the geometries described above was carried out by evaluating a ratio between Volume and Capacity of each single intersection equal to 0.8, thus equal to 80% of the maximum capacity of each scheme in a double lane examined.

In general, the performances of the road geometry are measured through its level of service while the safety measures can consider surrogate parameters such us the time to collision (TTC). This parameter is obtained considering the trajectories of the individual vehicles and the potential conflicts. Through the TTC it is possible to provide interesting information in advance, suggesting that the new layouts should be considered where justified by considerations on costs and benefits. The VISSIM output includes vehicle trajectory files that were then imported into the Surrogate Safety Assessment Model-SSAM, a software application designed to perform analysis of vehicle trajectory data output from microscopic traffic simulation models to derive proximal measures such as conflicts, based on time to collision (TTC) or postencroachment time (PET). The estimated conflicts were then applied considering conflict models already analyzed by other sources of literature. The approach of using proximal safety indicators such as conflicts has been suggested as an alternative to the use of crash data, especially where the facility being evaluated does not exist and thus does not have a crash record. In particular the flow values used are equal to 80% of the saturation flux of each of the two rotary geometries. In particular, reference was made to a symmetric O/D matrix for both roundabouts and shown as on Table 5.

The analyzed traffic mix refers to observed traffic distributions and consists of the following percentages: 70%of LPV

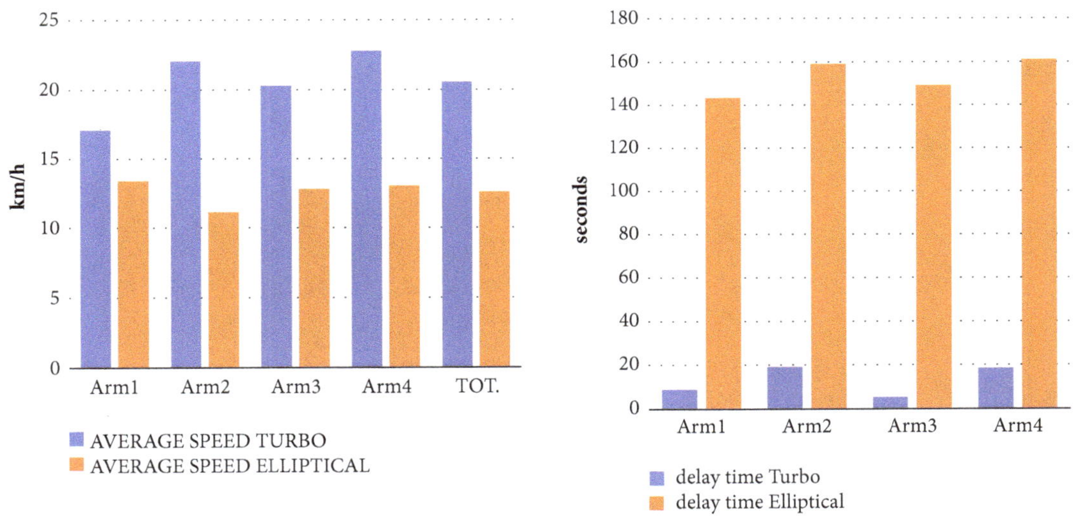

FIGURE 2: Comparison of average speed and delay time for investigated road scheme on saturated conditions.

(Light Passenger Vehicle), 20%of HPV (Heavy Passenger Vehicle, 5%of Buses and the same percentage is dedicated to Motorcycles. The comparison of the geometries was carried out by evaluating the parameters characterizing the LOS service level of the road intersections and in particular it was estimated: maximum and average length of the queues, average delay, average speed, and number of STOPs. In order to make the two geometries comparable, reference is made to normalized values.

4. Elliptical and Turbo Roundabout Operational Safety Results

Two schemes with two-lane circulatory ring were presented in this work. The choice of one or the other scheme depends on several factors. In general, an elliptical roundabout is a good choice when constraints such as precedence, existing road alignments, buildings, and/or wetlands affect the shape. As regards the turbo roundabout scheme, the separation of directional flows must be underlined through specific vertical signs or curbs [15].

In general, the physical separation of traffic lanes is interrupted only at entry points in the internal circulating carriageway. Since the weaving on the roundabout is no longer possible, drivers should be assisted by clear signage and lane signage.

4.1. Comparison of VISSIM Results. The simulation carried out with the VISSIM software, made it possible to compare the intersections described above through the evaluation of parameters relating to the service level of an infrastructure. In particular, the length of the queues, delays and the stops number were evaluated. These parameters make it possible to evaluate how much a road geometric scheme has, for a given vehicular flow, the propensity to be subject to congestion phenomena and therefore to a low level of service. This status implies the possible collapse of the mobility of the vehicles inside it. Graphs on Figure 2 shows a sharp difference in the delay values along the two geometries with a very high possibility of congestion in the case of the elliptic roundabout compared to the turbo type.

The graph therefore shows a higher average speed along the turbo-rotary because the preselection of the manoeuvres does not compromise the travel speed of the entire roundabout. It shows a total average speed value of about 20-23km/h for the turbo-type roundabout and about a reduced value of 40% for the elliptical roundabout. Likewise, it assumes almost zero or reduced delay values in the turbo-type intersection, while elevating values for the elliptical.

As far as the number of stops is concerned, there is a greater value in the arms B and D of the turbo roundabout, while they are practically of equal value along the elliptical roundabout in accordance with Figure 3. This phenomenon provides an increase in congestion and possible impacts in the aforementioned points.

An operational analysis is conducted to determine if the turbo roundabout will accommodate projected traffic volumes at an acceptable level of service (LOS). Roundabout LOS is measured in control delay consistent with the elliptical roundabout.

It is also fundamental to be able to understand the service level of the analyzed intersections, to evaluate the LOS parameter overall and/or for each arm. As can be seen from the graph shown in Figure 4, the value of LOS afferent to the elliptical roundabout remains almost constant along the arms for the traffic conditions previously discussed. In particular, the LOS=F=6 shows a bad condition of internal movement of this scheme in the roundabout, suggesting a widespread congestion of vehicular traffic to the flow analyzed. On the contrary, as regards the turbo-type roundabout, there is a more acceptable value of LOS, especially in arms 1 and 3 where the accumulated delays and the phenomenon of congestion seem to decrease by about 40% compared to two arms, 2 and 4.

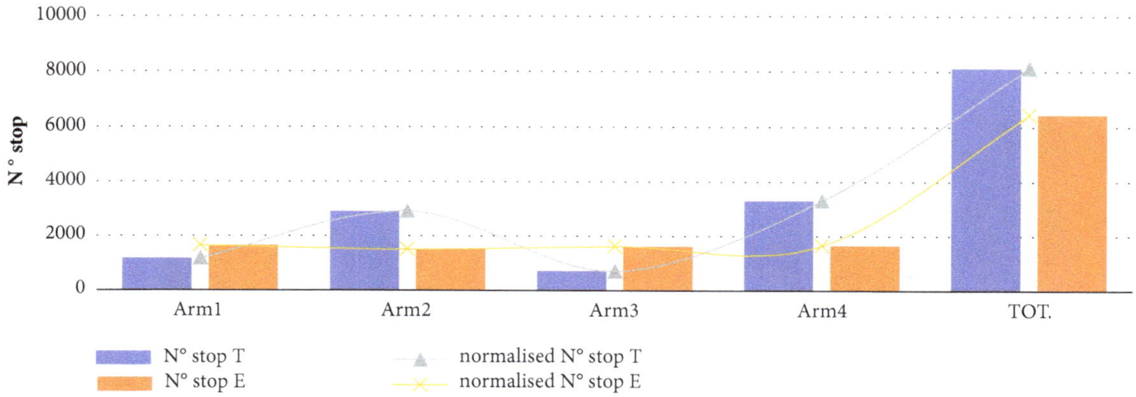

FIGURE 3: Comparison of number of stop related to turbo and elliptical scheme on congested conditions.

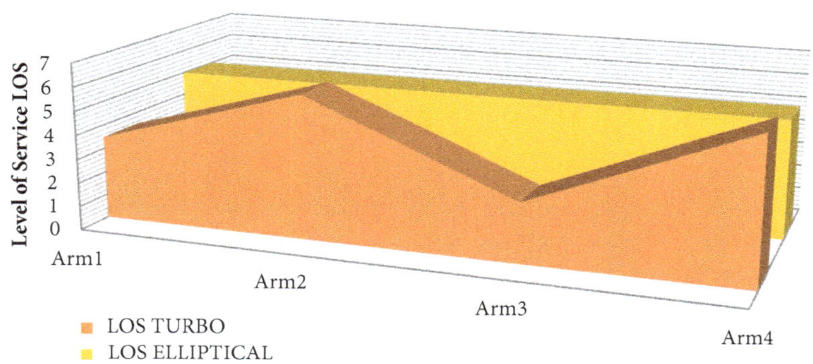

FIGURE 4: Level of service of investigated road schemes at grade.

The greater extension of the diameter along the main direction Arm1_Arm3 of the elliptical roundabout shows through the layouts produced by the microsimulator on Figure 5 a free flow condition near the circulatory carriageway while there is a greater thickening of vehicles along the arms for both the examined configurations. The lack of separation of the lanes on the main ring generates more collisions in the elliptical roundabout as explained in Figure 5.

4.2. Comparison of SSAM Results. TTC and PET are both indicators of proneness to establish a vehicle collision.

The probability of a conflict occurring is greater than the value of these parameters is lower [15]. The research was founded on specific range for both safety parameters in order to establish the possibility of defining a potential conflict, considering 0.1 s < TTC <1.5 s and 0.5 s < PET < 2.5 s. Among the data associated with each conflict is a MinTTC variable, which is the simulation time where the minimum TTC value for that conflict was observed [16]. That variable was used to filter out all conflicts that occurred during the simulation time. Two comparative safety evaluations were conducted based on different measures like described on Figure 6:

(i) The estimated conflicts from SSAM: this is done for entry flows of 0.8 V/C (absolute and normalized per 1000 vehicles) and for 2400 veh/h and 3040veh/h.

(ii) Estimated crashes are based on applying estimated conflicts (for entry flows of 0.8 V/C and for 3040 veh/h) in crash-conflict models; as can be seen from Figure 6, it assists a maximum PET value for both the turbo and elliptic type scheme below the value of 5. The average value, on the contrary, of results for the elliptic configuration is about 30% of the value average relative to the turbo. In terms of PET, there is a clear variability between the two geometries in terms of both maximum value and average value. In particular, the maximum value of the elliptical roundabout reaches a value equal to 5 determining an increase in the possible conflicts within it. Figure 6 shows the distribution of TTC and PET values considering turbo and elliptical scheme. In terms of average values it is possible to consider the high value of TTC for turbo and the same history for PET value.

We can notice a greater distribution in the arms, especially in the main direction with regard to the turbo-type roundabout, while an increase in yellow at the rotary crown couplings as regards the elliptical roundabout, equally distributed along the two directions. Equal distribution in the elliptic is denoted with regard to the conflicts shown in blue which instead are lower in the roundabout of the turbo type and bordering only in part of the lanes dedicated to the manoeuvre of turning to the right. Table 6 shows the comparison of the points

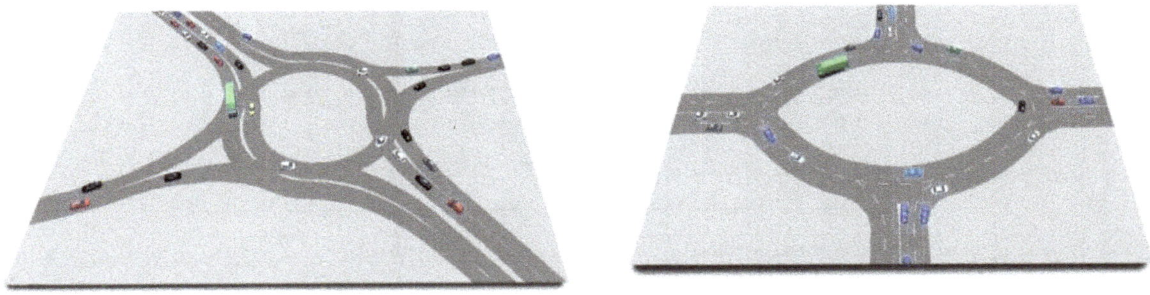

FIGURE 5: Microsimulator layout.

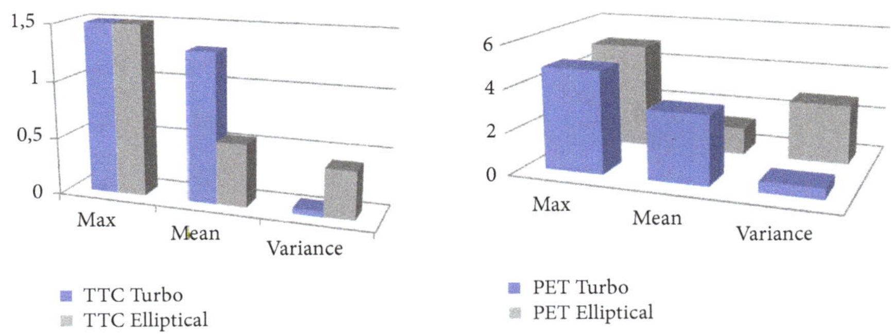

FIGURE 6: Surrogate safety parameters comparison: TTC and PET.

TABLE 6: Collision type distribution.

Intersection scheme	Collision Type		
	crossing	rear end	lane change
Turbo	7	12021	145
Total conflict point 12173			
Elliptical	453	8002	3073
Total conflict point 11528			

of conflict for the various manoeuvres on the two types of intersections investigated.

Through the graphic representation of the values mentioned in Table 6 along the geometry, it is better to see Figure 7 in the arms of direct turning of the turbo the types of conflict called rear end while in the elliptical roundabout the same types of conflict are thickened inside the rotary crown near the access arms.

In the elliptical roundabout, also in Figure 7, there is a concentration of the conflicts of the wool change type at the coupling between the circulatory carriageway and the entry arms.

5. Conclusions

This research investigated two alternatives type of double-lane roundabout, turbo and elliptical, in order to define the safest solution considering direct and surrogate parameters.

To date, mutual comparisons in terms of surrogate safety between these two schemes are not discussed in the sector literature but can be compared to other roundabout geometries and or in terms of capacity-delay/geometrical parameters.

The work highlights the potential for estimating road safety through the replacement solution. Therefore the research work carried out wanted to highlight the negativity of one or the other scheme in order not only to compare two but to provide more comprehensive data that can be incorporated into the comparison of more geometries and also to help to technicians or local authorities during the preliminary selection phase of infrastructure projects.

The assessment of the critical aspects related to road safety and to the possible collision of vehicles has been addressed through the use of traffic microsimulation and a dedicated tool.

The hypotheses analyzed consider the presence of a double lane on the ring that if it does not have the curb of separation between them, it allows a greater percentage of collisions due to the lane change manoeuvres along the central ring. In fact the preselection of the manoeuvring lane along the turbo entry arm limits the collision formations for lane change. The evaluation of safety surrogate parameters such as TTC and PET allows analyzing in a preliminary phase the criticalities found in particular geometries and providing the bases to local authorities in order to choose, with the same geometry as a satin, which can generate the maximum reliable value of collisions even in critical conditions of saturation.

Finally, it should be noted that the presence of more points of collision within the circulatory ring can make traffic

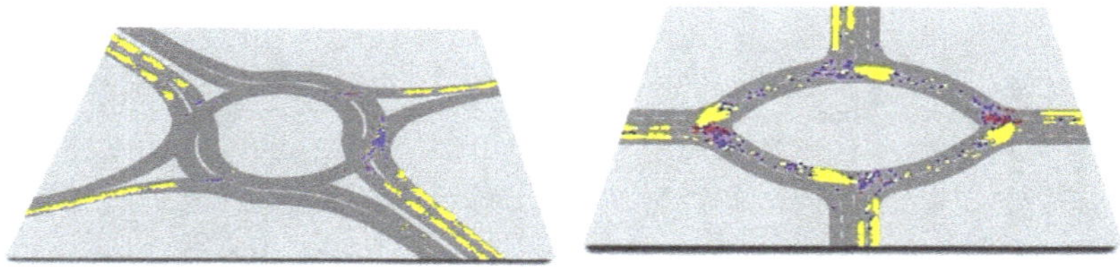

Legend	
Graphical representation	Conflict type
Blue dots	Lane-change
Yellow dots	Rear-end
Red dots	Crossing

FIGURE 7: Layout related to the types of collisions made by SSAM in the two roundabouts examined.

less easy, especially if more lanes are invaded. These reasons are to evaluate in advance preventive scenarios and possible alternatives and solutions to the problems in progress.

6. Limitations and Future Development

This research is focused on the first phase of comparison in terms of safety of turbo and elliptical roundabout geometries in critical conditions of vehicular outflow with symmetrical distribution along the main and secondary directions. Future developments will be conducted varying not only transport matrices but also percentages of heavy vehicles.

Disclosure

This research work is not a public or private expense for its elaboration. No research project of the Kore University of Enna, Italy, and the University of Maribor, Slovenia, has financed this work.

Conflicts of Interest

The authors declare that there are no conflicts of interest regarding the publication of this paper.

References

[1] J. Archer and I. Kosonen, "The Potential of Micro-Simulation Modelling in Relation to Traffic Safety Assessment," in *Proceedings of the ESS Conference*, Hamburg, Germany, 2000.

[2] T. Campisi, G. Tesoriere, and A. Canale, "Microsimulation approach for BRT system: The case study of urban turbo roundabout," in *Proceedings of the International Conference of Computational Methods in Sciences and Engineering 2017, ICCMSE 2017*, Greece, April 2017.

[3] H. Chin and S. Quek, "Measurement of Traffic Conflicts," *Safety Science*, vol. 26, pp. 169–185, 1997.

[4] R. Souleyrette and J. Hochstein, *Development of a Conflict Analysis Methodology Using SSAM; Final Report*, Center for Transportation Research and Education, Iowa State University, Ames, IA, USA, 2012.

[5] O. Feldman, "The GEH measure and quality of the highway assignment models," in *Proceedings of the European Transport Conference*, Glasgow, UK, 2012.

[6] L. G. H. Fortuijn, "Turbo roundabouts: Design principles and safety performance," *Transportation Research Record: Journal of the Transportation Research Board*, no. 2096, pp. 16–24, 2009.

[7] D. Gettman, L. Pu, T. Sayed, and S. Shelby, "Surrogate safety assessment model and validation," FHWA-HRT-08-051, 2008.

[8] T. Giuffrè, S. Trubia, A. Canale, and B. Persaud, "Using microsimulation to evaluate safety and operational implications of newer roundabout layouts for European Road networks," *Sustainability*, vol. 9, no. 11, 2017.

[9] R. Guide, "Wisconsin Department of Transportation," 2006.

[10] H. Hatami and I. Aghayan, "Traffic efficiency evaluation of elliptical roundabout compared with modern and turbo roundabouts considering traffic signal control," *Promet – Traffic & Transportation*, vol. 29, no. 1, pp. 1–11, 2017.

[11] F. Huang, P. Liu, H. Yu, and W. Wang, "Identifying if VISSIM simulation model and SSAM provide reasonable estimates for field measured traffic conflicts at signalized intersections," *Accident Analysis & Prevention*, vol. 50, pp. 1014–1024, 2013.

[12] PTV-Vision, VISSIM 5.30-05 user manual, 2011.

[13] L. Rodegerdts, "Roundabouts: An informational guide," *Transportation Research Board*, vol. 672, 2010.

[14] T. Sayed and S. Zein, "Traffic conflict standards for intersections," *Transportation Planning and Technology*, vol. 22, pp. 309–323, 1999.

[15] T. Tollazzi, R. Mauro, M. Guerrieri, and M. Renčelj, "Comparative analysis of four new alternative types of roundabouts: "Turbo", "flower", "target" and "four-flyover" roundabout," *Periodica Polytechnica Civil Engineering*, vol. 60, no. 1, pp. 51–60, 2016.

[16] T. Tollazzi, G. Tesoriere, M. Guerrieri, and T. Campisi, "Environmental, functional and economic criteria for comparing "target roundabouts" with one- or two-level roundabout intersections," *Transportation Research Part D: Transport and Environment*, vol. 34, pp. 330–344, 2015.

14

Capacity-Constrained Contraflow Adaption for Lane Reconfiguration in Evacuation Planning

Wu Ni,[1] Wenbo Li,[2] Hailei Wang,[2] Chengxin Xiong,[1] and Dan Guo[1]

[1] *School of Computer and Information, Hefei University of Technology, Hefei 230009, China*
[2] *Institute of Technology Innovation, Hefei Institutes of Physical Science, Chinese Academy of Sciences, Hefei 230088, China*

Correspondence should be addressed to Wenbo Li; wbli@iim.ac.cn

Academic Editor: Ludovic Leclercq

This paper presents a heuristic contraflow-based reconfiguration evacuation algorithm, which is named Capacity-Constrained Contraflow Adaption (CC-Adap). First, it effectively calculates optimal candidate routes for evacuation. Second, an evaluation method is proposed for estimating these candidate routes. Third, CC-Adap utilizes a contraflow-based method to reconfigure the evacuation routes to improve capacity constraints. Fourth, traffic conditions are updated in real time. Fifth, CC-Adap reuses historical evacuation routes to reduce the computational cost and accelerate the evacuation process. Experimental results show that CC-Adap generates high-performing evacuation strategies and can be used to tackle large-scale evacuation planning.

1. Introduction

In recent years, both natural and man-made disasters have posed serious threats to humans, such as Hurricane Andrew [1], Hurricane Katrina [2], the "9/11" terrorist attack event [3], and 2011 Tohoku Earthquake Tsunami [4]. Effective evacuation plans are necessary for saving lives and minimizing casualties. The most important objective of an effective evacuation plan is to transfer evacuees to safe areas as quickly as possible. However, dynamic factors make this objective complicated to achieve. For example, the original intention in designing a transportation network is not to address the sharp increase of traffic flow when an emergency occurs [5]. The number of evacuees might far exceed the road capacity, thereby making evacuation route planning a computational challenge.

A well-known routing algorithm, namely, Capacity-Constrained Route Planner (CCRP) [1], was proposed for tackling evacuation route planning. CCRP models the capacity constraints of roads and intersections and selects the shortest paths for evacuation. Later, two heuristics, namely, Intelligent Load Reduction (ILR) and Incremental Data Structure (IDS), were introduced to improve the algorithm's performance [6]. Although routing algorithms can effectively solve mid-size traffic evacuation problems [6], they might not be effective for large-scale evacuation scenarios with massive number of evacuees, which might lead to heavy traffic congestion, owing to capacity constraints. Contraflow-based methods are considered effective techniques for alleviating traffic congestion [2]. Contraflow-based methods reconfigure the edges in the ideal direction and reallocate edge capacity to reduce the evacuation time. For example, Kim et al. presented a heuristic approach that combines a Greedy method and a contraflow-based method for implementing transportation network reconfiguration and, to some extent, minimizing the total evacuation time [2]. However, finding an optimal contraflow plan for reconfiguring network is computationally challenging because we must enumerate all possible combinations of edges in the exhaustive search space [2]. Moreover, it takes a considerable amount of time to evaluate these contraflow candidates by calculating the evacuation time.

In this paper, we propose a Capacity-Constrained Contraflow Adaption (CC-Adap) algorithm for effective evacuation. CC-Adap consists of five steps. First, we utilize an evacuation algorithm to generate optimal candidate routes for evacuees, which finds the routes with maximal flow rate among all the available routes from source vertices to sink vertices. Second, we propose an evaluation method for heuristically evaluating the candidate routes' capabilities based on the traffic conditions and determining the appropriate routes for evacuation. Then, we implement a contraflow-based method for optimizing the routes' performances and reducing traffic congestion. Next, traffic conditions are updated in real time to make the route planning more practical. Finally, CC-Adap reuses the historical evacuation routes before calculating the available routes at each new time step. Experimental results indicate that CC-Adap boosts performance in evacuation route planning.

The remainder of this paper is organized as follows: Section 2 introduces some related works. Section 3 defines the transportation network and contraflow problem. Section 4 details the CC-Adap evacuation algorithm. Experimental results and analysis are presented in Section 5. Section 6 presents the conclusions of this paper.

2. Related Work

To the best of our knowledge, the existing methods can be divided into descriptive and prescriptive approaches [5]. Descriptive approaches aim at mimicking real emergency evacuations based on the simulation of the traffic situations and drivers' behaviours, while prescriptive methods mainly focus on providing a high-performing evacuation plan with minimum evacuation time.

2.1. Descriptive Methods. The aim of descriptive methods is to simulate traffic evacuation situations as vividly as possible. Many traffic evacuation simulation tools have been proven to be capable of solving complex traffic evacuation problems, including MITSIMLab [7], VISSIM [8], MATSim [9], DynusT and DYNASMART [10], and AIMSUN [11]. In addition, some simulation methods utilize multimodel integration methods to simulate evacuees' behaviours during evacuation and combine various existing evacuation methods for more realistic traffic simulation. For example, Pan et al. implemented a simulation framework for classifying evacuees' behaviours into locomotion, steering, and social for evacuation analysis [12]. Wu and Huang combined a control volume model and a flow merging hypothesis to simulate evacuees' behaviours in high-rise building evacuation situations [13]. Noh et al. considered different behaviours of heterogeneous evacuees in building evacuation [14]. They presented a partially dedicated evacuation strategy that divides evacuees into heterogeneous groups, assigns corresponding routes based on a flow model and a simulation optimization approach, and, to some extent, minimizes the average evacuation time [14]. Beloglazov et al. studied the behaviours of people in wildfire evacuation and integrated traffic simulator, wildfire simulator, and behaviour simulator to define the risk metric and provided detailed plan of how evacuation unfold [15]. Yuan et al. constructed a multilevel agent decision model for simulating driver's behaviours and determining actions for each agent [16]. These descriptive methods can achieve satisfactory performance in traffic evacuation for local communities. However, most exhibit poor generalization performance owing to high computational complexity.

2.2. Prescriptive Methods. Prescriptive methods typically provide suggested schedules for evacuation planners, aiming at reducing traffic congestion and minimizing total evacuation time. Zeng and Wang made a small modification to CCRP to improve the performance, which gives priority to longer evacuation routes for evacuating evacuees [17]. Kang et al. calculated the Dijkstra shortest evacuation paths and chose the best one for evacuees based on current traffic conditions [18]. Shahabi and Wilson proposed a Capacity-Aware Shortest Path Evacuation Routing (CASPER) framework for emergency traffic evacuation, which utilizes an advanced traffic model and an intelligent routing algorithm to calculate the optimal evacuation plan for evacuees [5]. Khan et al. used the Intelligent Transportation System (ITS) to compute the maximum traffic flow and the routes with least traffic congestion towards safe places for evacuees based on real-time traffic conditions [19]. Pourrahmani et al. considered the uncertain evacuee demand at the pick-up source points by utilizing fuzzy credibility theory to present a genetic algorithm (GA) for handling the stated evacuation problem [20]. Chen et al. implemented a distribution Load-balancing Emergency Guiding System (LEGS) to assign the fastest paths to transfer evacuees to the exits according to the capacity constraints and concurrent movement of evacuees [21]. Ikeda and Inoue proposed a Multi-Objective Genetic Algorithm (MOGA) to evaluate evacuation routes by considering route distance and evacuation time to select optimal evacuation routes [22]. Heydar et al. divided the transportation network into pedestrian network and vehicular network, where pedestrians were assigned the shortest routes to the safe areas in pedestrian network and vehicles transferred evacuees to designated shelters in vehicular network [23]. These prescriptive methods have demonstrated the feasibility of calculating evacuation strategies for traffic evacuation. However, the performances of these prescriptive methods are usually limited owing to the capacity constraints of road segments and intersections.

Previous works [24–27] have considered the benefits of contraflow-based methods on evacuation planning. In addition, in the literatures, the optimality of evacuation planning by integrating contraflow-based methods has been validated. For example, Xie et al. designed a bilevel framework for optimizing network evacuation performance, in which the lower level solves the traffic assignment problem and the upper level integrates a contraflow method with crossing elimination strategies to optimize the traffic assignment [28]. Wang et al. implemented a multiobjective optimization model to determine the evacuation priorities and setup time for doing the contraflow operations [3]. Even et al. implemented a conflict-based path-generation algorithm to simultaneously guide the evacuation and select the to-be-reversed roads [29]. Kim et al. presented a Bottleneck Relief (BR) approach that iteratively flips min-cut edges based on the theorem in [30] for solving massive traffic evacuation problems [2].

Above-mentioned contraflow-based methods are selected to generate contraflow strategy that can tackle capacity constraints effectively [2, 3, 24–29]. However, there are two difficulties in calculating the appropriate contraflow strategy for the complex traffic system. The first difficulty is that it is a time-consuming task to determine the appropriate to-be-reversed edges in the complex traffic network [2, 24, 25]. The second difficulty is that it is difficult to determine the ideal direction of these edges after being reversed [2, 3, 28, 29]. Thus, it is vital for the planners to find an appropriate contraflow strategy that could overcome these two difficulties.

So, this paper presents a prescriptive algorithm to generate appropriate contraflow strategy and optimal evacuation plan. For the first difficulty, the presented algorithm utilizes Greedy method and iterative optimization technique to select the to-be-reversed edges on the evacuation routes. For the second difficulty, the ideal direction of these to-be-reversed edges is determined by the traffic flow on the evacuation routes.

3. Modeling and Problem Formulation

Various methods, such as simulation [13–15], classification network [23], and mathematical modeling [1, 2], are used to formulate the evacuation situations. In this paper, the mathematical graph is used to describe the traffic evacuation situations [1, 2, 5, 6, 17].

3.1. Network Modeling. Suppose we are given a multisource and multisink transportation network $G(V, E)$, where $V = \{V_S, V_M, V_D\}$ and E, respectively, represent the sets of vertices and edges. Here, V_S is the source vertex set, V_M is the transition vertex set, and V_D is the sink vertex set. The definitions of the vertices and edges are given as follows:

(i) Each vertex $V_i \in V$ has a maximal vertex capacity $V_{i,MCap}$ and each source vertex $V_s \in V_S$ has an extra initial occupancy $V_{s,Occ}$, which represents the number of evacuees. For each sink vertex $V_d \in V_D$, the value of $V_{d,MCap}$ can be set according to the actual demand. In this paper, $V_{d,MCap}$ of each sink vertex is set to *infinity*.

(ii) Each edge $e^{\langle V_i,V_j\rangle} \in E$ has a maximal edge capacity $e_{MCap}^{\langle V_i,V_j\rangle}$, a travel time $e_{traT}^{\langle V_i,V_j\rangle}$, and an initial direction. $e_{MCap}^{\langle V_i,V_j\rangle}$ is the number of evacuees (e.g., residents or vehicles) that can pass per unit period. The edge direction is initialized according to the real road conditions. $e_{traT}^{\langle V_i,V_j\rangle}$ is the time cost when evacuees pass edge $e^{\langle V_i,V_j\rangle}$. Various factors should be considered when formulating edge travel time, such as road length, traffic flow, and drivers' behaviours. Many works have studied the relationship between these factors and edge travel time [13–15, 17]. This paper focuses on the impact of road length and traffic flow on the edge travel time. Considering the emergency situations, the traffic flow on the edges should be limited by the maximal edge capacity. Thus, the evacuees can use maximal speed passing one given edge when the traffic flow does not exceed the maximal edge capacity. In other words, more evacuees can be evacuated to safe area using this strategy in the minimum time. According to this, the edge travel time in the transportation network model is determined by the length and the maximal edge capacity of one given road.

There are different traffic elements in transportation network including buildings, crossroads, playgrounds, shelters, and parks. Buildings, crossroads, and playgrounds are the most common traffic elements in transportation network. Figure 1 illustrates the modeling results of the buildings, the crossroad, and the playground. Firstly, the buildings are modeled as source vertex. Secondly, the crossroad is modeled as transition vertex. Thirdly, the playground is modeled as sink vertex. The edges of the vertices are modeled according to the traffic configuration. As for the other traffic elements, they are modeled similar to these three common traffic elements shown in Figure 1.

The objective of transportation evacuation is to find an optimal plan with minimum evacuation time. The objective function is defined in

$$EvaT = \min \left[\max_{r\in RS} (ts + c(r))\right]$$
$$\text{s.t.} \quad \begin{cases} ① \ \sum_{r\in RS} f(r) = V_{s,Occ} \quad V_s \in V_S, V_s \in r \\ ② \ \sum_{V_d\in V_D} \sum_{r\in RS} f(r) = \sum_{V_s\in V_S} V_{s,Occ} \quad V_d \in r \\ ③ \ V_{i,MCap} \geq 0; e_{MCap}^{\langle V_i,V_j\rangle} \geq 0; V_{i,cap}^{t_i} \geq 0; e_{cap}^{(\langle V_i,V_j\rangle,t_i)} \geq 0 \end{cases} \tag{1}$$

$EvaT$ is the total evacuation time, ts is the initial time step when a group of evacuees leaves the source vertex, RS is the to-be-solved evacuation route set, and r is an evacuation route in RS.

$c(r)$ is used to calculate the travel time of route r when evacuees pass through the edges along route r and is expressed in (2). For route $r\langle V_0 - V_1 - \cdots - V_n - \cdots - V_N\rangle$, where V_n is the nth vertex of route r, V_0 is the source vertex of route r, and V_N is the sink vertex of route r. $e_{traT}^{\langle V_{n-1},V_n\rangle}$ means the travel time from vertex V_{n-1} to vertex V_n along route r.

$$c(r) = \sum_{n=1}^{N} e_{traT}^{\langle V_{n-1},V_n\rangle} \quad \forall e^{\langle V_{n-1},V_n\rangle} \in r \tag{2}$$

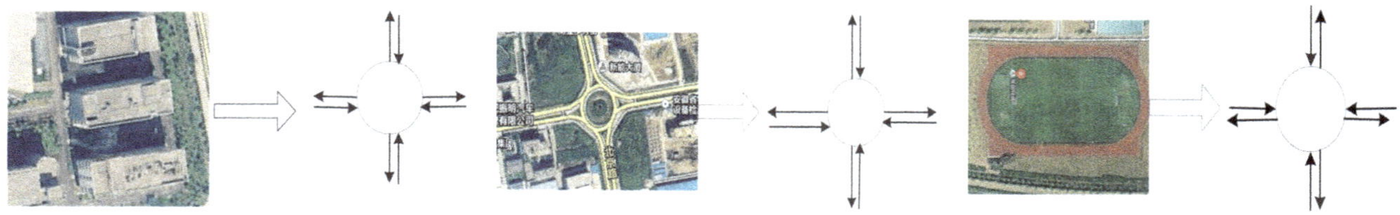

(a) Source vertex (b) Transition vertex (c) Sink vertex

FIGURE 1: Illustration of the vertices and edges.

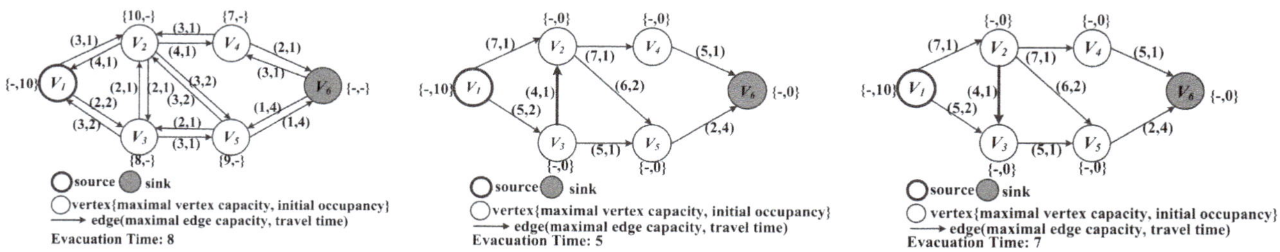

(a) Original network (b) Evacuation planning 1 (c) Evacuation planning 2

FIGURE 2: An evacuation scenario and two contraflow-based evacuation plans.

$f(r)$ is used to calculate the maximal traffic flow of route r under capacity constraints of vertices and edges and is expressed in (3). $V_{n,cap}^{t_n}$ denotes the available residual capacities of vertex V_n on route r at its arrival time t_n, where $t_n = ts + \sum_{l=1}^{n} e_{traT}^{\langle V_{l-1},V_l\rangle}$, and $e_{cap}^{(\langle V_{n-1},V_n\rangle,t_{n-1})}$ denotes the available residual capacity of edge $e^{\langle V_{n-1},V_n\rangle}$ on route r at its arrival time t_{n-1}.

$$f(r) = \min \begin{cases} V_{0,Occ} \\ V_{n,cap}^{t_n} \\ e_{cap}^{(\langle V_{n-1},V_n\rangle,t_{n-1})} \end{cases} \qquad (3)$$

$$\text{s.t.} \quad \forall e^{\langle V_{n-1},V_n\rangle} \in r,\ V_n \in r,\ 1 \le n \le N$$

Formula (1) illustrates the capacity constraints along the time dimension during evacuation. Constraint ① ensures that the outflow of source vertex V_s does not exceed the initial occupancy of source vertex V_s. Constraint ② enforces that the inflow of the sink vertices must be equal to the initial occupancy of the source vertices. In addition, the model capacity of vertices and edges cannot be negative during evacuation due to Constraint ③.

3.2. Contraflow Problem. In a large transportation network, long distance traffic jams usually cover the roads when emergencies occur. Contraflow-based methods are considered effective techniques for remedying traffic congestion [2]. However, an appropriate contraflow reconfiguration is difficult to calculate. Generally, two main difficulties are encountered in calculating optimal contraflow reconfiguration: (1) it is NP-hard to find all possible combinations of edge directions and calculate the evacuation time of these contraflow combinations and (2) it is time-consuming to evaluate the effects of dynamic traffic flow on these contraflow reconfiguration candidates.

Figure 2 illustrates an evacuation scenario and two contraflow-based evacuation planning results. The evacuation time is 8 time units based on the original network configuration in Figure 2(a). Figures 2(b) and 2(c) illustrate two contraflow reconfigurations, which combine all the two-way edge directions and merge the edge capacities to increase the upper bound on the evacuation capacity. The contraflow reconfiguration in Figure 2(b) reduces the evacuation time to 5 time units, while the contraflow reconfiguration in Figure 2(c) reduces the evacuation time to 7 time units. The only difference between the two contraflow reconfigurations is the edge direction between vertex V_2 and vertex V_3. Thus, it is vital for an optimal contraflow reconfiguration to select the critical edges that affect evacuation performance and determine their ideal direction after reversal.

4. Proposed Algorithm

This paper proposes the CC-Adap algorithm, which heuristically evaluates the evacuation routes, and integrates it with a contraflow-based method for distributing the evacuation routes with the least traffic congestion and maximal flow rate to evacuees. In addition, CC-Adap reuses historical evacuation routes to reduce computational costs and accelerate the evacuation process. Algorithm 1 gives the pseudocode of the CC-Adap algorithm.

Step 0 (initialization). For a given network $G(V, E)$, let C be the to-be-reversed edge set, ts be the initial time step

```
Input: G(V, E), C ← ∅, ts = 0, RS ← ∅, HR ← ∅
Output: RS and C
while there are still evacuees in source vertices do
    r ← GetEvaRoute(G, ts);
    if r is not available then
        ts++; HRouteReuse(G, HR, ts);
    else
        δ = −1;        //Set initial value of δ
        δ =RouteEvaluation(G, HR, ts, r);
        if δ == −1 then δ = 1;        //Exceptional case
        else
            if δ == 1 then
                c ← ContraflowReconfiguration(G, r);
                NetworkUpdate(G, ts, r);
                C ← C ∪ c;
                r.ρ = 0;
                RS ← RS ∪ r;
                HR ← HR ∪ r;
            end if
        end if
    end if
end while
```

ALGORITHM 1: CC-Adap algorithm.

when evacuees leave source vertices, *RS* be the to-be-solved evacuation planning route set, and *HR* be the historical evacuation route set.

Step 1 (GetEvaRoute). CC-Adap utilizes the evacuation route generation method in the Max-Flow Rate Priority (MFRP) algorithm [31] to select optimal candidate route r from among the available routes from source vertices to the nearest sink vertices. MFRP introduces parameter *FlowRate* to describe the route capacity. The formula of *FlowRate* is shown in (4).

$$FlowRate = \frac{f(r)}{c(r)} \tag{4}$$

The main strategy of the MFRP algorithm [31] is to calculate the available candidate routes by Dijkstra algorithm and balance the number of transferred evacuees with the travel time to select evacuation routes with maximal *FlowRate*. The performance of MFRP has been analyzed in [31]. CC-Adap guarantees the shortest possible evacuation time by employing MFRP [31] to select the optimal candidate routes.

Step 2 (RouteEvaluation). The use of low-quality routes might aggravate traffic congestion and increase the total evacuation time. CC-Adap introduces a parameter rq for quantizing route r, which is generated in Step 1, and historical evacuation route to avoid selecting lower-quality evacuation routes. The formula of rq is presented in

$$rq = \min\left(\frac{V_{s,Occ}}{f(r')}, c(r) - c(r')\right) * \left(f(r) - f(r')\right) \tag{5}$$

where r' is the historical evacuation route in *HR*, which has the same source vertex as r. Equation (5) represents the evacuation capability gap between routes r and r' when they evacuate evacuees at same times under the same traffic conditions.

The *RouteEvaluation* function operates as follows: First, we introduce parameter δ for reserving the evaluation result and set $\delta = -1$ initially. For each historical evacuation route r' in *HR*, we evaluate rq between r and r' by employing (5), where r and r' have the same source vertex, and reserve the evaluation result using δ as follows, which is expressed in (6): If rq is less than zero, then we set $\delta = 0$ and calculate the next historical evacuation route r' in *HR*. Otherwise, r is suitable for evacuation and we set $\delta = 1$ and terminate the evaluation procedure. If there is no route with the same source vertex as r in *HR*, we assume that r can also be used for evacuation and set $\delta = 1$.

$$\delta = \begin{cases} 0 & rq < 0 \\ 1 & rq \geq 0 \end{cases} \tag{6}$$

Step 3 (ContraflowReconfiguration). Greedy method can be used in evacuation route planning [2]. The appropriate contraflow strategy can be determined using Greedy method which has the advantage of obtaining the global optimal solution by local optimal solution. Thus, CC-Adap integrates Greedy method and iterative optimization technique to reverse the two-way edges on the evacuation routes iteratively in this paper. However, the edges on the evacuation routes cannot be reversed repeatedly as long as the edges have been flipped. The modified edge direction of the to-be-reversed edges depends on the traffic flow along the evacuation routes. The upper bound on the capacity of the flipped edges is increased by merging the two-way edges. The travel time of

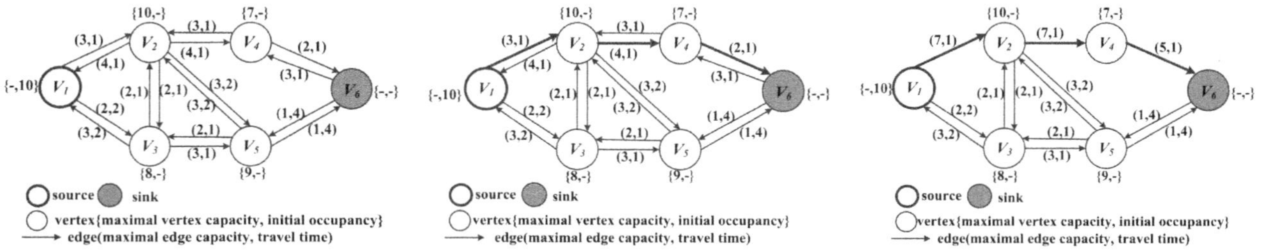

(a) Original network (b) Select evacuation route (c) Contraflow result

FIGURE 3: Example of the proposed contraflow-based method.

the reversed edges is equal to that of the two-way edges, which have the same edge direction as the flipped edges.

Figure 3 illustrates the steps for implementing the proposed contraflow reconfiguration method. Figure 3(a) is the original network configuration. In Figure 3(b), we select evacuation route $\langle V_1 - V_2 - V_4 - V_6 \rangle$. Its traffic flow is $\min\{10, 3, 10, 4, 7, 2, \infty\} = 2$. As shown in Figure 3(c), the two-way edges on this route are merged and the maximal capacity of the flipped edges is increased. After contraflow reconfiguration, the traffic flow is increased to $\min\{10, 7, 10, 7, 7, 5, \infty\} = 5$. Thus, the evacuation capability of route $\langle V_1 - V_2 - V_4 - V_6 \rangle$ is improved.

Step 4 (NetworkUpdate). After evacuees are transferred from source vertex to sink vertex, the available residual capacities of transition vertices and edges along route r are updated in (7) and (8), respectively, and the source vertex's residual evacuees of route r are updated in (9).

$$V_{n,cap}^{t_n} = V_{n,cap}^{t_n} - f(r) \quad V_n \in r \tag{7}$$

$$e_{cap}^{(\langle V_{n-1}, V_n \rangle, t_{n-1})} = e_{cap}^{(\langle V_{n-1}, V_n \rangle, t_{n-1})} - f(r) \quad e^{\langle V_{n-1}, V_n \rangle} \in r \tag{8}$$

$$V_{0,Occ} = V_{0,Occ} - f(r) \quad V_0 \in r \tag{9}$$

Based on (7), (8), and (9), the traffic conditions can be monitored in real time for the next-iteration route calculation.

Step 5 (HRouteReuse). When there is no available evacuation route at the current time step, CC-Adap proceeds to the next time step (ts++) and returns to Step 1 to calculate routes. Before calculating routes, CC-Adap reuses historical evacuation routes in HR to reduce the computational cost and accelerate the evacuation process. To avoid evacuating along the same evacuation routes continually, CC-Adap introduces a parameter ρ that marks historical evacuation route hr in HR if hr is reserved at the last time step. ρ is defined in

$$\rho = \begin{cases} 0 & hr \text{ is reserved last time step} \\ 1 & hr \text{ isn't reserved last time step} \end{cases} \tag{10}$$

The *HRouteReuse* function operates as follows: For each historical evacuation route hr in HR, if $f(hr) = 0$ under the current traffic conditions, it will be excluded from HR. Otherwise, if the corresponding ρ is equal to 0, hr is reserved at the last time step and is not suitable for evacuation at the current time step, we set $\rho = 1$ for the next iteration. If ρ is equal to 1, then hr is used for evacuation at the current time step and traffic conditions are updated by Step 4.

In each iteration, the generated route r in Step 1 is reserved in HR and RS when r is used to transfer evacuees and ρ of r is set to 0. Meanwhile, the to-be-reversed edges on route r are reserved in C. To complete the evacuation, Steps 1–5 can be repeated until there is no evacuee who needs to be evacuated.

5. Experiment

5.1. Experimental Design. There are many evacuation algorithms that can provide evacuation route planning, such as the classical CCRP algorithm [1], LRP algorithm [17], MR algorithm [31], MFRP algorithm [31], and Greedy algorithm [2]. Reference [31] has validated that the performance of MFRP algorithm is better than LRP and MR. Besides, Greedy algorithm is a classical contraflow evacuation algorithm. CC-Adap algorithm integrates MFRP algorithm and Greedy algorithm in this paper. Thus, CCRP, MFRP, and Greedy algorithms are selected to validate the performance of CC-Adap algorithm.

5.2. Experiment Data. In this paper, we use the following notations to describe evacuation situations: n is the number of vertices, s is the number of source and sink vertices, m is the number of edges, and p is the number of evacuees. The experimental data that are used in this paper is generated by the network simulator tool NETGEN [32]. However, NETGEN does not generate the maximal vertex capacity for transition vertices. This paper uses a random number generation function, namely, *rand()*, to generate the maximal vertex capacity for transition vertices in (11). The seed of the *rand()* function is set to the vertex number to avoid generating the same networks. α and β are two coefficients in (11), which imply that the lower bound on the edge capacity is β and the upper bound on the edge capacity is $(rand()\%((\alpha * p)/s)) + \beta$. α and β can be reset according to the practical requirements. In this paper, we set $\alpha = 3$ and $\beta = 20$ [31].

$$V_{i,MCap} = \left(rand()\% \frac{\alpha * p}{s} \right) + \beta \quad V_i \in V_M \tag{11}$$

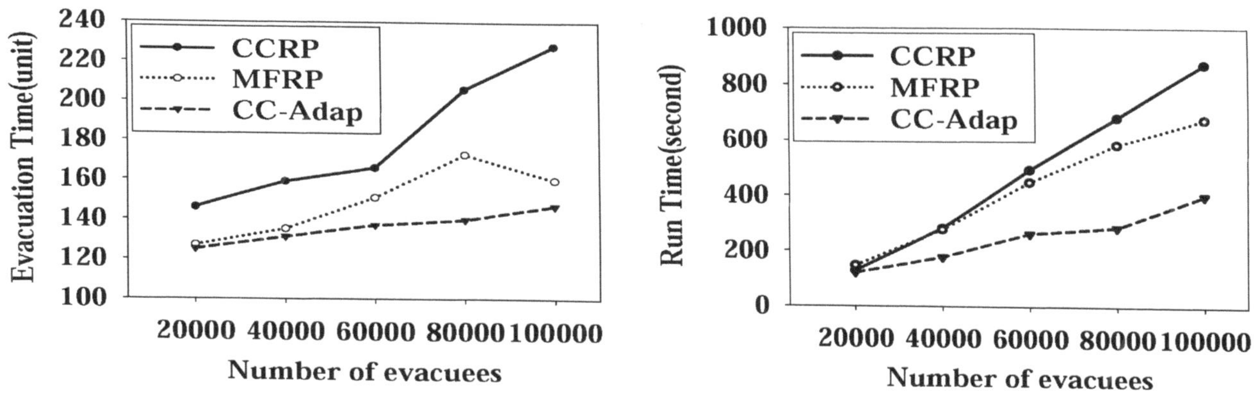

(a) $n = 256$; $m = 768$; $s = 26$

(b) $n = 256$; $m = 768$; $s = 26$

FIGURE 4: Evacuation quality in a small transportation network.

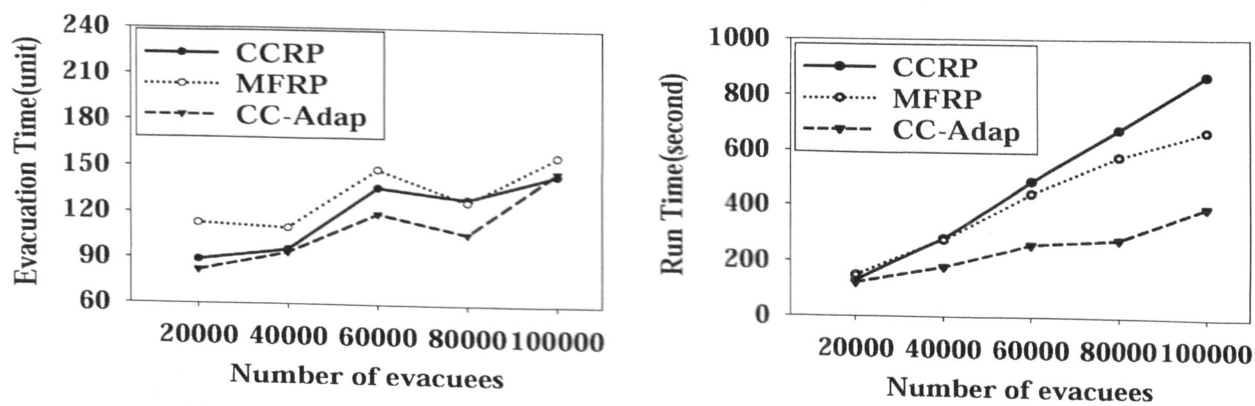

(a) $n = 1024$; $m = 5258$; $s = 102$

(b) $n = 1024$; $m = 5258$; $s = 102$

FIGURE 5: Evacuation quality in a large transportation network.

5.3. Experimental Result

5.3.1. Feasibility Verification. The purpose of this section is to evaluate the feasibility of using CC-Adap in evacuation planning. CC-Adap is compared with CCRP and MFRP to determine whether CC-Adap can be used for evacuation planning and to optimize evacuation plans by applying a contraflow-based method.

In this section, two experiments are carried out. Each experiment has five test groups and the number of vertices, edges, source, and sink vertices is fixed. We vary the number of evacuees from 20000 to 100000. Figures 4 and 5 depict the results on the evacuation time and run time of the three algorithms in small-scale and large-scale transportation networks, respectively. According to Figure 4(a), CC-Adap has the smallest and nearly linearly growing evacuation time with the increase of the number of evacuees, while CCRP and MFRP have steeply increasing evacuation time and the evacuation time of CCRP is almost double that of CC-Adap. The run time of CCRP and MFRP grows rapidly as the number of evacuees increases, while CC-Adap has a stable run time and the run time of CCRP and MFR than twice that of CC-Adap when the number of 100000 in Figure 4(b). As shown in Figure 5, C the shortest evacuation time with the increase i of evacuees in Figure 5(a), followed by CCRP, a the longest evacuation time. In terms of the Adap, CCRP, and MFRP all show nearly lin the increase in the number of evacuees, an run time that is nearly half those of CCRP

5.3.2. Scalability Verification. The purp to evaluate the scalability of CC-Adap t situations and CC-Adap is compared Greedy algorithms.

Comparison on Networks with Di Figure 6 plots the experimental r tion time and run time of the f number of edges. This experim subexperiments. The number

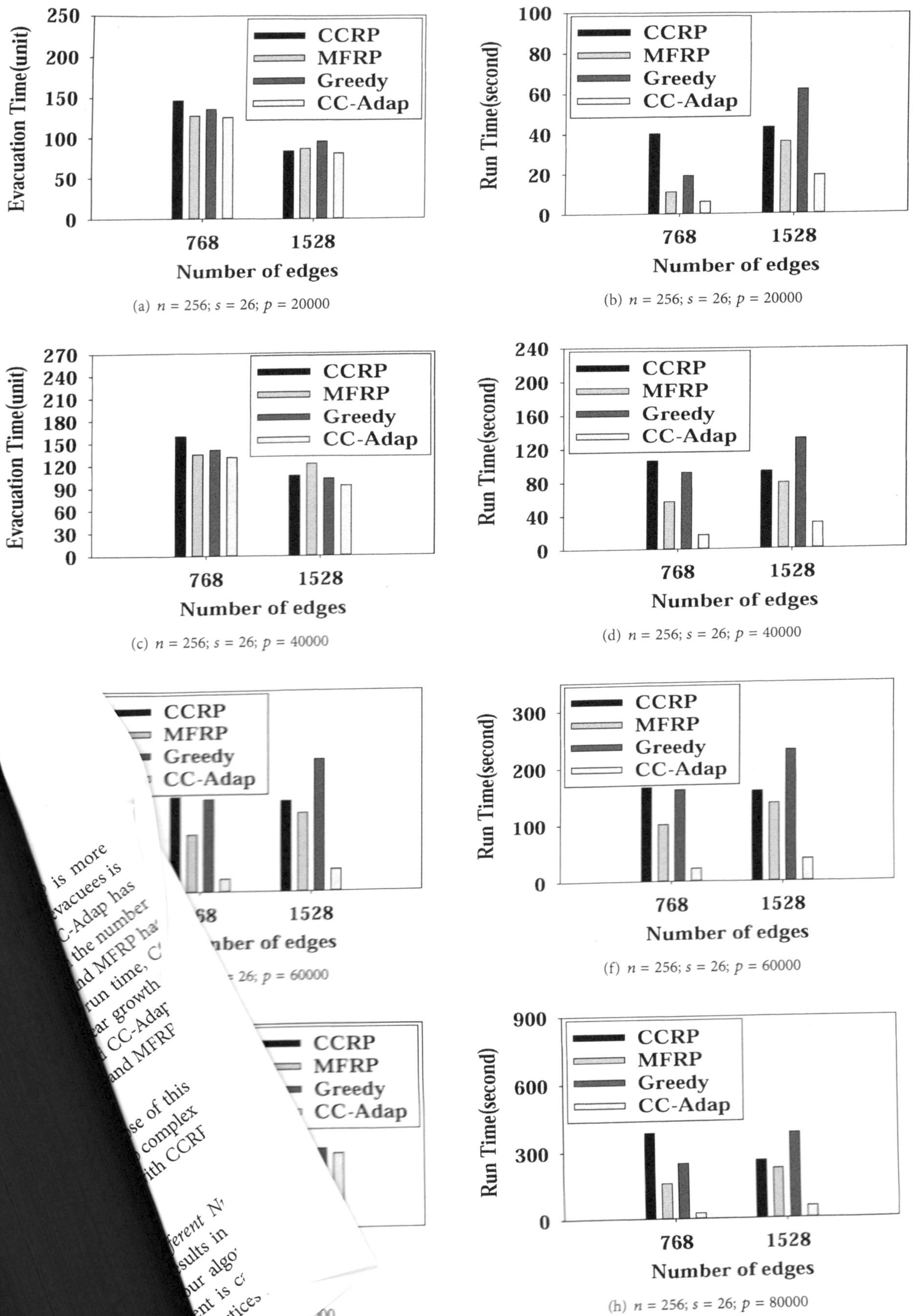

(a) $n = 256$; $s = 26$; $p = 20000$

(b) $n = 256$; $s = 26$; $p = 20000$

(c) $n = 256$; $s = 26$; $p = 40000$

(d) $n = 256$; $s = 26$; $p = 40000$

(f) $n = 256$; $s = 26$; $p = 60000$

(h) $n = 256$; $s = 26$; $p = 80000$

FIGURE 6: Continued.

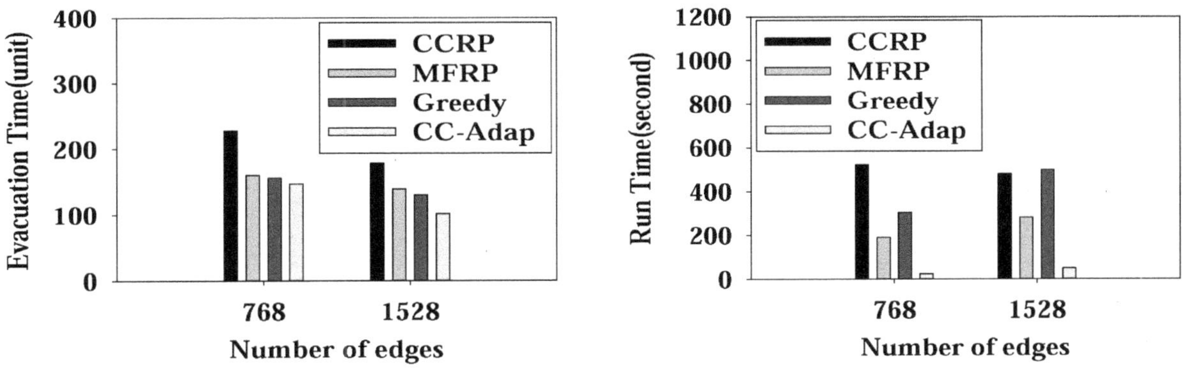

(i) $n = 256$; $s = 26$; $p = 100000$

(j) $n = 256$; $s = 26$; $p = 100000$

FIGURE 6: Evacuation quality with respect to the number of edges.

the number of source and sink vertices is fixed at 26. The number of evacuees is varied from 20000 to 100000. Each subexperiment has two test groups and the number of edges is varied from 768 to 1528. According to Figure 6, CC-Adap has the shortest evacuation time in each subexperiment as the number of edges varies. In addition, CC-Adap shows the minimum change in evacuation time when the number of edges varies, followed by Greedy, MFRP, and CCRP. Regarding the run time of these four algorithms when the number of edges varies, as shown in Figure 6, CC-Adap has the shortest run time comparing with the CCRP, MFRP, and Greedy algorithms; the run time of CC-Adap is less than 20 percent of those of CCRP and Greedy. In addition, the number of edges has little effect on the run time of CC-Adap, while the CCRP, MFRP, and Greedy algorithms are affected by the variation of the number of edges.

Comparison on Different Number of Evacuees. Figure 7 plots the experimental results on evacuation time and run time of the four algorithms against the number of evacuees. This experiment is carried out with three subexperiments, and the numbers of vertices, edges, and source and sink vertices are fixed. Each subexperiment has five test groups and the number of evacuees is varied from 20000 to 100000. According to Figure 7, the evacuation time of CC-Adap increases (perhaps linearly) with the increase in the number of evacuees, while CCRP has a steeply increasing evacuation time, MFRP and Greedy show volatility growth of evacuation time with the increase in the number of evacuees. In particular, according to Figures 7(c) and 7(e), the evacuation time of CC-Adap is almost 50 percent those of CCRP, MFRP, and Greedy when tackling a massive number of evacuees. Moreover, CC-Adap has a short and stable run time with the increase in the number of evacuees, while the CCRP, MFRP, and Greedy algorithms have steeply, perhaps superlinearly, increasing run time. The run time of CC-Adap is almost 10 percent that of the Greedy algorithm in Figures 7(b), 7(d), and 7(f) and that of CCRP in Figures 7(b) and 7(d) when tackling a massive number of evacuees.

Comparison on Different Numbers of Source and Sink Vertices. Figure 8 plots the experimental results on the evacuation time and run time of the four algorithms against the number of source and sink vertices. This experiment is carried out with five subexperiments. The number of vertices is fixed at 256, the number of edges is fixed at 1528, and the number of evacuees is varied from 20000 to 100000. Each subexperiment has two test groups and the number of source and sink vertices is varied from 26 to 52. According to Figure 8, CC-Adap produces the solutions with shortest evacuation time comparing with the CCRP, MFRP, and Greedy algorithms when the number of source and sink vertices is varied. In addition, the evacuation time of CC-Adap has the slowest increase with the increase in the number of source and sink vertices comparing with the other three algorithms. For different number of source and sink vertices, CC-Adap has the fastest run time comparing with the CCRP, MFRP, and Greedy algorithms. As the number of source and sink vertices varies from 26 to 52, the run time of the MFRP and Greedy algorithms rises steeply, CCRP shows a small growth in run time, and CC-Adap shows the smallest increase in run time.

Comparison on Large Transportation Networks. Figure 9 depicts the experimental results of the four algorithms in terms of evacuation time and run time on a large-scale transportation network. The experiment is carried out with five test groups and has a fixed network configuration of 1024 vertices, 5258 edges, and 102 source and sink vertices. We vary the number of evacuees from 20000 to 100000 to test the algorithms' performances. As shown in Figure 9(a), CC-Adap can generate high-quality evacuation solutions with the shortest evacuation time when the number of evacuees grows in a large-scale transportation network. Moreover, the run time of CC-Adap shown in Figure 9(b) also has the smallest and most stable growth with different number of evacuees in large-scale transportation network evacuation scenarios. In addition, CC-Adap has the minimum growth ratio of run

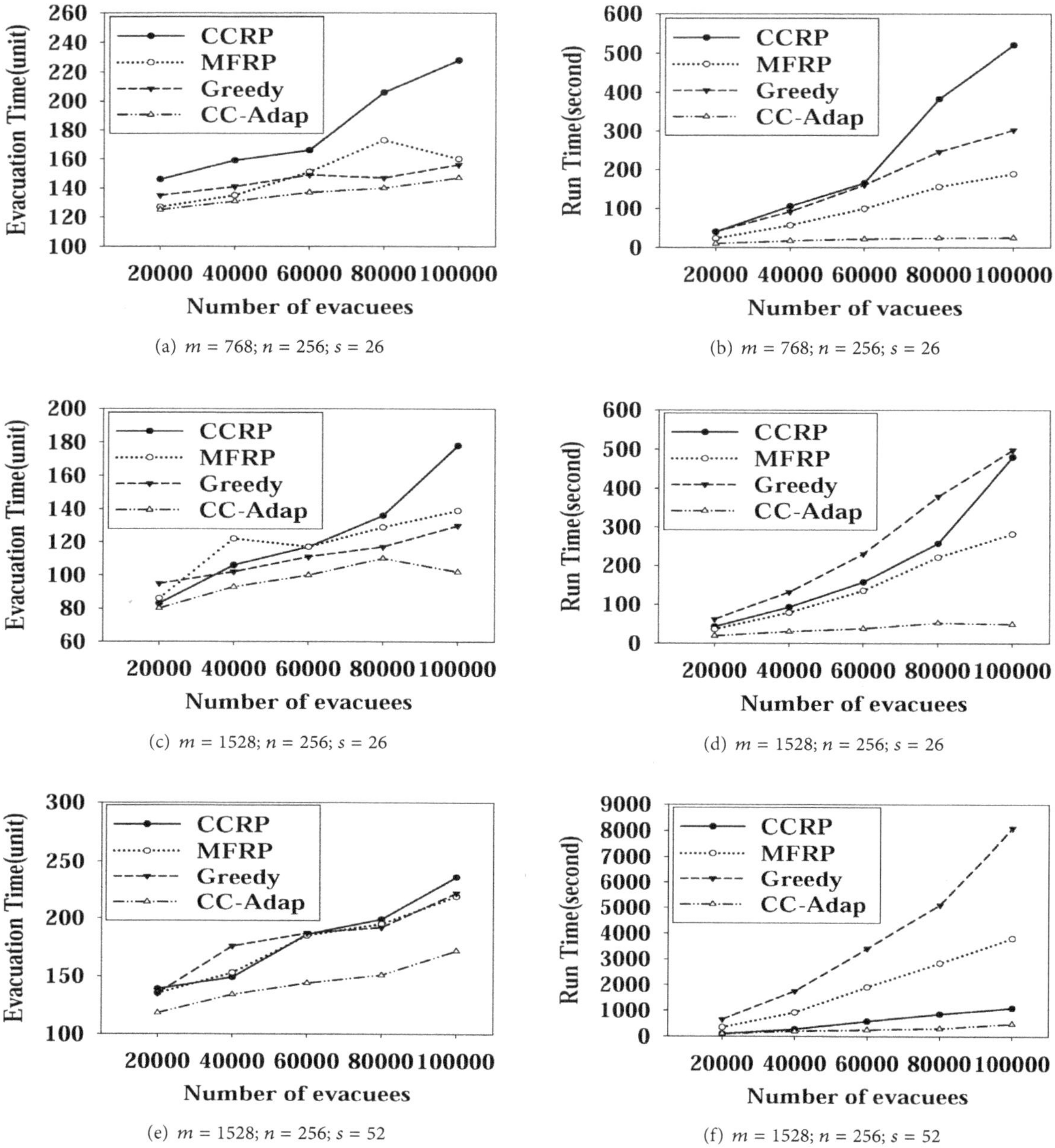

(a) $m = 768; n = 256; s = 26$

(b) $m = 768; n = 256; s = 26$

(c) $m = 1528; n = 256; s = 26$

(d) $m = 1528; n = 256; s = 26$

(e) $m = 1528; n = 256; s = 52$

(f) $m = 1528; n = 256; s = 52$

FIGURE 7: Evacuation quality with respect to the number of evacuees.

time with the increase in the number of evacuees, comparing with CCRP, MFRP, and Greedy (which have growth ratios of nearly 50 percent).

5.4. Summary of Experiments. There are many dynamic factors that are critical to the transportation network, such as number of evacuees and affected areas, which makes evacuation management a daunting task. These dynamic factors can be evaluated systematically by building appropriate evacuation models [33]. Only by taking these factors into consideration we can optimize evacuation management. This section will discuss these dynamic factors and evaluate the optimality of the CC-Adap algorithm based on the experimental results.

(i) **Number of edges**: the more edges there are, the more difficult it is to find appropriate routes to evacuate evacuees. CC-Adap can intelligently evaluate the routes to choose the optimal one for evacuation, thereby making it more scalable with the number of edges.

(ii) **Number of source and sink vertices**: an increase in the number of source and sink vertices in a transportation network means that the evacuees are

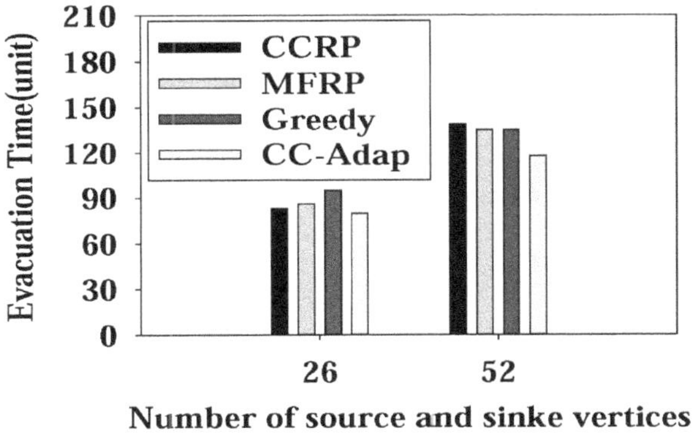

(a) $n = 256$; $m = 1528$; $p = 20000$

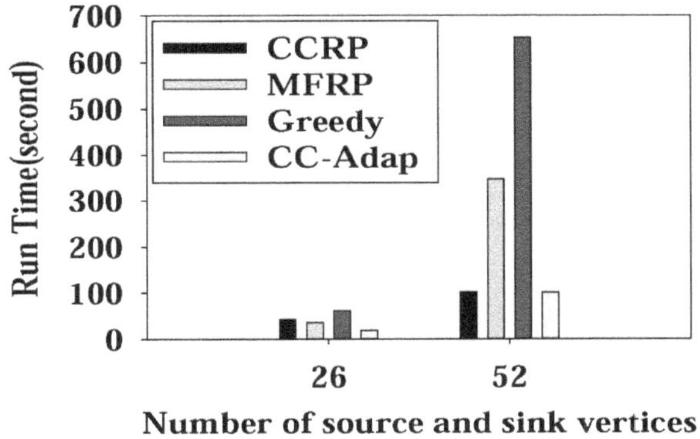

(b) $n = 256$; $m = 1528$; $p = 20000$

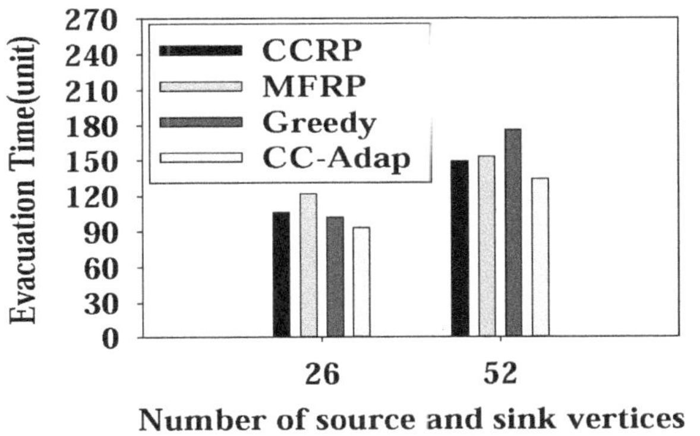

(c) $n = 256$; $m = 1528$. $p = 40000$

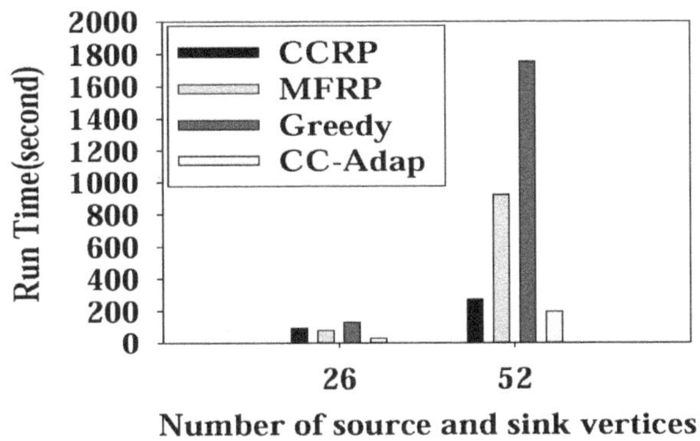

(d) $n = 256$; $m = 1528$; $p = 40000$

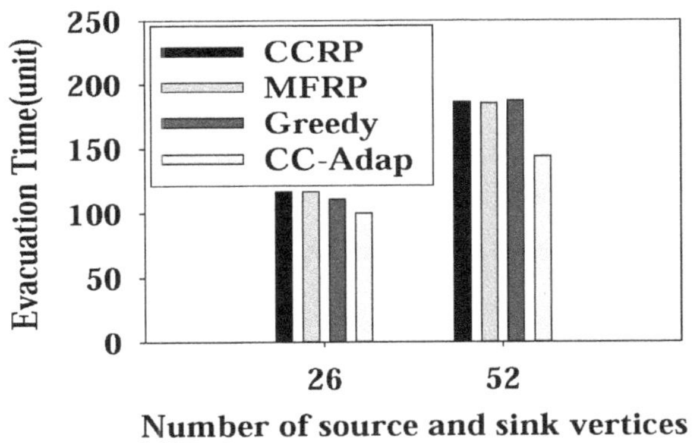

(e) $n = 256$; $m = 1528$; $p = 60000$

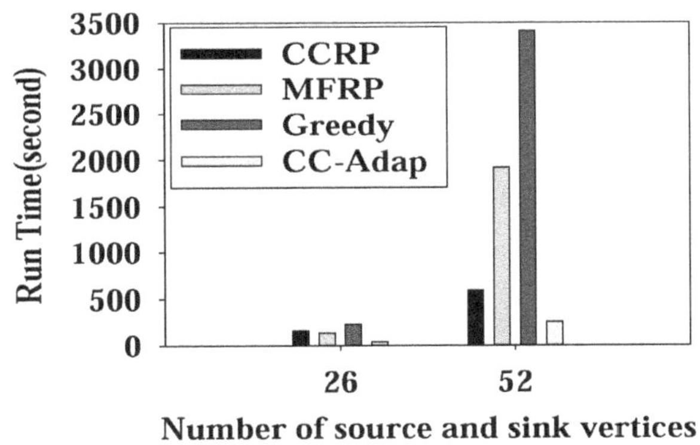

(f) $n = 256$; $m = 1528$; $p = 60000$

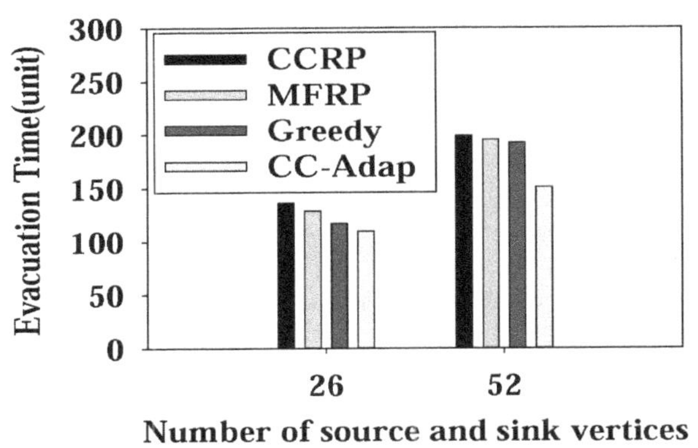

(g) $n = 256$; $m = 1528$; $p = 80000$

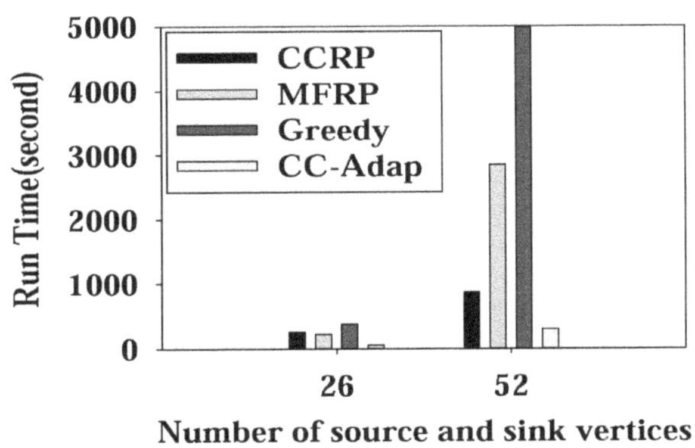

(h) $n = 256$; $m = 1528$; $p = 80000$

FIGURE 8: Continued.

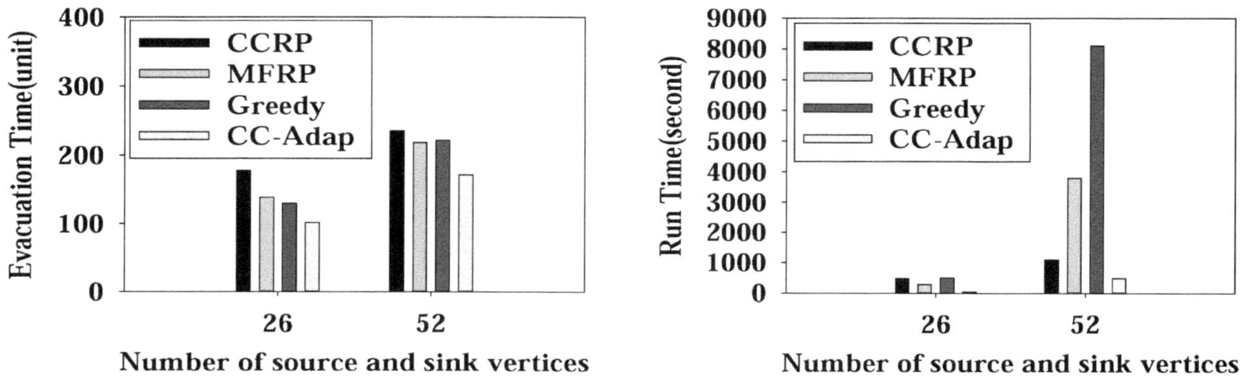

(i) $n = 256$; $m = 1528$; $p = 100000$

(j) $n = 256$; $m = 1528$; $p = 100000$

FIGURE 8: Evacuation quality with respect to the number of source and sink vertices.

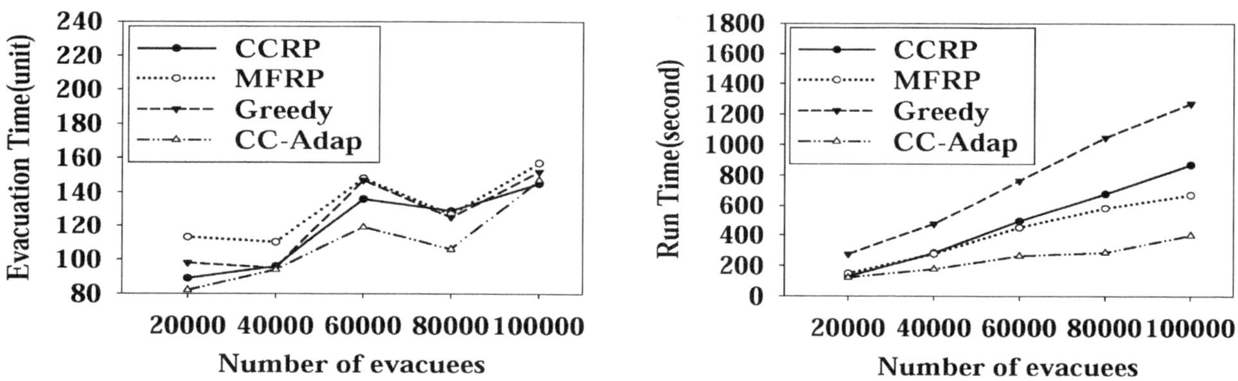

(a) $n = 1024$; $s = 102$; $m = 5258$

(b) $n = 1024$; $s = 102$; $m = 5258$

FIGURE 9: Evacuation quality of large-scale transportation network evacuation.

more scattered in the network [1]. It might lead to difficulty in calculating the evacuation routes. CC-Adap reuses the historical evacuation routes to reduce the computational costs, thereby making CC-Adap more scalable with the number of source and sink vertices.

(iii) **Number of evacuees**: capacity constraints are the major limiting factor in optimizing an evacuation strategy. Traffic congestion occurs easily when there are too many evacuees. CC-Adap applies a contraflow-based method to improve the performance of evacuation routes, which makes CC-Adap more scalable with the number of evacuees.

(iv) **Traffic scale**: large and complex urban evacuation problems cause most evacuation route planning approaches to fail. CC-Adap integrates an iterative optimization technique and a contraflow-based method to optimize an evacuation plan, thereby making CC-Adap more effective in addressing urban evacuation planning.

6. Conclusions

This paper presents a heuristic contraflow reconfiguration evacuation algorithm, which is called Capacity-Constrained Contraflow Adaption algorithm. CC-Adap can heuristically evaluate evacuation routes to assign high-quality routes to evacuees. Moreover, the proposed contraflow-based method is applied to reconfigure evacuation routes to reduce traffic congestion and improve evacuation performance. Meanwhile, CC-Adap reuses historical evacuation routes to reduce computational costs and accelerate the evacuation process. Experimental results have shown that the CC-Adap algorithm can optimize evacuation route planning in terms of evacuation time and run time.

However, more research is required for CC-Adap. First, a more effective evacuation route generation method should be constructed. Second, the evaluation method should be improved to evaluate the evacuation routes more effectively.

Conflicts of Interest

The authors declare that they have no conflicts of interest.

Acknowledgments

This work is supported by the National Key Research and Development Program of China (no. 2016YFB0502600) and the National Natural Science Foundation of China (no. 41701594).

References

[1] Q. Lu, B. George, and S. Shekhar, "Capacity constrained routing algorithms for evacuation planning: A summary of results," in *Proceedings of the International Symposium on Spatial and Temporal Databases*, pp. 291–307, Berlin, Germany, 2005.

[2] S. Kim, S. Shekhar, and M. Min, "Contraflow transportation network reconfiguration for evacuation route planning," *IEEE Transactions on Knowledge and Data Engineering*, vol. 20, no. 8, pp. 1115–1129, 2008.

[3] J. W. Wang, H. F. Wang, W. J. Zhang, W. H. Ip, and K. Furuta, "Evacuation planning based on the contraflow technique with consideration of evacuation priorities and traffic setup time," *IEEE Transactions on Intelligent Transportation Systems*, vol. 14, no. 1, pp. 480–485, 2013.

[4] T. Takabatake, T. Shibayama, M. Esteban, H. Ishii, and G. Hamano, "Simulated tsunami evacuation behavior of local residents and visitors in Kamakura, Japan," *International Journal of Disaster Risk Reduction*, vol. 23, pp. 1–14, 2017.

[5] K. Shahabi and J. P. Wilson, "CASPER: Intelligent capacity-aware evacuation routing," *Computers, Environment and Urban Systems*, vol. 46, pp. 12–24, 2014.

[6] S. Kim, B. George, and S. Shekhar, "Evacuation route planning: scalable heuristics," in *Proceedings of the 15th annual ACM international symposium on Advances in geographic information systems*, p. 20, 2007.

[7] M. Jha, K. Moore, and B. Pashaie, "Emergency evacuation planning with microscopic traffic simulation," *Transportation Research Record*, no. 1886, pp. 40–48, 2004.

[8] M. Fellendorf and P. Vortisch, "Microscopic traffic flow simulator VISSIM," in *Fundamentals of traffic simulation*, pp. 63–93, Springer, New York, NY, USA, 2010.

[9] G. Lämmel, D. Grether, and K. Nagel, "The representation and implementation of time-dependent inundation in large-scale microscopic evacuation simulations," *Transportation Research Part C: Emerging Technologies*, vol. 18, no. 1, pp. 84–98, 2010.

[10] Y.-T. Hsu and S. Peeta, "Clearance Time Estimation for Incorporating Evacuation Risk in Routing Strategies for Evacuation Operations," *Networks and Spatial Economics*, vol. 15, no. 3, pp. 743–764, 2015.

[11] V. Zyryanov and A. Feofilova, "Simulation of Evacuation Route Choice," *Transportation Research Procedia*, vol. 20, pp. 740–745, 2017.

[12] X. Pan, C. S. Han, K. Dauber, and H. H. Law, "A multi-agent based framework for the simulation of human and social behaviors during emergency evacuations," *Ai & Society*, vol. 22, no. 2, pp. 113–132, 2007.

[13] G.-Y. Wu and H.-Ch. Huang, "Modeling the emergency evacuation of the high rise building based on the control volume model," *Safety Science*, vol. 73, no. 5, pp. 62–72, 2015.

[14] D.-J. Noh, J. Koo, and B.-I. Kim, "An efficient partially dedicated strategy for evacuation of a heterogeneous population," *Simulation Modelling Practice and Theory*, vol. 62, pp. 157–165, 2016.

[15] A. Beloglazov, M. Almashor, E. Abebe, J. Richter, and K. C. B. Steer, "Simulation of wildfire evacuation with dynamic factors and model composition," *Simulation Modelling Practice and Theory*, vol. 60, pp. 144–159, 2016.

[16] S. Yuan, S. A. Chun, B. Spinelli, Y. Liu, H. Zhang, and N. R. Adam, "Traffic evacuation simulation based on multi-level driving decision model," *Transportation Research Part C: Emerging Technologies*, vol. 78, pp. 129–149, 2017.

[17] M. Zeng and C. Wang, "Evacuation route planning algorithm: Longer route preferential. International Symposium on Neural Networks," in *Proceedings of the International Symposium on Neural Networks*, pp. 1062–1071, Berlin, Germany, 2009.

[18] W. Kang, F. Zhu, Y. Lv, G. Xiong, L. Xie, and B. Xi, "A heuristic implementation of emergency traffic evacuation in urban areas," in *Proceedings of the 2013 IEEE International Conference on Service Operations and Logistics, and Informatics, SOLI 2013*, pp. 40–44, chn, July 2013.

[19] M. U. S. Khan, O. Khalid, Y. Huang et al., "MacroServ: A Route Recommendation Service for Large-Scale Evacuations," *IEEE Transactions on Services Computing*, vol. 10, no. 4, pp. 589–602, 2017.

[20] E. Pourrahmani, M. R. Delavar, and M. A. Mostafavi, "Optimization of an evacuation plan with uncertain demands using fuzzy credibility theory and genetic algorithm," *International Journal of Disaster Risk Reduction*, vol. 14, pp. 357–372, 2015.

[21] L.-W. Chen, J.-H. Cheng, and Y.-C. Tseng, "Distributed Emergency Guiding with Evacuation Time Optimization Based on Wireless Sensor Networks," *IEEE Transactions on Parallel and Distributed Systems*, vol. 27, no. 2, pp. 419–427, 2016.

[22] Y. Ikeda and M. Inoue, "An Evacuation Route Planning for Safety Route Guidance System after Natural Disaster Using Multi-objective Genetic Algorithm," in *Proceedings of the 20th International Conference on Knowledge Based and Intelligent Information and Engineering Systems, KES 2016*, pp. 1323–1331, gbr, September 2016.

[23] M. Heydar, J. Yu, Y. Liu, and M. E. H. Petering, "Strategic evacuation planning with pedestrian guidance and bus routing: a mixed integer programming model and heuristic solution," *Journal of Advanced Transportation*, vol. 50, no. 7, pp. 1314–1335, 2016.

[24] U. Pyakurel, T. N. Dhamala, and T. N. Dhamala, "Models and algorithms on contraflow evacuation planning network problems," *International Journal of Operations Research*, vol. 12, no. 2, pp. 36–46, 2015.

[25] U. Pyakurel and T. N. Dhamala, "Continuous Dynamic Contraflow Approach for Evacuation Planning," *Annals of Operations Research*, vol. 253, no. 1, pp. 573–598, 2017.

[26] U. Pyakurel and T. N. Dhamala, "Continuous Time Dynamic Contraflow Models and Algorithms," *Advances in Operations Research*, vol. 2016, Article ID 7902460, 7 pages, 2016.

[27] J. Zhao, Y. Liu, and P. Li, "A network enhancement model with integrated lane reorganization and traffic control strategies," *Journal of Advanced Transportation*, vol. 50, no. 6, pp. 1090–1110, 2016.

[28] C. Xie, D.-Y. Lin, and S. Travis Waller, "A dynamic evacuation network optimization problem with lane reversal and crossing elimination strategies," *Transportation Research Part E: Logistics and Transportation Review*, vol. 46, no. 3, pp. 295–316, 2010.

[29] C. Even, V. Pillac, and P. Van Hentenryck, "NICTA evacuation planner: Actionable evacuation plans with contraflows," *Frontiers in Artificial Intelligence and Applications*, vol. 263, pp. 1143–1148, 2014.

[30] D. R. Fulkerson, "Flows in networks," in *Recent advances in mathematical programming*, pp. 319–331, McGraw-Hill, New York, NY, USA, 1963.

[31] D. Guo, C. Gao, W. Ni, and X. Hu, "Max-Flow rate priority algorithm for evacuation route planning," in *Proceedings of the 1st IEEE International Conference on Data Science in Cyberspace, DSC 2016*, pp. 275–283, chn, June 2016.

[32] D. Klingman, A. Napier, and J. Stutz, "NETGEN: a program for generating large scale capacitated assignment, transportation, and minimum cost flow network problems," *Management Science*, vol. 20, no. 5, pp. 814–821, 1974.

[33] E. Ronchi, F. N. Uriz, X. Criel, and P. Reilly, "Modelling large-scale evacuation of music festivals," *Case Studies in Fire Safety*, vol. 5, pp. 11–19, 2016.

15

A Methodology for Choosing between Route Deviation and Point Deviation Policies for Flexible Transit Services

Yue Zheng[1] **Wenquan Li**[1] **and Feng Qiu**[2]

[1] *Jiangsu Key Laboratory of Urban ITS, Jiangsu Province Collaborative Innovation Center of Modern Urban Traffic Technologies, School of Transportation, Southeast University, Nanjing, China*
[2] *Department of Computer Science, University of Victoria, Victoria, Canada*

Correspondence should be addressed to Wenquan Li; wenqli@seu.edu.cn

Academic Editor: Antonino Vitetta

Flexible transit services, which bring together the characteristics of fixed-route transit and demand-responsive transit, have been proven to be cost-efficient in low-density residential areas. In this paper, a methodology is proposed to assist planners in making better decisions when choosing between route deviation policy and point deviation policy, which are two promising types of flexible transit services. A user cost function is developed to measure the service quality of the transit systems, and analytical models are constructed to compare the system performance under both expected and unexpected demand levels. Based on the experiments for various scenarios over a real-life transit example, the critical demands, which represent the switching point between the two competing service policies, have been derived. Our findings show that point deviation policy is more efficient at low-demand levels, while route deviation policy is a better choice at low-to-moderate demand levels. At unexpectedly high demand levels, route deviation policy is better able to accommodate rejected passengers than point deviation policy.

1. Introduction

With the acceleration of urbanization in recent decades, an increasing number of suburban areas with low population density have arisen. The increasing dispersion of population results in a low degree of resource sharing for traditional fixed-route transit, which has a rigid route and schedule structure. Demand-responsive transit, which can provide much of the desired flexibility with curb-to-curb services, is considered costly to deploy and is therefore mostly limited to specialized operations such as paratransit for people with reduced mobility [1].

To address these issues, researchers and practitioners seek to introduce flexible transit services to meet the needs of new travel patterns in suburban and rural areas. All these flexible transit systems combine the efficiency of fixed-route transit with the flexibility of demand-responsive transit. They are regarded as fixed-route services in that they have a set of stops with predetermined schedules. On the other hand, they are considered as demand-responsive systems because passengers can ask for service at optional locations. Previous works based on actual operational data have shown that flexible transit services are more cost-efficient than pure demand-responsive transit [2] and conventional bus systems [3] in low-demand areas.

Based on the investigation by Koffman [4], flexible operating policies can be categorized into six main types: route deviation, point deviation, demand-responsible connector (DRC), request stops, flexible route segments, and zone route. Among these flexible services, DRC and zone route are implemented as feeder lines that have only one transfer point. Request stops can be regarded as a route deviation where only a few deviations are available. Flexible route segments operate in point deviation mode in some portion of the route. Hence, a comparison between route deviation and point deviation can generally reflect the difference in system structure for non-feeder services.

According to a survey by Potts et al. [5], nearly 40% of transit agencies in the United States and Canada operate flexible transit services. Among these, route deviation is by far the most common form (up to 63.9%), while point deviation only accounts for 16%. One of the main factors hindering the widespread use of point deviation service is the sophisticated support systems required to schedule the curb-to-curb requests [6]. However, with the rapid development of intelligent transportation systems, point deviation is becoming a promising operating policy from a long-term perspective.

Most of these studies revealed that flexible transit services are promising operating policies for shaping new travel patterns in low-demand areas and passengers are generally willing to use these innovative transit systems [7, 8]. However, how to choose the most appropriate policy in a certain service area, especially between route deviation policy and point deviation policy, remains a challenge for transit operators.

In this paper, we present a methodology to assist decision makers in choosing between route deviation and point deviation policies. Analytical models are developed to evaluate the system performance of two competing services. Using the analytical models, the critical demand, in which the two transit services have the same performance, can be identified, and the reliabilities of the two systems are further discussed. Our work serves as a meaningful step towards selecting between flexible transit services in areas with low and fluctuating travel demand.

2. Literature Review

Route deviation and point deviation policies differ in terms of degree of flexibility. More specifically, the route deviation systems are much more constrained than those allowed for point deviation; they must operate along a well-defined path and deviate to serve curb-to-curb requests within a service area around the path [9]. On the other hand, route deviation systems have a lower administrative/dispatch requirements and are easier to operate than point deviation services. Some transit agencies (for example, Tennessee Department of Transportation [10]) find it is a hard to choose between route deviation and point deviation policies when deploying new flexible transit services.

Although route deviation has been the most popular type of flexible transit services [4], it has been seldom studied by researchers. Most existing studies focused on improving the planning and scheduling of point deviation services. Quadrifoglio et al. [11–13] evaluated the upper and lower bounds of the maximum longitudinal velocity of service vehicles and proposed scheduling algorithms of point deviation service for both static and dynamic scenarios. Fu [14] developed a model to reveal the relationship between feasible deviations, slack time, zone size, and dwell time. Zhao and Dessouky [15] analyzed the relationship between the service cycle time and the length and width of the service area. Nourbakhsh and Ouyang [16] developed a new structured point deviation system that outperforms fixed-route transit under low-to-moderate demand levels. Qiu et al. [17] explored the feasibility of replacing fixed-route transit with point deviation policies and derived the switching demand densities between the two policies. Qiu et al. [18] presented a new demi-flexible transit system and compared its performance with fixed-route and point deviation services. Yang et al. [19] proposed a GIS-based method to select the most cost-effective route for deviated fixed-route transit (DFRT) based on the lowest operating cost per passenger. Chen and Nie [20] developed a new hybrid system that integrates the fixed-route service with point deviation service. The point deviation system serves as a connector between passengers' origin/destination and the fixed-route service to improve accessibility.

Most previous studies are conducted in operating environments with expected and predictable demand. Only a very limited number of studies focus on the uncertainty of travel demand in low-demand areas. Qiu et al. [21] proposed a dynamic station strategy to accept more passengers when the actual travel demand is higher than demand. Zheng et al. [22] proposed a slack arrival strategy to reduce the number of rejected passengers through redistributing the slack time among segments. In practice, the actual demand frequently deviates from the expected demand levels in low-demand service areas and the uncertain travel demand makes it hard to serve all curb-to-curb requests, which definitely degrade the service level of the systems.

In this paper, we aim to investigate and establish the conditions that are fit for the implementation of route deviation or point deviation policy. The system performances are compared not only at expected demand levels, but also at unexpected demand levels to test the reliability of the two transit services. To our knowledge, this paper is the first to develop a methodology for solving this problem.

3. System Description

3.1. Service Areas and Demand. The service area in our study represents a residential community and is modeled as a rectangle of width W and length L, delimited by two terminal checkpoints located at connection centers (see Figures 1 and 2). The transit system serves as community buses to transport passengers between their houses and connection centers, where they can transfer to the major transit network.

In general, the number of transit lines passing by the connection centers is very high; therefore, the temporal demand distribution at the connection centers is assumed to follow a Poisson distribution. It is also reasonable to assume that the trip origins and destinations outside checkpoints are uniformly and independently distributed in the service area according to a homogeneous spatial Poisson process. There are three main types of passengers in the service area with proportions of γ_1, γ_2, and γ_3 defined as follows ($\gamma_1 + \gamma_2 + \gamma_3 = 1$).

Type I: pick up and drop off both at checkpoints.

Type II: pick up at checkpoints, and drop off not at checkpoints.

Type III: pick up not at checkpoints, and drop off at checkpoints.

3.2. Competing Transit Policies. To compare the performance of the two flexible transit systems, both route deviation policy and point policy are implemented with the same fleet size M and passenger distribution. Bus cruising speed control strategy [23] is assumed to be implemented to eliminate the variation of demand locations.

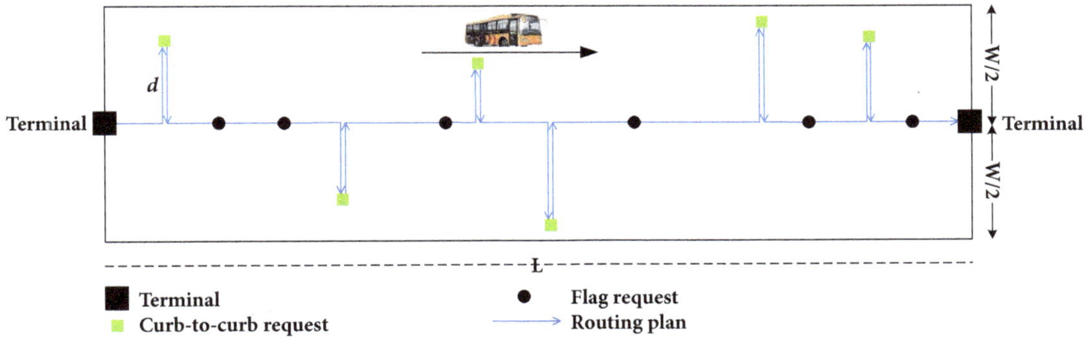

FIGURE 1: Route deviation operating policy.

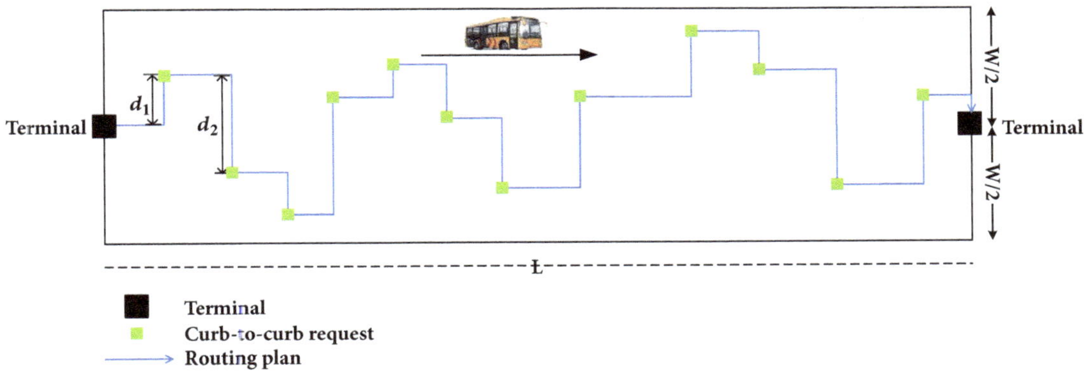

FIGURE 2: Point deviation operating policy.

3.2.1. Route Deviation Policy. As shown in Figure 1, the service vehicle operates a regular schedule along the base route and deviates to serve the curb-to-curb requests on either side of the base route with a maximum allowable deviation width of $W/2$. Types II and III passengers can request the transit in two ways. The first kind of request is called a flag request in which passengers have a vertical walk between the base route and their house and the service vehicles stop to board or discharge passengers at any location requested along the base route. Another kind of request is a curb-to-curb request which requires passengers to make advance notice to schedule their noncheckpoint stops. After serving the curb-to-curb passengers, the vehicle must return to the base route at the exact point where it left to accommodate the deviation.

The slack time for route deviation service is allotted based on the expected demand levels of the flag requests and curb-to-curb requests. A no-rejection policy is applied for flag requests as the variation of travel demand has limited impact on the service cycle. The deviation service operates on a first come, first available basis. When the slack time is unable to accommodate the actual demand, some of the curb-to-curb customers who reserve later must be rejected due to the slack time limitation.

3.2.2. Point Deviation Policy. As another kind of flexible transit operating policy, service vehicles in point deviation system are not constrained to follow any predefined base route. Except for the time constraints of the checkpoints, the remainder of the service is demand-responsive and vehicles serve curb-to-curb requests within the service area (see Figure 2).

Given that the path between the scheduled stops might change in each ride, flag requests are not possible. Instead, all types II and III passengers are required to make reservations to schedule their noncheckpoint stops through smartphones or the Internet. Similar to route deviation system, there is also a predefined slack time based on the expected demand level. Accepted curb-to-curb requests have no walking in their trip. If the actual demand exceeds the expected demand level, rejections may occur for curb-to-curb passengers.

4. Modeling

In our study, passengers are assumed to be transit-dependent. Vehicle follows a rectilinear metric, which has been proven to be a good approximation of reality [24]. The services are provided by vehicles with the following characteristics: the cruising speed considering stops due to traffic and pedestrian interference, V_b; the dwelling time of a flag stop or a request stop, T_d^r; and dwelling time of a checkpoint, T_d^f. The walking speed of passenger is V_{wk}. θ and θ_c represent the expected demand rate (passenger/h) and actual demand rate (passenger/h) in the service area, respectively.

In practice, the actual travel demand frequently deviates from the expected demand level in low-demand service

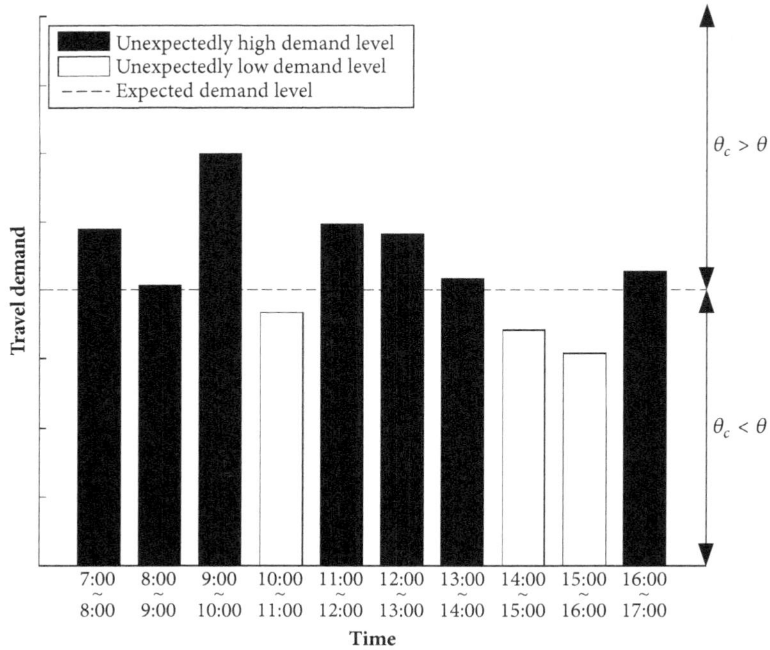

FIGURE 3: Uncertainties in travel demand.

areas due to uncertainties (see Figure 3). To investigate the performance of the two flexible transit systems, we measure the system performance at three demand levels, namely, the expected demand level, an unexpectedly high demand level, and an unexpectedly low-demand level.

4.1. Measure of Performance. There is no designed idle time between trips, and we keep the same fleet under different transit policies. If we disregard the difference in complexity of the reservation and scheduling systems, which only account for a small fraction of long-term transit operation, the two transit policies are considered have the same operating cost. Thus, in our analysis, the performance measure of the transit system is defined as the user cost function F, which reflects the service quality and greatly impacts the passengers' attitude towards the transit services. Similar to the previous study by Li and Quadrifoglio [25], the user cost function is constructed as the weighted sum of three indicators, the expected walking time K, the expected waiting time A, and the expected riding time R, with weights of ω_K, ω_A, and ω_R, respectively. A lower value of F clearly indicates a better transit service policy.

$$F = \omega_K K + \omega_A A + \omega_R R \quad (1)$$

4.2. Analytical Model of Route Deviation Service. For route deviation service, we assume the proportions of the curb-to-curb type II and type III passengers are α and β, respectively. Correspondingly, the proportions of the flag request type II and type III passengers are 1-α and 1-β.

4.2.1. At the Expected Demand Level. At the expect demand level, all curb-to-curb requests can be accepted. The expected vertical distance between the base route and the request stops (see Figure 1) is

$$d = \frac{W}{4} \quad (2)$$

Assuming a vehicle serves k passenger in a single-trip time T_r, the travel distance D of a vehicle in one trip is

$$D = L + (k\alpha\gamma_2 + k\beta\gamma_3) \times 2d = L + \frac{W(k\alpha\gamma_2 + k\beta\gamma_3)}{2} \quad (3)$$

In route deviation service, another two relationships can be obtained:

$$T_r = \frac{D}{V_b} + k(\gamma_2 + \gamma_3) T_d^r + T_d^f \quad (4)$$

$$T_r \times \theta = k \times M \quad (5)$$

Combining (3)-(5), we can derive the single-trip time T_r as follows:

$$T_r = \frac{2ML + 2V_b M T_d^f}{2MV_b - W\theta(\alpha\gamma_2 + \beta\gamma_3) - 2\theta V_b T_d^r (\gamma_2 + \gamma_3)} \quad (6)$$

In the system, type I passengers do not walk as part of their trips ($K_I = 0$). Curb-to-curb type II and type III passengers also have no waking time $K_{II}^c = K_{III}^c = 0$. Flag request passengers walk vertically between their houses and base route, and their expected walking time is

$$K_{II}^f = K_{III}^f = \frac{W}{4V_{wk}} \quad (7)$$

Thus, the expected walking time per passenger is

$$K = \gamma_1 K_{\mathrm{I}} + \gamma_2 \left(\alpha K^{\mathrm{c}}_{\mathrm{II}} + (1-\alpha) K^{\mathrm{f}}_{\mathrm{II}}\right) + \gamma_3 \left(\beta K^{\mathrm{c}}_{\mathrm{III}} + (1-\beta) K^{\mathrm{f}}_{\mathrm{III}}\right) = (\gamma_2 + \gamma_3 - \alpha\gamma_2 - \beta\gamma_3) \frac{W}{4V_{wk}} \quad (8)$$

The headway of route deviation system is $2T_r/M$, and the expected waiting time of the type I, type II, and flag request type III passengers is

$$A_{\mathrm{I}} = A_{\mathrm{II}} = A^{\mathrm{f}}_{\mathrm{III}} = \frac{T_{\mathrm{r}}}{M} \quad (9)$$

Curb-to-curb type III passengers can receive a scheduled pick-up time from the system when they make advanced notice. They are more likely to spend their time at their house or comfortable locations instead of waiting outside before the scheduled pick-up time. Therefore, their waiting time is defined as the interval between the scheduled pick-up time and the actual pick-up time. The waiting time for a curb-to-curb type III passenger i consists of two parts. The first part is the sum of service time of the flag requests before the vehicle picks up i. The second part is the follow-up curb-to-curb customers who make reservations later than i causing additional stops before the vehicle picks up i. Based on mathematical probability theory, the expected value of the waiting time can be expressed as

$$A^{c}_{\mathrm{III}} = \frac{\sum_{i=1}^{k(\alpha\gamma_2+\beta\gamma_3)-1} i}{k(\alpha\gamma_2+\beta\gamma_3)} \left(\frac{W}{2V_b} + T^r_d\right)\frac{1}{2} + \frac{k((1-\alpha)\gamma_2 + (1-\beta)\gamma_3)}{2} T^r_d = \frac{Wk(\alpha\gamma_2+\beta\gamma_3)}{8V_b} + \frac{T^r_d k(\alpha\gamma_2+\beta\gamma_3)}{4} - \frac{W}{8V_b} - \frac{T^r_d}{4} + \frac{T^r_d k((1-\alpha)\gamma_2 + (1-\beta)\gamma_3)}{2} \quad (10)$$

Thus, the expected waiting time per passenger is

$$A = \gamma_1 A_{\mathrm{I}} + \gamma_2 A_{\mathrm{II}} + \gamma_3 \left(\beta A^{\mathrm{c}}_{\mathrm{III}} + (1-\beta) A^{\mathrm{f}}_{\mathrm{III}}\right) = \frac{\left(2L + 2V_b T^f_d\right)(\gamma_1 + \gamma_2 + \gamma_3(1-\beta))}{2MV_b - W\theta(\alpha\gamma_2+\beta\gamma_3) - 2\theta V_b T^r_d(\gamma_2+\gamma_3)} + \frac{\gamma_3\beta W k(\alpha\gamma_2+\beta\gamma_3) - \gamma_3\beta W}{8V_b} + \frac{\gamma_3\beta T^r_d k(\alpha\gamma_2+\beta\gamma_3) - \gamma_3\beta T^r_d}{4} + \frac{\gamma_3\beta T^r_d k((1-\alpha)\gamma_2 + (1-\beta)\gamma_3)}{2} \quad (11)$$

Type I passengers travel from one terminal to another, so their expected riding time is

$$R_{\mathrm{I}} = T_{\mathrm{r}} \quad (12)$$

Types II and III passengers can be dropped off or picked up uniformly anytime in the trip between two terminals, and their expected riding time can be expressed as

$$R_{\mathrm{II}} = R_{\mathrm{III}} = \frac{T_{\mathrm{r}}}{2} \quad (13)$$

Then, the expected riding time per passenger is

$$R = \gamma_1 R_{\mathrm{I}} + \gamma_2 R_{\mathrm{II}} + \gamma_3 R_{\mathrm{III}} = \frac{\left(ML + V_b M T^f_d\right)(1+\gamma_1)}{2MV_b - W\theta(\alpha\gamma_2+\beta\gamma_3) - 2\theta V_b T^r_d(\gamma_2+\gamma_3)} \quad (14)$$

4.2.2. At Unexpectedly Low-Demand Level. When $\theta_c < \theta$, there will be idle time at the downstream checkpoint. For this condition, the expected walking time is equal to the result from (8). The expected waiting of the type I, type II, and flag request type III passengers can be derived by (9).

For each ride, the actual served passengers can be expressed as

$$k_c = \frac{\theta_c \times T_r}{M} = \frac{2\theta_c L + 2V_b\theta_c T^f_d}{2MV_b - W\theta(\alpha\gamma_2+\beta\gamma_3) - 2\theta V_b T^r_d(\gamma_2+\gamma_3)} \quad (15)$$

The expected waiting time of curb-to-curb type III passengers can be obtained by replacing k with k_c in (10). Then, we can obtain the expected waiting time at unexpectedly low-demand levels by $A = \gamma_1 A_{\mathrm{I}} + \gamma_2 A_{\mathrm{II}} + \gamma_3(\beta A^c_{\mathrm{III}} + (1-\beta)A^f_{\mathrm{III}})$.

At unexpectedly low-demand levels, there will be idle time at the terminal checkpoints. The expected riding times of different types of passengers can be obtained as follows:

$$R_{\mathrm{I}} = \frac{L + W(k_c\alpha\gamma_2 + k_c\beta\gamma_3)/2}{V_b} + k_c(\gamma_2+\gamma_3)T^r_d + T^f_d \quad (16)$$

$$R_{\mathrm{II}} = R_{\mathrm{III}} = \frac{L + W(k_c\alpha\gamma_2 + k_c\beta\gamma_3)/2}{2V_b} + \frac{k_c(\gamma_2+\gamma_3)}{2}T^r_d + \frac{T^f_d}{2} \quad (17)$$

Thus, the expected riding time per passenger at unexpectedly low-demand levels can be expressed as

$$R = \gamma_1 R_{\mathrm{I}} + \gamma_2 R_{\mathrm{II}} + \gamma_3 R_{\mathrm{III}} = (1+\gamma_1)\left[\frac{L + V_b T^f_d}{2V_b} + \left(\frac{W(\alpha\gamma_2+\beta\gamma_3)}{4V_b} + \frac{T^r_d(\gamma_2+\gamma_3)}{2}\right)k_c\right] \quad (18)$$

4.2.3. At Unexpectedly High Demand Level. When the actual demand exceeds the service capacity of deviation services, some of the curb-to-curb requests have to be rejected due to slack time limitation. These rejected customers may need to wait more than one operating cycle before pick-up. However, waiting for the next available vehicle makes it difficult for the rejected passengers to follow their schedules. Thus, we assume that rejected curb-to-curb passengers become flag request passengers for desired transit service.

When $\theta_c > \theta$, the first θ passengers can be accepted, and the values of the passenger cost indicators are the same as those at the expected demand level. For the last θ_c-θ passengers, type I and flag request types II and III passengers can be accepted and the actual number of rejected request is $(\theta_c - \theta)(\alpha\gamma_2 + \beta\gamma_3)$. For these θ_c-θ passengers, $K_{\text{I}}^{\theta_c-\theta} = 0$, and the expected walking time of types II and III passengers is

$$K_{\text{II}}^{\theta_c-\theta} = K_{\text{III}}^{\theta_c-\theta} = \frac{W}{4V_{wk}} \tag{19}$$

We define $\delta = K$ in (8), and the expected walking time of all passengers can be calculated as

$$K = \frac{\theta}{\theta_c}\delta + \frac{(\theta_c - \theta)(\gamma_2 + \gamma_3) W}{4V_{wk}\theta_c} \tag{20}$$

Because of the no-rejection policy, the last $(\theta_c - \theta)(\gamma_2 + \gamma_3)$ flag request passengers prolong the single-trip time. Thus, the actual single-trip time T_r^a is

$$T_r^a = T_r + (k_c - k)(\gamma_2 + \gamma_3) T_d^r \tag{21}$$

At unexpectedly high demand levels, the expected waiting of the type I, type II, and flag request type III passengers can be calculated by replacing T_r^a with T_r in (9). The expected waiting time of curb-to-curb type III passengers can be expressed as

$$\begin{aligned} A_{\text{III}}^c &= \frac{\sum_{i=1}^{k(\alpha\gamma_2+\beta\gamma_3)-1} i}{k(\alpha\gamma_2 + \beta\gamma_3)}\left(\frac{W}{2V_b} + T_d^r\right)\frac{1}{2} \\ &+ \frac{k((1-\alpha)\gamma_2 + (1-\beta)\gamma_3) + (k_c - k)(\gamma_2 + \gamma_3)}{2} \\ &\cdot T_d^r = \frac{Wk(\alpha\gamma_2 + \beta\gamma_3)}{8V_b} + \frac{T_d^r k(\alpha\gamma_2 + \beta\gamma_3)}{4} - \frac{W}{8V_b} \\ &- \frac{T_d^r}{4} + \frac{T_d^r(k_c(\gamma_2 + \gamma_3) - k\alpha\gamma_2 - k\beta\gamma_3)}{2} \end{aligned} \tag{22}$$

Thus, the expected waiting time of all passengers is

$$A = \gamma_1 A_{\text{I}} + \gamma_2 A_{\text{II}} + \gamma_3 \left(\frac{k\beta}{k_c} A_{\text{III}}^{\text{c}} + \frac{k_c - k\beta}{k_c} A_{\text{III}}^{\text{f}}\right) \tag{23}$$

Similarly, the expected riding time of type I passengers is T_r^a and the expected riding time of types II and III passengers is $T_r^a/2$; thus, the expected riding time per passenger is

$$\begin{aligned} R &= \gamma_1 R_{\text{I}} + \gamma_2 R_{\text{II}} + \gamma_3 R_{\text{III}} \\ &= (1 + \gamma_1) \frac{T_r + (k_c - k)(\gamma_2 + \gamma_3) T_d^r}{2} \end{aligned} \tag{24}$$

4.3. Analytical Model of Point Deviation Service. For point deviation service, a "no-backtracking constraint policy" is applied, which forbids backward movements of the service vehicles; therefore, passengers are served in order of horizontal coordinates [26]. This routing policy may not be optimal but it is easy to implement, especially in complex road networks.

4.3.1. At the Expected Demand Level. At the expected demand level, we assume an ideal case in which no rejection occurs ($K = 0$) and the downstream terminal checkpoint features no idle time. As shown in Figure 2, let d_1 indicate the vertical distance between the terminal checkpoints and the fist/last on-request stop. We have that $\text{E}(d_1) = W/4$. Let d_2 indicate the vertical distance between any pair of on-request stops; we obtain that $\text{E}(d_2) = W/3$.

Similar to the previous analytical method for point deviation service, the following relationships can be obtained:

$$\begin{aligned} D &= L + 2 \times d_1 + [k(\gamma_2 + \gamma_3) - 1] \times d_2 \\ &= L + \frac{W}{6} + \frac{kW(\gamma_2 + \gamma_3)}{3} \end{aligned} \tag{25}$$

$$T_{\text{r}} = \frac{D}{V_b} + k(\gamma_2 + \gamma_3) T_d^r + T_d^f \tag{26}$$

$$T_{\text{r}} \times \theta = k \times M \tag{27}$$

Combining (25)-(27), the single-trip time T_{r} can be derived as follows:

$$T_{\text{r}} = \frac{6ML + MW + 6MV_b T_d^f}{6MV_b - 2W\theta(\gamma_2 + \gamma_3) - 6\theta V_b T_d^r(\gamma_2 + \gamma_3)} \tag{28}$$

The expected waiting time of different types of passenger can be expressed as

$$A_{\text{I}} = A_{\text{II}} = \frac{T_{\text{r}}}{M} \tag{29}$$

$$\begin{aligned} A_{\text{III}} &= \frac{\sum_{i=1}^{k(\gamma_2+\gamma_3)-1} i}{k(\gamma_2 + \gamma_3)}\left(\frac{W}{3V_b} + T_d^r\right)\frac{1}{2} \\ &= \frac{Wk(\gamma_2 + \gamma_3)}{12V_b} + \frac{T_d^r k(\gamma_2 + \gamma_3)}{4} - \frac{W}{12V_b} - \frac{T_d^r}{4} \end{aligned} \tag{30}$$

Thus, the expected value of waiting time per passenger is

$$\begin{aligned} A &= \gamma_1 A_{\text{I}} + \gamma_2 A_{\text{II}} + \gamma_3 A_{\text{III}} \\ &= \frac{(6L + W + 6V_b T_d^f)(\gamma_1 + \gamma_2)}{6MV_b - 2W\theta(\gamma_2 + \gamma_3) - 6\theta V_b T_d^r(\gamma_2 + \gamma_3)} \\ &+ \frac{\gamma_3 Wk(\gamma_2 + \gamma_3) - \gamma_3 W}{12V_b} \\ &+ \frac{\gamma_3 T_d^r k(\gamma_2 + \gamma_3) - \gamma_3 T_d^r}{4} \end{aligned} \tag{31}$$

Just as in the route deviation service, the expected riding time of type I passengers is T_{r} and the expected riding time of types II and III passengers is $T_r/2$. Thus, the expected riding time per passengers can be calculated as

$$R = \gamma_1 R_{\mathrm{I}} + \gamma_2 R_{\mathrm{II}} + \gamma_3 R_{\mathrm{III}} = \frac{\left(6ML + MW + 6MV_b T_d^f\right)(1+\gamma_1)}{12MV_b - 4W\theta(\gamma_2+\gamma_3) - 12\theta V_b T_d^r (\gamma_2+\gamma_3)} \quad (32)$$

4.3.2. At Unexpectedly Low-Demand Level. If $\theta_c < \theta$, no passengers are rejected by the system ($K = 0$). In this case, the expected waiting time of types I and II passengers can be calculated by (29). The expected waiting of the types III passengers can be calculated by replacing k with k_c ($k_c = \theta_c T_r/M$) in (30). Then, we can determine the expected waiting time per passenger as $A = \gamma_1 A_{\mathrm{I}} + \gamma_2 A_{\mathrm{II}} + \gamma_3 A_{\mathrm{III}}$.

As the actual demand is lower than expected, the vehicle will arrive earlier than the schedule arrival time. The expected riding times of different types of passengers are

$$R_{\mathrm{I}} = \frac{L}{V_b} + \frac{W}{6V_b} + \frac{k_c W(\gamma_2+\gamma_3)}{3V_b} + k_c(\gamma_2+\gamma_3)T_d^r + T_d^f \quad (33)$$

$$R_{\mathrm{II}} = R_{\mathrm{III}} = \frac{L}{2V_b} + \frac{W}{12V_b} + \frac{k_c W(\gamma_2+\gamma_3)}{6V_b} + \frac{k_c(\gamma_2+\gamma_3)T_d^r}{2} + \frac{T_d^f}{2} \quad (34)$$

Thus, the expected riding time per passenger at unexpectedly low-demand levels can be expressed as

$$R = \gamma_1 R_{\mathrm{I}} + \gamma_2 R_{\mathrm{II}} + \gamma_3 R_{\mathrm{III}} = (1+\gamma_1)\left[\frac{L}{2V_b} + \frac{W}{12V_b} + \frac{k_c W(\gamma_2+\gamma_3)}{6V_b} + \frac{k_c(\gamma_2+\gamma_3)T_d^r}{2} + \frac{T_d^f}{2}\right] \quad (35)$$

4.3.3. At Unexpectedly High Demand Level. Unlike route deviation service, in point deviation service, the rejected passengers cannot issue a flag request since there is no base route for the service. Therefore, we assume that rejected customers utilize the nearest checkpoint for transit. Generally, curb-to-curb passengers are required to request service at least two hours in advance. Thus, the rejected passengers are considered to have sufficient time to reach the checkpoint before pick up.

For the last $\theta_c - \theta$ passengers, type I passengers can be accepted and the rejected curb-to-curb request is $(\theta_c - \theta)(\gamma_2 + \gamma_3)$. The expected walking time of types II and III passengers is

$$K_{\mathrm{II}}^{\theta_c-\theta} = K_{\mathrm{III}}^{\theta_c-\theta} = \frac{L}{4V_{wk}} + \frac{W}{4V_{wk}} \quad (36)$$

TABLE 1: Parameter values.

Parameter	Value
L	3 miles
W	1 miles
M	1
V_b	25 miles/h
V_{wk}	3 miles/h
T_d^r	12 sec
T_d^f	15 sec
$\gamma_1/\gamma_2/\gamma_3$	0.2/0.4/0.4

Thus, the expected walking time per passenger can be expressed as

$$K = \frac{(\theta_c - \theta)(\gamma_2+\gamma_3)(L+W)}{4V_{wk}\theta_c} \quad (37)$$

For the last $\theta_c - \theta$ passengers, the expected waiting time of type I passengers is $A_{\mathrm{I}}^{\theta_c-\theta} = T_r/M$. As the rejected type III passengers are informed about the transit schedule, $A_{\mathrm{III}}^{\theta_c-\theta} = 0$. Rejected type II passengers either walk directly to their destinations or obtain a ride to the downstream checkpoint; thus the expected waiting time can be derived as

$$A_{\mathrm{II}}^{\theta_c-\theta} = \frac{1}{2}\times 0 + \frac{1}{2}\times\frac{T_r}{M} = \frac{T_r}{2M} \quad (38)$$

We define $\lambda = A$ in (31), and the expected waiting time of all passengers can be derived as

$$A = \frac{\theta}{\theta_c}\lambda + \frac{\gamma_1(\theta_c-\theta)T_r}{M\theta_c} + \frac{\gamma_2(\theta_c-\theta)T_r}{2M\theta_c} \quad (39)$$

Similarly, in the last $\theta_c - \theta$ passengers, the expected riding time of type I passengers is $R_{\mathrm{I}}^{\theta_c-\theta} = T_r$. For types II and III passengers, the expected riding time can be expressed as

$$R_{\mathrm{II}}^{\theta_c-\theta} = R_{\mathrm{III}}^{\theta_c-\theta} = \frac{1}{2}\times 0 + \frac{1}{2}\times T_r = \frac{T_r}{2} \quad (40)$$

We can observe that the expected riding time of the $\theta_c - \theta$ passengers remains unchanged; therefore the expected riding time of all passengers can be calculated by (32).

5. Result Analysis

5.1. Parameter Values. In this section, a case study is conducted based the Route 289 service in the suburban area of Zhengzhou City in China which currently operates under a pure flag stop policy [18]. In this paper, we use the system parameters and operating data to test system performance under both route deviation policy and point deviation policy. The parameter values for the analysis are presented in Table 1. The weights values (ω_K, ω_A, and ω_R) in the user cost function may be varied according to a wide range of socioeconomic and environment factors. Similar to the previous research, we assume that $\omega_A = 1$ and $\omega_R = 1$ [16] and the default weight of the walking time is set to $\omega_K = 2$ [27]. Cases with different weights will be discussed in sensitivity analysis.

TABLE 2: Passenger cost indicators in route deviation service (α=0.1 and β=0.1).

Indicator (min)	Demand (passengers/h)						
	26	30	34	38	42	46	50
T_r	8.38	8.54	8.71	8.89	9.08	9.27	9.47
K	3.60	3.60	3.60	3.60	3.60	3.60	3.60
A	8.04	8.20	8.37	8.54	8.72	8.91	9.11
R	5.03	5.13	5.23	5.33	5.45	5.56	5.68
F	20.27	20.53	20.80	21.08	21.37	21.67	21.99

TABLE 3: Passenger cost indicators in point deviation service.

Indicator (min)	Demand (passengers/h)						
	26	30	34	38	42	46	50
T_r	12.02	13.08	14.36	15.91	17.84	20.30	23.55
K	0.00	0.00	0.00	0.00	0.00	0.00	0.00
A	7.53	8.27	9.17	10.25	11.60	13.33	15.60
R	7.21	7.85	8.62	9.55	10.70	12.18	14.13
F	14.73	16.12	17.78	19.80	22.31	25.51	29.73

5.2. Comparison of System Performance at the Expected Demand Level. At the expected demand levels, the passenger cost indicators from the analytical models of both route deviation policy and point deviation policy are presented in Tables 2 and 3, respectively. The results illustrate that point deviation service outperforms route deviation service when the travel demand is low because there is no walking ($K = 0$) in the system. However, point deviation service is much more sensitive to the demand levels. The single-trip time T_r sharply increases with increasing demand under point deviation policy, which likely the result of long detours required by the curb-to-curb services. In contrast, the demand has limited influence on route deviation service ($\alpha = 0.1$ and $\beta = 0.1$). Consequently, the passenger cost function of point deviation service enlarged dramatically due to the rapidly increasing values of A and R and overtakes route deviation service with the critical demand $\theta = 42$ passengers/h.

To test the influence of the curb-to-curb passengers' proportions, the user cost functions with different α and β are presented in Figure 4. The results indicate that, for any proportions of α and β, point deviation service has a clear advantage in relatively low demand (e.g., $\theta < 38$ passengers/h). With an increasing proportion of curb-to-curb passengers (α and β), the user function F becomes more sensitive to the travel demand and the critical demand between the two policies becomes larger. When $\alpha = \beta = 0.7$, point deviation service outperforms route deviation service under all demand levels. This is because, under route deviation policy, the vehicle has to return to the base route after serving a curb-to-curb passenger which becomes inefficient with a large proportion of curb-to-curb passengers. Besides, route deviation service with a large proportion of curb-to curb type III passengers ($\alpha = 0.1$, $\beta = 0.7$) has a lower cost function than its counterparts ($\alpha = 0.4$, $\beta = 0.4$). Although the two cases have the same number of curb-to curb passengers ($\alpha\gamma_2 + \beta\gamma_3$), curb-to curb type III passengers have shorter waiting time than those waiting at the terminals because of the advanced scheduled pick-up time notice. We can also observe that growth trends of the user cost for the two cases ($\alpha = 0.1$, $\beta = 0.7$, and $\alpha = \beta = 0.4$) are almost the same which indicate that, for any other combinations of α and β, the growth trend is within the range of these three selected combinations. Hence, we use $\alpha = \beta = 0.1$, $\alpha = \beta = 0.4$, and $\alpha = \beta = 0.7$ as typical cases in this study.

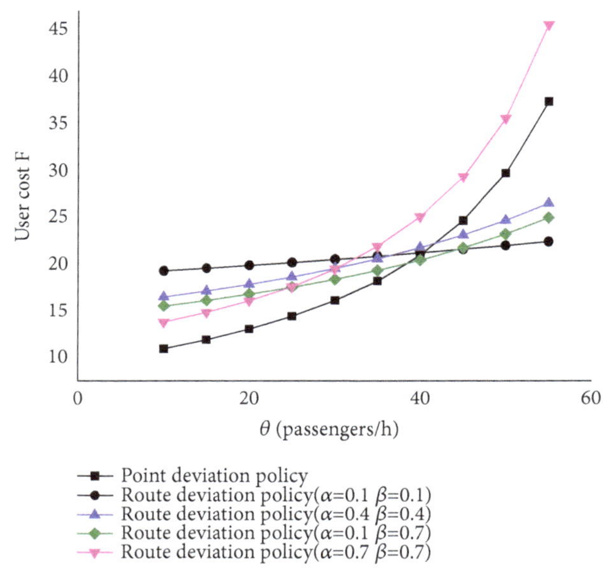

FIGURE 4: Comparison of system performance at the expected demand levels.

The above results confirm that, in the designed demand range, point deviation service is desirable at low-demand levels. The difference in system performance narrows between the two policies with the increasing demand. Route deviation service is more suitable for the relatively higher demand levels. With an increase in the proportion of curb-to-curb

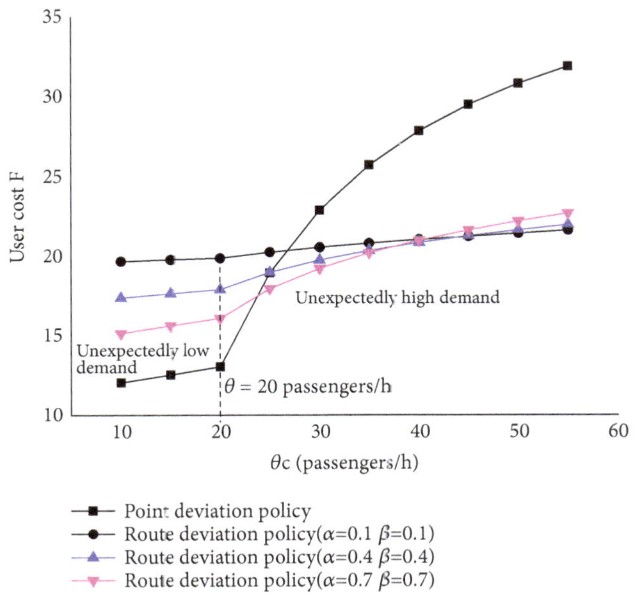

FIGURE 5: Comparison of system performance at unexpected demand levels.

passengers, route deviation service becomes more flexible but more sensitive to the demand. When this proportion is considerably large, point deviation service always performs better than route deviation service.

5.3. Comparison of System Performance at Unexpected Demand Levels. To test the reliability of the two transit services, the system performance at unexpected demand levels is presented in Figure 5 assuming an expected demand rate of $\theta = 20$ passengers/h. The results illustrate that both services experience a slight decrease at unexpectedly low-demand levels. The decreasing trend becomes more apparent with an increasing proportion of on-demand requests.

However, at unexpectedly high demand levels, the user cost of point deviation service surges with increasing actual demand θ_c. The system performance of route deviation service is relatively more stable especially when α and β are small. This result occurs because the passengers rejected in point deviation service have to walk to the nearest checkpoints while in route deviation service, rejected curb-to curb passengers can simply walk directly to the base route and issue flag requests.

Based on these observations, we can conclude that route deviation service is more reliable than point deviation service and is more suitable to apply in the service areas with great demand variations.

5.4. Sensitivity Analysis. Sensitivity analysis is conducted under both expected and unexpected demand levels to examine how the performance of the two competing transit polices is affected by key input parameters. The results are presented in Figure 6. The weights of waiting time and walking time may vary in different external operation environments. To study the influence of these weights on the system performance, we consider a case where $\omega_K = 3$, $\omega_A = 3$, $\omega_R = 1$. This may occur in cases where walking and waiting are highly undesirable, such as under extreme weather or in unsafe neighborhood. As shown in Figure 6(a), the user costs increase under both policies, and the margin becomes more remarkable than in the base case. The critical demand between the two systems becomes larger, which suggests that point deviation service has a larger suitable demand range in a bad operation environment. At unexpectedly high demand levels, the difference in the user cost between the two policies becomes larger since walking is more unendurable in this case (see Figure 6(b)).

Another experiment with a long length ($L = 8$ miles) is conducted (see Figures 6(c) and 6(d)). In this case, the waiting time and riding time are enlarged because of the long travel distance. The critical demand is smaller than that in the base case which indicates that route deviation service has a large application demand range for long distance route. At unexpectedly high demand levels, the difference becomes more remarkable because the rejected curb-to-curb passengers in point deviation service have to spend more time walking to the nearest checkpoints.

In real-life cases, the permitted deviation ranges from 0.25 to 1.5 miles [4]. A larger width ($W = 1.5$ miles) is investigated, and the results are presented in Figures 6(e) and 6(f). In this case, route deviation service with a small proportion of curb-to-curb passenger has a slight increase in the user cost. In contrast, the user cost of point deviation service experiences a dramatic increase with increasing demand due to the longer travel distance needed to travel to serve curb-to-curb requests. The switching point of the two transits is at a lower demand level, which demonstrates that point deviation service is quite sensitive to the width of the service area and it is more favorable to operate in narrowed service area.

Furthermore, a two-vehicle case ($M = 2$) is investigated, and the results are displayed in Figures 6(g) and 6(h). The two vehicles run along the base route in opposite directions, and they leave the two terminals at the same time. The outcomes indicate that point deviation service is more sensitive to the fleet size and has a better performance under low-to-moderate demand levels. At unexpectedly high demand levels, route deviation service shows better reliability than that in the base case. This result illustrates that more available vehicles can improve the robustness of route deviation service.

6. Conclusion

This paper helps planners to select the most appropriate flexible transit services in low-demand areas. Route deviation service and point deviation service are chosen as typical flexible policies in this study. The former is the most widely used in practice, while the latter has been extensively investigated by researchers.

The two flexible systems differ in operation rules and degree of flexibility offered. Generally, route deviations systems are much more constrained than those allowed for point deviation systems. By modeling a real-life transit service, we compare the system performance of the two competing transit policies under both expected and unexpected demands.

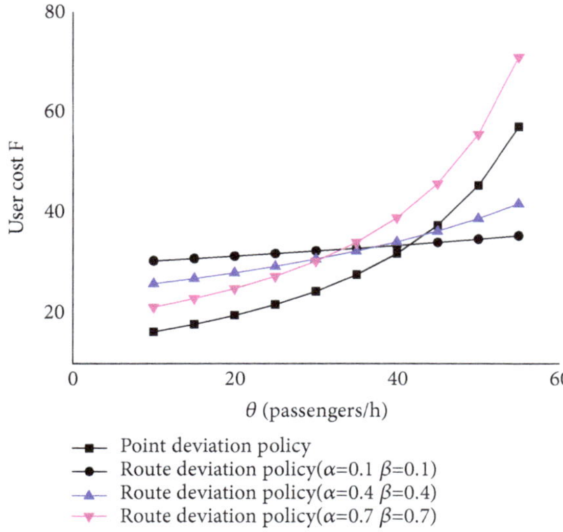

(a) At expected demand ($\omega_K = 3$, $\omega_A = 3$, $\omega_R = 1$)

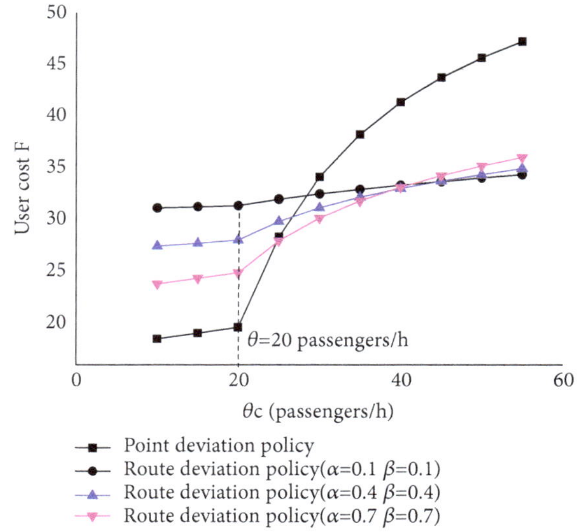

(b) At unexpected demand ($\omega_K = 3$, $\omega_A = 3$, $\omega_R = 1$)

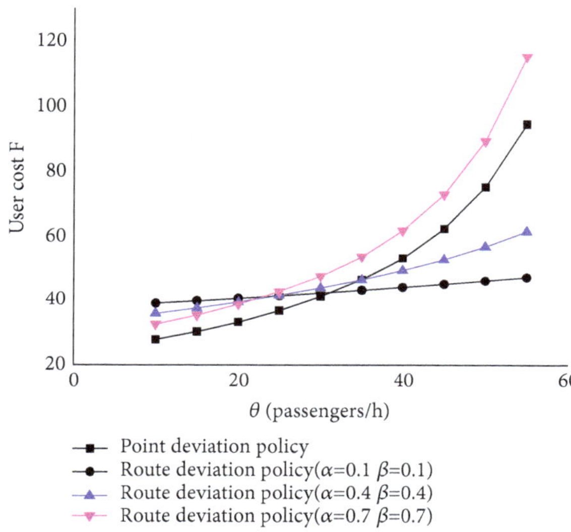

(c) At expected demand ($L = 8$ mile)

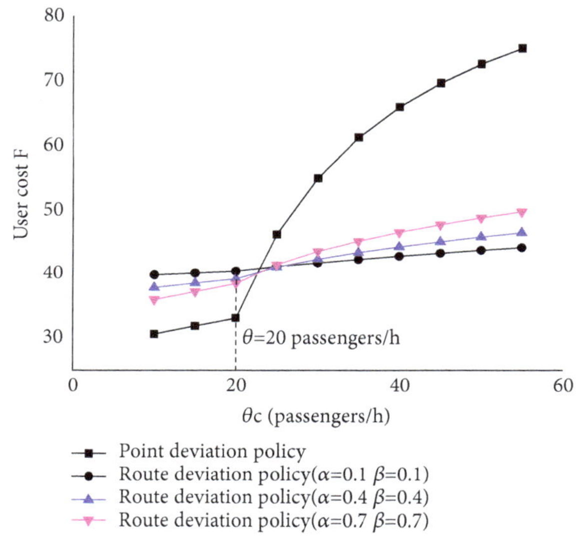

(d) At unexpected demand ($L = 8$ mile)

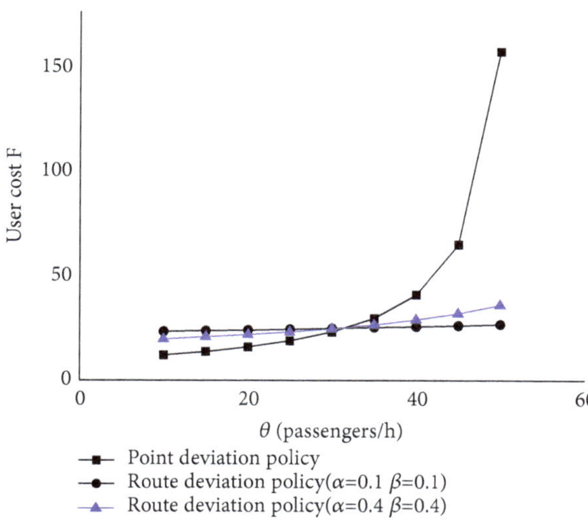

(e) At expected demand ($W = 1.5$ mile)

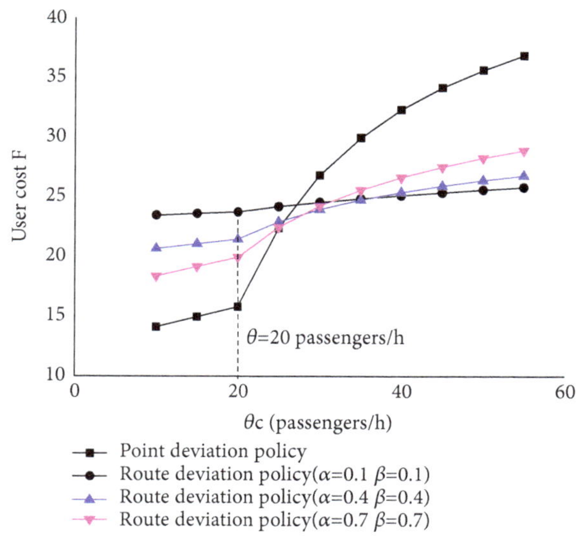

(f) At unexpected demand ($W = 1.5$ mile)

Figure 6: Continued.

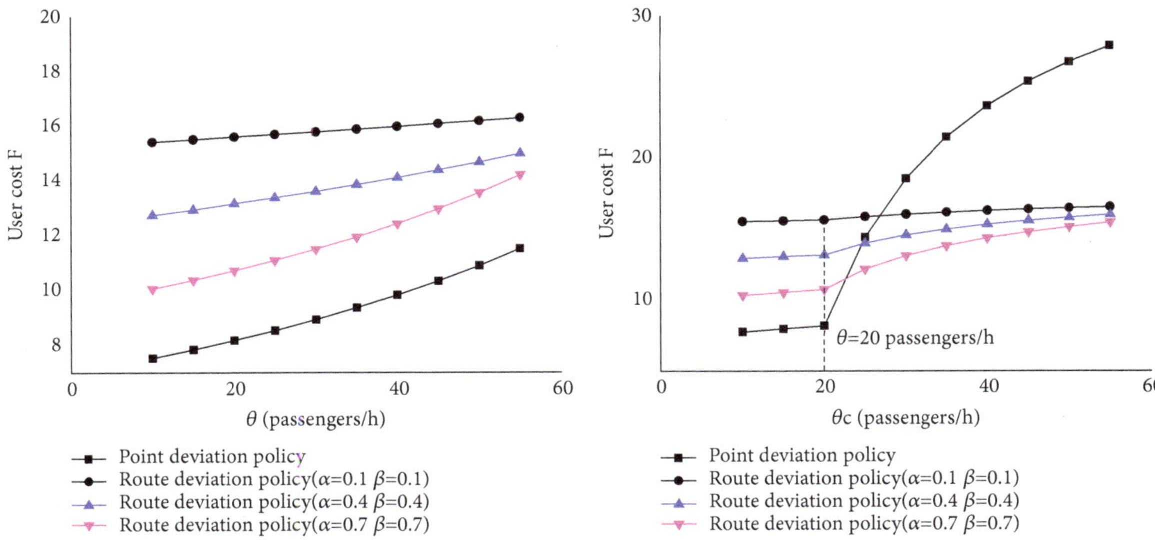

(g) At expected demand ($M = 2$)

(h) At unexpected demand ($M = 2$)

FIGURE 6: Sensitivity analysis for different parameters.

The results suggest that point deviation policy is preferable at low-demand levels and that route deviation policy is better at low-to-moderate demand levels. When the proportion of on-demand requests is considerably large, route deviation is not preferred even at relatively high demand. Route deviation policy is considered to be more reliable because the rejected curb-to-curb customers can still get access to transit with a shorter walking distance. In contrast, point deviation policy might not be suitable when the demand in the service area fluctuates significantly.

In practical use, point deviation service is more complex to implement than route deviation service because it requires a more sophisticated navigation and scheduling system. In addition, the travel time between checkpoints is more difficult to determine in point deviation service since the path might change for each ride. These difficulties explain why this promising operating policy has been applied by only a small percentage of transit agencies compared to route deviation policy. Therefore, an automated dispatch system with high navigation accuracy and an imbedded routing algorithm are needed to promote the widespread use of point deviation service. For future study, we plan to develop simulation models to reproduce the operation of these two competing service types more accurately considering the vehicle capacity and passengers' travel mode choice. Another possible research direction is to compare other kinds of flexible transit policies and define their application conditions.

Disclosure

This paper is an extended version of paper "Flexible Transit Services Choosing between Route Deviation and Point Deviation Policy" presented in 2018 TRB Annual Meeting.

Conflicts of Interest

The authors declare that they have no conflicts of interest.

Acknowledgments

The research reported in this paper was supported by the National Natural Science Foundation of China (Grant no. 61573098).

References

[1] D. Gupta, H. W. Chen, L. A. Miller, and F. Surya, "Improving the efficiency of demand-responsive paratransit services," *Transportation Research Part A: Policy and Practice*, vol. 44, no. 4, pp. 201–217, 2010.

[2] J. Becker, R. Teal, and R. Mossige, "Metropolitan transit agency's experience operating general-public demand-responsive transit," *Transportation Research Record*, no. 2352, pp. 136–145, 2013.

[3] S. R. Fittante and A. Lubin, "Adapting the Swedish service route model to suburban transit in the United States," *Transportation Research Record*, vol. 2536, pp. 52–59, 2015.

[4] D. Koffman, "Operational Experiences with Flexible Transit Services: A Synthesis of Transit Practice," TCRP Synthesis 53, Washington, Wash, USA, 2004.

[5] J. F. Potts, M. A. Marshall, E. C. Crockett, and J. Washington, "A Guide for Planning and Operating Flexible Public Transportation Services," Tech. Rep. 140, Transportation Research Board, Washington, Wash, USA, 2010.

[6] N. R. Velaga, J. D. Nelson, S. D. Wright, and J. H. Farrington, "The potential role of Flexible Transport Services in enhancing rural public transport provision," *Journal of Public Transportation*, vol. 15, no. 1, pp. 111–131, 2012.

[7] J. Yu, X. Lu, S. Pan, and C. Guo, "Traveler willingness to use flexible transit services in China: Case study of Qilu Software Park," *Journal of Urban Planning and Development*, vol. 143, no. 2, 2017.

[8] C. Frei, M. Hyland, and H. S. Mahmassani, "Flexing service schedules: Assessing the potential for demand-adaptive hybrid transit via a stated preference approach," *Transportation Research Part C: Emerging Technologies*, vol. 76, pp. 71–89, 2017.

[9] F. Errico, T. G. Crainic, F. Malucelli, and M. Nonato, "A survey on planning semi-flexible transit systems: Methodological issues and a unifying framework," *Transportation Research Part C: Emerging Technologies*, vol. 36, pp. 324–338, 2013.

[10] Tennessee Department of Transportation (TDOT), *City of Gallatin Transit Feasibility Study*, Tennessee Department of Transportation (TDOT), 2005, https://www.gallatintn.gov/DocumentCenter/View/1208/Gallatin-Transit-Feasibility-Study-December-2005-PDF.

[11] L. Quadrifoglio, R. W. Hall, and M. M. Dessouky, "Performance and design of mobility allowance shuttle transit services: Bounds on the maximum longitudinal velocity," *Transportation Science*, vol. 40, no. 3, pp. 351–363, 2006.

[12] L. Quadrifoglio, M. M. Dessouky, and K. Palmer, "An insertion heuristic for scheduling mobility allowance shuttle transit (MAST) services," *Journal of Scheduling*, vol. 10, no. 1, pp. 25–40, 2007.

[13] L. Quadrifoglio, M. M. Dessouky, and F. Ordóñez, "Mobility allowance shuttle transit (MAST) services: MIP formulation and strengthening with logic constraints," *European Journal of Operational Research*, vol. 185, no. 2, pp. 481–494, 2008.

[14] L. Fu, "Planning and design of flex-route transit services," *Transportation Research Record*, no. 1791, pp. 59–66, 2002.

[15] J. Zhao and M. Dessouky, "Service capacity design problems for mobility allowance shuttle transit systems," *Transportation Research Part B: Methodological*, vol. 42, no. 2, pp. 135–146, 2008.

[16] S. M. Nourbakhsh and Y. Ouyang, "A structured flexible transit system for low demand areas," *Transportation Research Part B: Methodological*, vol. 46, no. 1, pp. 204–216, 2012.

[17] F. Qiu, W. Li, and A. Haghani, "A methodology for choosing between fixed-route and flex-route policies for transit services," *Journal of Advanced Transportation*, vol. 49, no. 3, pp. 496–509, 2015.

[18] F. Qiu, J. Shen, X. Zhang, and C. An, "Demi-flexible operating policies to promote the performance of public transit in low-demand areas," *Transportation Research Part A: Policy and Practice*, vol. 80, pp. 215–230, 2015.

[19] H. Yang, C. R. Cherry, R. Zaretzki, M. S. Ryerson, X. Liu, and Z. Fu, "A GIS-based method to identify cost-effective routes for rural deviated fixed route transit," *Journal of Advanced Transportation*, vol. 50, no. 8, pp. 1770–1784, 2016.

[20] P. W. Chen and Y. M. Nie, "Analysis of an idealized system of demand adaptive paired-line hybrid transit," *Transportation Research Part B: Methodological*, vol. 102, pp. 38–54, 2017.

[21] F. Qiu, W. Li, and J. Zhang, "A dynamic station strategy to improve the performance of flex-route transit services," *Transportation Research Part C: Emerging Technologies*, vol. 48, pp. 229–240, 2014.

[22] Y. Zheng, W. Li, and F. Qiu, "A slack arrival strategy to promote flex-route transit services," *Transportation Research Part C: Emerging Technologies*, vol. 92, pp. 442–455, 2018.

[23] C. F. Daganzo and J. Pilachowski, "Reducing bunching with bus-to-bus cooperation," *Transportation Research Part B: Methodological*, vol. 45, no. 1, pp. 267–277, 2011.

[24] L. Quadrifoglio, M. M. Dessouky, and F. Ordóñez, "A simulation study of demand responsive transit system design," *Transportation Research Part A: Policy and Practice*, vol. 42, no. 4, pp. 718–737, 2008.

[25] X. Li and L. Quadrifoglio, "Feeder transit services: Choosing between fixed and demand responsive policy," *Transportation Research Part C: Emerging Technologies*, vol. 18, no. 5, pp. 770–780, 2010.

[26] L. Quadrifoglio and X. Li, "A methodology to derive the critical demand density for designing and operating feeder transit services," *Transportation Research Part B: Methodological*, vol. 43, no. 10, pp. 922–935, 2009.

[27] M. Wardman, "Public transport values of time," *Transport Policy*, vol. 11, no. 4, pp. 363–377, 2004.

16

Driver's Lane Selection Model Based on Phase-Field Coupling and Multiplayer Dynamic Game with Incomplete Information

Xiaoyuan Wang,[1,2] Fang Wang,[1,3] Dong Kong,[3] Yaqi Liu,[1,3] Liping Liu,[3] and Chen Chen[3]

[1]*College of Electromechanical Engineering, Qingdao University of Science & Technology, Qingdao 266000, China*
[2]*State Key Laboratory of Automotive Safety and Energy, Tsinghua University, Beijing 100084, China*
[3]*Institute of Intelligent Transportation, School of Transportation and Vehicle Engineering, Shandong University of Technology, Zibo 255049, China*

Correspondence should be addressed to Xiaoyuan Wang; wangxiaoyuan@sdut.edu.cn

Academic Editor: Rakesh Mishra

Intelligent driving is an effective means to achieve the active safety of automobile, and the accurate prediction of vehicle group situation is the premise to achieve the intelligent driving of vehicle. Lane selection and lane changing are not only the most fundamental reasons for the transformation of vehicle group situation, but also the basic contents for the research on driver behavior of traffic flow theory. In this paper, with a view to the background of Internet of Things, the vehicle group situation was given a comprehensive consideration on the basis of the factors which influence driver's behavior. The driver's lane selection behavior was analyzed under the condition of incomplete information, and lane selection model based on phase-field coupling and multiplayer dynamic game with incomplete information was constructed considering the time-varying character of driving propensity. The means of actual driving experiment, virtual driving experiment, and microscopic simulation of traffic flow were used to verify the model. The verification results showed that the model built in this paper can objectively reflect the actual operation characteristic of traffic flow on road section and the process of lane selection. The theoretical basis of the research on lane selection can be provided for intelligent driving especially anthropomorphic driving under the condition of Internet of Things.

1. Introduction

With the rapid development of the transportation industry, the car ownership has increased rapidly. The contradiction among people, vehicle, and environment has become increasingly conspicuous, and the traffic safety is becoming more and more serious. Lane changing is one of the main factors affecting traffic safety, and lane selection which need drivers to analyze all kinds of information and consider a variety of factors in an instant is an important premise for lane changing, and the rationality of lane selection is directly related to driving safety. The multisource information of people-vehicle-environment can be collected and shared through Internet of Things, which can provide decision-making basis for driver's lane selection process and improve the active safety of driver and vehicle. Therefore, it is of great practical significance to deeply study the lane selection behavior under the condition of Internet of Things. And the theoretical basis of the research on lane selection can be provided for intelligent driving, especially the anthropopathic driving.

Many scholars have made sustained efforts to the research on lane changing model. Gipps [1] built a lane changing model to cover the urban driving situation, where traffic signals, obstructions, and heavy vehicles all exerted an influence. QI YANG et al. [2] and Ahmed K.I [3] divided the lane changing behavior into three parts, namely, producing an intention to change lane, detecting the conditions, and executing a lane change. M.Rickert et al. [4] presented a lane changing model based on cellular automata approach under two-lane traffic condition. The basic scheme of the model is to determine whether there is an obstacle in the lane where the target vehicle is located and whether there is enough gap in the adjacent lane. Hidas [5] classified the maneuvers into free, forced, and cooperative lane changes under congested traffic conditions. And the lane changing model based on intelligent agent concept was developed. Mehdi Keyvan-Ekbatani et al. [6] investigated the decision process of lane changing maneuvers for a variety of drivers based on a two-stage test drive on freeways. Four distinct strategies were empirically found, namely, speed leading, speed leading and overtaking, lane leading, and traffic leading. Esmaeil Balal et al. [7] introduced a Fuzzy Inference System (FIS) which models a driver's binary decision to or not to execute a discretionary lane changing move on freeways. Yugong Luoet al. [8] proposed a dynamic automated lane change maneuver. The key technologies for this maneuver were trajectory planning and trajectory tracking. The planning problem was converted into a constrained optimization problem in the trajectory planning method, and a reference trajectory satisfying the demands of safety, comfort, and traffic efficiency was calculated and updated to complete the automated lane change and avoid the potential collisions. Shengbo Li et al. [9] presented a four-component framework to model, analyze, and synthesize a platoon of CAVs from the perspective of multiagent consensus control. The modeling technique of four-component framework was systematically introduced, the major performance metrics were discussed, and the design of distributed controller was presented.

Driving on the road is a dynamic game process, and many scholars have studied driver's lane changing behavior from the perspective of game theory. Hideyuki Kita [10] developed a two-person non-zero-sum noncooperative game theoretical model to describe the interactions of cars in merging or weaving sections and only the influence of through car on the behavior of a merging car was considered in the analysis. Both the merging and through cars would attempt to take the best actions for themselves by forecasting the other's action, respectively. Peng Jinshuan et al. [11] discussed a noncooperative mixed strategy game between the target vehicle and the following vehicle in the target lane and deeply analyzed the payoffs of the players and the Nash equilibrium. Alireza Talebpour et al. [12] put forward a two-person non-zero-sum noncooperation game model to analyze the interaction between the target vehicle and the following vehicle which is in the target lane. The lane changing models with complete and incomplete information were, respectively, established according to whether they was equipped with V2V communication. Zhang Yuanliang [13] proposed a discretionary lane changing model based on Stackelberg game and studied the game process of the target vehicle and the following vehicle in the target lane. Based on receding horizon optimal control and dynamic game theory, Meng Wang et al. [14] put forward a predictive approach for unified lane changing and car-following control. In the proposed approach, discrete lane change decisions and continuous accelerations were evaluated jointly to minimize the payoff function and obtain the Nash equilibrium solution.

The shortcomings of the previous studies on lane changing are mainly as follows. First, the driving propensity, especially the time-varying driving propensity is not fully considered in the previous studies. Driving propensity would change with road environment and vehicle group situation. Different drivers are different in the aspects of expected speed, expected distance, weight preference of payoffs, and even the extent of information perception and mastery on traffic environment. Second, the vehicle types and multilane environment are seldom taken into consideration in the existing studies. Different vehicle types have different influences on drivers. And in the multilane environment, the composition of vehicle type proportion would be more complicated, the propensity type of driver would be more various, and the frequency of lane changing would be higher. Third, in the existing lane, changing model based on game theory, only the two-person static game model is constructed, which is between the vehicle to change lane and the following vehicle on target lane. The interactive influence of the two vehicles was analyzed, but the influences of lane selection strategies of surrounding vehicles were ignored. Fourth, the completeness degree of information the driver obtained is seldom considered in the previous lane changing model based on game theory. The completeness degree for drivers to learn and master the information would be different under the different development level of Internet of Things.

Aimed at the issues of the previous studies, the grouping vehicles driving on the urban expressway were chosen to be the research object and the vehicle group situation was analyzed under the background of Internet of Things. On this basis, driver's lane selection model based on phase-field coupling and multiperson dynamic game with incomplete information was constructed considering the influence of time-varying driving propensity. The payoff of every player was analyzed to obtain driver's lane selection strategy.

2. Method

Before constructing the lane selection model, the vehicle group situation was defined and its mathematical description was built. And the vehicle group situation types were identified after the division and reduction. Based on the game analysis of driver's lane selection behavior, the driver's lane selection model based on phase-field coupling and multiperson dynamic game with incomplete information was constructed.

2.1. Vehicle Group Situation. Traffic situation refers to the states and situations of all traffic entity deployment and behavior in the driver's interest-sensitive area (interest-sensitive area refers to an area which has a greater impact on vehicle safety and allocates more driver's attention), and it contains all the information of traffic entity [15, 16]. In this paper, vehicle group situation in which driver can perceive in the interest-sensitive area, including the characteristics of traffic entity and vehicle group relationship and also considering the macroscopic characteristics of density and speed, was taken as an example to research the driver's lane selection behavior.

2.1.1. Definition of Vehicle Group Situation. As shown in Figure 1, the three-lane condition on urban expressway was taken as an example, target vehicle n_1 is chosen as the research

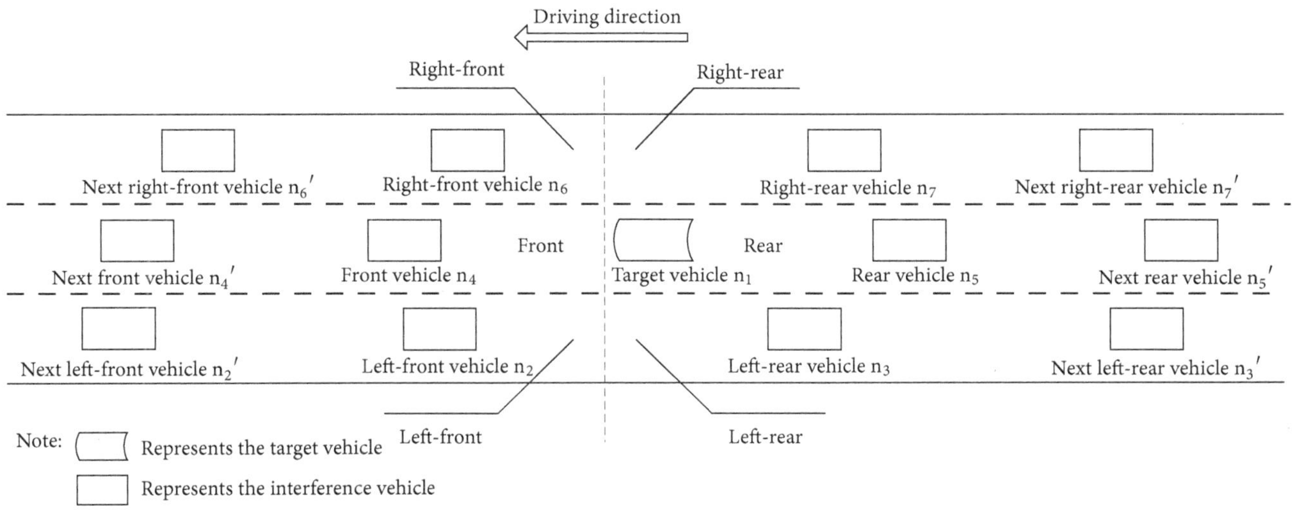

FIGURE 1: Vehicle group situation when target vehicle is located in the middle lane.

subject (unless otherwise specified, the case of target vehicle n_1 located in middle lane is taken as an example), and the interest-sensitive area is divided into left-front, left-rear, front, rear, right-front, and right-rear subarea according to the position of target vehicle's front bumper. The target vehicle n_1, left-front vehicle n_2, left-rear vehicle n_3, front vehicle n_4, rear vehicle n_5, right-front vehicle n_6, and right-rear vehicle n_7 were chosen as the game players to analyze and construct the driver's lane selection model. The next front or next rear vehicles would not be included in game player and they were only considered when calculating the driving payoffs.

2.1.2. Mathematical Description of Vehicle Group Situation. During driver's lane selection, generally, radical drivers focus on traffic efficiency, tend to change left, and mainly consider the average speed of vehicle; conservative drivers focus on comfort, tend to change right, and mainly consider the traffic flow density. So it is necessary for developing the mathematical description of vehicle group situation to take the two macroscopic parameters of density and speed into consideration. Meanwhile, in the microscopic aspect, the interactive influences with surrounding vehicles would be considered when the driver makes a decision on whether to change lane. The concept of "force" in the physics can be referenced and the description of interactive relationships among vehicles can be abstracted into interaction force. Thus, the vehicle group situation the target vehicle located in can be abstractly described through the set of lane force the target vehicle received in the vehicle group relationship. And if the vehicle in a subarea has a positive impact on the target vehicle to choose the lane in which the subarea is located, then the force is considered to be an attraction, and the converse effect is defined as repulsion. The effect size was used to describe the magnitude of force, and the effect size of the greatest repulsion was represented by -1, whereas the effect size of the greatest attraction was represented by 1 (Table 1).

The relationship among traffic flow parameters of traffic volume q, velocity v, and density ρ is $q = v \cdot \rho$. Combining the actual research, the traffic state of urban expressway is divided into three levels: smooth, slow, and congestion. And the corresponding classification of level of service is shown as Table 2.

The fuzzy logic method [17, 18] was used to grade the effect size, in which the vehicle type (Small, Middle, Large), relative distance (Danger, Near, Medium, Far), relative speed (Negative Large, Negative Small, Zero, Positive Small, Positive Large), traffic flow density of each lane (Small, Middle, Large), average speed of each lane (Large, Middle, Small), and the driver's propensity of target vehicle (Radical, Common, Conservative) were comprehensively considered. In the fuzzy logic method, the fuzzy variables, corresponding fuzzy sets, and the calculation of fuzzy subset boundary value can reference the literature [19]. Limited to the space, only one of the typical language fuzzy rules was listed as follows.

If the type of target vehicle is small, the type of left-rear vehicle is small, the driving propensity of target vehicle driver is radical, the relative distance between target vehicle and left-rear vehicle is far and their relative speed is positive large, the traffic flow density in left lane is small, and the average speed in left lane is large, then the effect size of left-rear vehicle acted on target vehicle is 1.

The force that left-rear vehicle acted on target vehicle can be obtained based on the above method. Similarly, the force that each vehicle in the interest-sensitive area acted on target vehicle and the force it received also can be obtained. The force set for target vehicle n_1 can be expressed as f_1^0 = [front vehicle force, rear vehicle force, left-front vehicle force, left-rear vehicle force, right-front vehicle force, right-rear vehicle force] $= [F_1^q, F_1^h, F_1^{zq}, F_1^{zh}, F_1^{yq}, F_1^{yh}]$. Similarly, the force set f_i^0 for other vehicles in interest-sensitive area also can be expressed in the same way.

Considering that the contribution rate of vehicles located in different areas was different, the contribution rate that

Table 1: Corresponding effect size of different forces.

Force	Strong repulsion	Middle repulsion	Weak repulsion	Zero	Weak attraction	Middle attraction	Strong attraction
Effect size	$[-1, -0.7)$	$[-0.7, -0.3)$	$[-0.3, 0)$	0	$(0, 0.3]$	$(0.3, 0.7]$	$(0.7, 1]$

Table 2: Classification of level of service for basic section on urban expressway.

Level of service	Density (pcu/km/lane)	Velocity (km/h)	Maximum service volume (pcu/h/lane)
Smooth	$\rho \leqslant 32$ (Small)	$v \geqslant 54.5$ (Large)	1600
Slow	$32 < \rho \leqslant 50$ (Middle)	$40 \leqslant v < 54.5$ (Middle)	2100
Congestion	$\rho > 50$ (Large)	$v < 40$ (Small)	2100

target vehicle driver perceived can be regarded as the weight of corresponding force to calculate the force that each lane acted on target vehicle through the way of weighted summation. The contribution rates of vehicle in each subarea that the driver with different propensity types perceived could be obtained through the analytic hierarchy process based on the questionnaire survey. The result was shown as Table 4. In Table 4, ε_i^k represented the contribution rate of vehicle which is located in subarea k for target vehicle n_i $(i = 1, 2, 3, 4, 5, 6, 7)$. For example, ε_i^{ql} represented the contribution rate of the vehicle in front area for the target vehicle n_i when it located in the left lane. Therefore, the force of the lane that target vehicle n_1 located was $F_1^{lol} = f_1^q \varepsilon_1^q + f_1^h \varepsilon_1^h$, the force of left lane was $F_1^{zl} = f_1^{zq} \varepsilon_1^{zq} + f_1^{zh} \varepsilon_1^{zh}$, the force of right lane was $F_1^{yl} = f_1^{yq} \varepsilon_1^{yq} + f_1^{yh} \varepsilon_1^{yh}$, and then the mathematical description of vehicle group situation that the target vehicle n_1 located was $F_1^0 = [F_1^{zl}, F_1^{lol}, F_1^{yl}]$. Similarly, the vehicle group situation that other vehicles located also can be calculated.

2.1.3. Division and Reduction of Vehicle Group Situation. When the target vehicle is located on the left lane or right lane, the forces from the adjacent lane and the separated lane can be merged to obtain the force of side lane. Therefore, the vehicle group situation would be $2^2 = 4$, $2^3 = 8$, and $2^2 = 4$ types, respectively, when target is vehicle located in left, middle, and right lane, and totally be 16 types. The attraction is represented by "+", and the repulsion is represented by "-". Thus the 16 types of vehicle group situation are shown as Figure 2.

2.1.4. Identification of Vehicle Group Situation. The types of vehicle group situation at different moment needs to be identified due to the time-varying characteristics. The bases for the identification of vehicle group situation are spatial-time location of vehicles in interest-sensitive area and the force that target vehicle received. The identification process of the vehicle group situation that target vehicle located in under three-lane condition is shown as Figure 3.

2.2. Driver's Lane Selection Model Based on Phase-Field Coupling and Multiperson Dynamic Game with Incomplete Information. Game theory can analyze the situation with interactive influences among multiple persons. First, driver's lane selection is a process of pursuing interests, such as driving safety, efficiency, and comfort. It is necessary for target vehicle driver to take the influences of lane selection strategies that its surrounding vehicle drivers have taken into consideration. And the multiperson dynamic game model needs to be constructed. Second, in the actual situation, there exist differences in the completeness degree of information and the usage of the interconnected devices the vehicle equipped. Therefore, the completeness degree of information that drivers could be obtained will be different. So it is necessary to study driver's lane selection behavior under the condition of incomplete information. Third, under normal circumstances, multiperson dynamic game is relatively complex and there not necessarily exists an equilibrium solution, often reaching the satisfaction state of each player. Thus, the satisfactory solution of the model can be solved. Fourth, driving preference, namely, the driving propensity, refers to the decision preference for lane selection under different vehicle group situations. Decision preference would be different for different types of driver and would change with the dynamic changes of vehicle group situation state. Therefore, in order to more accurately simulate the driving situations, the time-varying driving propensity should be considered during the model construction.

In conclusion, the model based on phase-field coupling and multiperson dynamic game with incomplete information can be constructed with the consideration of vehicle group situation and time-varying driving propensity to explore the driver's lane selection model adapted to the Internet of Things condition, especially the lane selection model for intelligent vehicles.

2.2.1. Game Analysis of Lane Selection Behavior. In Figure 1, vehicles n_1, n_4, and n_5 in middle lane had three behavior strategies, namely, *change left* (*CL*), *no changing* (*NC*), and *change right* (*CR*). Vehicles n_2 and n_3 in left lane had two behavior strategies, namely, *NC* and *CR*. And vehicles n_6 and n_7 in right lane had two behavior strategies, namely, *NC* and *CL*. In the actual game of lane selection, vehicles only equipped with partial interconnected devices, and only partial information about vehicle type, relative distance, relative speed, and the corresponding strategy space can be obtained. However, the information about the characteristics of other drivers and the corresponding driving payoffs cannot be obtained. Therefore, it is assumed in this paper that drivers

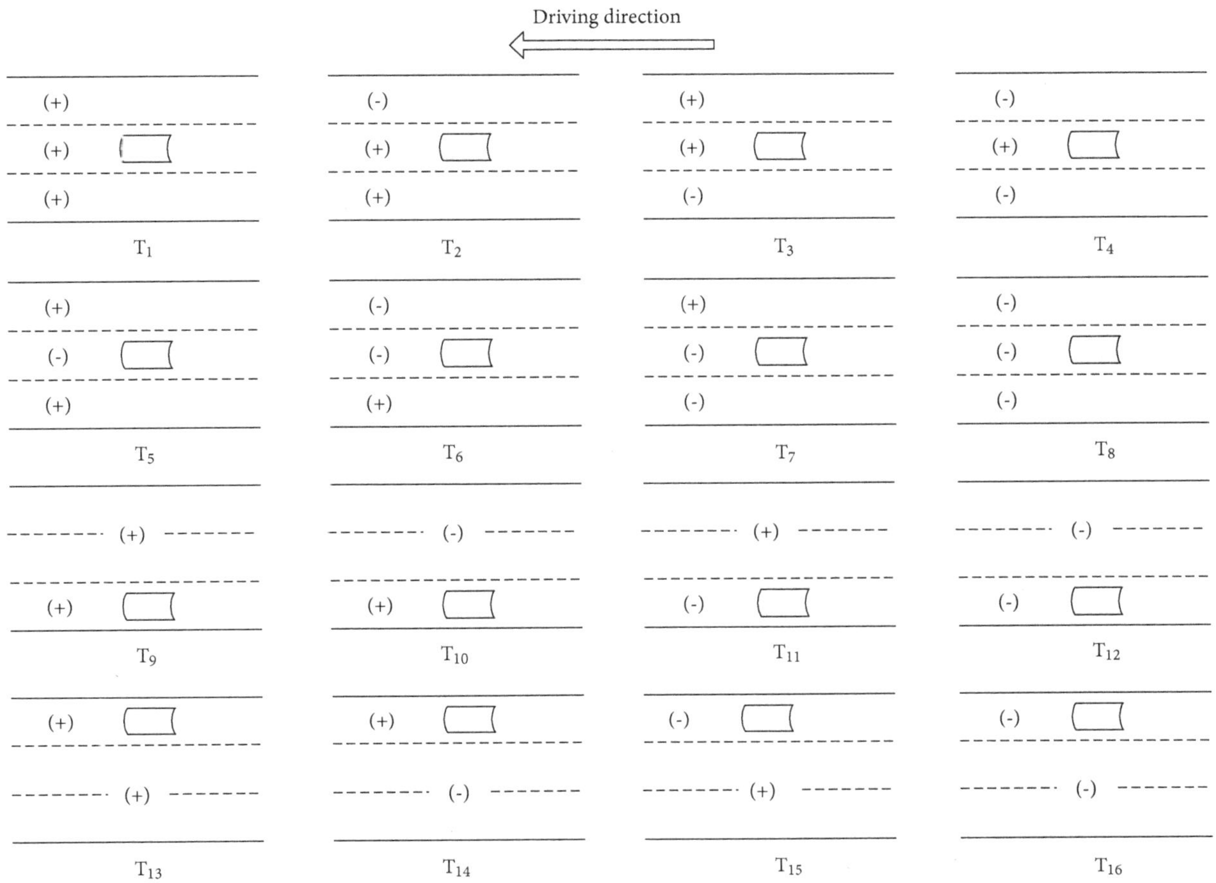

FIGURE 2: Reduction of vehicle group situation under three-lane condition.

involved in the lane selection game all get the incomplete information.

(1) Determination on the Order for Drivers to Take Action. In the process of dynamic game, the order for players to take action is successive, and the player who takes action later can obtain the information about the players who take action before him through observing their actions. And on this basis, the types of the players who take action before him are inferred and the prior probability distribution of their types is corrected, so as to choose his optimal action. Due to the fact that radical drivers are more likely to choose their own actions sooner than conservative drivers, and, compared to the vehicle at a rear position, the vehicle at a front position has the priority to take action. Thus, the order for drivers to take action is determined based on two indexes of the front or rear position in vehicle group relationship and radical degree of driver's propensity. Generally, drivers would have a higher priority to take action when their vehicles located at a front position and the driving propensity is more radical. The method in literatures [20, 21] was referenced to identify the driving propensity; the propensity type with the largest probability was chosen as the driver's propensity. The specific process of the determination on the order for drivers to take action was shown in Figure 4.

(2) Embedding of Hidden Markov Model. The driver's propensity would be considered in every round of the game, and it will change with the environment, which would affect driver's consciousness and manipulation behavior and further affect the vehicle group situation. Therefore, the Hidden Markov model was established against the double stochastic process of vehicle group situation and driving propensity to explore the transformation mechanism of driving propensity. The detailed process refers to literature [22].

Finally, the formula for the optimized transformation probability of vehicle group situation is as follows:

$$\hat{a}_{ij} = \frac{\sum_{t=1}^{L-1} P\left(x_t = T_i, x_{t+1} = T_j \mid u_1, \cdots, u_L\right)}{\sum_{t=1}^{L-1} P\left(x_t = T_i \mid u_1, \cdots, u_L\right)} = \frac{\sum_{t=1}^{L-1} \varepsilon_t(i, j)}{\sum_{t=1}^{L-1} \gamma_t(i)} = \frac{\sum_{t=1}^{L-1} a_{ij} b_j(u_{t+1})}{\sum_{t=1}^{L-1} \sum_{j=1}^{16} a_{ij} b_j(u_{t+1})} \quad (1)$$

The optimization algorithm for the probability distribution of driver's propensity is as follows:

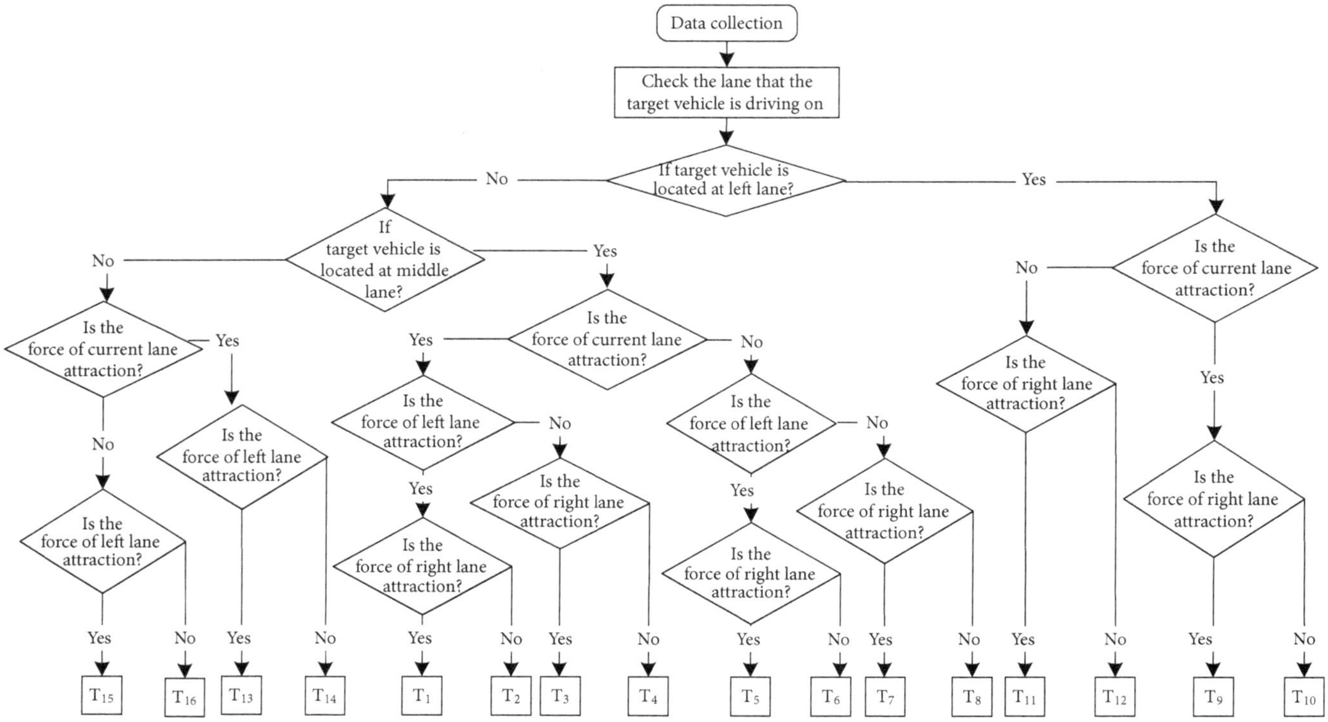

FIGURE 3: Identification flowchart of vehicle group situation under three-lane condition.

$$\hat{b}_j(u_k) = \frac{\sum_{t=1,\ y_t=u_k}^{L} P(x_t = T_j \mid u_1, \cdots, u_L)}{\sum_{t=1}^{L} P(x_t = T_j \mid u_1, \cdots, u_L)} = \frac{\sum_{t=1,\ y_t=u_k}^{L} \gamma_t(j)}{\sum_{t=1}^{L} \gamma_t(j)} = \frac{\sum_{t=1,\ y_t=u_k}^{L} \alpha_t(j)\beta_t(j)}{\sum_{t=1}^{L} \alpha_t(j)\beta_t(j)} \quad (2)$$

The optimization algorithm for the probability distribution of the initial vehicle group situation state π_i is as follows:

$$\hat{\pi}_1 = \gamma_1(i) = \frac{\alpha_1(i)\beta_1(i)}{\sum_{i=1}^{16} \alpha_1(i)\beta_1(i)} \quad (3)$$

(3) Calculation of Driving Payoff. Once the game strategies of players were selected, the situation of the game and the corresponding payoff function would be determined. The driving payoffs of driver's selection can be measured from the aspects of driving safety, efficiency, and comfort.

(a) Calculation of Payoff on Driving Safety. The payoff on driving safety (expressed by $u_{n_i}^1$) refers to the change of the integrated force (the weighted summing of the forces that vehicle in each subarea acted on the target vehicle) that the vehicle got before and after executing the selected operation. The target vehicle n_1 was chosen as an example, and the contribution rate (as shown in Table 3) that target vehicle perceived was donated as the weight of corresponding force that vehicles in each subarea acted on it. Before executing the selected operation, the integrated force that target vehicle received was $F_1 = f_1^q\varepsilon_1^q + f_1^h\varepsilon_1^h + f_1^{zq}\varepsilon_1^{zq} + f_1^{zh}\varepsilon_1^{zh} + f_1^{yq}\varepsilon_1^{yq} + f_1^{yh}\varepsilon_1^{yh}$, and, after executing the selected operation, the integrated force F_1' also could be obtained in the same way. Then, the payoff on driving safety that target vehicle n_1 obtained was $u_{n_1}^1 = F_1' - F_1$. Similarly, the integrated forces that vehicle in each subarea received before and after executing the selected operation, respectively, were F_i and F_i', and the corresponding payoff on driving safety was $u_{n_i}^1 = F_i' - F_i$.

(b) Calculation of Payoff on Driving Efficiency. Payoff on driving efficiency (expressed by $u_{n_i}^2$) can be represented in terms of the improvement on road traffic condition of the front area, and it can be measured through the difference between the integrated forces from the front area that vehicle received before and after executing the selected operation. The contribution rate (as shown in Table 3) that target vehicle perceived also was donated as the weight of corresponding force that vehicles in front area acted on it. Target vehicle n_1 was chosen as an example; the integrated forces that target vehicle received before and after all drivers executed the selected operation, respectively, were $F_f^1 = \sum \varepsilon_1^{qc} f_1^{qc}$ and $F_{f'}^1 = \sum \varepsilon_1^{qc'} f_1^{qc'}$. Then, the payoff on driving efficiency that target vehicle n_1 obtained was $u_{n_1}^2 = F_{f'}^1 - F_f^1$, where f_1^{qc}, ε_1^{qc}, $f_1^{qc'}$, and $\varepsilon_1^{qc'}$, respectively, represented the force and the contribution rate of the vehicles in front subareas before and after executing the selected operation. Similarly, the integrated forces that vehicle in each subarea received before and after executing the selected operation, respectively, were F_f^i

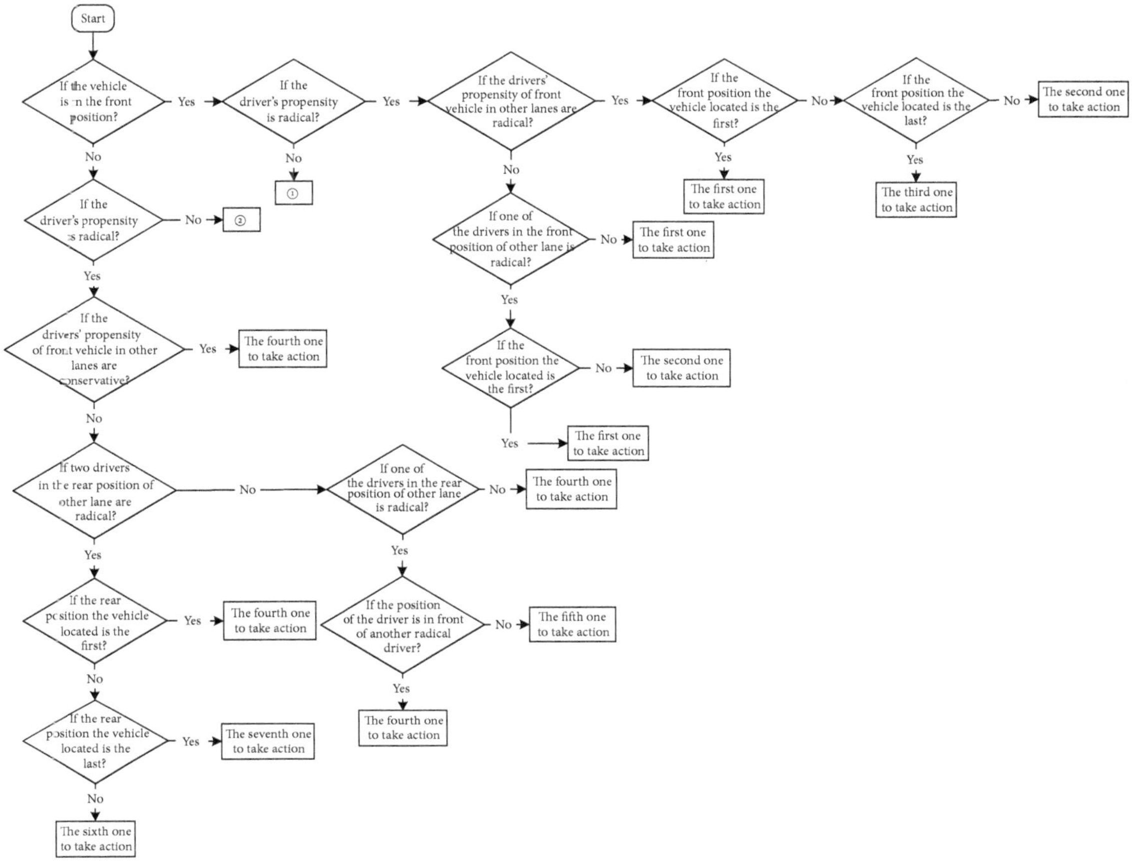

(a) Partial flow chart of the determination on the order for drivers to take action

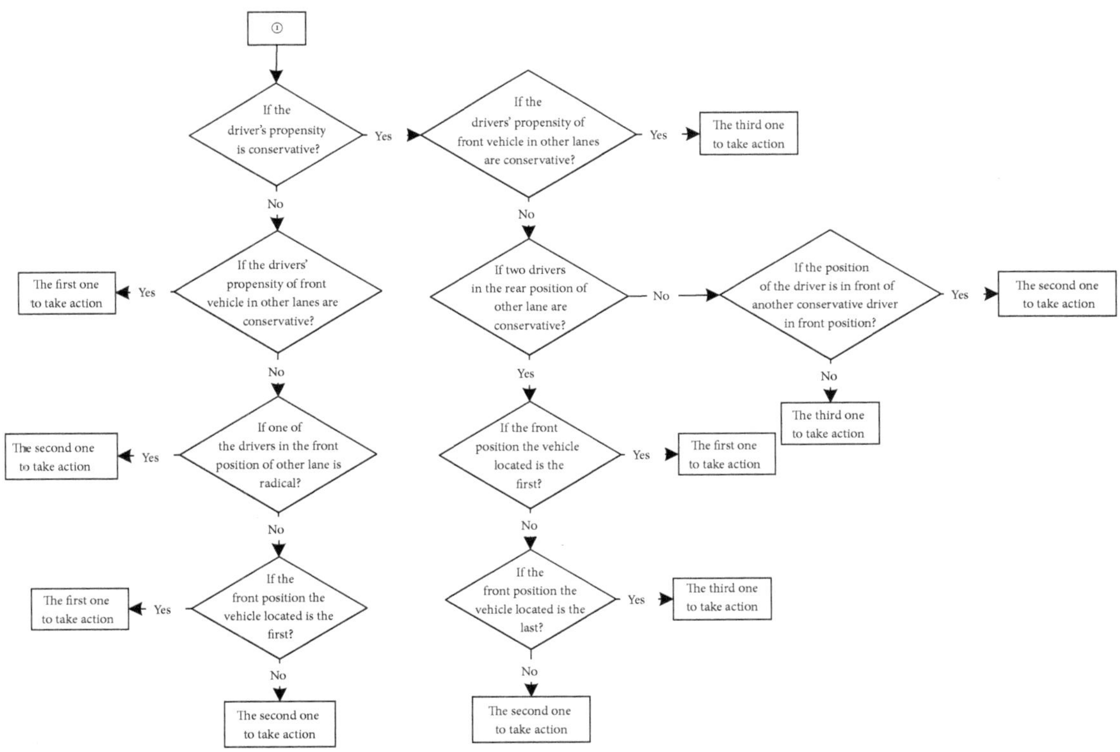

(b) Partial flow chart of the determination on the order for drivers to take action

FIGURE 4: Continued.

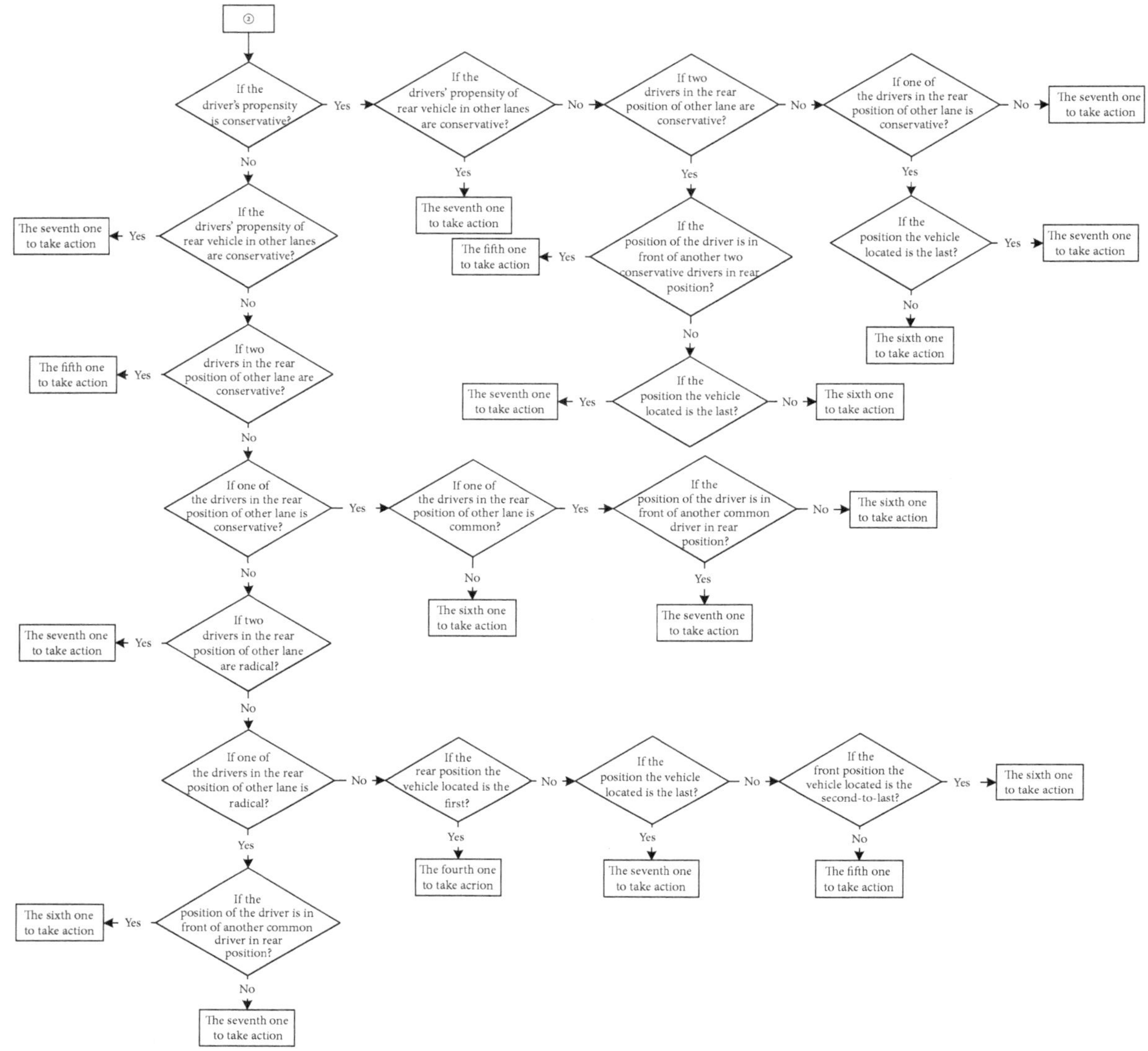

(c) Partial flow chart of the determination on the order for drivers to take action

FIGURE 4: Flow chart of the determination on the order for drivers to take action.

and $F^i_{f'}$, and the corresponding payoff on driving efficiency was $u^2_{n_i} = F^i_{f'} - F^i_f$.

(c) Calculation on Payoff of Driving Comfort. Payoff on driving comfort (expressed by $u^3_{n_i}$) can be represented in terms of the difference between the forces that the lane the vehicle located acted on it before and after executing the selected operation, where the lane force included the cumulative force of the two vehicles in front of the vehicle in game and the force of the rear vehicle behind it, and the cumulative force referred to the sum of the forces that the front vehicle acted on the vehicle in game and the next front vehicle acted on the front vehicle. The contribution rate (as shown in Table 3) that target vehicle perceived also was donated as the weight of corresponding force that vehicles in front and rear areas acted on it. Therefore, the target vehicle n_1 was taken as an example, the forces that the target vehicle received from the lane it located before and after all drivers executed the selected operation, respectively, were $F^1_o = \varepsilon^q_1(f^q_1 + f^q_4) + \varepsilon^h_1 f^h_1$ and $F^1_{o'} = \varepsilon^{q'}_1(f^{q'}_{1''} + f^{q'}_{4''}) + \varepsilon^{h'}_1 f^{h'}_1$. Then, the payoff on driving efficiency that target vehicle n_1 obtained after executing the selected operation was $u^3_{n_1} = F^1_{o'} - F^1_o$, where, ε^q_1, ε^h_1 and $\varepsilon^{q'}_1$, $\varepsilon^{h'}_1$ represented the contribution rate of the vehicles in front and rear subareas before and after executing the selected operation, respectively. f^h_1 and $f^{h'}_1$ represented the force that the vehicle in rear subarea acted on target vehicle n_1 before and after executing the selected operation, respectively. $f^{q'}_{1''}$ represented the force of the front vehicle $n_{1''}$ acted on the target vehicle n_1 after executing the selected operation and $f^{q'}_{4''}$ represented the force of the next front vehicle $n_{4''}$ on the front vehicle $n_{1''}$. Similarly, the forces

TABLE 3: Contribution rates from vehicles in each subarea, the different propensity types driver perceived.

Target vehicle located in left lane			Contribution rate			
Propensity type	ε_i^{ql}	ε_i^{hl}	ε_i^{lyq}	ε_i^{lyq}	ε_i^{gyq}	ε_i^{gyh}
Radical	0.246	0.153	0.186	0.228	0.088	0.099
Common	0.254	0.198	0.181	0.195	0.081	0.091
Conservative	0.265	0.226	0.175	0.171	0.078	0.085
Target vehicle located in middle lane			Contribution rate			
Propensity type	ε_i^{q}	ε_i^{q}	ε_i^{q}	ε_i^{zh}	ε_i^{yq}	ε_i^{yh}
Radical	0.217	0.217	0.217	0.184	0.125	0.153
Common	0.224	0.224	0.224	0.164	0.136	0.134
Conservative	0.239	0.193	0.112	0.144	0.138	0.174
Target vehicle located in right lane			Contribution rate			
Propensity type	ε_i^{qr}	ε_i^{hr}	ε_i^{lzq}	ε_i^{lzh}	ε_i^{gzq}	ε_i^{gzh}
Radical	0.241	0.157	0.192	0.235	0.082	0.093
Common	0.245	0.189	0.189	0.218	0.075	0.084
Conservative	0.254	0.215	0.183	0.199	0.071	0.078

TABLE 4: Fuzzy weight of different payoffs that drivers with different types of propensity perceived.

Type of driver		Fuzzy weight	
	Payoff on driving safety ω_i^1	Payoff on driving efficiency ω_i^2	Payoff on driving comfort ω_i^3
Radical	0.35	0.5	0.15
Common	0.4	0.35	0.25
Conservative	0.5	0.2	0.3

that the lane the vehicle located acted on it before and after executing the selected operation, respectively, were F_o^i and $F_{o'}^i$, and the corresponding payoff on driving comfort was $u_{n_i}^3 = F_{o'}^i - F_o^i$.

Due to the differences of drivers with different types of propensity in the importance of cognition in terms of driving payoffs, for example, radical drivers pay more attention to the payoff on driving efficiency, and conservative drivers pay more attention to the payoff on driving safety. Therefore, the importance of cognition in terms of payoffs on driving safety, efficiency, and comfort can be donated as the fuzzy weight to give the corresponding payoffs, so as to measure the driving payoffs uniformly. The fuzzy weight of different payoffs that drivers with different types of propensity perceived was shown as Table 4. Therefore, the driving payoff for each driver in the game was $u_{n_i} = \omega_i^1 u_{n_i}^1 + \omega_i^2 u_{n_i}^2 + \omega_i^3 u_{n_i}^3$ $(i = 1, 2, \cdots, 7)$.

2.2.2. Model Construction. According to the game analysis of driver's lane selection, the lane selection model of the grouping vehicles based on multiperson dynamic game with incomplete information was constructed under three-lane condition. n_1, n_2, n_3, n_4, n_5, n_6, and n_7 were chosen as the game players, and the dynamic game process of the grouping vehicles was as follows.

At time t, the vehicle group situation was identified according to the method in part 1.1.4, and driver's propensity was identified according to the method in [19, 22]. In the first round, the order for drivers in the vehicle group situation to take action was determined according to the identified front or rear position of vehicle and the radicalness of the driver's propensity. Then, in each secondary round, the driver who was to take action would choose his own lane selection strategy according to the current vehicle group situation, the strategies of drivers who had taken action, and the prediction of the strategies for drivers who would take action after him. After driver's lane selection in every secondary round, the vehicle group situation needed to be reidentified according to the method in part 1.1.4. And the transformation of driver's propensity needed to be studied based on the embedded Hidden Markov model due to the interaction relationship between vehicle group situation and driving propensity. Meanwhile, according to the order, the driver who was to take action would choose his own lane selection strategy according to the current vehicle group situation, the strategies of drivers who had taken action, and the prediction of the strategies for drivers who would take action after him. Repeat the process of driver's lane selection, identification of vehicle group situation, and transformation of driving propensity until all drivers in vehicle group situation have selected the lane to drive on. After all drivers in the vehicle group situation have chosen their lane selection strategies, it was judged according to the requirements of game stopping time whether the present game met the requirements, where the requirements of game stopping time include three aspects: safety requirement, namely, judging whether the current integrated force the vehicle received from the vehicles in each subarea is greater than or equal to zero; efficiency requirement, namely, judging whether the difference of the integrated force from the front area at current and initial moment is greater than or

TABLE 5: Main microscopic parameters obtained from data preprocessing.

	Relative distance between target vehicle and surrounding vehicle (m)					
Data	Left-front vehicle	Left-rear vehicle	Front vehicle	Rear vehicle	Right-front vehicle	Right-rear vehicle
Parameter	Δd_2	Δd_3	Δd_4	Δd_5	Δd_6	Δd_7
	Relative speed between target vehicle and surrounding vehicle (m/s)					
Data	Left-front vehicle	Left-rear vehicle	Front vehicle	Rear vehicle	Right-front vehicle	Right-rear vehicle
Parameter	Δv_2	Δv_3	Δv_4	Δv_5	Δv_6	Δv_7

TABLE 6: Parameter calibration of lane selection model.

Parameter	V_1	V_2	V_3	V_4	D_1	D_2	D_3	D_4
Radical	-3.3	-1.6	1.6	3.3	8.2	22.5	36.7	51.3
Common	-4.8	-2.5	2.5	4.8	12.6	25.8	40.1	60.5
Conservative	-8.4	-4.5	4.5	8.4	22.3	38.3	54.4	70.4

equal to zero; comfort requirement, namely, judging whether the force the vehicle received from the lane it located is greater than zero. When it cannot meet the requirements of game stopping time, the vehicle group situation at current moment would be reidentified and the transformation of driver's propensity would be studied based on the embedded Hidden Markov model, so as to redetermine the order for drivers to take action. And then, a new round lane selection game would be conducted according to the game rules of the previous round until the requirements of stopping time can be satisfied after the end of the game at round N. Then, the lane selection strategy and the corresponding driving payoff of drivers at time t would be obtained, namely, the lane selection strategy at each round of game. After that, the simulation clock would be promoted to the time $t + 1$, and a new round lane selection game would be conducted. The process of driver's lane selection game was shown in Figure 5.

3. Results

The model parameters were calibrated after processing the experiment data. And on this basis, the model solution was obtained and so the driver's lane selection strategy.

3.1. Data Processing and Model Calibration. The vehicle trajectory sets of the section I-80 at the duration of 4:00-4:15 pm and the section US-101 at the duration of 7:50-8:05 am in the Next Generation Simulation (NGSIM) data set were selected to calibrate and validate the model constructed in this paper, where, the vehicles running on the lanes with the lane IDs 2, 3, and 4 were chosen as study subjects when entering into the date collection area (the most left lane ID was 1 on the driving direction and the number would increase successively to the right). The vehicles in HOV lane and other auxiliary lanes (the driving behavior is different from that in other lanes) and the continuous lane changing (the continuous lane changing is more likely the mandatory lane changing) would not be taken into consideration. Lane distribution diagrams of the studied sections I-80 and US-101 were shown in Figure 6.

3.1.1. Data Processing. Each record of NGSIM trajectory data contains not only the status information of a single vehicle at a moment including the lane driving on, vehicle location, instantaneous speed, instantaneous acceleration, and so on but also the ID of preceding and following vehicle at the current moment. Therefore, the status information of the surrounding vehicle at the current moment and a certain interval moment can be queried according to the combination of video, vehicle ID, and time information. The vehicle type, relative distance, and relative speed in the vehicle group situation can be got through the combination of the obtained information, and the microscopic parameters which meet the calculation need of vehicle group situation can be obtained finally. The main microscopic parameters obtained from the data preprocessing of NGSIM are shown in Table 5 [23].

3.1.2. Calibration of Lane Selection Model. Based on the empirical value and the scientific analysis and process of the vehicle trajectory data set for I-80 section, the model parameters were calibrated through the repeated training and expert opinion. The calibration of the partial model parameters was shown in Table 6.

3.2. Solution of Lane Selection Model. According to the game analysis and the dynamic game process of driver's lane selection, at time, when the game reached the end of the round N, if all drivers in vehicle group situation got their own relatively satisfied driving states and the requirements of game stopping time can be reached from the aspects of safety, efficiency, and comfort. Then, vehicles in the grouping situation would keep a relatively stable driving state at this time, and driver's lane selection game would end at this time. The game analysis of driver's lane selection was referenced and the calculation about the requirements of game stopping time was as follows.

Safety requirement is to judge whether the integrated force $F_i^n = \sum f_i^{k^n} \varepsilon_i^{k^n}$ that vehicle n_i received from the vehicles in each subarea at the end of round n ($n = 1, 2, \cdots, N$) is greater than or equal to zero. Efficiency requirement is to judge whether the difference $F_{f^n}^i - F_f^i$ of the integrated force vehicle n_i received from the front area at the end of round n ($n = 1, 2, \cdots, N$) and the initial moment (namely,

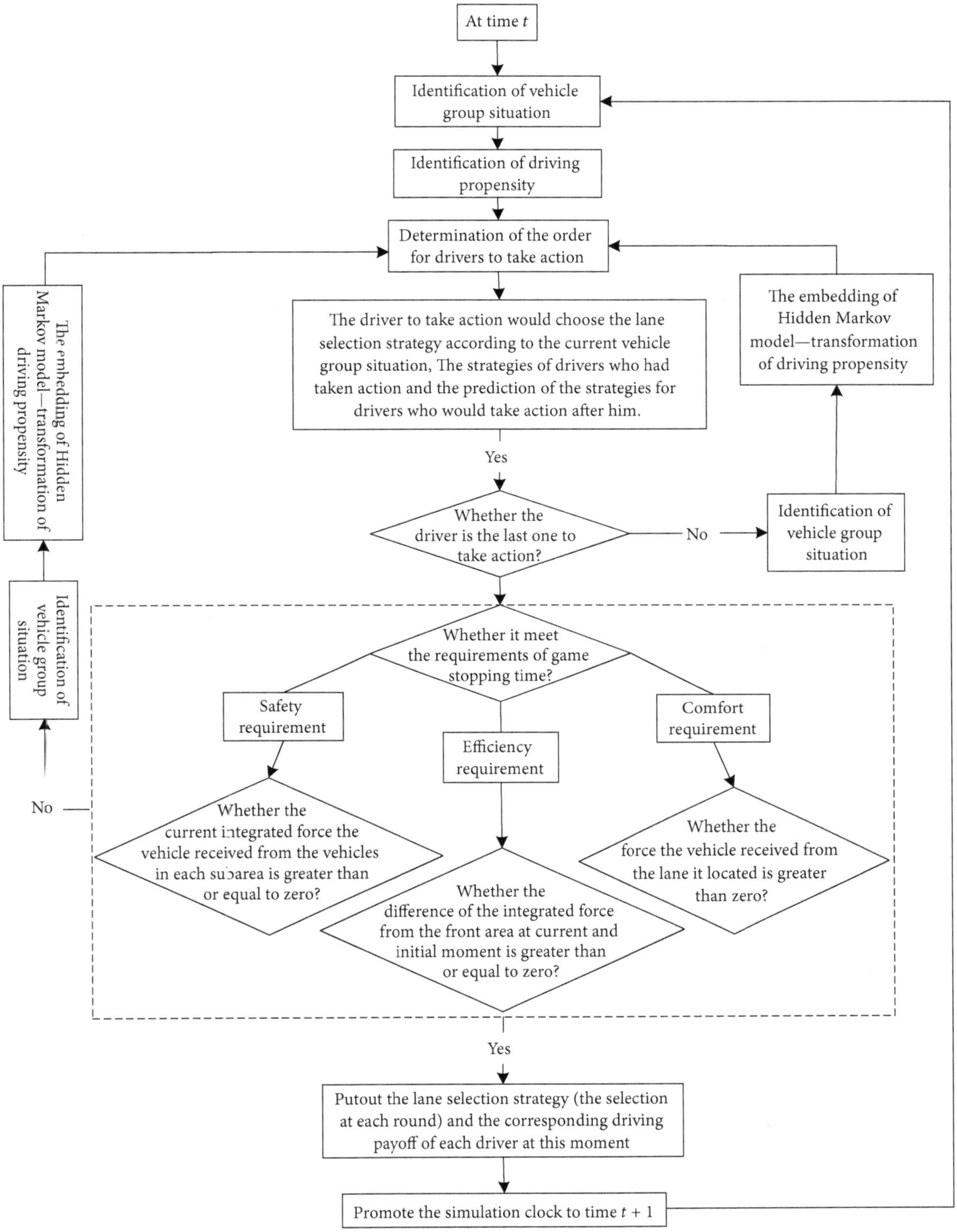

FIGURE 5: Process of driver's lane selection game.

$F^i_{f^n} = \sum \varepsilon_i^{qc^n} f_i^{qc^n}$ and $F^i_f = \sum \varepsilon_i^{qc} f_i^{qc}$) is greater than or equal to zero. Comfort requirement is to judge whether the force $F^i_{o^n} = \varepsilon_i^{q^n}(f_{i'}^{q^n} + f_{i''}^{q^n}) + \varepsilon_i^{h^n} f_i^{h^n}$ that vehicle n_i received from the lane it located at the end of round n ($n = 1, 2, \cdots, N$) is greater than zero.

When the game was over, the lane selection strategy and the corresponding payoff of each driver in vehicle group situation can be obtained according the process of game, and the lane selection strategy was the choice in each round of game. According to the game analysis, payoff of driving safety

TABLE 7: Model verification results.

Lane ID	Prediction times	Comparison between prediction and actual results		Accuracy
		Agree times	Disagree times	
2	50	41	9	82.00%
3	150	127	23	84.67%
4	150	131	19	87.33%
Total	350	299	51	85.43%

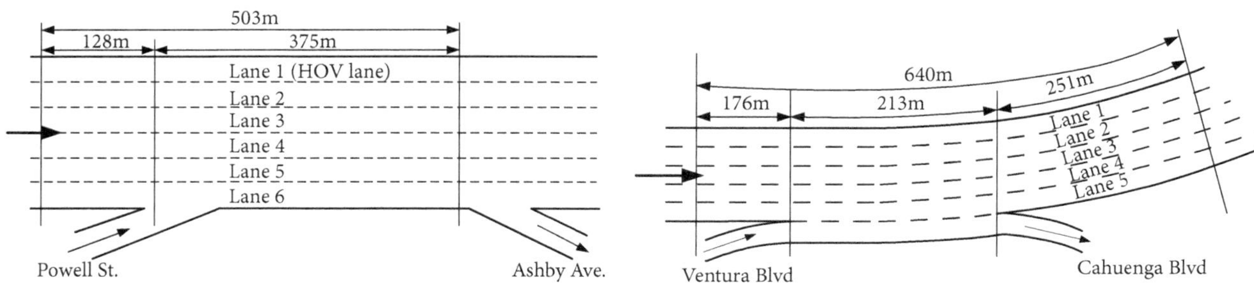

(a) Northward section of I-80 (California-Emeryville)
(b) Southward section of US-101 (California-Los Angeles)

FIGURE 6: Lane distribution diagrams of the studied sections.

was $u^1_{n_i} = F_i^N - F_i^0$, payoff of driving efficiency was $u^2_{n_i} = F^i_{f^N} - F^i_f$, and payoff of driving comfort was $u^3_{n_i} = F^i_{o^N} - F^i_o$. The driving payoff of each driver at the end of the game after round N was $u_{n_i} = \omega_i^1 u^1_{n_i} + \omega_i^2 u^2_{n_i} + \omega_i^3 u^3_{n_i}$ based on the importance of cognition in terms of driving payoffs.

4. Discussions

The efficiency of the model was verified through experiment and simulation. And based on the verification results, the strength and weakness of the constructed model over existing methods were compared.

4.1. Verification of Lane Selection Model. The means of actual driving experiment, virtual driving experiment, and microscopic simulation of traffic flow were used to verify the constructed model.

4.1.1. Verification of Lane Selection Model Based on NGSIM Trajectory Data. The vehicle trajectory data of US-101 collected through NGSIM program was used to verify the model. Driver's lane selection model constructed in this paper was applied to analyze and process the video information and vehicle trajectory data collected from the studied road section and further to predict the driver's lane selection on the target lanes. The lane changing of vehicle was selected as the index to examine the prediction results of the driver's lane selection because that it is obvious and easy to observe for vehicle's status when driver selected the strategy of "*CL*" or "*CR*". The prediction results were compared with the actual lane changing situations to verify the validity of the model and the comparison results were shown in Table 7.

4.1.2. Verification of Lane Selection Model Based on Actual Driving Experiment. The road section of Yuanshan Road between Xincun West Road intersection and Renmin West Road of Zibo city, Shandong province, was selected as the experimental route. The experiment was conducted between 7:30 am and 9:30 am of normal working day with the good condition of weather and road and the traffic flow was nonfree flow. The overall length of the studied road section was 1.2km, as shown in Figure 7. Based on the video monitoring, 30 drivers with different driving propensities were chosen to conduct the experiment, and the experimental equipment was shown in Figure 8. During the experiment, the information of road, traffic, and environment was collected and the experimental data was stored. After the experiment, the lane selection model based on multiplayer game built in this paper was used to analyze and process the collected data. The effectiveness and validity of the model could be judged through comparing the prediction results with the actual lane selections recorded in the video (Figure 9). The results of actual driving experiment were shown in Table 9.

4.1.3. Verification of Lane Selection Model Based on Interactive Parallel Driving Simulated Experiment. In order to verify the model, the multichannel interactive parallel driving simulation system was used to conduct the virtual driving experiment, as shown in Figure 10. In this confirmatory experiment, the same traffic scene of virtual reality was constructed according to the road environment of actual driving experiment. Before the experiment, 20 drivers who participated in the experiment were trained in the driving simulator. During the experiment, the experimental data was stored and the whole video was recorded with the interference

TABLE 8: Comparison between prediction and actual lane selection result for driving simulation experiment.

Driver's number	Prediction times (time)	Comparison between prediction and actual results		Accuracy
		Agree times (time)	Disagree times (time)	
1	90	75	15	83.33%
2	90	79	11	87.78%
...	...	...	...	...
9	90	76	14	84.44%
10	90	78	12	86.67%
...	...	...	...	...
19	90	75	15	83.33%
20	90	77	13	85.56%
Average	90	77.25	12.75	85.83%

TABLE 9: Comparison of microscopic simulation result and measured data.

Evaluation index	Measured value	Simulation value	error (%)
Average speed (m/s)	12.36	11.78	4.69
Average density (veh/km)	35.41	36.93	4.29

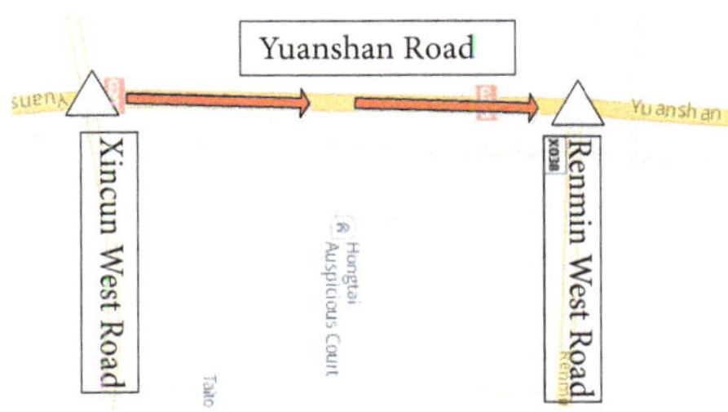

FIGURE 7: Experimental route.

for driver was avoided. The lane selection model based on dynamic game with incomplete information built in this paper was used to analyze and process the collected data, and the optimal strategy result for the driver was calculated and output. And then the calculation result was compared with the actual lane selection result of the driver in experiment, and the result of comparative analysis was shown in Table 8.

It can be seen from Table 8 that the integrating degree of prediction and actual lane selection result was high with the built lane selection model in driving simulation experiment, and the average accuracy of the model prediction could reach 85.83%.

4.1.4. Verification of Lane Selection Model Based on Microscopic Simulation of Traffic Flow. According to actual driving experiment, car-following model and lane changing decision-making model were built, respectively, based on the optimal control theory and fuzzy multiobjective decision theory. The experimental data of drivers with different driving propensity was input to the microscopic simulation model of lane selection that did consider (simulation 1) or did not consider (simulation 2) the dynamic game among the drivers. To verify the effectiveness and validity of the model built in this paper, the simulation results in microscopic regularity (such as lane changing frequency and lane changing times) and macroscopic regularity (such as speed, density, traffic volume, and lane occupancy) of traffic flow were compared with the actual situations of actual driving experiment.

The verification results of traffic flow in microscopic aspect were shown in Figure 11, and it described the distribution of lane changing times under different traffic volume in three-lane traffic scene. As mentioned, simulation 1 was the case that considered the driver's lane selection game for real-time simulation, while simulation 2 did not.

The verification of traffic flow in macroscopic aspect mainly included the average speed, average density and lane utilization ratio, and the verification results were shown in Table 9 and Figure 12.

In Table 9, the average speed and average density were selected as the evaluation index to compare the simulation and measured values based on the data of actual driving experiment, and the model error was in the acceptable range.

Figure 12 is the simulation result of the relationship between traffic volume and lane utilization ratio under three-lane condition, where the full line represented the actual result of the relationship between traffic volume and lane utilization ratio and the dot represented the simulation result obtained from the microscopic simulation model considering the dynamic game of driver's lane selection. It can be seen from Figure 12 that the microscopic simulation model which considered the dynamic game of driver's lane selection could simulate the relationship between traffic volume and lane utilization ratio much better, which indicated that the model built in this paper had a higher accuracy and feasibility.

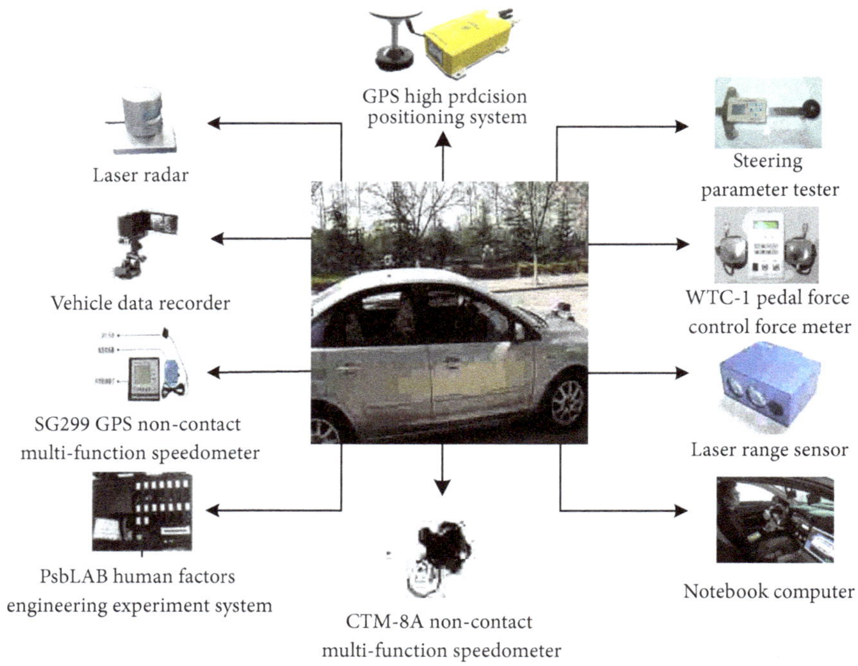

FIGURE 8: Information collection system of actual driving experiment.

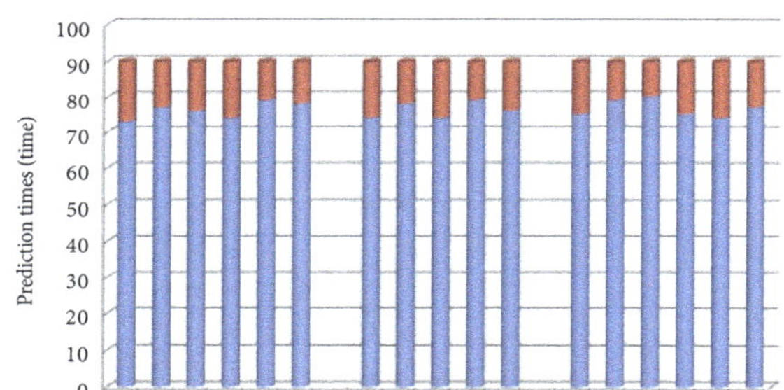

	1	2	3	4	5	6	...	13	14	15	16	17	...	25	26	27	28	29	30
Disagree times	17	13	14	16	11	12		16	12	16	11	14		15	11	10	15	16	13
Agree times	73	77	76	74	79	78		74	78	74	79	76		75	79	80	75	74	77

FIGURE 9: Comparison between prediction and actual results of actual driving experiment.

4.2. Discussion. The dynamic game model with incomplete information established in this paper comprehensively considered the factors which influence driver's decision-making behavior, such as vehicle group situation, driving propensity, and the changing of preference. And it mainly analyzed driver's lane selection behavior under the current condition of Internet of Things, especially the Internet of Vehicles.

Compared with the existing research on driver's lane selection model, the proposed method is superior in the following aspects. First of all, the driving propensity was considered in the proposed method. And furthermore, the Hidden Markov model was embedded to study the transformation of driving propensity in order to take the influence of time-varying driving propensity into consideration. Second, the influence of the lane selection strategies for surrounding vehicles on target vehicle was considered rather than only the influence of following vehicle on target lane. Third, the dynamic game model was constructed rather than static game model. Fourth, driving payoff was measured considering three aspects of driving safety, driving efficiency, and driving comfort, rather than only one aspect of driving safety.

However, the proposed method still has the following shortcomings. First, only one coupling relationship between vehicle group situation and driving propensity was considered when studying the transformation of driving propensity. However, different coupling relationship has different influence on the transformation of driving propensity. And the state of driving propensity (vehicle group situation) at time $t + 1$ would be influenced by the state of vehicle group situation (driving propensity) at time t. Second, only the situation that vehicles equipped with partial interconnected devices and all drivers obtained the incomplete information was studied. And the problems of driver's lane selection should be studied under the different penetration rate for vehicles equipped (or not equipped) with interconnected devices. Third, because of the differences of China and the United States in road condition and driving environment, there will be some error in model itself when applying the natural driving data (NGSIM data) for model calibration under the road condition of China; thus the model accuracy will be influenced. Fourth, to adapt to more complex traffic environments, the model needs to be extended to four or more lane scenarios and the game behavior of drivers in different road segments, such as intersections or ramps, needs to be comprehensively considered. In addition, driver's moods and the infectivity of driving behaviors should be taken into consideration when constructing the model.

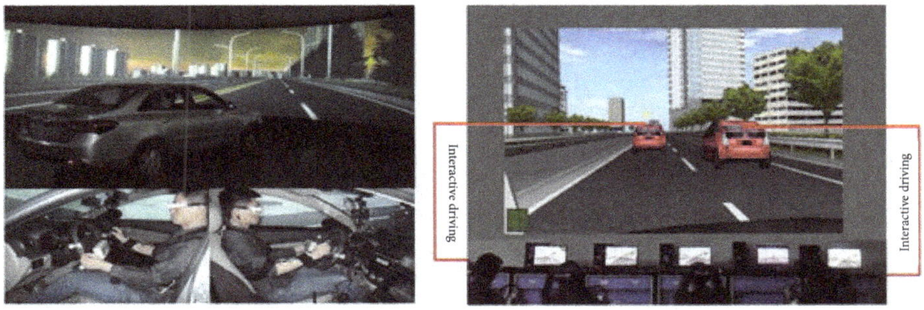

FIGURE 10: Driving simulation experiment.

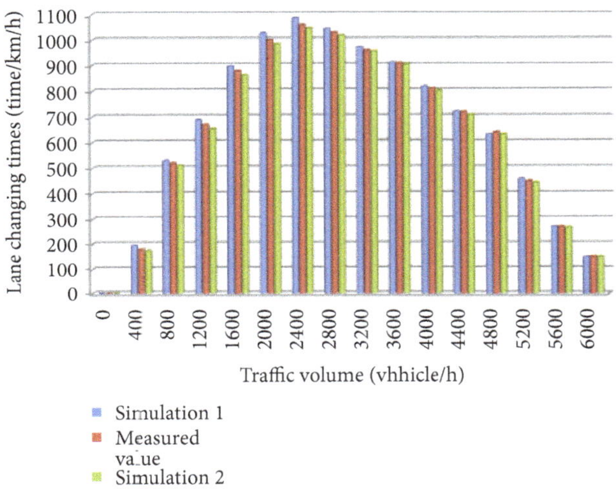

FIGURE 11: Comparison of lane changing times in simulated value and measured value under three-lane condition.

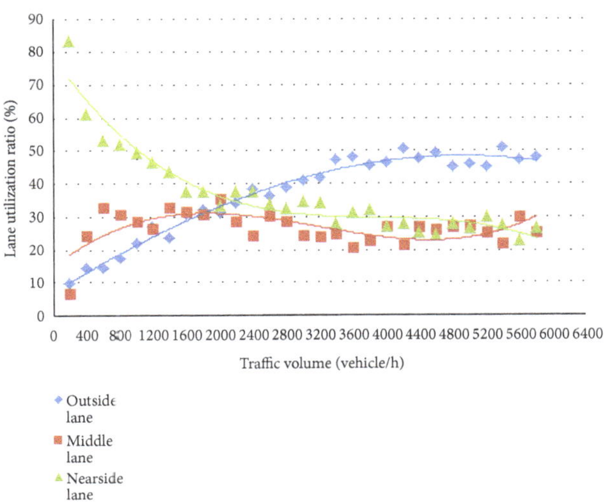

FIGURE 12: Simulation result of relationship between traffic volume and lane utilization ratio under three-lane condition.

5. Conclusion

In this paper, the game behavior of the grouping vehicles running on the urban expressway road section was analyzed, and the factors which influence driver's decision-making behavior such as vehicle group situation and driving propensity were comprehensively considered. The driver's lane selection behavior was analyzed under the condition of incomplete information, and lane selection model based on phase-field coupling and multiplayer dynamic game with incomplete information was constructed considering the time-varying character of driving propensity. The driver's lane selection strategy and the corresponding payoff were obtained through judging the stopping time and analyzing the process of the game. The means of actual driving experiment, virtual driving experiment, and microscopic simulation of traffic flow were used to verify the model. The verification results showed that the model built in this paper can objectively reflect the actual operation characteristic of traffic flow on road section and the process of lane selection. The theoretical basis of the research on lane selection can be provided for intelligent driving especially anthropomorphic driving under the condition of Internet of Things.

Conflicts of Interest

The authors declare that they have no conflicts of interest.

Acknowledgments

This study was supported by the State Key Laboratory of Automotive Safety and Energy under Project no. KF16232, Natural Science Foundation of Shandong Province (Grants nos. ZR2014FM027 and ZR2016EL19), Social Science Planning Project of Shandong Province (Grant no. 14CGLJ27), and the National Natural Science Foundation of China (Grants nos. 61074140, 61573009, and 1508315).

References

[1] P. G. Gipps, "A model for the structure of lane-changing decisions," *Transportation Research Part B: Methodological*, vol. 20, no. 5, pp. 403–414, 1986.

[2] Q. Yang and H. N. Koutsopoulos, "A microscopic traffic simulator for evaluation of dynamic traffic management systems," *Transportation Research Part C: Emerging Technologies*, vol. 4, no. 3, pp. 113–129, 1996.

[3] K. I. Ahmed, *Modeling Drivers' Acceleration and Lane Changing Behaviors*, Department of Civil and Environmental Engineering, MIT, 1999.

[4] M. Rickert, K. Nagel, M. Schreckenberg, and A. Latour, "Two lane traffic simulations using cellular automata," *Physica A: Statistical Mechanics and its Applications*, vol. 231, no. 4, pp. 534–550, 1996.

[5] P. Hidas, "Modelling vehicle interactions in microscopic simulation of merging and weaving," *Transportation Research Part C: Emerging Technologies*, vol. 13, no. 1, pp. 37–62, 2005.

[6] M. Keyvan-Ekbatani, V. L. Knoop, and W. Daamen, "Categorization of the lane change decision process on freeways," *Transportation Research Part C: Emerging Technologies*, vol. 69, pp. 515–526, 2016.

[7] E. Balal, R. L. Cheu, and T. Sarkodie-Gyan, "A binary decision model for discretionary lane changing move based on fuzzy inference system," *Transportation Research Part C: Emerging Technologies*, vol. 67, pp. 47–61, 2016.

[8] Y. Luo, Y. Xiang, K. Cao, and K. Li, "A dynamic automated lane change maneuver based on vehicle-to-vehicle communication," *Transportation Research Part C: Emerging Technologies*, vol. 62, pp. 87–102, 2016.

[9] S. E. Li, Y. Zheng, K. Li et al., "Dynamical modeling and distributed control of connected and automated vehicles: challenges and opportunities," *IEEE Intelligent Transportation Systems Magazine*, vol. 9, no. 3, pp. 46–58, 2017.

[10] H. Kita, "A merging-giveway interaction model of cars in a merging section: a game theoretic analysis," *Transportation Research Part A: Policy and Practice*, vol. 33, no. 3-4, pp. 305–312, 1999.

[11] P. Jin-shuan, F. Rui, S. Jing-jing, and Z. Qiong, "Reaearch of drivers lane change decision-making mechanism," *Journal of Wuhan University of Technology*, vol. 33, no. 12, pp. 46–50, 2011.

[12] A. Talebpour, H. S. Mahmassani, and S. H. Hamdar, "Modeling lane-changing behavior in a connected environment: a game theory approach," *Transportation Research Part C: Emerging Technologies*, vol. 59, pp. 216–232, 2015.

[13] Y.-L. Zhang, "Discretionary lane changing modeling based on stackelberg game theory," *Jiaotong Yunshu Xitong Gongcheng Yu Xinxi/Journal of Transportation Systems Engineering and Information Technology*, vol. 14, no. 5, pp. 67–73, 2014.

[14] M. Wang, S. P. Hoogendoorn, W. Daamen, B. van Arem, and R. Happee, "Game theoretic approach for predictive lane-changing and car-following control," *Transportation Research Part C: Emerging Technologies*, vol. 58, pp. 73–92, 2015.

[15] W. Lei, W. Xiaohui, Y. Xinyue, and W. Xiaoyuan, "Study on the recognition of traffic situation and its state transition mechanism," *Communications Standardization*, vol. 2, no. 3, pp. 61–65, 2007.

[16] J. Zhang, X. Wang, J. Wang, and X. G. Ban, "Vehicle group relationship transformation mechanism under dynamic and complex three-lane conditions," *Advances in Mechanical Engineering*, vol. 7, no. 2, Article ID 687517, 2014.

[17] B. Karatop, C. Kubat, and Ö. Uygun, "Talent management in manufacturing system using fuzzy logic approach," *Computers & Industrial Engineering*, vol. 86, pp. 127–136, 2015.

[18] D. V. Petrović, M. Tanasijević, V. Milić, N. Lilić, S. Stojadinović, and I. Svrkota, "Risk assessment model of mining equipment failure based on fuzzy logic," *Expert Systems with Applications*, vol. 41, no. 18, pp. 8157–8164, 2014.

[19] W. Kegang, *Study on the Coupling Mechanism of Vehicle Cluster Situation and Drivers Propensity*, Shandong: Shandong University of Technology, Shandong, China, 2015.

[20] X. Wang, J. Zhang, X. Ban, and D. Tan, "Dynamic feature extraction method of driver's propensity under complicated vehicle group," *Advances in Mechanical Engineering*, vol. 2013, Article ID 287653, 10 pages, 2013.

[21] X. Wang, J. Liu, and J. Zhang, "Dynamic recognition model of driver's propensity under multilane traffic environments," *Discrete Dynamics in Nature and Society*, vol. 2012, Article ID 309415, 15 pages, 2012.

[22] X. Y. Wang, J. L. Zhang, X. G. Ban, and S. Gao, "Study on transformation mechanism of driver's propensity under two-lane conditions," *Applied Mechanics and Materials*, vol. 411–414, pp. 1911–1918, 2013.

[23] J. Zhang, X. Wang, J. Wang, and J. Wang, "Decision-making model of lane-change behavior based on integrated cognitive vehicle cluster situations," in *Green Intelligent Transportation Systems*, vol. 419 of *Lecture Notes in Electrical Engineering*, pp. 77–94, Springer Singapore, Singapore, 2018.

17

Airworthiness Support Measures Analogy to the Prospective Roundabouts Alternatives: Theoretical Aspects

Andriy Viktorovich Goncharenko

Aircraft Airworthiness Retaining Department, Academic and Research Aerospace Institute, National Aviation University, 1, Kosmonavta Komarova Avenue, Kyiv 03058, Ukraine

Correspondence should be addressed to Andriy Viktorovich Goncharenko; andygoncharenco@yahoo.com

Guest Editor: Raffaele Mauro

The goal of this paper is to investigate theoretically the possible directions of some specified methods for the alternative roundabouts effectiveness modeling and optimization. The out-coming criteria have an economical interpretation. Those are the objective functionals of the alternative roundabouts effectiveness as the profit gained in the course of the traffic flow changes in the view of the integral form. This is modeled taking into consideration the transport infrastructure functioning elements such as the traffic flow of a capacity model. It takes into account two major components of the transportation services which are the alternative roundabouts business' incomes and expenses relating to the roundabouts transportation worthiness support. The prototypic approach is that one from the aircraft airworthiness support measures models. Corresponding managerial influences with respect to environmental, safety, utility, and other issues, as well as probable impacts, are modeled with the construction of the relevant under-integral expressions, equations, and appropriate coefficients and parameters of the mathematical models. The achieved theoretical results, on the basis of the Euler-Lagrange equation and accepted assumptions, have been checked for the sufficiency of the objective functional maximum presence at the "point" with the use of the conducted computer simulation. The necessary diagrams are plotted in order to illustrate the theoretical contemplations and speculations, as well as to check the correctness of the applied mathematical derivations and visualize the models' preciseness and abilities. The theoretically constructed mathematical models have a significance of the prognostic values applicability required at the alternative roundabouts effectiveness modeling and optimization ensuring their design progress and evolutions.

1. Introduction

The complexity of the modern roundabouts design, their diversities in designations, and functional, economic, and environmental criteria structures elaboration [1] are predetermining the multialternativeness of the new types of the roundabouts evaluation.

Rational economic activity is the foundation providing the development in all areas and spheres of life [2]. Therefore, theoretical researches with the help of the plausibly developed hypothetical mathematical models are of a great importance [2]; moreover, the usefulness of such models application for the alternative roundabouts effectiveness evaluations is beyond any doubts and it can be traced with the most up-to-date publications [3–15]. The brief literature survey of the presented paper just is highlighting the principal possibilities of the proposed methods implementations in those areas of science that relate with the multiattribute assessment of road design solutions [3], geometric design of turbo roundabouts [4], or the journal paper [5] dedicated to the turbo roundabouts, design principles, and safety performance, as well as alternativeness [6] and transportation infrastructure [7] aspects.

It is also necessary to underline the importance of the developed here approach for other transportation facilities, like railway [8]; however the methods are aimed at modeling alternative roundabouts effectiveness and in its sense continue the researches of the assessment, analysis, and simulation discussed in references [9–15]. Some of those publications emphasize the evaluation of the environmental and functional benefits of "innovative" roundabouts situational features, like [11] or the study of the vehicle speed [14] for the design of the roundabouts.

Remarkable in this context is that all the publications either directly or indirectly have dealt or might have dealt with mathematical modeling and simulation in regard to economical parameters.

One of such approaches and corresponding mathematical modeling is the objective of the presented paper.

There are numerous strategies to cope with the problems related with transportation services in general. The diversity of such strategies is stipulated by different types of transportation means, other managerial factors, including financial levers of influence.

For example, in Belgium, as that was reported by the TV news and broadcasted by the mass media, the local subway service becomes free of charge; no one will have to pay in fares, when the on-ground going vehicles' pollutant emission exceeds the accepted allowed limits.

The presented paper theoretical concept is centered upon the issues related with the support of the alternative roundabouts worthiness (vehicle worthiness, riding worthiness, transportation worthiness, etc.). The prototypic approach is adopted from the aircraft airworthiness support measures concepts of [16–20] developed from subjective analysis [21] on the basis of the Jaynes' principle [22–24] in the framework of the calculus of variations theory [25].

2. Mathematical Modeling and Developed Methods

There is an irresistible temptation to model and investigate the process of a roundabout functioning under the prism of its effectiveness. It will depend upon the number of factors; specialists, who have an extensive experience in the field of roundabouts, distinguish the most significant from them [1, 3–15]. It is proposed to consider theoretically the modeling and optimizing of a few, for example, competing roundabouts. The supposed roundabouts worthiness is deemed to be supported in an analogous to airworthiness way [16–20].

2.1. Basic Model. Every aspect of a roundabout functioning is being considered under the prism of multialternativeness, if it deals with someone's choice or individual subjective preferences distribution in regard to the achievable set of alternatives [21], on one hand. And on the other hand, there is a multi-"optionality" phenomenon when the objectively existing options of the ongoing processes are under consideration.

All the issues of a roundabout operation or exploitation relating transportation facilities, network, entire and in particular infrastructure capacity, functional and economical criteria, environmental impacts, safety and precaution means and measures are based upon operational incomes and costs analysis [2].

Profitableness of a roundabout construction along with the roundabout's design, creation, building, and management costs subtracted from the incomes ensures a good reaction of the transportation business upon the challenges of the functioning [1–15].

Herewith, it is proposed a model based upon the principle approach expressed with the following formula:

$$J = \int_{p_0}^{p_1} \left(apcc' - bpc' \right) dp. \tag{1}$$

Here, in equation (1), J is the objectives functional depending upon the variated traffic flow p within the traffic flow's possible range of alterations $[p_0 \dots p_1]$. The under-integral function of

$$apcc' - bpc' \tag{2}$$

expresses the specific profitableness of the roundabout, reduced per a vehicle in the roundabout's varied traffic flow. In the expressions of (1) and (2), a has a meaning of the proportionality coefficient for the supposed model of the incomes formation, whereas b bears a sense of such a coefficient for the accepted costs model. The function of c takes its own role in the roundabout's operational income versus cost balance formation. For the stated problem setting, it is investigated the model constructed of the expressions of (1) and (2), being also dependent upon the first complete derivative of

$$c' = \frac{dc}{dp} = c'_p. \tag{3}$$

Mathematically, formulated in the view of the equations of (1) – (2), this particularly given problem setting is stated as the simplest problem of the calculus of variations for the objective functionals likewise [25]:

$$J = \int_{p_0}^{p_1} F\left(p, c, c'_p\right) dp, \tag{4}$$

where $F(p, c, c'_p)$ is the under-integral function of the stated problem objective functional (1).

That is,

$$F\left(p, c, c'_p\right) = apcc'_p - bpc'_p. \tag{5}$$

The purpose is to maximize the objective functional (1) on conditions of (2) and (3) by finding such function of c that delivers the wanted maximum of the profit formation; and for the general view integral of (4) there are the necessary conditions for the extremum existence in the view of the well-known Euler-Lagrange equation [25]:

$$\frac{\partial F}{\partial c} - \frac{d}{dp}\left(\frac{\partial F}{\partial c'_p}\right) = 0. \tag{6}$$

2.2. Solution to the Stated Basic Profit Optimization Problem. In previous speculations (1) – (6) framework mentioned above, the sought after solution will be next.

In accordance with the Euler-Lagrange equation (6) [25], in case of (1) – (5)

$$\frac{\partial F}{\partial c} = apc'_p. \tag{7}$$

Then, for the second member of equation (6) that condition yields

$$\frac{\partial F}{\partial c_p'} = apc - bp. \tag{8}$$

Differentiating the partial derivative of the right hand part of equation (8) for the second time with respect to the independent variable p in the complete form derivative it can be obtained that

$$\frac{d}{dp}\left(\frac{\partial F}{\partial c_p'}\right) = \frac{\partial}{\partial p}\left(\frac{\partial F}{\partial c_p'}\right)\frac{dp}{dp} + \frac{\partial}{\partial c}\left(\frac{\partial F}{\partial c_p'}\right)\frac{dc}{dp} + \frac{\partial}{\partial c_p'}\left(\frac{\partial F}{\partial c_p'}\right)c_p'', \tag{9}$$

where

$$\partial c_p'' = \frac{d^2 c}{dp^2}. \tag{10}$$

Thus, with taking into account the conditions written with the equations of (9) and (10), for equations (6) and (8) it yields

$$\frac{d}{dp}\left(\frac{\partial F}{\partial c_p'}\right) = ac - b + apc_p'. \tag{11}$$

Here in (11) the complete derivative (9) has got an abridged form since

$$\frac{\partial}{\partial c_p'}\left(\frac{\partial F}{\partial c_p'}\right) \equiv 0. \tag{12}$$

Hence, the second derivative (10) is omitted in the final form of the Euler-Lagrange equation (6); and equation (6) now does not depend upon the second derivative of the sought function, i.e.,

$$\frac{\partial}{\partial c_p'}\left(\frac{\partial F}{\partial c_p'}\right)c_p'' \equiv 0. \tag{13}$$

Therefore, substituting the relevant expressions of (7) and (11) for their values into the equation (6), it becomes possible to rewrite it in the view of

$$apc_p' - \left(ac - b + apc_p'\right) = 0. \tag{14}$$

Finally, after cancelling the similar members in equation (14), it yields even not a differential but algebraic equation with respect to the function of c, that determines the considered model structure (1):

$$ac - b = 0. \tag{15}$$

Furthermore, the solution in the view of function c has happened to be not depending upon the independent variable p at all:

$$c = \frac{b}{a}. \tag{16}$$

2.3. Another Model a Roundabout Functioning. It is definitely the simplest model described with expressions of (1) – (5) and yielding the solution in the view of equation (16) obtained through the procedures of (6) – (16) should be checked for the suspected extremum existence since all mathematical constructions and derivations have been performed on the basis of the necessity conditions (6).

Another theoretical development of the model of (1) – (5) applicable for an alternative roundabout functioning is as the following one.

Suppose, there is a nonlinear (raised to a certain power) dependence of the profitableness function (under-integral function) of F upon the traffic flow p. Herein, it is proposed the next up modification (see and compare with the equation of (5)):

$$F\left(p, c, c_p'\right) = ap^n cc_p' - bpc_p', \tag{17}$$

where n is the power index magnifying the impact of the traffic flow p exerted upon the component of the vehicle contribution in the traffic flow.

The theoretical solution to the objective functional similar to (1) or (4), in the framework of the approach analogous to (6) – (16), results in the other view optimal function.

However, it is obvious, in the proposed model modification expressed with the equation of (17) versus (5), the principle of the wanted solution finding will not differ from the traditional method (6).

2.4. Developed Model Solution. The substitution of the relevant interrelationship of (17) pertaining with the stated problem setting for its value into the objective functional (1) gives for members of equation (6) such formulae.

For the partial derivative with respect to the sought function, likewise (7), the expression will be in accordance with the model equation (17):

$$\frac{\partial F}{\partial c} = ap^n c_p'. \tag{18}$$

For the complete derivative, similar to (9), with respect to the independent variable, the partial derivative with respect to the complete derivative of the sought function with respect to the independent variable, in analogous way to (8), now gets the view of the following equation:

$$\frac{\partial F}{\partial c_p'} = ap^n c - bp. \tag{19}$$

In its turn the complete form derivative (9) in such case yields similar to (11)

$$\frac{d}{dp}\left(\frac{\partial F}{\partial c_p'}\right) = anp^{n-1}c - b + ap^n c_p'. \tag{20}$$

After that collecting corresponding members of equations (18) and (20) into the necessary condition Euler-Lagrange equation (6), there is a possibility to obtain the other equation than (14), i.e.,

$$ap^n c_p' - \left(anp^{n-1}c - b + ap^n c_p'\right) = 0. \tag{21}$$

Again, making obvious transformations, (21) yields the needed result as the following equation:

$$anp^{n-1}c - b = 0. \tag{22}$$

From (22) it inevitably means formula (23):

$$c = \frac{b}{anp^{n-1}} = c(p). \tag{23}$$

This time, unlike in the case resulting with the constant value of (16), the sought solution (23) of the model (17) – (23) is a certain function of the independent variable, traffic flow p: $c(p)$.

2.5. Computer Simulation and Check of the Suspected Solutions Optimality. As the solutions in the view of equations (16) and (23) are found on just the necessary condition (6) for the extremums existence, it is required to check the solutions for being really optimal subject to the stated problem settings constraints.

The numerical data for the conducted calculation experimentation are as follows. For the basic model described with the procedures of expressions (1) – (16): $a = 2$; $b = 20$ of such Conditional Units (CU) that are correspond to the measurement units of the fractional members of the under-integral function (5) of the objective functional (1). The range of the traffic flow p alteration has been considered from $p_0 = 1$ up to $p_1 = 200$ of CU (possible unit is vehicles per an hour (v/h)). The result of such primitive, so far, experimentation is $c = 10$ CU.

In order to make sure of the optimality of the obtained function $c = 10$ value, it is possible to let the function $c = 10$ some definite variation with the fixed boundaries.

Such effect is modelled with the help of the matrix-vector data description.

The traffic flow matrix at the next three specified points is

$$\mathbf{P} = \begin{bmatrix} p_0^2 & p_0 & 1 \\ \left(\frac{p_0 + p_1}{2}\right)^2 & \frac{p_0 + p_1}{2} & 1 \\ p_1^2 & p_1 & 1 \end{bmatrix}. \tag{24}$$

The column-vector of the free function values difference is given in the view of

$$\mathbf{C} = \begin{bmatrix} 0 \\ -0.0005 \cdot c_{\text{opt}}\left(\frac{p_0 + p_1}{2}\right) \\ 0 \end{bmatrix}, \tag{25}$$

where $c_{\text{opt}}((p_0 + p_1)/2)$ depicts the modeled variated function interrelationships between the alternating traffic flow p in v/h and the function's optimal value of equation (16) $c_{\text{opt}} = 10$ CU in the middle of the accepted range of the traffic flow p variation.

Now, from the matrix-vector equation engaging the equations of (24) and (25)

$$\mathbf{C} = \mathbf{P} \cdot \mathbf{K}, \tag{26}$$

where

$$\mathbf{K} = \begin{bmatrix} a_1 \\ b_1 \\ c_1 \end{bmatrix}, \tag{27}$$

where a_1, b_1, and c_1 are the coefficients, in CU, converting the traffic flow values of the quadratic forms of (24) via (26) with (27) into the free function values of (25).

Solving for (27) it returns

$$\mathbf{K} = \mathbf{P}^{-1} \cdot \mathbf{C}. \tag{28}$$

And for the inverse matrix it yields

$$\mathbf{P}^{-1} = \begin{bmatrix} \frac{2}{39{,}601} & -\frac{4}{39{,}601} & \frac{2}{39{,}601} \\ -\frac{601}{39{,}601} & \frac{804}{39{,}601} & -\frac{203}{39{,}601} \\ \frac{40{,}200}{39{,}601} & -\frac{800}{39{,}601} & \frac{201}{39{,}601} \end{bmatrix}. \tag{29}$$

This gives

$$\mathbf{K} = \begin{bmatrix} a_1 = 5.05 \cdot 10^{-7} \\ b_1 = -1.015 \cdot 10^{-4} \\ c_1 = 1.01 \cdot 10^{-4} \end{bmatrix}. \tag{30}$$

In accordance with the expressions of the procedure of (24) – (30) the free function variations calculated by

$$c_o(p) = a_1 p^2 + b_1 p + c_1 + c_{\text{opt}}(p) \tag{31}$$

for a decreasing function variation and

$$c_O(p) = -\left(a_1 p^2 + b_1 p + c_1\right) + c_{\text{opt}}(p) \tag{32}$$

for an increasing function variation, illustrated in Figure 1, give the results shown in Figures 2 and 3.

In Figure 1 $c_{\text{opt}}(p)$ is standing for the formula (16) extremal value of the optimal function; $c_o(p)$ has been calculated by formula (31) and $c_O(p)$ by formula (32).

In Figure 2 $Z_{\text{opt}}(p)$, $Z_o(p)$, and $Z_O(p)$ are the objective functionals values plotted as a result of the calculations performed by the formula (1) with the variated upper limit of integration for the corresponding determining functions of $c_{\text{opt}}(p)$ by the formula (16): extremal value of the optimal function; $c_o(p)$ by formula (31); $c_O(p)$ by formula (32).

And in Figure 3 the corresponding objective functionals values are plotted for the marginal traffic flow value $p_1 = 200$ v/h.

The nonlinear no constant model (17) – (23) numerical investigation is illustrated in Figures 4–6.

The optimal solution now is being calculated by the extremal formula of (23) and it has been variated in the style of the one described with the expression of (24) – (32).

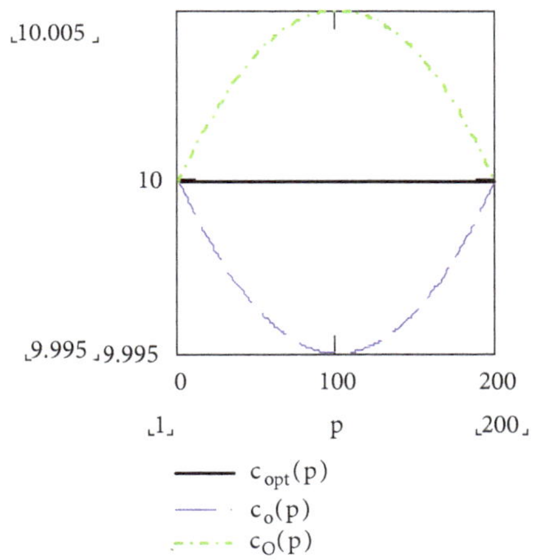

FIGURE 1: Free function variations for the traffic flow in comparison with the extremal solution (optimal function).

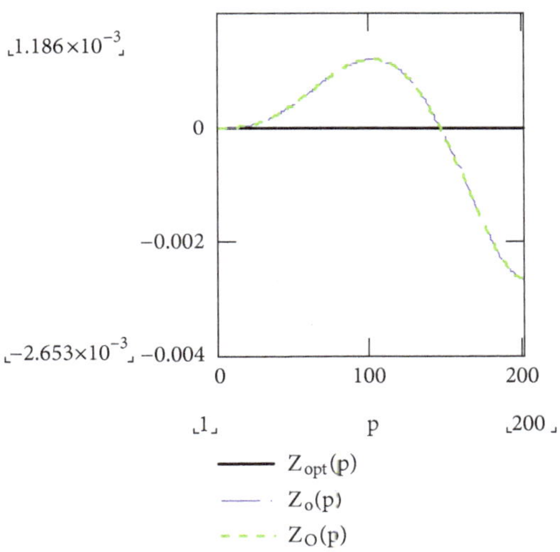

FIGURE 2: Objective functional value.

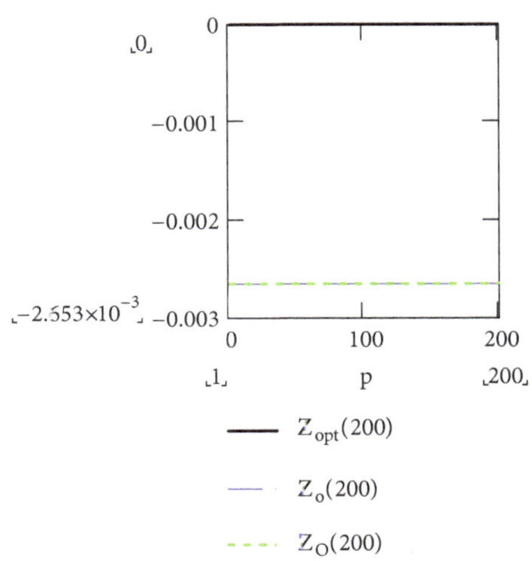

FIGURE 3: Objective functional value at the final boundary.

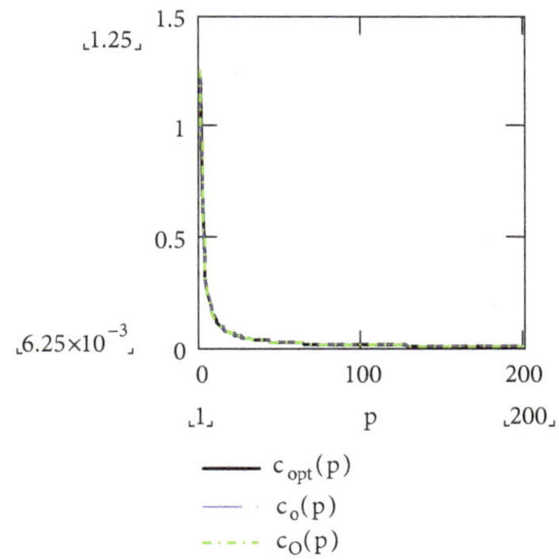

FIGURE 4: Free function variations for the traffic flow in comparison with the extremal solution (optimal function).

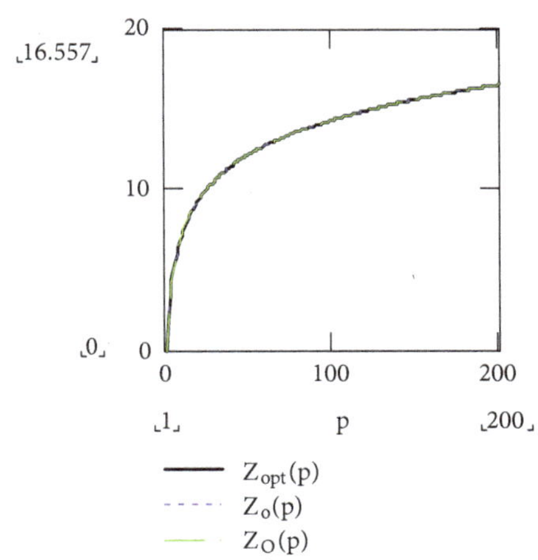

FIGURE 5: Objective functional value.

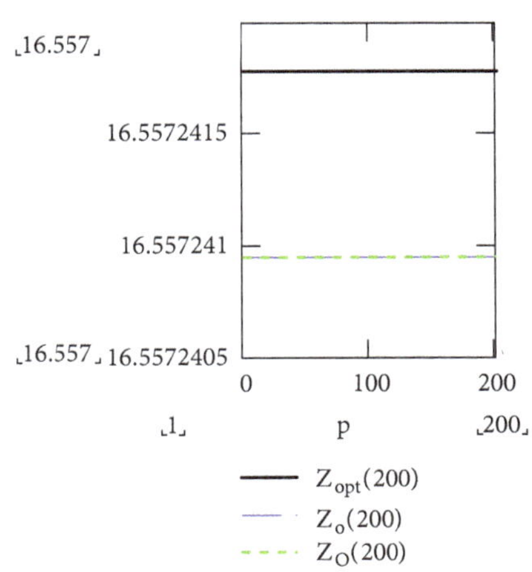

FIGURE 6: Objective functional value at the final boundary.

$c_o(p) =$

	0
0	1.25
1	0.624999876
2	0.416666419
3	0.312499631
4	0.24999951

$c_{opt}(p) =$

	0
0	1.25
1	0.625
2	0.416666667
3	0.3125
4	0.25

$c_O(p) =$

	0
0	1.25
1	0.625000124
2	0.416666914
3	0.312500369
4	0.25000049

$c_o(p) =$

	0
195	$6.377061058 \cdot 10^{-3}$
196	$6.344808309 \cdot 10^{-3}$
197	$6.31288382 \cdot 10^{-3}$
198	$6.28128266 \cdot 10^{-3}$
199	$6.25 \cdot 10^{-3}$

$c_{opt}(p) =$

	0
195	$6.37755102 \cdot 10^{-3}$
196	$6.345177665 \cdot 10^{-3}$
197	$6.313131313 \cdot 10^{-3}$
198	$6.281407035 \cdot 10^{-3}$
199	$6.25 \cdot 10^{-3}$

$c_O(p) =$

	0
195	$6.378040982 \cdot 10^{-3}$
196	$6.345547021 \cdot 10^{-3}$
197	$6.313378807 \cdot 10^{-3}$
198	$6.28153141 \cdot 10^{-3}$
199	$6.25 \cdot 10^{-3}$

FIGURE 7: Initial and final values of the variated determining free function.

The difference is for the accepted in the computerized numerical experimentations vector solution (30), which makes

$$\mathbf{K} = \begin{bmatrix} a_1 = 6.282 \cdot 10^{-10} \\ b_1 = -1.263 \cdot 10^{-7} \\ c_1 = 1.256 \cdot 10^{-7} \end{bmatrix}. \tag{33}$$

3. Results and Discussion

3.1. Analysis of the Obtained Results. The obtained theoretical results represented with the diagrams shown in Figures 1–6 brightly distinguish the described above two variants of the roundabouts functional strategies.

The maximal value of the objective functional is clearly seen in Figures 2 and 3 for the basic case (1) – (16) simulated with the procedures (24) – (32).

For a more complicated case study (17) – (23) calculated with (24) – (29) and (31) – (33) the effect of the maximization is hardly noticeable (see Figure 5) because of the tiny variations of the determining function values of $c_{opt}(p)$: found with the use of the extremal solution equation (23) (see Figure 4).

Nevertheless, even such negligibly small variations as proposed herewith prove the maximum value deliverance to the objective functional (1) with the under-integral function (17) and corresponding extremal (23) (see Figure 6).

3.2. Discussion. Theoretically, the obtained results are for a rather simple modeling. Nonetheless, the principle is fundamental: the profit is the difference between the incomes and costs of any object (including roundabouts) functioning. And the amount of the profit provides all necessary factors, which are numerous, with their implementation procedures and further outcomes.

The other fact of the simplifications accepted at the modeling is that the problem setting is stated for the fixed boundaries calculus of variations problem.

It is perfectly visible in Figure 1. This approach is applied in the (17) – (23) model too. To make a proof for such statement a couple of the fragmental parts pieces of the calculation tables are presented in Figure 7.

Moreover, all the theoretically considered factors [1–15] have a subjective analysis [21], based upon [22–24], component worth of being studied as a developmental idea of [16–20].

4. Conclusions

The theoretical concepts case study, analyzed in this paper on the alternative roundabouts worthiness in the way analogous to the aircraft airworthiness supporting measures effectiveness research, proves that the described calculus of variations approach allows obtaining the objectively existing optimal values of the operational (functioning) purpose functional with the help of the specially introduced determining functions.

Conditions of multialternativeness for the considered theoretically different types of the roundabouts are embodied with the corresponding parameters of the proposed models. The conditions of uncertainty related with the roundabouts alternatives can be taken into account with the subjective entropy of the individuals' preferences functions optimal distributions.

The preferences functions give the possibility of the considered alternative roundabouts assessment with respect to the preferences entropy uncertainty measure, which should be a part of the further research dedicated to the theory of subjective analysis since it would be useful to find more theoretical results and applicable areas as well as some other effectiveness functions and their variables.

Conflicts of Interest

The author declares that there is no conflict of interests regarding the publication of this paper.

References

[1] T. Tollazzi, G. Tesoriere, M. Guerrieri, and T. Campisi, "Environmental, functional and economic criteria for comparing "target roundabouts" with one- or two-level roundabout intersections," *Transportation Research Part D: Transport and Environment*, vol. 34, pp. 330–344, 2015.

[2] E. Silberberg and W. Suen, *The Structure of Economics. A Mathematical Analysis*, McGraw-Hill Higher Education, 2001.

[3] E. K. Zavadskas, "Multi-attribute assessment of road design solutions by using the COPRAS method," *The Baltic Journal of Road and Bridge Engineering*, vol. 2, no. 4, pp. 195–203, 2007.

[4] T. Džambas, S. Ahac, and V. Dragčević, "Geometric design of turbo roundabouts," *Tehnički vjesnik*, vol. 24, no. 1, pp. 309–318, 2017.

[5] L. G. H. Fortuijn, "Turbo roundabouts: Design principles and safety performance," *Journal of the Transportation Research Board*, vol. 2096, no. 1, pp. 16–24, 2009.

[6] J. C. Engelsman and M. Uken, "Turbo roundabouts as an alternative to two lane roundabouts," in *Proceedings of the 26th Annual Southern African Transport Conference, SATC '07*, pp. 581–589, Pretoria, South Africa, 2007.

[7] Z. Tesoriere, "Infrastructure as interface: Thinking the urban and the high-speed railway station: Italian case-studies," *Spaces & Flows: An International Journal of Urban & Extra Urban Studies*, vol. 3, no. 1, pp. 71–87, 2013.

[8] A. Cazzani, M. Cattani, R. Mauro, and F. Stochino, "A simplified model for railway catenary wire dynamics," *European Journal of Environmental and Civil Engineering*, vol. 21, no. 7-8, pp. 936–959, 2017.

[9] O. Giuffrè, A. Granà, and S. Marino, "Comparing performances of turbo-roundabouts and double-lane roundabouts," *Modern Applied Science*, vol. 6, no. 10, pp. 70–79, 2012.

[10] A. B. Silva, P. Mariano, and J. P. Silva, "Performance Assessment of Turbo-roundabouts in Corridors," *Transportation Research Procedia*, vol. 10, pp. 124–133, 2015.

[11] M. Guerrieri, F. Corriere, B. Lo Casto, and G. Rizzo, "A model for evaluating the environmental and functional benefits of "innovative" roundabouts," *Transportation Research Part D: Transport and Environment*, vol. 39, pp. 1–16, 2015.

[12] L. Vasconcelos, A. Silva, Á. Seco, P. Fernandes, and M. Coelho, "Turboroundabouts: multicriterion assessment of intersection capacity, safety, and emissions," *Transportation Research Record: Journal of the Transportation Research Board*, vol. 2402, pp. 28–37, 2014.

[13] D. Ticali and F. Corriere, "Turbo roundabouts: geometric design parameters and performance analysis," *GSTF Journal on Computing (JoC)*, vol. 2, no. 1, pp. 227–232, 2012.

[14] H. Pilko, D. Brčić, and N. Šubić, "Study of vehicle speed in the design of roundabouts," *Građevinar- Journal of the Croatian Association of Civil Engineers*, vol. 66, no. 5, pp. 407–416, 2014.

[15] J. Chalmovský, J. Štefaňák, L. Miča et al., "Statistical-numerical analysis for pullout tests of ground anchors," *The Baltic Journal of Road and Bridge Engineering*, vol. 12, no. 3, pp. 145–153, 2017.

[16] A. V. Goncharenko, "Aeronautical and aerospace material and structural damages to failures: theoretical concepts," *International Journal of Aerospace Engineering*, vol. 2018, Article ID 4126085, 7 pages, 2018.

[17] A. V. Goncharenko, "Multi-optional hybrid effectiveness functions optimality doctrine for maintenance purposes," in *Proceedings of the 2018 IEEE 14th International Conference on Advanced Trends in Radioelecrtronics, Telecommunications and Computer Engineering (TCSET '18)*, pp. 771–775, IEEE, Lviv-Slavske, Ukraine, 2018.

[18] A. Goncharenko, "Aircraft operation depending upon the uncertainty of maintenance alternatives," *Aviation*, vol. 21, no. 4, pp. 126–131, 2017.

[19] A. V. Goncharenko, "Optimal UAV maintenance periodicity obtained on the multi-optional basis," in *Proceedings of the 2017 IEEE 4th International Conference Actual Problems of Unmanned Aerial Vehicles Developments (APUAVD '17)*, pp. 65–68, IEEE, Kiev, Ukraine, 2017.

[20] A. Goncharenko, "Several models of artificial intelligence elements for airctaft control," in *Proceedings of the 2016 IEEE 4th International Conference on Methods and Systems of Navigation and Motion Control (MSNMC '16)*, pp. 224–227, IEEE, Kiev, Ukraine, 2016.

[21] V. Kasianov, *Subjective Entropy of Preferences. Subjective Analysis: Monograph*, Institute of Aviation, Warsaw, Poland, 2013, http://kasianovv.wixsite.com/entropyofpreferences.

[22] E. T. Jaynes, "Information theory and statistical mechanics," *Physical Review A: Atomic, Molecular and Optical Physics*, vol. 106, pp. 620–630, 1957.

[23] E. T. Jaynes, "Information theory and statistical mechanics II," *Physical Review A: Atomic, Molecular and Optical Physics*, vol. 108, pp. 171–190, 1957.

[24] E. T. Jaynes, "On The Rationale of Maximum-Entropy Methods," *Proceedings of the IEEE*, vol. 70, no. 9, pp. 939–952, 1982.

[25] I. M. Gelfand and S. V. Fomin, *Calculus of Variations*, State Publishing House of Physical and Mathematical Literature, Moscow, Russia, 1961.

18

Driver's Eco-Driving Behavior Evaluation Modeling Based on Driving Events

Chen Chen,[1] **Xiaohua Zhao**,[1] **Ying Yao**,[1] **Yunlong Zhang**,[2] **Jian Rong**,[1] **and Xiaoming Liu**[1]

[1]*Beijing Key Laboratory of Traffic Engineering, Beijing Engineering Research Center of Urban Transport Operation Guarantee, Beijing University of Technology, Beijing 100124, China*
[2]*Texas A&M University, College Station, TX 77843, USA*

Correspondence should be addressed to Xiaohua Zhao; zhaoxiaohua@bjut.edu.cn

Academic Editor: Yue Liu

Eco-driving is an effective means to reduce vehicle fuel consumption. Although many researches and devices have been developed to introduce eco-driving, quantitative effects of driver behaviors on fuel consumption are still unclear, as well as quantitative eco-driving advices. To solve this problem and promote the application of eco-driving in China, a driving-events-based eco-driving behavior evaluation model was proposed in this paper. First, based on taxicab operating data, the relationship between three vehicle operating parameters (speed, acceleration, and driving mode duration) and fuel consumption was analyzed. Then, nine fuel-consumption-involved driving events (including Accelerating Sharply, Decelerating Sharply, and Long-Time Accelerating) were proposed and defined. Using the frequency of each driving event in a certain distance as independent variable and vehicle fuel consumption as dependent variable, principal component analysis (PCA) and multiple linear regression were applied to establish driver's eco-driving behavior evaluation model. The model was proved to be highly accurate (96.72%). At last, based on the evaluation model, corresponding quantitative eco-driving advices were provided to help driver to improve their driving skills.

1. Introduction

Road transportation consumes large amounts of oil resources (about 40%) in 2015 [1] and contributes nearly 80% in air pollutant emission [2]. Some technological breakthroughs are potential to solve this problem, for instance, the usage of advanced engine control system and electric powered vehicles. However, due to existing technological challenges and public attitudes, the popularization of such methods is likely to take a long time.

On a short-term horizon, change in driver behavior, namely, the "eco-driving," arises as a promising method to reduce energy use. Eco-driving includes several driving methods that could achieve higher energy efficiency, so as to reduce fuel consumption and exhaust emission. It has been verified that different driving method could result in different fuel consumption and emission. De Vlieger [3] found that fuel consumption and emission increased by 12%–40% and 20%–50%, respectively, when driving aggressively. Van Mierlo et al. [4] indicated that a considerable reductions of 5%~25% in fuel consumption can be achieved after fuel-saving driving practice.

According to previous studies, the application of eco-driving can mainly be classified into three categories: route planning, trajectory optimization, and driver behavior improvement.

Route planning mainly deals with the questions on macroscopic level, for example, the traffic congestion avoidance. Pham et al. [5] developed a participatory sensing service FUEOOGLE, which allowed drivers to find the most fuel-efficient routes by plotting vehicle fuel consumption on city road network. Barth et al. [6] developed an environmentally friendly navigation algorithm, which took advantage of mobile-source energy and emission models and minimized fuel consumption and pollutant emissions. Further, Boriboonsomsin et al. [7] presented another eco-routing navigation system by taking real-time traffic information into consideration. Yao and Song [8] collected emission and fuel consumption data, based on which a set of mesoscopic vehicle emission and fuel consumption model is established. What is more, an eco-route planning algorithm is proposed.

Trajectory optimization focuses on mesoscopic level. By collecting and utilizing road gradients data, traffic signal information, leading car speed, and so on, the most fuel-efficiency trajectory is designed. Barth et al. [9] developed a dynamic eco-driving velocity planning algorithm, which used signal phasing and timing information of an arterial corridor. The algorithm then was tested and showed 12% reduction of fuel consumption and CO2 emission. Mandava et al. [10] developed arterial velocity planning algorithms that provided dynamic speed advice to drivers by using the signal information when approaching an intersection. It was verified that energy/emission savings were 12%–14% when using velocity planning. Chen [11] developed an optimization model to determine the optimal eco-driving trajectory at signalized intersection using signal phases information. The case study showed that eco-driving could reduce emissions by 50%.

Driver behavior feedback aims at the improvement of individual driver's vehicle maneuvering from microscopic level. By providing real-time/trip-summarized eco-driving advices during/after travels, drivers adjust their driving behavior to achieve a higher fuel economy. Boriboonsomsin et al. [12] introduced an eco-driving feedback device that could provide both real-time and detained trip emission information. The real-time feedback information included fuel consumption and CO_2 emission in a color scheme from red (poor) to green (good). The detained trip information displayed total travel time, average speed, total fuel consumption, and CO_2 emission. Application effects from 20 samples showed that average fuel economy improved by 6% in city streets. Frank et al. [13] implemented an android application that gathered relevant data from vehicle's CAN bus, providing a driver with second-by-second representative eco-score. The eco-score is classified into four categories: BAD, AVERAGE, GOOD, and VERY GOOD, derived by three fuzzified variables (velocity, road slope, and power consumption) based on a fuzzy system. By introducing drivers the basic eco-driving concepts and recommendations after travel, a seven-volunteers-participated experiment showed that the android application significantly reduced overall energy consumption. Araújo et al. [14] presented a smartphone application that could detect driving pattern and then suggested new behaviors in real-time to reduce fuel consumption. Variables used to classify driving pattern included speed, acceleration, road slope, fuel consumption, throttle signal, and engine rotations. Also, the fuzzy system is used in the classification and feedback model. Eight hints such as "shift gear earlier" and "accelerating too high" were provided to drivers during travel. Wu et al. [15] trained 22 recruited drivers based on a driving simulator by coaching them how to take fuel economy manipulations. The monitoring of fuel consumption before and after training showed a 5.45% reduction. Besides, Wu et al. [16] developed an Internet + based eco-driving support platform with a cellphone as feedback device to provide fuel consumption and behavior deficiency to drivers after travel. A validation of 50 taxicab drivers' fuel consumption during a month decreases by 4.5%~13%.

Among the three eco-driving application categories, route planning and trajectory optimization highly rely on the integration of detailed traffic information and road alignment data. These methods may play tremendous roles under connected vehicle or automatic vehicle circumstance. At present, the developing foundation of connected vehicle and automatic vehicle in China limits their application. Conversely, driver behavior feedback is not subjected to these limitations and has huge room for improvement. Thus, the improvement of driver behavior is paid more attention.

Generally, the keys of improving driver eco-driving behavior are (1) evaluation of drivers' eco-driving behavior and (2) quantitative and operational driving advices feedback. The former is to inform drivers why they are fuel-guzzling by identifying drivers' eco-driving behavior deficiency and the latter is to help drivers to improve driving skills.

As to the first aspect (evaluation of drivers' eco-driving behavior), the detection of high-fuel-consumption behavior use in applications mentioned above is mostly achieved by monitoring abnormal fuel usage and consequently provides warning (red/green colors, voices, etc.) messages through on-board screen or smartphone. Some developers have proposed several behavior-based identification algorithms, for example, the fuzzy-system-based eco-score assessment. Nevertheless, the thresholds used in their models mainly derive from experience judgment, other than quantitative analysis based on driving behavior. Also, in the research field of driver behavior evaluation, there are some researches concentrating on the effect of driving styles on fuel consumption. Ferreira et al. [17] studied the impact of driving styles on fuel consumption, exploiting parameters including speed, acceleration, engine rotation speed, and so on, which were collected from vehicle CAN bus and GPS devices. Naive Bayes' algorithm was used to identify the main influence factors that had major effects on fuel consumption. Eleven events like excessive acceleration were classified into five classes (namely, extreme below average, below average, on average, above average, and extreme above average), according to the frequency of events per 100 km. Then, probabilities for fuel consumption of each class were provided. However, this study indicated the probability that how driver behavior impacts fuel consumption, while the accurate quantitative relationship between driving behavior and fuel consumption was still unclear. Other studies [3, 18, 19] mainly focused on the influence of different driving styles classes (aggressive, normal, moderate, conservative, etc.) on energy usage, which was too abstract in practical application.

As to the second aspect (driving advices feedback), most of eco-driving advices which are provided to driver are qualitative at present, for example, "Avoid sharp acceleration." Drivers need to adjust their manipulation cyclically (a process like "driving, feedback, driving with low acceleration, feedback, driving with lower acceleration...") to approach a better fuel economy. What is more, it seems that process-based eco-driving advice may be easier to understand and operate by drivers than instantaneous-based advice. For example, the message "Accelerate XX km/h within Y s" may be better than "Acceleration should be less than XX m/s^2." Therefore, the concept, driving events, which are defined as typical driving action during a period of time such as accelerating and decelerating, is proposed and used in this paper.

In fact, eco-driving deficiency detection and corresponding advices are integrated. When one knows where and why he/she makes mistake, he/she can correct it quickly. Therefore, the objectives of this paper are as follows:

(1) To quantitatively explore how driver behaviors (driving events) affect fuel consumption, thus evaluating driver's eco-driving behavior.

(2) To propose more quantitative and clearer driving advices on drivers' operating behaviors, thus guiding drivers to improve fuel economy.

In summary, to promote the spreading and application of eco-driving in China, this paper is proposed to establish a driving-event-based eco-driving behavior evaluation model and provide quantitative eco-driving advices. The detection of these driving events and the development of the model can help drivers to realize their driving behavior deficiency and then correct it.

The remainder of the paper is organized as follows. Section 2 introduces the data collection and processing. Then, in Section 3, the proposing of driving events is presented. In Section 4 the establishment of the model is built and eco-driving advices are provided. What is more, in Section 5 the method and results of this paper are discussed and directions of future work are provided. Finally, in Section 6 the conclusions of this paper are summarized.

2. Data Collection and Processing

2.1. Vehicle Fuel Consumption and Operating Data Collection. In this paper, 307 taxicabs' fuel consumption and operating data in one day were collected second by second through Beidou/Global Positioning System (GPS) dual-mode positioning device and on-board diagnostic (OBD) device mounted on vehicles. Collected data included time, vehicle instantaneous fuel consumption (L/h), speed (km/h), engine speed (rounds per minute, RPM), engine torque (N·m), cumulative travel distance (km), and vehicle's location information (i.e., the latitude and longitude). All taxicabs used in the study were Hyundai Elantra, with 4-cylinder 1.6 liter engine, and their fuel consumption was certified by the National Level IV emission standard. All of these cabs were put into operation in 2013 and were regularly maintained every 5000 km complied with taxicab company requirement. So it was assumed that all of taxicabs had almost the same working condition and fuel consumption characteristics. The amount of the collected data was 1.14 million, and the total driving time was 3176 hours. On average, every driver's daily driving time was about 10.3 hours.

2.2. Data Screening and Processing. In this paper, the evaluation of driver's behavior under normal condition is mainly concerned. Thus, data in following cases are not taken into consideration.

(1) Speed > 100 km/h. Taxicabs used in this paper are mainly operated on Beijing urban road, with the highest speed limit of 80 km/h. Due to the strict management of taxicab company, speeding is rare. Therefore, data with vehicle speed more than 100 km/h (illegal drive) are ignored.

(2) Absolute Value of Acceleration > 20 km/h/s. Acceleration can be calculated through vehicle speed. Data with vehicle acceleration more than 20 km/h/s or lower than −20 km/h/s is judged beyond engine's physical capabilities [20, 21].

Besides, data with values missing more than consecutive 2 seconds in time sequence are labeled. In the following analysis, data above and below the labeled data item are regarded as two different driving processes.

Figure 1 plots the distributions of daily speed and fuel consumption per hundred kilometers (FPH, L/100 km, calculated as (1)) of all the 307 taxicabs in the data-collected day. The results show that taxis' average speed mainly distributes in 20–30 km/h in one day. FPHs are generally in the range of 7–10 L/100 km with a nearly normal distribution.

$$\text{FPH}\ \ (\text{L}/100\,\text{km}) = \frac{\sum(\text{instantaneous fuel consumption (L/h)}/3600)}{\sum(\text{speed (km/h)}/3600)} \times 100, \tag{1}$$

where FPH is FPH of every driver in the data-collected day.

3. Defining Driving Events

3.1. Influence of Vehicle Operating Parameters on Fuel Consumption. Drivers' different operating behaviors result in various vehicle operating states. In this paper, several operating parameters including speed (km/h), acceleration (km/h/s), and driving mode duration (s) are used to evaluate driver's eco-driving behavior. Speed and acceleration have been widely proved to have significant influence on vehicle fuel consumption [22], while few studies concern about driving mode duration. Driving mode duration refers to the duration time of one driving mode. For example, if drivers maintain accelerating for 3 seconds, the driving mode duration of this process is 3. The reason why driving mode duration is taken into consideration is the positive correlation between vehicle instantaneous fuel consumption (IFC, L/h) and the duration time of acceleration when preprocessing the data, as shown in Figure 2. When continuously accelerating, IFC grows continuously. However, when acceleration process is interrupted by cruising or deceleration, IFC would decrease and IFC in the beginning of next accelerating process is lower than that when continuously accelerating.

The influence of three parameters on vehicle fuel consumption is shown in Figure 3. FPH (its calculation method is similar to (1), while for each speed value, only the samples that have the same speed are used, which can represent the average FPH under each speed case) is used as the measurement for speed, and IFC (for each acceleration value, average IFC is used) is chosen as the measurement of acceleration and driving mode duration.

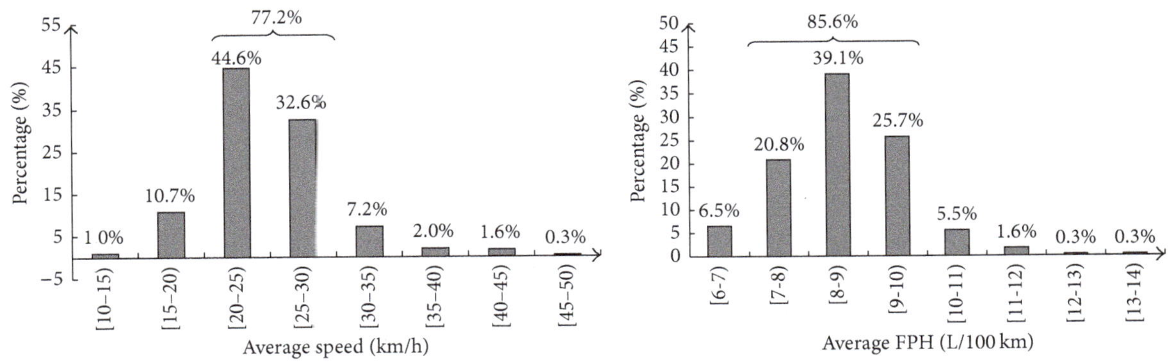

(a) Average speed (b) Average FPH

FIGURE 1: Distributions of average daily speed and FPH.

Time (yyyymmddhhmmss)	Vehicle ID	Speed (km/h)	Instantaneous fuel consumption (L/h)
20140815124154	11BP8350	0	0.903221
20140815124155	11BP8350	0	0.866404
20140815124156	11BP8350	0	1.157593
20140815124157	11BP8350	1	5.73186
20140815124158	11BP8350	10	5.15094
20140815124159	11BP8350	22	5.79366
20140815124200	11BP8350	27	6.60942
20140815124201	11BP8350	29	7.46226
20140815124202	11BP8350	36	7.80216
20140815124203	11BP8350	38	7.8516
20140815124204	11BP8350	40	2.475
20140815124205	11BP8350	40	2.078204
20140815124206	11BP8350	41	5.12004
20140815124207	11BP8350	42	5.24982
20140815124208	11BP8350	44	5.5341
20140815124209	11BP8350	48	6.36222
20140815124210	11BP8350	50	6.26334
20140815124211	11BP8350	51	2.91996
20140815124212	11BP8350	50	0.961562
20140815124213	11BP8350	50	1.198728
20140815124214	11BP8350	49	1.908836
20140815124215	11BP8350	50	4.61328
20140815124216	11BP8350	50	3.202268
20140815124217	11BP8350	51	4.84812
20140815124218	11BP8350	52	5.50938

Raw data

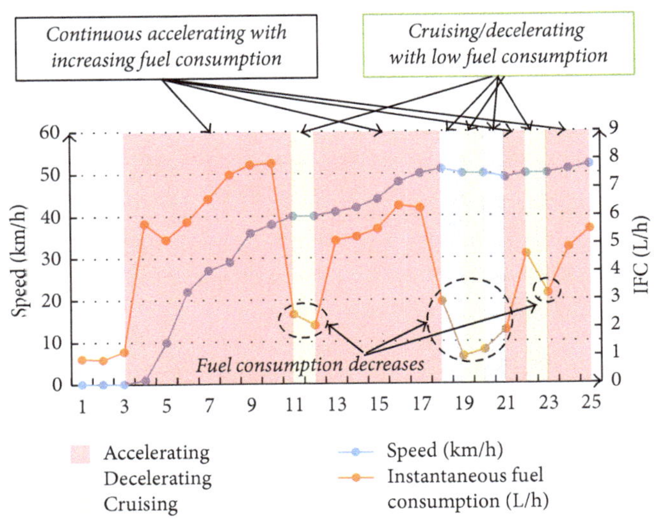

Line chart of raw data

FIGURE 2: Positive correlation between vehicle instantaneous fuel consumption and the duration time of acceleration.

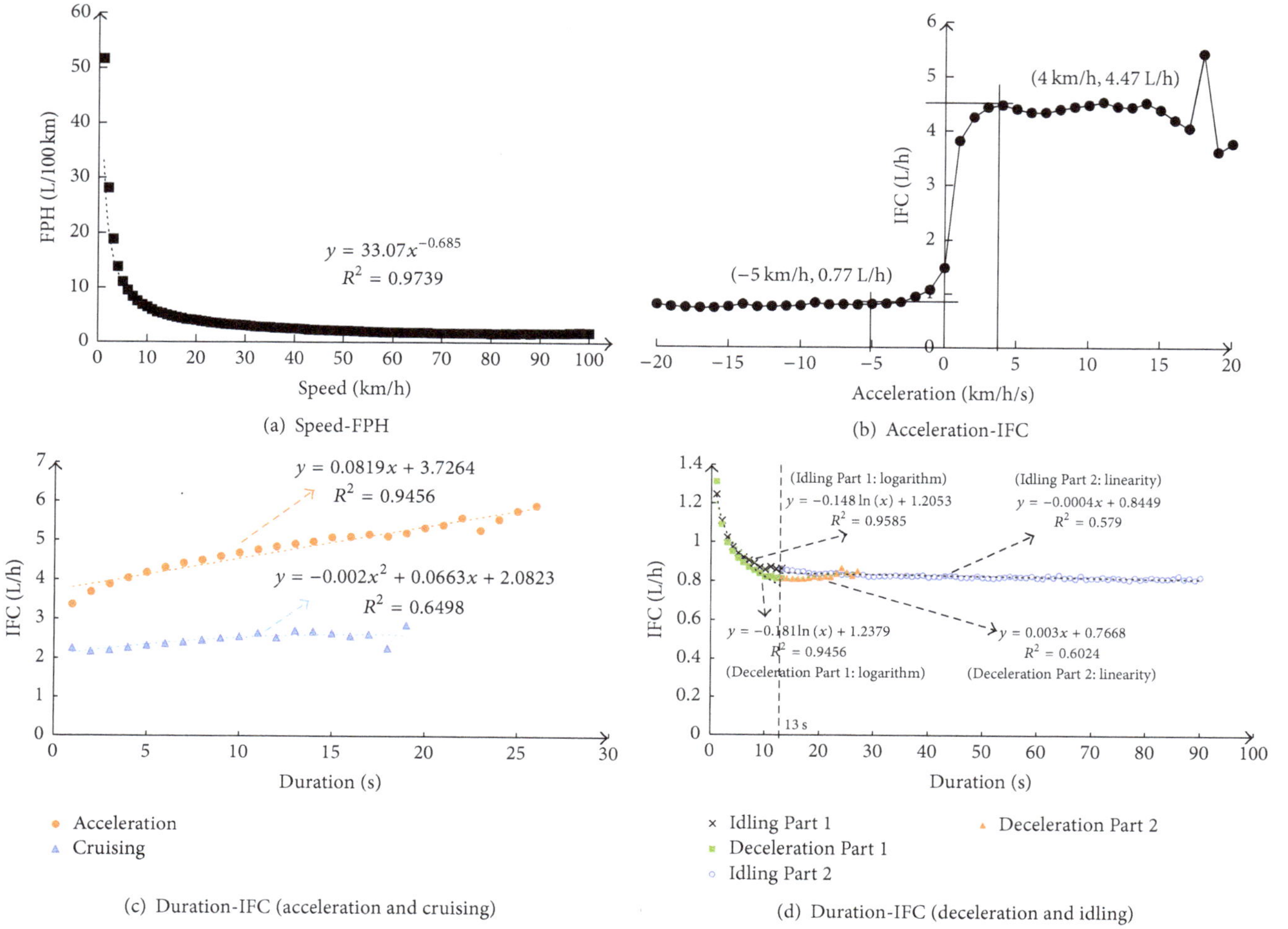

FIGURE 3: Influence of three parameters on vehicle fuel consumption. *Note.* Unit of acceleration is km/h/s instead of m/s^2. The reason is that the values using km/h/s as unit are in accordance with the ones displayed on the dashboard, which are easy to understand for drivers.

Figure 3(a) shows that FPH decreases rapidly when speed is lower than 30 km/h and then tends to be stable with the increase of speed. Previous researches suggest that wind resistance increases sharply when speed is greater than 96 km/h (60 mph) and consequently results in huge fuel consumption [23]. The data with speed greater than 100 km/h are overlooked; therefore this situation is not included nor discussed in this research.

Figure 3(b) shows that IFC in acceleration is apparently higher than in deceleration. When accelerating, IFC first increases rapidly and then remains stable, which may be caused by the engine feature limitation. When decelerating, IFC decreases with the increase of deceleration. Specifically, IFC is maintained at about 0.77 L/h when deceleration is greater than 5 km/h/s, nearly the same as idling (seen in Figure 3(d)). To explain this phenomenon, it is probably that clutch pedal is completely depressed and the fuel is only used to maintain the engine operation at high deceleration. Although IFC shows a decrease in high deceleration, FPH may have a different tendency. Thus, in this paper sharply decelerating is also considered as poor driving behavior.

Figure 3(c) shows that, with the increase of driving mode duration, IFC of acceleration linearly increases, while IFC of cruising increases slowly.

Figure 3(d) plots the relationship between IFC and duration time of deceleration and idling. It is revealed that IFC of idling and deceleration has similar performances. To better describe the tendency, the curves of deceleration and idling are separated as two parts at the point 13 s: the logarithm curve (Part 1) and linearity line (Part 2). As can be seen, IFC of deceleration and idling firstly decreases with logarithm curve and then tends to be stable.

3.2. Propose and Define Driving Events. By reviewing eco-driving advices mentioned in previous researches [24], nine driving events that are closely related to vehicle fuel consumption and can be recognized simply by the combination of three parameters above (speed, acceleration, and driving mode duration) are proposed. The nine driving events include Accelerating Sharply (AS), Decelerating Sharply (DS), Long-Time Accelerating (LA), Long-Time Idling (LI), Running with Low Speed (LS), Cruising with Higher Speed

(HS), Starting Moderately (SM), Frequently Stop and Start (SS), and Braking Moderately (BM).

Definitions of every driving events are list as follows. The determination of thresholds used in definitions is explained later.

(1) AS (Accelerating Sharply): $a(t) > 4$ km/h/s. A period with continuous instantaneous ASs is identified as one AS event.

(2) DS (Decelerating Sharply): $a(t) < 5$ km/h/s. A period with continuous instantaneous DS is identified as one DS event.

(3) LA (Long-Time Accelerating): $T_a > 5/s$. Accelerating that lasts more than 5 s is identified as one LA event.

(4) LI (Long-Time Idling): $T_i > 60/s$. Idling that lasts more than 60 s is identified as one LI event.

(5) LS (Running with Low Speed). Average speed during 60 seconds is no more than 23 km/h (see (2)), and a period with continuous LSs is identified as one LS event.

$$\text{ave}(s(t), s(t-1), \ldots, s(t-59)) \leq 23\,\text{km/h}. \tag{2}$$

(6) HS (Cruising with Higher Speed). When driving, during 5 seconds, one has the following:

(i) Average speed is no less than 60 km/h; see (3)

(ii) Instantaneous acceleration is no more than 1 km/h; see (4)

(iii) Speed standard deviation is no more than 1.5 km/h; see (5)

(iv) Speed variation is no more than 1 km/h; see (6)

(v) A period with continuous HS is identified as one HS event.

$$\text{ave}(s(t), s(t-1), s(t-2), s(t-3), s(t-4)) \geq 60\,\text{km/h} \tag{3}$$

$$\max(a(t), a(t-1), a(t-2), a(t-3), a(t-4)) \leq 1\,\text{km/h/s} \tag{4}$$

$$\text{std}(s(t), s(t-1), s(t-2), s(t-3), s(t-4)) \leq 1.5\,\text{km/h} \tag{5}$$

$$\text{abs}(s(t) - s(t-4)) \leq 1\,\text{km/h}. \tag{6}$$

(7) SM (Starting Moderately). When vehicle accelerates from idling, during 5 seconds, one has the following:

(i) Speed variation $\in [10, 20]$ km/h/s; see (7)

(ii) Product of driving mode codes equals 1 or 3; see (8)

(iii) Maximum acceleration is no more than 4 km/h/s; see (9)

(iv) A period with continuous SM is identified as one SM event.

$$10\,\text{km/h} \leq s(t) - s(t-4) \leq 20\,\text{km/h} \tag{7}$$

$$m(t) \times m(t-1) \times m(t-2) \times m(t-3) \times m(t-4) = 1 \parallel= 3 \tag{8}$$

$$\max(a(t), a(t-1), a(t-2), a(t-3), a(t-4)) \leq 4\,\text{km/h/s}. \tag{9}$$

(8) SS (Frequently Stop and Start). A vehicle idles within 3 seconds after starting from an idling. A period with continuous SS is identified as one SS event.

(9) BM (Braking Moderately). When decelerating, during 5 seconds, one has the following:

(i) Deceleration $\in [-25, -15]$ km/h/s; see (10)

(ii) Product of driving mode codes $\in (31, 49)$; see (11)

(iii) Minimum acceleration is no less than -5 km/h; see (12)

(iv) A period with continuous BM is identified one BM event.

$$-25\,\text{km/h/s} \leq s(t) - s(t-4) \leq -15\,\text{km/h/s} \tag{10}$$

$$31 < m(t) \times m(t-1) \times m(t-2) \times m(t-3) \times m(t-4) < 49 \tag{11}$$

$$\min(a(t), a(t-1), a(t-2), a(t-3), a(t-4)) \geq -5\,\text{km/h}, \tag{12}$$

where $a(t)$ is acceleration on t second, km/h/s, T_a is accelerating duration time, s, T_i is decelerating duration time, s, $s(t)$ is speed on t second, km/h, ave, std, abs, max, and min are calculating the sum, standard deviation, maximum value, and minimum value of given arguments, and $m(t)$ is driving mode code of t second, 1, 2, 3, 4 index accelerating, decelerating, cruising, and idling, respectively.

3.3. Threshold Determination

3.3.1. Acceleration and Deceleration Threshold. Acceleration threshold is used to define AS, and it is also used in the definition of SM (see (7), (9)). Besides, deceleration threshold is used in definition of DS and BM (see (10), (12)). According to models shown in Figure 3(b), the rate of mutation of instantaneous fuel consumption (RMF) between two contiguous acceleration values is used as a measurement to determine the threshold. RMF of every acceleration value is calculated as (13) and then plotted in Figure 4.

$$\text{RMF}_a = \frac{g(a) - g(a-1)}{g(a-1)}, \quad a \in [1, 20], \tag{13}$$

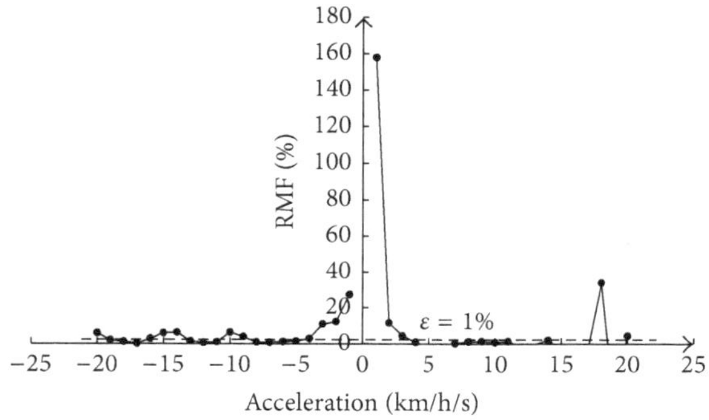

FIGURE 4: RMF of acceleration.

where RMF_a is RMF of given acceleration value and $g(a)$ is instantaneous fuel consumption when acceleration is a.

Figure 4 shows that as acceleration (deceleration) increases, RMF firstly decreases rapidly and then maintains stable. In terms of accelerating, IFC is thought of achieving maximum when RMF is less than a threshold ε. The corresponding acceleration value is determined as acceleration threshold because it is better not to accelerate with so big acceleration for its higher IFC. As for decelerating, IFC is thought of achieving minimum (nearly the same as idling) when RMF is less than a threshold ε. Obviously, drivers have not taken engine brake in this case. Taking $\varepsilon = 1\%$, acceleration and deceleration thresholds are 4 km/h/s and −5 km/h/s, respectively.

3.3.2. Low Speed Threshold and High Speed Threshold. Low speed threshold and high speed threshold are used in LS (see (2)) and HS (see (3)), respectively, to determine if a vehicle is in high-fuel-consumption state or in fuel economy state. Similarly with acceleration, RMF between two contiguous speed values is used in the determination of speed threshold. RMF of every speed values is calculated as (14) and then plotted in Figure 5.

$$RMF_s = \frac{g(s) - g(s-1)}{g(s-1)}, \quad a \in [1, 20], \tag{14}$$

where RMF_s is RMF of given speed value and $g(s)$ is instantaneous fuel consumption when speed is s.

According to Figure 5, a vehicle is thought of getting rid of low speed state when RMF is less than a threshold ε_1 and running in a high speed state when RMF is less than a threshold ε_2. Taking $\varepsilon_1 = 3\%$ and $\varepsilon_2 = 1\%$, low speed threshold and high speed threshold are 23 km/h/s and 60 km/h/s, respectively.

3.3.3. Time Threshold in LA. The period of 5 s is long enough for drivers to accelerate to a higher speed (if accelerating with 4 km/h/s, speed increases 20 km/h). Meanwhile, operation complexity of laddering acceleration every 5 s is acceptable for driver.

3.3.4. Time Threshold in LS and LI. Considering that both stopping time at intersections and passengers getting on/off time in Beijing are generally less than 60 s, thresholds used in LS and LI are set to 60 s to avoid misjudging these two cases. Moreover, in terms of LI, Union of Concerned Scientists [25] also suggests that idling time is better not beyond 60 s.

3.3.5. Time Threshold in HS, SM, and BM. For HS, it seems difficult for drivers to cruise with stable speed. For SM and BM, the period of 5 s is long enough for drivers to accelerate/decelerate to a higher/lower speed (if accelerating/decelerating with 4/-5 km/h/s, speed increases/decreases by 20/25 km/h).

3.3.6. Product of Driving Mode Codes Threshold SM and BM. Ranges of product of driving mode codes used in SM (see (8)) and BM (see (11)) are calculated to ensure that the whole driving events process is an acceleration or deceleration process (allowing temporal cruising). Enumeration method is used to determine the range, as shown in Table 1.

3.3.7. Speed Standard Deviation Threshold in HS. Speed standard deviation used in LS (see (5)) is used to ensure that vehicle speed oscillation is less than 1 km/h. Similar to product of driving mode codes, threshold 1.5 of speed standard deviation is determined by enumeration method as well.

4. Eco-Driving Behavior Evaluation Modeling and Advices Providing

307 drivers were randomly divided into five groups. Four of the five groups ($61 * 4 = 244$ samples) were used to establish the model (training set). The remaining group (63 samples) was used to verify the model (test set). As preparations for modeling, two tasks were completed at first: (1) identifications of nine driving events of every samples and (2) fuel-related hundred-mark system establishment.

4.1. Nine Driving Events Identification of Every Sample and Description. According to the definition listed above, driving events were identified and the average frequency every 50 km of nine driving events was calculated to normalize the travel distance. In the data collection day, average frequency of driving events every 50 km was shown in Table 2. Meanwhile, average frequency of driving events every 1 km was also shown to have a better view.

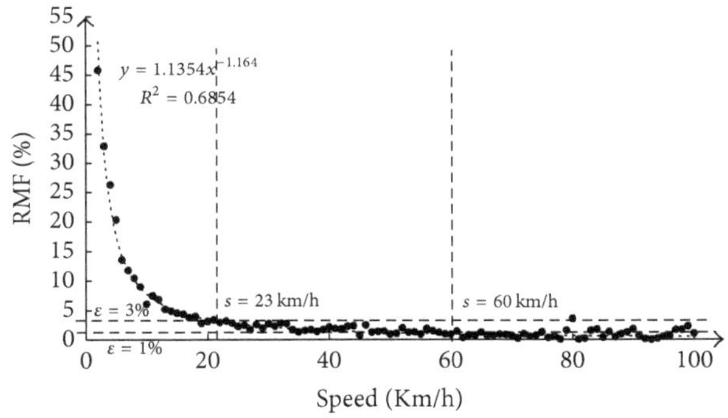

FIGURE 5: RMF of speed.

TABLE 1: Enumeration of the product of driving mode codes in SM and BM.

Cases	Driving mode code on every second					Product	Remarks
	1 s	2 s	3 s	4 s	5 s		
				SM			
Case 1	*1*	*1*	*1*	*1*	*1*	*1*	*Accelerating continuously*
Case 2	1	1	1	1	2	2	Having decelerating
Case 3	*1*	*1*	*1*	*1*	*3*	*3*	*Accelerating with temporal cruising*
Case 4	1	1	1	1	4	4	Having idling
Case 5	1	1	1	3	3	9	Too much cruising
				BM			
Case 1	2	2	2	2	1	16	Having accelerating
Case 2	*2*	*2*	*2*	*2*	*2*	*32*	*Decelerating continuously*
Case 3	*2*	*2*	*2*	*2*	*3*	*48*	*Decelerating with temporal cruising*
Case 4	2	2	2	3	3	72	Too much cruising

Note. (1) Italicized cases stand for the cases that could be identified as an SM process or BM process; (2) the order of driving mode codes is not necessarily ranked as list in table. For example, 2-2-3-2-2 is also identified as case 3 (BM); (3) driving mode code 1, 2, 3, 4 index accelerating, decelerating, cruising, and idling, respectively.

TABLE 2: Average frequency of driving events every 50 km and 1 km.

Driving events	AS	DS	LA	LI	LS	HS	SM	SS	BM
Frequency/50 km	122.4	88.8	136.4	8.1	28.1	34.0	53.8%	1.2	32.9%
Frequency/km	2.44	1.77	2.72	0.16	0.56	0.68	1.08%	0.02	0.65%

It should be noted that unit of SM and BM was the ratio of SM/All-Staring and BM/All-Braking instead of the frequency of SM and BM. The reason was that the more the SM or BM occurred, the better the fuel economy in general is. But in fact, more SM and BM were always accompanied with more starting and braking processes, which would increase fuel consumption compared with cruising. Thus, the ratio of SM/All-Staring and BM/All-Braking might be better as indicators.

Table 2 indicates that during a certain travel distance, drivers are prone to AS and LA (nearly 2.5 times per kilometer), while being less to SS and LI. LS occurs about every 2 km on average, which may be caused by the close distance between intersections and frequent traffic jams in Beijing.

4.2. Fuel-Related Hundred-Mark System. The evaluation of driver's eco-driving behavior was aimed at helping drivers to promote eco-driving behavior. Hence, a simple and easy-to-understand indicator was needed. In this paper, a hundred-mark system was used to measure driver's eco-driving behavior. Scores of every driver was calculated by the following linear equation:

$$\text{SCORE}_a = 40 + \left(1 - \frac{(\text{FPH} - 6)}{14 - 6}\right) \times 60, \qquad \text{FPH} \in [6, 14], \tag{15}$$

where SCORE_a is scores of every driver and FPH is FPH of every driver, and it is calculated using (1).

TABLE 3: Correlation between arguments.

Driving events	1	2	3	4	5	6	7	8	9
(1)	1	.905**	.680**	.607**	.769**	−.566**	−.010	.271**	.410**
(2)	.905**	1	.747**	.586**	.745**	−.550**	.007	.315**	.565**
(3)	.680**	.747**	1	.684**	.854**	−.736**	.040	.312**	.238**
(4)	.607**	.586**	.684**	1	.787**	−.613**	.013	.366**	.037
(5)	.769**	.745**	.854**	.787**	1	−.775**	.009	.326**	.191**
(6)	−.566**	−.550**	−.736**	−.613**	−.775**	1	.006	−.279**	.018
(7)	−.010	.007	.040	.013	.009	.006	1	−.198**	.199**
(8)	.271**	.315**	.312**	.366**	.326**	−.279**	−.198**	1	.002
(9)	.410**	.565**	.238**	.037	.191**	.018	.199**	.002	1

Note. (a) **Correlation is significant at the 0.01 level; (b) (1)–(9) represent AS, DS, LA, LI, LS, HS, SM, SS, and BM, respectively.

TABLE 4: Results of PCA.

Component	Total	% of variance	Cumulative%
F1	4.793	53.258	53.258
F2	1.377	15.302	68.560
F3	1.059	11.763	80.324
F4	.715	7.940	88.264
F5	.391	4.349	92.613
F7	.303	3.369	95.982
F8	.190	2.110	98.092
F9	.118	1.310	99.402

Obviously, the higher the fuel consumption, the lower the scores. Driver's minimum score was 40 and maximum was 100 corresponding to the maximum fuel consumption of 14 L/100 km and the minimum of 6 L/100 km. When fuel consumption was lower than 6 L/100 km, score was 100. When fuel consumption was higher than 14 L/100 km, score was 40.

4.3. Establishment of Eco-Driving Behavior Evaluation Model. Taking numbers of every driving events occurred during a certain distance as independent variables and vehicle fuel consumption as dependent variable, driver's eco-driving behavior evaluation model was established. Multiple linear regression was chosen as modeling method due to the fact that change of the arguments (i.e., the change of driver behavior to better or worse) could directly influence evaluation score. In this case, the reason why driver had high fuel cost could be intuitively provided, as well as advices that which vehicle manipulation should be improved or eliminated.

When modeling, strong collinearity between arguments was observed, as shown in Table 3.

In this case, directly using linear regression might cause model distortion. Therefore, principal component analysis (PCA) was used to eliminate collinearity.

First, data of 244 samples were normalized using Z standardization, as follows:

$$x^* = \frac{x - \mu}{\sigma}, \tag{16}$$

where x^* is normalized value, x is raw value, μ is mean, and σ is standard deviation.

Second, using PCA, the result was shown in Table 4.

The first three principal components were extracted since their Eigen values were greater than 1. The expressions of every principal components were listed as follows:

$$\begin{aligned} F_1 = {} & 0.181N_{\mathrm{AS}} + 0.184N_{\mathrm{DS}} + 0.187N_{\mathrm{LA}} + 0.167N_{\mathrm{LI}} \\ & + 0.195N_{\mathrm{LS}} - 0.172N_{\mathrm{HS}} - 0.01N_{\mathrm{SM}} \\ & + 0.088N_{\mathrm{SS}} + 0.06N_{\mathrm{BM}} \end{aligned} \tag{17}$$

$$\begin{aligned} F_2 = {} & 0.155N_{\mathrm{AS}} + 0.227N_{\mathrm{DS}} - 0.038N_{\mathrm{LA}} - 0.175N_{\mathrm{LI}} \\ & - 0.068N_{\mathrm{LS}} + 0.138N_{\mathrm{HS}} + 0.375N_{\mathrm{SM}} \\ & - 0.322N_{\mathrm{SS}} + 0.592N_{\mathrm{BM}} \end{aligned} \tag{18}$$

$$\begin{aligned} F_3 = {} & 0.145N_{\mathrm{AS}} + 0.217N_{\mathrm{DS}} - 0.129N_{\mathrm{LA}} - 0.183N_{\mathrm{LI}} \\ & - 0.136N_{\mathrm{LS}} + 0.270N_{\mathrm{HS}} - 0.681N_{\mathrm{SM}} \\ & + 0.367N_{\mathrm{SS}} + 0.369N_{\mathrm{BM}}, \end{aligned} \tag{19}$$

where F_1 is the first principal component, F_2 is the second principal component, F_3 is the third principal component, N_{AS} is number of AS occurring every 50 km, N_{DS} is number of DS occurring every 50 km, N_{LA} is number of LA occurring every 50 km, N_{LI} is number of LI occurring every 50 km, N_{LS} is number of LS occurring every 50 km, N_{HS} is number of HS occurring every 50 km, N_{SM} is number of SM occurring every 50 km, N_{SS} is number of SS occurring every 50 km, and N_{BM} is number of BM occurring every 50 km.

Third, taking three principal components as arguments and vehicle fuel consumption as depend variable, multiple

Table 5: Influence on score when driving event occurs once (every 50 km).

Driving events	AS	DS	LA	LI	LS	HS	SM	SS	BM
Coefficients	−0.03	−0.04	−0.05	−0.44	−0.18	0.07	0.02	−1.05	0.11
Rank	8	7	6	2	3	5	9	1	4

linear regression was used to establish driver's eco-driving behavior evaluation model. The model was expressed as follows:

$$\begin{aligned} \mathrm{SCORE}_M &= 80.549 - 6.927F_1 + 1.455F_2 + 0.565F_3, \\ R^2 &= 0.823, \end{aligned} \tag{20}$$

where SCORE_M is model score of every driver.

In this model, the relationship between the three principal components and score was presented. Meanwhile, the principal components are the linear combination of nine arguments (i.e., the frequency of nine driving events occurs every 50 km). Thereby the score could be expressed by nine arguments also with linear form. Coefficients of every argument were shown in Table 5. The coefficients could also indicate how score will change when driver performs one driving event.

Table 5 suggests that, in every 50 km, once SS and LI occur, scores decrease by 1.05 and 0.44, respectively, which indicates that the occurrences of these two driving events will result in obvious increase of fuel consumption. Thus, drivers should pay more attention to avoid SS and LI to achieve higher fuel efficiency.

4.4. Model Validation. The remaining group (63 samples) was used to verify the eco-driving behavior evaluation model. Vehicle fuel consumption during one day was calculated using vehicle operation data and then converted to scores. Nine driving events were also identified and their frequencies (every 50 km) were counted. Substituting these nine arguments into the evaluation model, model scores were obtained.

Relative error was used to measure the model accuracy.

$$E_i = \frac{\left|\mathrm{SCORE}_M - \mathrm{SCORE}_A\right|}{\mathrm{SCORE}_A}. \tag{21}$$

E_i indicates the relative error between fuel consumption-calculating score and model score of ith driver. Taking average relative error as measurement, validation results showed that model error was 3.28%; that is, model accuracy was 96.72%.

Relationship between fuel consumption-calculating score and model score was plotted in Figure 6.

4.5. Driving Advices. Based on the definition of the nine driving events, nine eco-driving advices described as several processes were proposed as follows.

(1) AS. Avoid sharp acceleration. When accelerating, it is better to accelerate to 15 km/h within 4 s.

(2) DS. Avoid sharp deceleration. When decelerating, it is better to decelerate to 20 km/h with 4 s.

(3) LA. Avoid long-time acceleration. Take laddering accelerating operation to have vehicle instantaneously cruising or decelerating within accelerating process. It is better not to continuously accelerate beyond 5 s.

(4) LI. Avoid long-time idling. Anticipating traffic condition and signal phase in intersection, drivers should switch off engine when predicting stopping more than 60 s.

(5) LS. Make travel planning and try to bypass jam roads.

(6) HS. Try to maintain higher speed but not over speed.

(7) SM. When starting, it is better to accelerate to 20 km/h within 5 s.

(8) SS. Under congestion condition, it is better to start when the space from the front car is more than half a car to avoid frequently stopping and starting.

(9) BM. When decelerating, it is better to decelerate to 15–20 km/h within 5 s.

5. Discussion

In our research, the database used in the definition of driving events and establishment of evaluation model is collected in only one day. It seems that the dataset might not be large enough, but the longer average operation time of taxicabs (10.3 hours) is able to cover various traffic conditions and road types.

In some previous studies, external circumstances such as road slope, weather condition, traffic condition, and vehicle load have influences on car's fuel consumption. However, in this paper, these factors are not taken into consideration for the following reasons:

(1) The information is difficult to collect and the integration between different data sources is complex. Thus, the application of paper results may be limited.

(2) The general improvement of driver' manipulation skills is mainly focused on, no matter how the external conditions change. For example, the consideration of traffic condition before travels can help drivers to avoid traffic congestions and yield a lower energy usage. Nevertheless, if drivers engage in slow traffic flow, eco-driving manipulation still can save fuel compared with normal driving. Similarly, weather condition can influence vehicle fuel consumption, while the overall performance of driving manipulation for

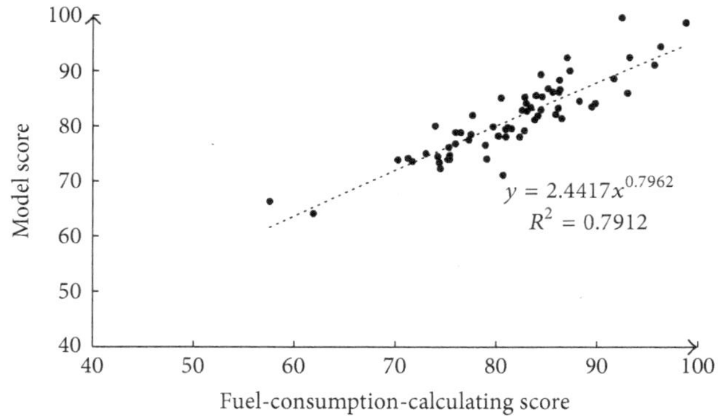

FIGURE 6: Relationship between fuel consumption-calculating score and model score.

specific driver is steady. In other words, if vehicle operating parameters (speed, acceleration, and driving mode duration), which can be regarded as the results of driver manipulation, are in line with eco-driving advices, fuel consumption will be lower than that when driving normally under the same external circumstance.

In Section 2, data with vehicle speed more than 100 km/h (illegal drive) are ignored. One may doubt that this treatment may influence modeling. However, this paper mainly deals with the eco-driving under normal driving conditions. Moreover, data that speed is more than 100 km/h only occupy 1% in overall database. Thus the ignorance is irrelevant for the modeling and result.

According to the model coefficients, it appears that AS, DS, LA, LI, LS, and SS (bad manipulations) will result in high fuel consumption for their negative coefficients in the model expression. On the contrary, HS, SM, and BM (good manipulations) will improve the evaluation scores for their positive coefficients in evaluation model. This phenomenon is in accordance with previous literature and common sense, which verifies the methodology of this paper from other sides.

In Section 3, consecutive same driving events are regarded as one driving event; that is, the frequency is taken into the model but the duration time of every event is ignored. This is because of the fact that the goal of this research is to inform drivers that how many times they take bad/good manipulations, and then they are expected to avoid/take more these manipulations. In other words, numbers of these events are advices to decrease instead of the duration time of every event.

The limitation of this research is about the thresholds used when defining driving events. According to data observation and driving experience, a series of parameter thresholds used in evaluation model are proposed and they work well in the modeling process. Nevertheless, these thresholds could not be assured to be optimal. On the other hand, the robustness of this model was not fully analyzed for the data limitation. But as a method for eco-driving behavior evaluation, the model could be applied to the practice after calibration and optimization with new data.

The foundation of this research is based on the taxicabs' operation data. It will be beneficial to promote these research findings to other types of vehicles. What is more, the nine driving events could be identified by only speed parameter. So when practically applying, cellphone could be used as both data source and feedback media after further validation.

Furthermore, driving-events-based evaluation model could only help drivers improve their driving behavior in the nine aspects. In future work, machine learning could be introduced into the evaluation modeling to explore hidden factors that influence fuel consumption and evaluate driver's eco-driving behavior more accurately. A fuel consumption model that fits the fuel consumption characteristics of Chinese vehicles will be developed too.

6. Conclusion

This paper establishes a driver eco-driving behavior evaluation model based on vehicle operating and fuel consumption data. First, the influence of three vehicle operating parameters including speed, acceleration, and driving mode duration on fuel consumption is analyzed. Then, nine driving events that are fuel consumption related are proposed. Using PCA and multiple linear regression, an eco-driving behavior evaluation model is established and nine corresponding eco-driving advices are provided.

The establishment of this model provides a quantitative and intuitive way to evaluate the effect of drivers' behavior on fuel consumption and several procedural-driving-events-based eco-driving advices. The outcomes of this paper could be applied directly to the development of eco-driving support device for its simple but accurate algorithms, laying the foundation for the eco-driving promotion in China.

Notation

FPH: Fuel consumption per hundred kilometers (L/100 km)
IFC: Instantaneous fuel consumption (L/h)
AS: Accelerating Sharply (driving event)
DS: Decelerating Sharply (driving event)
LA: Long-Time Accelerating (driving event)
LI: Long-Time Idling (driving event)
LS: Running with Low Speed (driving event)
HS: Cruising with Higher Speed (driving event)

SM: Starting Moderately (driving event)
SS: Frequently Stop and Start (driving event)
BM: Braking Moderately (driving event)
RMF: Mutation of instantaneous fuel consumption (%)
PCA: Principal component analysis (algorithm).

Conflicts of Interest

The authors declare that they have no conflicts of interest.

Acknowledgments

The authors would like to acknowledge the project of National Natural Science Foundation of China no. 61672067 for supporting this research.

References

[1] China Statistical Yearbook, "National Bureau of Statistics of the People's Republic of China," 2016, http://www.stats.gov.cn/tjsj/ndsj/2015/indexch.htm.

[2] Y. B. Cui, "Study of automobile emissions saving countermeasures," *Science & Technology Vision*, vol. 4, p. 190, 2014.

[3] I. De Vlieger, "On-board emission and fuel consumption measurement campaign on petrol-driven passenger cars," *Atmospheric Environment*, vol. 31, no. 22, pp. 3753–3761, 1997.

[4] J. Van Mierlo, G. Maggetto, E. Van De Burgwal, and R. Gense, "Driving style and traffic measures - Influence on vehicle emissions and fuel consumption," *Proceedings of the Institution of Mechanical Engineers, Part D: Journal of Automobile Engineering*, vol. 218, no. 1, pp. 43–50, 2004.

[5] Pham and D. Nam et al., Fueoogle: a participatory sensing fuel-efficient maps application, 2009.

[6] M. Barth, K. Boriboonsomsin, and A. Vu, "Environmentally-Friendly navigation," in *Proceedings of the 10th International IEEE Conference on Intelligent Transportation Systems, ITSC 2007*, pp. 684–689, October 2007.

[7] K. Boriboonsomsin, M. J. Barth, W. Zhu, and A. Vu, "Eco-routing navigation system based on multisource historical and real-time traffic information," *IEEE Transactions on Intelligent Transportation Systems*, vol. 13, no. 4, pp. 1694–1704, 2012.

[8] E. Yao and Y. Song, "Study on eco-route planning algorithm and environmental impact assessment," *Journal of Intelligent Transportation Systems: Technology, Planning, and Operations*, vol. 17, no. 1, pp. 42–53, 2013.

[9] M. Barth, S. Mandava, K. Boriboonsomsin, and H. Xia, "Dynamic ECO-driving for arterial corridors," in *Proceedings of the IEEE Forum on Integrated and Sustainable Transportation System (FISTS '11)*, pp. 182–188, IEEE, Vienna, Austria, July 2011.

[10] S. Mandava, K. Boriboonsomsin, and M. Barth, "Arterial velocity planning based on traffic signal information under light traffic conditions. intelligent transportation systems," in *Proceedings of the 2009 12th International IEEE Conference on Intelligent Transportation Systems*, pp. 160–165, October 2009.

[11] Z. Chen, *An Optimization Model for Eco-Driving at Signalized Intersection*, AM University, 2013.

[12] K. Boriboonsomsin, A. Vu, and M. Barth, *Eco-driving: pilot evaluation of driving behavior changes among us drivers*, University of California Transportation Center, 2010.

[13] R. Frank, G. Castignani, R. Schmitz, and T. Engel, "A novel eco-driving application to reduce energy consumption of electric vehicles," in *Proceedings of the 2013 2nd IEEE International Conference on Connected Vehicles and Expo, ICCVE 2013*, pp. 283–288, IEEE, December 2013.

[14] R. Araújo, Â. Igreja, R. De Castro, and R. E. Araújo, "Driving coach: a smartphone application to evaluate driving efficient patterns," in *Proceedings of the 2012 IEEE Intelligent Vehicles Symposium, IV 2012*, pp. 1005–1010, IEEE, Alcala de Henares, Spain, June 2012.

[15] Y. Wu, X. Zhao, J. Rong, Y. Liu, and L. Xu, "Potential of Eco-driving in Reducing Fuel Consumption and Emissions Based on a Driving Simulator," *Journal of Beijing University of Technology*, vol. 41, pp. 1211–1218, 2015.

[16] Y. Wu, X. Zhao, C. Chen, J. Rong, and Y. Zhang, "Development of an internet + based eco-driving support platform and its application: a case study in beijing taxicabs," Transportation Research Board 17-05681, 2017.

[17] J. C. Ferreira, J. De Almeida, and A. R. Da Silva, "The impact of driving styles on fuel consumption: a data-warehouse-and-data-mining-based discovery process," *IEEE Transactions on Intelligent Transportation Systems*, vol. 16, no. 5, pp. 2653–2662, 2015.

[18] D. Vangi and A. Virga, "Evaluation of energy-saving driving styles for bus drivers," *Proceedings of the Institution of Mechanical Engineers, Part D: Journal of Automobile Engineering*, vol. 217, no. 4, pp. 299–305, 2003.

[19] G. Kedar-Dongarkar and D. Manohar, "Driver classification for optimization of energy usage in a vehicle," *Procedia Computer Science*, vol. 8, pp. 388–393, 2012.

[20] Z. Yunming, *Experimental and Simulation Investigation and Operating Characteristic of a Hybrid Hydraulic Passenger Car with an Output Coupled Power-Split Transmission*, University of Technology, Beijing , China, 2014.

[21] GB 7528-2012, "Safety specifications for power-driven vehicles operating on roads".

[22] K. Ahn, H. Rakha, A. Trani, and M. van Aerde, "Estimating vehicle fuel consumption and emissions based on instantaneous speed and acceleration levels," *Journal of Transportation Engineering*, vol. 128, no. 2, pp. 182–190, 2002.

[23] M. Barth and K. Boriboonsomsin, "Traffic congestion and greenhouse gases," *ACCESS Magazine*, vol. 1, 2009.

[24] K. Jakobsen, S. C. H. Mouritsen, and K. Torp, "Evaluating eco-driving advice using GPS/CANBus data," in *Proceedings of the 21st ACM SIGSPATIAL International Conference on Advances in Geographic Information Systems, ACM SIGSPATIAL GIS 2013*, pp. 44–53, November 2013.

[25] Union of Concerned Scientists, "How to Maximize Your Vehicle's Fuel Economy," 2016, http://www.ucsusa.org/clean-vehicles/fuel-efficiency/how-to-maximize-fuel-economy#.VuEQ8_19671.

19

Analysis on Topological Properties of Dalian Hazardous Materials Road Transportation Network

Pengyun Chong,[1] Bin Shuai,[1] Shaowei Deng,[1] Jianting Yang,[2] and Hui Yin[3]

[1] *School of Transportation and Logistics, Southwest Jiaotong University, Chengdu 610031, China*
[2] *Faculty of Geosciences and Environmental Engineering, Southwest Jiaotong University, Chengdu 610031, China*
[3] *Kunming Hydroelectric Investigation Design and Research Institute, China Hydropower Engineering Consulting Group Corporation, Kunming 650051, China*

Correspondence should be addressed to Bin Shuai; bsh67@126.com

Academic Editor: Hong Chen

To analyze the topological properties of hazardous materials road transportation network (HMRTN), this paper proposed two different ways to construct the cyberspace of HMRTN and constructed their complex network models, respectively. One was the physical network model of HMRTN based on the primal approach and the other was the service network model of HMRTN based on neighboring nodes. The two complex network models were built by using the case of Dalian HMRTN. The physical network model contained 154 nodes and 238 edges, and the statistical analysis results showed that (1) the cumulative node degree of physical network was subjected to exponential distribution, showing the network properties of random network and that (2) the HMRTN had small characteristic path length and large network clustering coefficient, which was a typical small-world network. The service network model contained 569 nodes and 1318 edges, and the statistical analysis results showed that (1) the cumulative node degree of service network was subjected to power-law distribution, showing the network properties of scale-free network and that (2) the relationship between nodes strength and their descending order ordinal and the relationship between nodes strength and cumulative nodes strength were both subjected to power-law distribution, also showing the network properties of scale-free network.

1. Introduction

Complexity science is the kind of science of the 21st century. Complex network has developed rapidly in the past few years as an important part of the complexity science [1]. Its main idea is to convert the various elements within the research system into nodes of the network, and the relationship between the elements into the edges to form a network in order to describe the relationship among the various elements within the research system. Exploring the essential properties of the real system through the studies of the topological properties of the network structure is an effective method to research complexity science.

Mathematical theory basis of complex network theory is graph theory. Euler's study on "seven-bridge problem" of Gurney Myers in 1736 was the beginning of the studies into complex network theory. Since then people abstracted a number of complex systems into networks to describe and research from the perspective of the network topology, such as the Internet network [2], the food chain network [3], neural networks [4], and the citation network [5]. Milgram first proposed small-world phenomenon of network in 1967 in the analysis of network structure [6]. Based on this, Watts and Strogatz proposed WS small-world network in the study of small-world phenomenon generation process mechanisms and presented the network generation algorithm for the first time in 1998 [7]. Barabási and Albert found the scale-free properties of real-world networks in 1999, which called BA model, and presented the network generation algorithm [8]. The research results of Watts and Barabási et al. subverted people's awareness of the traditional network, causing many scholars in the field of physics, economics, and computer communications to focus on complex network [9–11].

The normal operation of traffic and transportation system is an important prerequisite for the steady and rapid development of the national economic, and the complex network theory has attracted wide attention of scholars in transportation field. In current studies, scholars research the topological properties of the transportation network mainly by means of graph theory and network topology theory and so on. Transportation network has growth and irregularity; therefore, these theories can describe networks of simple structure well, but not competent for complex transportation network. The research results of application of complex network theory to the transportation network topological properties are mainly concentrated on the railway network [12–14], airport network [15–20], and urban traffic network [10, 21].

In the studies on topological properties of railway network, Sen et al. [12] studied the Indian railway network in the P space, which showed that it had a shorter average path length and large clustering coefficient, presenting small-world network properties. Li and Cai [13] researched China railway passenger transportation network, and the results showed that the physical railway network had the tree structure and the traffic flow network was a typical small-world network. Kurant and Thiran [14] made an empirical study on the three kinds of space forms P, R, and L of European and Swiss railway networks and analyzed the similarities and differences among the railway network topological properties of different space forms.

In the studies on topological properties of airport network, Guimerà et al. [15, 16] researched global airport networks, and results showed that the airline network was a small-world network, with average shortest path length of 4.4 and the network average clustering coefficient of 0.62. Chi et al. [17] constructed a directed and weighted airport network of USA, with similar research results to Guimerà et al, that is, American airport network also presenting small-world properties, with average shortest path length of 2.4 and the network average clustering coefficient of 0.618. Other scholars have conducted similar studies, such as Bagler [18] on India airport network, Guida and Maria [19] on Italy airport network, and Li and Cai [20] on Chinese airport network. Their results showed that the airport network was small-world network, and the network average clustering coefficient was similar. The only difference was that the average shortest path length of global airport network was greater than that of each national airport networks.

In the studies on topological properties of urban traffic network, Latora and Marchiori [10] researched the complex network properties of the Boston subway network by constructing an undirected and weighted subway network, and results showed that the Boston subway network presented small-world properties. Jiang and Claramunt [21] researched urban road network, and results showed that it presented small-world network properties.

In general, the studies on complex network are mainly focused on the following areas: (1) to simulate the geometrical statistics (such as the number of nodes, the number of edges, the network average node degree) of real world by using generation algorithm of random network, scale-free network, and small-world network [7, 8]; (2) to research dynamic problems of network topology structure [22]; (3) to observe the change of its properties after the implementation of different operation on the network, that is, to research network survivability at different attack strategies [23]. Studies of applying complex network theory to researching the transportation network can be divided roughly into two categories: (1) to research the topological properties of the transportation network, which mostly concentrated on the railway network [12–14], airport network [15–20], urban rail transit network [10], and urban road network [21]; (2) to study how to improve the reliability and survivability of the network under a deliberate or random attack, optimizing network structure when the topological properties of the network are known [24].

Compared with the common freight transportation, hazardous materials transportation objects are flammable, explosive, corrosive, and radioactive such as gasoline, explosive, strong acid, and peroxide. So once an accident happens during the transport process, it will cause serious damage to the environment, the surrounding residents, and the normal operation of economic activity. At present, China railway, airport, and shipping are not perfect in hazardous materials transportation infrastructure and management mechanism, so road transportation is the main way of the hazardous materials transportation. According to statistics, road transports cover about 80% of our hazardous materials [25]. In 2009, China's annual hazardous materials transport volume is more than 4 million tons, and in the first half of 2013, the total increase in main chemicals is about 6.3%, and more than 95% of the hazardous materials relate to offsite transportation problem [25], so hazardous materials transportation in our country presents properties of having a large volume, a high growth rate, a single mode of transportation, a low safety management level, and so on. Due to the increasing transport volume of hazardous materials, both road mileage and density of road network of hazardous materials transportation network increase significantly, and also the integration and the complexity of the network are more obvious.

Compared to other transportation networks, hazardous materials road transportation network (HMRTN) is a special kind of transit network, which mainly reflected on the following points. (1) Different network planning concept: for common freight transport, the general intention of road network planning is mainly reducing running costs or shortening the distance between two places. Because of the particularity of hazardous materials, transportation risk must be minimized. So the basic principle of HMRTN planning is to minimize risk, which also leads to differences between that and the other road transportation network. (2) Different network system composing elements: edges and nodes of HMRTN should be far away from area of high network density, such as population, schools, water, bridges, and government agencies, which leads to its network structure different from other transportation network. (3) Different network scale: edges and nodes of HMRTN are composed of the city's main roads and intersections, which is a part of urban road network but different from it. Considering the several factors above, HMRTN must present different topological properties from the other transportation network. When hazardous materials transportation vehicles encounter random attacks (such as traffic accidents) or deliberate attacks (like terrorist attacks) in the transport process, transportation network function will be impaired. In particular, when the accident locates at the network node or edge which is the most "fragile," it may paralyze the network. Therefore, applying complex network theory to researching topological properties of HMRTN helps to improve the understanding of the properties of network and reduce the damage of the unknown risks to network function.

Based on the above analysis, this paper researches the topological properties of HMRTN. On the basis of constructing hazardous materials road transportation physical network and service network structure models, and analyzing the interaction between them, this paper constructs the physical network model of HMRTN based on the primal approach and the service network model of HMRTN based on the neighboring nodes by applying complex network theory and puts forward the research process of network topological properties. The method is applied to the Dalian HMRTN. This paper researches the complex network properties of HMRTN through the statistical average shortest path length, the network diameter, average network clustering coefficient, node degree, strength, the strength parameters, and so on and provides scientific basis for the planning and optimization of HMRTN.

2. Methodology of Researching Topological Properties of HMRTN

In order to get the network topological properties, the following questions need to be focused on for the research method of hazardous materials road transportation network properties. (1) What is HMRTN? (2) What factors could be used to judge the topological properties of a network? (3) How to judge the type of a complex network? (4) How to construct a HMRTN based on complex network theory? (5) What is the research process of network topological properties? In view of the above questions, the main methods of this paper are explained through the following aspects.

2.1. The Structure Model of HMRTN. HMRTN system is composed of nodes of different nature, the road network of connection among the nodes, and the OD pairs' information with distribution tasks. Therefore, HMRTN includes physical network and service network, and it has the double attributes. Based on this, this paper constructs its network models, respectively.

(1) Physical Network Space. The network $G = (V, E)$ in urban road network represents hazardous materials road transportation physical network. V is a node set of network G that represents the places of actual meaning like intersections, and so forth. If $|V| = m_a$, say network G has m_a nodes, $V = \{v_1, v_2, v_3, \ldots, v_{m_a}\}$; E is the edge set of network G, which represents the connection path between two nodes, including expressway, national highway, and the urban road.

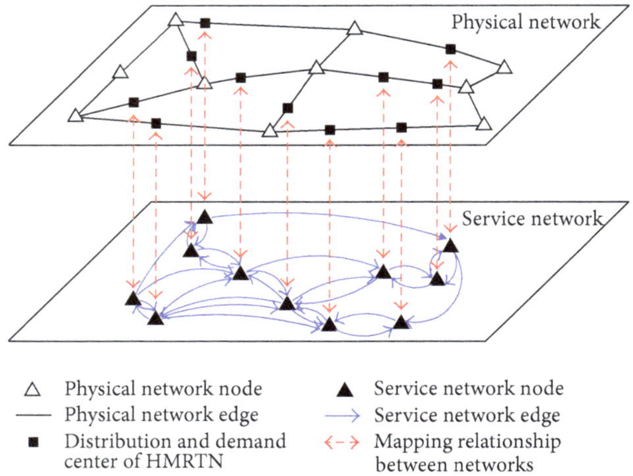

FIGURE 1: The spatial view of physical network/service network of HMRTN.

If $|E| = n_a$, representing n_a edges in network G, $E = \{e_1, e_2, e_3, \ldots, e_{n_a}\}$; W are the edge weights of network G, which represents a set of path lengths between any two nodes; that is, $W = \{w_1, w_2, w_3, \ldots, w_{n_a}\}$. Therefore, the physical network constructed in this paper is an undirected, weighted, and connected network.

(2) Service Network Space. In hazardous materials road transportation physical network $G = (V, E)$, where $s_{i,j}$ is the minimum risk path between any two nodes v_i and v_j, hazardous materials road transportation vehicles complete the distribution task along the minimum risk path. In the distribution process, transportation vehicles passing distribution center, demand center, and distribution route compose another huge service network $G' = (V', E')$, where V' is the places of actual meaning like hazardous materials distribution center, demand center, and so forth. If $|V'| = m_b$, which represents m_b nodes in network G', $V' = \{v_1, v_2, v_3, \ldots, v_{m_b}\}$. E' is the network edge set formed by distribution route passing two neighboring nodes. Because the distribution is divided into upstream and downstream, the edges in network G' are directed edges. If $|\vec{E}'| = n_b$, which represents n_b edges in network G', $\vec{E}' = \{e_{1,2}, e_{2,1}, e_{3,4}, \ldots, e_{i,j}\}$, $i, j \in m_b$, $i \neq j$. Therefore, the service network in this paper is a directed, weighted, and connected network.

HMRTN model and spatial interaction are shown in Figure 1.

2.2. Methodology of Establishing the Complex Network Model of HMRTN. Based on the above analysis, the physical network and service network of hazardous materials road transportation complex network models are constructed as follows.

2.2.1. The Physical Network Model of HMRTN Based on Primal Approach. To ensure abstract network reflecting the spatial relationship between the network topological properties and road intersection utmost well, in this paper, a hazardous materials road transportation physical network model based on the primal approach is constructed. The primal approach refers to the road path abstracted to network edges and the intersections to network nodes [26]. And the following basic definitions and assumptions are made to the network topology.

(1) Do not consider the properties and ranking differences of HMRTN edges and nodes.

(2) Do not consider the direction of the road, where the HMRTN is an undirected graph.

(3) Regard the intersections as network nodes and the paths between neighboring nodes as network edges.

(4) Regard the distance between two nodes as the network edge weight.

According to the above analysis, hazardous materials road transportation complex network model based on the primal approach has the following properties: (1) network model comes from the real world, so the network structure is subjected to real network. (2) Nodes in network model have clear spatial location and coordinate. (3) The number of edges that network nodes can connect is influenced by road condition and the surrounding environment, which influences the distribution of node degrees.

2.2.2. The Service Network Model of HMRTN Based on Neighboring Nodes. Service network is a complex network consisting of a large number of OD pairs intersecting with each other, which reflects the relationships among different nodes and properties of HMRTN well. In this paper, a hazardous materials complex network model based on the neighboring nodes is constructed, and the following basic definitions and assumptions are made to the network topology.

(1) Regard the places of actual meaning like hazardous materials distribution center, demand center, and so forth as network nodes and distribution routes between neighboring nodes as the network edges.

(2) Regard the number of transportation vehicles passing the edge as network edge weight.

(3) Vehicle distribution go-return routes affect the network edge weight. For example, for a path from the distribution center to the demand center passing neighboring nodes v_m and v_n, there is an edge $v_m \rightarrow v_n$ between them; in the same way, when returning from the demand center to the distribution center, there is another edge $v_n \rightarrow v_m$ passing neighboring nodes v_m and v_n.

(4) Do not consider the distribution volume of transportation vehicles, that is, no great difference in transport capacity of the transportation routes, and neighboring nodes have the same network edge weights.

(5) The tour-route distribution mode is used between the distribution center and demand center, and distribution center fully meets the demand of the transportation needs of demand center.

The properties of hazardous materials complex network model based on the neighboring nodes are as follows: (1) network nodes are a part of the physical network; they have accurate geographical location and spatial information; (2) network structure is subjected to physical network; (3) network node degree is the characterization of the service network, and the change of distribution task will directly affect the network degree distribution; (4) risk value, cost, and time need to be considered when choosing network distribution route, which has a direct influence on its complex network properties.

2.3. Methodology. Basic information for the study of HMRTN topological properties is network planning map and land-use map, but it cannot be directly used in calculation and analysis. Therefore, further calculation and processing are needed, and the specific steps are as follows.

Step 1. Complete the data processing using Arc-GIS software according to the city planning map and land-use map, and get the structure model of HMRTN.

Step 2. Construct topological structure of distribution network based on transportation network structure model according to hazardous materials distribution route data.

Step 3. Find the network topological adjacency matrix according to the distribution network topology data using the construction method of hazardous materials complex network model based on neighboring nodes.

Step 4. Calculate the network average node degree, average clustering coefficient, and degree distribution using topological adjacency matrix in the Matlab R2008a.

Step 5. Judge the complex network type and analyze the network topological properties.

3. The Eigenvalues Calculation and Topological Properties Analysis of Complex Network of Hazardous Materials Road Transportation

3.1. The Physical Network of HMRTN

3.1.1. The Basic Data of Physical Network of Dalian HMRTN. The Dalian network planning data in December 2013 is used in this paper, which contains 154 intersections and 238 paths. According to the construction method of hazardous materials road transportation physical network model based on the primal approach, Dalian hazardous materials road transportation physical complex network model is finally obtained by using Arc-GIS 10.0 to complete the process of spatial data, which is shown in Figure 2.

3.1.2. The Eigenvalues Calculation. Combined with Dalian hazardous materials road transportation physical complex network model, based on obtaining the network topological

Table 1: Parameters of Dalian HMRTN.

Network eigenvalues	m_a	n_a	$\langle k\rangle$	L (km)	T (km)	C	L_R (km)	C_R
Calculation results	154	238	3.0909	64.4485	193.3871	0.1714	4.1871	0.0216

m_a is the number of hazardous materials road transportation physical network's node, n_a is the number of hazardous materials road transportation physical network's edge, $\langle k\rangle$ is the average node degree of anetwork, L is the network characteristic path length, T is the network diameter, C is the network average clustering coefficient, L_R is the characteristic path length of random network, and C_R is network average clustering coefficient of random network.

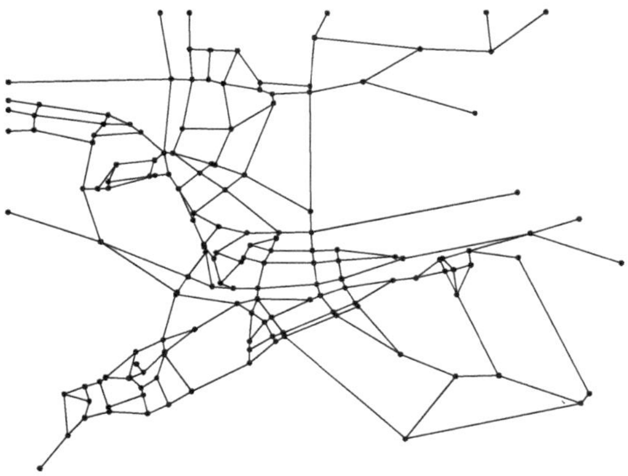

Figure 2: Physical network model of Dalian HMRTN based on primal approach.

adjacency matrix and distance adjacency matrix, the network eigenvalues such as the number of nodes, the number of edges, the network average node degree, characteristic path length, and network average clustering coefficient values, as well as the random network characteristic path length and network average clustering coefficient of the same scale, are calculated; the results are shown in Table 1.

The results in Table 1 show the following.

(1) There are 154 nodes and 238 edges in Dalian HMRTN, and the network average node degree is about 3.0909, between 3~4, which indicates that each intersection in the physical network is connected directly with an average of 3 paths. The average distance between any two nodes is 64.4485 km, and the largest shortest distance between two nodes is 193.3871 km.

(2) The physical network has a larger network average clustering coefficient, but it is much smaller than that of the urban road network [23].

(3) The characteristic path length L and network average clustering coefficient C of physical network are greater than those of random network of the same scale, which is consistent with the requirements $L \geq L_R$ and $C \gg C_R$ [27]. Therefore, hazardous materials road transportation physical network topological structure presents small-world properties.

3.1.3. *The Topological Properties Analysis*

(1) The Nodes Degree Distribution. Analyzing the network node degree distribution is helpful for understanding spatial distribution of network nodes and the access rules of new nodes. Through the statistics of Dalian hazardous materials road transportation physical network model, the statistical results of network node degree are shown in Table 2. By calculating the network node degree and node degree cumulative distribution probability, and using Matlab Curve Fitting toolbox for data fitting, the diagram is shown in Figure 3.

Table 2: The nodes degree statistics of physical network of Dalian HMRTN.

Node degree	1	2	3	4	5
Node	16	2	90	44	2
Ratio/%	10.39	1.30	58.44	28.57	1.30

In Table 2, nodes with node degree value below 3 in Dalian road transportation physical complex network account for 70.13% in the network, and nodes with node degree value above 5 only account for 1.30% in the network, which indicates that the network node degree distribution is not uniform, and the majority node degrees are generally small. The minimum value of node degree is 1, which is induced by some nodes in the boundary of network. The maximum value of node degree is 5, which shows that intersections connect with up to 5 paths in hazardous materials transportation network nodes.

In Figure 3, the network cumulative node degree distribution probability is subjected to exponential distribution, and the imitative effect is good. Known from [27], the connections among a new node and existing nodes in physical network are random, and network topological structure presents random network properties. This shows that the physical network is a tree network, and it is very easy to form isolated subgraphs after attack. And the low percolation point of tree network also means that the connectivity of physical network could be easily destroyed after attack.

(2) The Network Clustering Coefficient and Characteristic Path Length. The network average clustering coefficient and the characteristic path length are important indexes of network topological properties. In this paper, network topological structure properties of HMRTN are researched by observing the relationship among network characteristic path length, network average clustering coefficient, and network growth. The statistical results are shown in Figures 4 and 5.

In Figure 4, the linear fitting between network characteristic path lengths and the logarithm of nodes is good, and the fitting curve presents positive relationship, which shows that hazardous materials road transportation physical network presents property of small path length in the growth process.

In Figure 5, the ratio increases with the network node number increasing, and the fitting effect is very good.

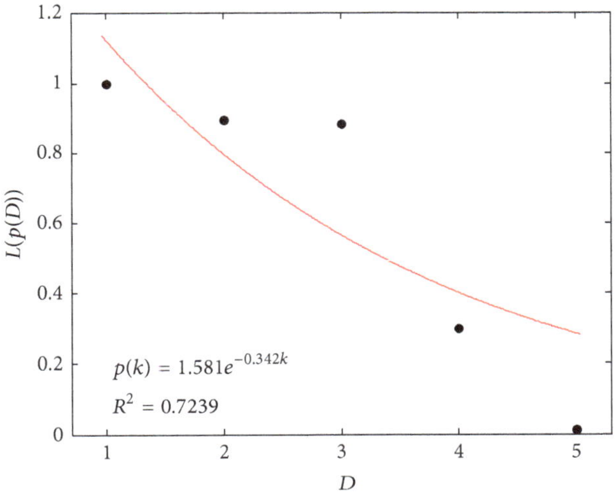

FIGURE 3: Probability distribution of cumulative node degree.

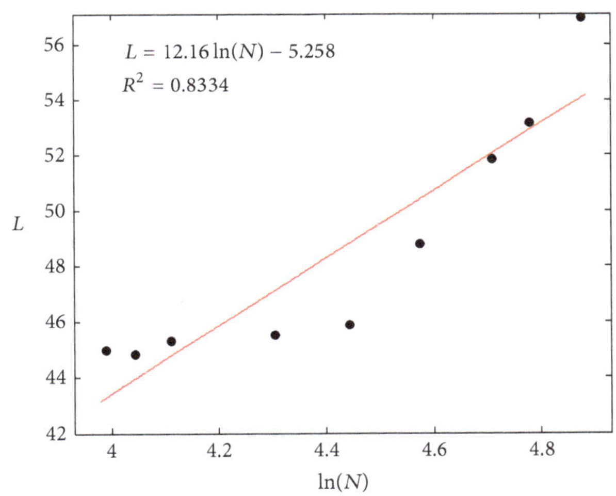

FIGURE 4: Curve of characteristic path length and $\ln(N)$.

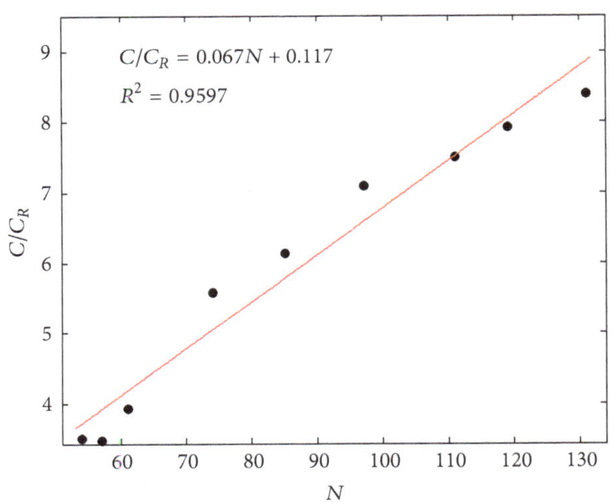

FIGURE 5: Curve of C/C_R and network nodes.

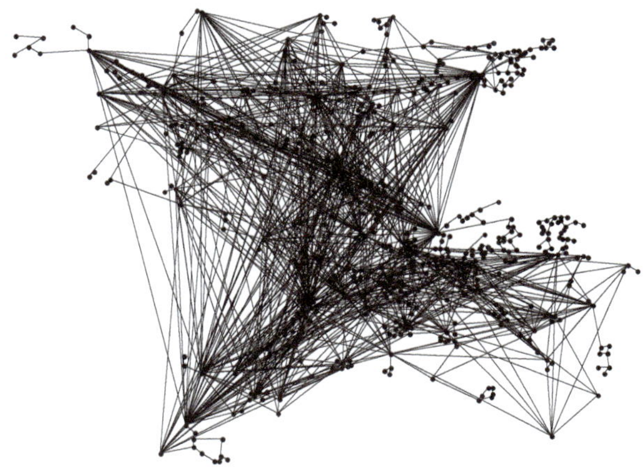

FIGURE 6: Service network model of Dalian HMRTN based on neighboring nodes.

It shows that the larger the scale of hazardous materials road transportation physical network is, the more obvious highly clustering properties of network topology are.

From the above results, Dalian hazardous materials road transportation physical network has small characteristic path length and large clustering coefficient, and it is a typical small-world network. Small characteristic path length shows that the amount of information the physical network contains is not complex although its scale may be large, which makes it possible for people's cognition of physical network space. Large clustering coefficient shows that the physical network has good connectivity.

3.2. The Service Network of HMRTN

3.2.1. The Basic Data of Service Network of Dalian HMRTN. In all the impact factors of the planning for hazardous materials road transportation distribution route, safety is the most important one, which is distribution route of minimum risk. Therefore, hazardous materials road transportation service network route is subjected to not only physical network but also the network risk.

Matching with the operation data of Dalian hazardous materials road transportation physical network, Dalian hazardous materials distribution data in December 2013 is used in this paper. The distribution network includes 1318 paths and 569 network nodes. According to the construction method of hazardous materials road transportation service network model based on the neighboring nodes, Dalian hazardous materials road transportation service complex network model is finally obtained by using Arc-GIS 10.0, which is shown in Figure 6.

3.2.2. The Eigenvalues Calculation and Topological Properties Analysis. Known from Section 2.2.2, the service network constructed in this paper is a directed network, so the node degree is divided into out degree and in degree. Due to the fact that the round-trip routes of hazardous materials transportation vehicles are generally the same, there are no

TABLE 3: The node degree statistics of service network of Dalian HMRTN.

Node degree	1	2	3	4	5	6	7	8	9	10	11	12
Node	180	175	70	30	11	6	6	5	13	3	5	3
Ratio/%	31.63	30.76	12.30	5.27	1.93	1.05	1.05	0.88	2.28	0.53	0.88	0.53
Node degree	13	14	15	16	17	18	19	20	21	22	23	24
Node	2	4	5	3	8	3	5	6	4	2	3	3
Ratio/%	0.35	0.70	0.88	0.53	1.41	0.53	0.88	1.05	0.70	0.35	0.53	0.53
Node degree	25	26	27	28	34	35	38	40	44	45	46	47
Node	2	2	1	1	1	1	1	1	1	1	1	1
Ratio/%	0.35	0.35	0.18	0.18	0.18	0.18	0.18	0.18	0.18	0.18	0.18	0.18

obvious differences between network node out degree and in degree in the service network. Based on this, network out degree will be regarded as the research object in this paper.

(1) The Node Degree and Node Degree Distribution. For hazardous materials road transportation service network based on the neighboring nodes, the node degree $D(v_i)$ of any node v_i represents the number of such nodes as distribution center, demand center, and so forth that are directly connected with v_i. Therefore, it does not include edge weight. The bigger $D(v_i)$ is, the stronger the connectivity of this node is. And in practice, this node may be demand center or secondary distributor in great demand of category and quantity. Through the statistics of Dalian hazardous materials road transportation service network model, the statistical results of network node degree are shown in Table 3, and by using Matlab Curve Fitting toolbox for data fitting, the network node degree and degree distribution probability diagram are shown in Figure 7.

From the statistical results in Table 3, the maximum node degree of Dalian hazardous materials road transportation service complex network is 47, and the average is about 4.63, which shows that one node connects with an average of 5 nodes in the distribution process. The nodes with node degrees below 3 account for 74.69% in the network, and the nodes with node degrees above 30 account for only 1.41% in the total number of nodes, presenting uniform distribution.

In Figure 7, the node degree distribution $p(D)$ decreases with the change of node degree D in power-law exponential forms, $p(D) \propto D^{-\beta}$, and $\beta \approx 1.156$. Known from [27], hazardous materials road transportation service network based on the neighboring nodes is subjected to power-law distribution, representing the properties of scale-free network.

For further study on complex network properties of the network nodes, the node degree and its cumulative distribution are dealt with by double logarithmic processing, and the fitting results are shown in Figure 8.

Through double logarithmic processing of node degree and its cumulative distribution probability and linear regression fitting of the results, the equation can be obtained as $\log(p(D)) = -1.481\log(D)+5.651$, where the coefficient $R^2 = 0.8639$. The statistical results show that hazardous materials road transportation service network is subjected to power-law distribution.

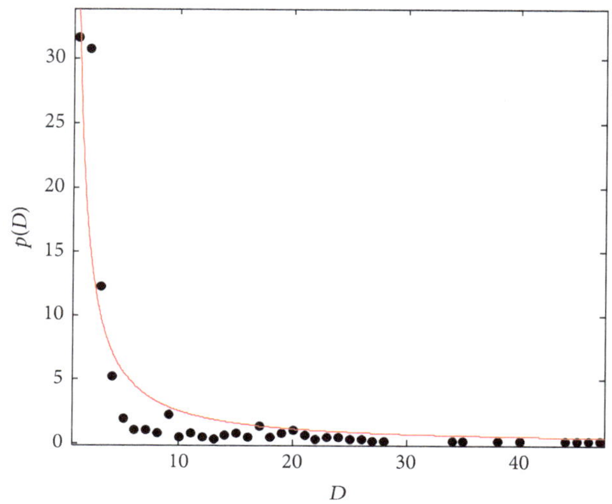

FIGURE 7: The relationship between nodes degree and node degree distribution of service network of Dalian HMRTN.

According to the above analysis, in hazardous materials road transportation service network based on the neighboring nodes, the relationship between node degree and degree distribution, and the double logarithmic relationship between node degree and cumulative node degree distribution probability are both subjected to power-law distribution and present scale-free network properties.

(2) The Node Strength. Analysis on network node strength distribution can help to understand not only the connected relationship among network nodes but also the network traffic density and the connectivity among network nodes. Service network is a network of weighted edges, and the weight represents the efficiency of paths between neighboring nodes. Referring to [28], hazardous materials road transportation service network node strength can be defined as follows in this paper:

$$D'(v_i) = \sum_{v_j \in V'_{v_i}} w_{v_i v_j}, \quad (1)$$

where $D'(v_i)$ is the node strength of node v_i, $v_j \in V'_{v_i}$ is node v_j directly connected with node v_i, and $w_{v_i v_j} = 0$ or 1, where

TABLE 4: The node strength statistics of service network of Dalian HMRTN.

Node strength	1	2	3	4	5	...	16	17	18	19	20
Node	50	18	16	11	9	...	6	4	3	2	6
Ratio/%	12.92	4.65	4.13	2.84	2.33	...	1.55	1.03	0.78	0.52	1.55
Node strength	68	69	70	71	72	...	155	194	195	223	289
Node	6	4	1	2	2	...	1	1	1	1	1
Ratio/%	1.55	1.03	0.26	0.52	0.52	...	0.26	0.26	0.26	0.26	0.26

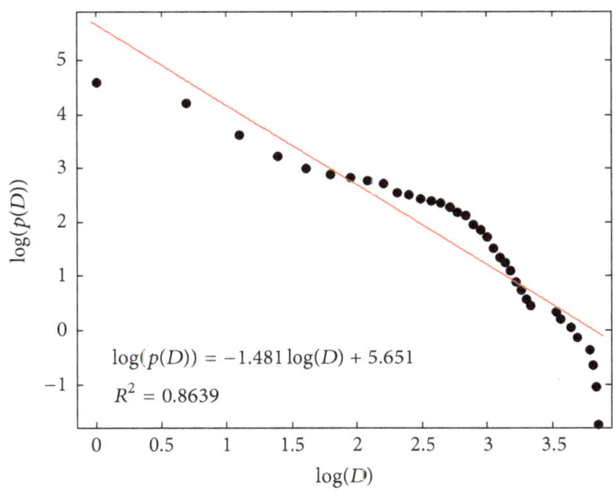

FIGURE 8: The relationship between nodes degree and cumulative node degree.

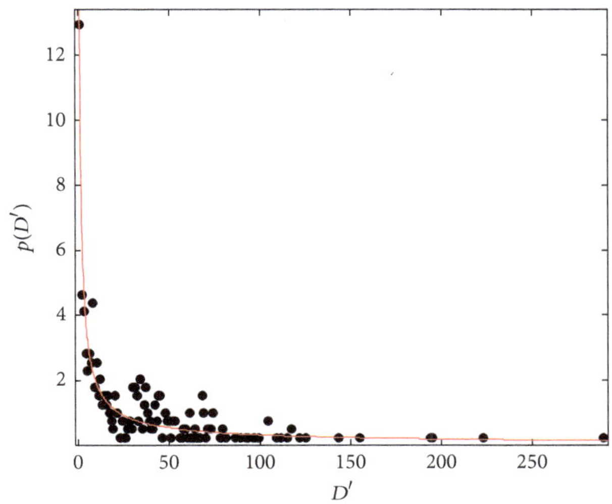

FIGURE 9: The relationship between nodes strength and nodes strength distribution of service network of Dalian HMRTN.

an edge exists between nodes $v_i \rightarrow v_j$, $w_{v_i v_j} = 1$, and otherwise $w_{v_i v_j} = 0$.

According to (1), the network node strength distribution can be defined as follows:

$$\prod \left(D'(v_i)\right) = \frac{D'(v_i)}{N}, \tag{2}$$

where $\prod(v_i)$ is the distribution probability of node strength equal to $D'(v_i)$ and N is the sum of network node strengths.

Through the statistics of Dalian hazardous materials road transportation service network, the statistical results of the network node strengths are shown in Table 4.

Through the analysis of the data in Table 4, the average node strength $\langle D' \rangle$ of the service network is 29.53, and the maximum node strength is 289. This shows that about 30 distribution routes go through each node. With an average of 1 node connected with 5 nodes; therefore, the average strength of two neighboring nodes is 6, which shows that an average of 6 distribution routes go through two neighboring nodes.

In Table 4, the top 20% nodes contribute 52.20% of accumulative node strength and the top 10% of cumulative node strength has reached 38.50% in the 569 nodes of Dalian hazardous materials road transportation service network. This shows that the normal operation of the whole service network is maintained by only a few nodes. Otherwise, the cumulative node strength of nodes with node strength below 5 is up to 26.87%, which indicates that most of the nodes in the service network are normal nodes. The distribution task through these nodes is relatively small, but their presence increases the coverage and density of the service network.

(3) The Node Strength Distribution. In order to research the complex network properties of the of network node strength, the statistics of Dalian hazardous materials road transportation service network node strengths are carried out in this paper, and the nodes strength and their strength distribution probability are shown in Figure 9.

In Figure 9, the node strength distribution $p(D')$ decreases with the change of node strength D' in power-law exponential forms, $p(D') \propto D'^{-\beta}$, and $\beta \approx 0.7751$. Known from [27], hazardous materials road transportation service network is generally subjected to power-law distributions and represents the properties of scale-free network.

For further study on internal dynamics properties of the network nodes, the relationship between service network node strength with intensity sequence number and cumulative node strength will be examined. Figure 10 shows the nodes strength changing with its descending order ordinal, and Figure 11 shows the node strength changing with the cumulative node strength increasing.

In Figure 10, the node strength and its descending order ordinal are dealt with by double logarithmic processing. Using Matlab Curve Fitting toolbox for linear regression fitting, the equation can be obtained as

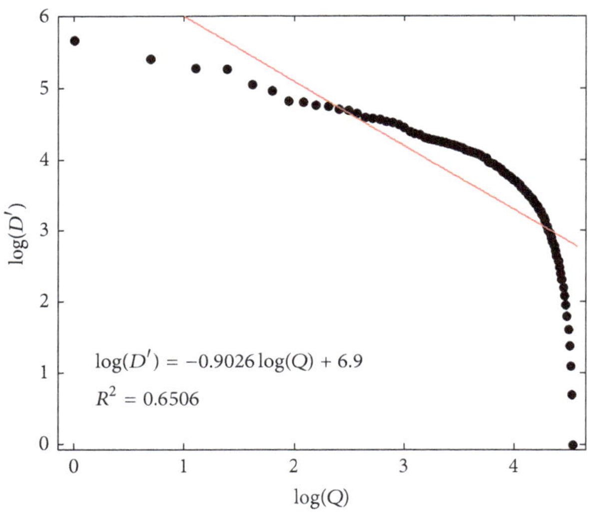

FIGURE 10: The relationship between nodes strength and its descending order ordinal of service network of Dalian HMRTN.

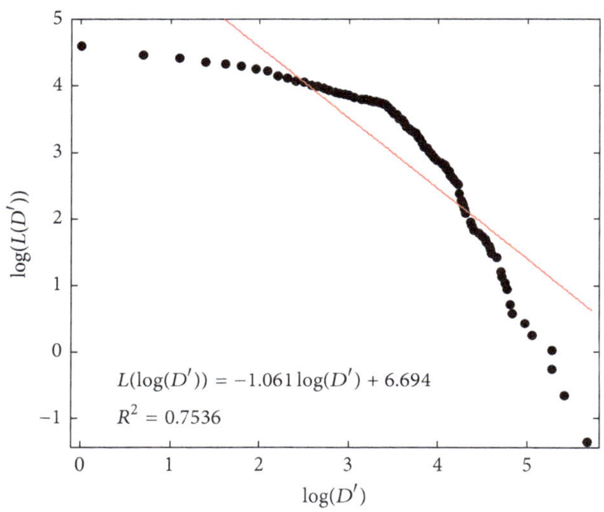

FIGURE 11: The relationship between nodes strength and cumulative nodes strength of service network of Dalian HMRTN.

$\log(D') = -0.9026\log(Q) + 6.9$ and the coefficient as $R^2 = 0.6506$. The above statistical results show that the node strength and its descending order ordinal are subjected to power-law distribution on the whole.

In Figure 11, the nodes strength and cumulative nodes strength are dealt with by double logarithmic processing. Using Matlab Curve Fitting toolbox for linear regression fitting, the equation can be obtained as $L(\log(D')) = -1.061\log(D') + 6.694$ and the coefficient as $R^2 = 0.7536$. The above statistical results show that the nodes strength and cumulative nodes strength are also subjected to power-law distribution on the whole.

According to the above analysis, in the hazardous materials road transportation service network based on neighboring nodes, node strength and its distribution probability, node strength and its descending order ordinal, and node strength and cumulative node strength distribution probability are all subjected to power-law distribution and represent properties of scale-free network. It presents that the production, circulation, and distribution of hazardous materials are completed by only a few nodes in the service network, which reflects the internal dynamic properties and essence of the service network.

4. Conclusions

To research the topological properties of HMRTN, this paper proposes two different ways to construct the cyberspace of HMRTN and constructs its complex network models, respectively. The Dalian HMRTN is regarded as a researching case. Using the complex network theory and statistical analysis, the main conclusions are as follows.

(1) By analyzing hazardous materials road transportation physical network, its node degree distribution in Dalian is very uneven. The cumulative node degree of physical network is subjected to exponential distribution, showing the network properties of random network; network characteristic path length and clustering coefficient change with the network scale, showing small-world network properties. These results reflect the complexity and the internal dynamic properties of hazardous materials road transportation physical network.

(2) The random network properties of physical network show that each node of physical connection in Dalian is random connection. It could be considered that, in HMRTN planning, urban long and wide main roads are typically preferred, and they should be away from populated areas with high road network density, which also leads to low connectivity between physical network nodes. Therefore, we must strengthen the construction of hazardous materials road transportation infrastructure and improve the network average node degree to ensure the connection and reliability of hazardous materials road transportation.

(3) The network average clustering coefficient represents the structure properties of the physical network to some extent; the greater its numerical value is, the better the connectivity of the network would be. Although the physical network presents small-world properties, the network average clustering coefficient is much smaller compared to other transportation networks. When the network paralysis occurs in some key nodes and edges, it would be very easy to make the nodes isolated. Therefore, improving the network average clustering coefficient of a physical network is an effective way to optimize the topological properties of hazardous materials road transportation physical network.

(4) By analyzing hazardous materials road transportation service network, its node degree and node strength distribution in Dalian are very uneven. The relationship between node degree and cumulative node degree of service network is subjected to power-law distribution, showing the network properties of scale-free network. The relationship among nodes strength, its descending order ordinal, and cumulative nodes strength are all subjected to power-law distribution, showing the network properties of scale-free network. These results reflect the complexity and the internal dynamic properties of hazardous materials road transportation service network. The results can also be used as an example of scale-free network growth model.

(5) The scale-free network properties presented by service network node strength show that hazardous materials in the whole process from production to consumption are always completed by only a few companies. And those companies' transport once blocked will cause the paralysis to the whole service network transportation efficiency. Therefore, we should try our best to reduce the pressure of the key companies, which can enhance the survivability of the whole network under the unexpected events.

(6) Service network node degree mainly emphasizes the connectivity between network nodes, while node strength presents distribution load. Therefore, there is difference between service network node degree and node strength ranking, and this difference reflects the contradiction between the planned network and distribution pressure. So service network node degree and node strength can be used as reference to the optimization of distribution.

This paper is only a preliminary study on hazardous materials road transportation complex network topological properties. HMRTN is a typical open complex network with road system and distribution system, which are mutually influenced and interdependent, and it is a super network including physical network and service network. Therefore, the coupling relationship between physical network and service network of hazardous material road transportation and network planning optimization will be the work of next step.

Conflict of Interests

The authors declare that there is no conflict of interests regarding the publication of this paper.

Acknowledgment

The authors would like to appreciate the financial support from the National Natural Science Foundation of China (no. 71173177).

References

[1] M. Rubinov and O. Sporns, "Complex network measures of brain connectivity: Uses and interpretations," *NeuroImage*, vol. 52, no. 3, pp. 1059–1069, 2010.

[2] R. Albert, H. Jeong, and A.-L. Barabási, "Error and attack tolerance of complex networks," *Nature*, vol. 406, no. 6794, pp. 378–382, 2000.

[3] J. A. Dunne, R. J. Williams, and N. D. Martinez, "Network structure and biodiversity loss in food webs: robustness increases with connectance," *Ecology Letters*, vol. 5, no. 4, pp. 558–567, 2002.

[4] L. A. N. Amaral, A. Scala, M. Barthélémy, and H. E. Stanley, "Classes of small-world networks," *Proceedings of the National Academy of Sciences of the United States of America*, vol. 97, no. 21, pp. 11149–11152, 2000.

[5] S. Redner, "How popular is your paper? An empirical study of the citation distribution," *The European Physical Journal B-Condensed Matter and Complex Systems*, vol. 4, no. 2, pp. 131–134, 1998.

[6] S. Milgram, "The small world problem," *Psychology Today*, vol. 1, no. 1, pp. 61–67, 1967.

[7] D. J. Watts and S. H. Strogatz, "Collective dynamics of "small-world" networks," *Nature*, vol. 393, no. 6684, pp. 440–442, 1998.

[8] A.-L. Barabási and R. Albert, "Emergence of scaling in random networks," *Science*, vol. 286, no. 5439, pp. 509–512, 1999.

[9] M. Metcalfe, "Knowledge sharing, complex environments and small-worlds," *Human Systems Management*, vol. 24, no. 3, pp. 185–195, 2005.

[10] V. Latora and M. Marchiori, "Is the Boston subway a small-world network?," *Physica A: Statistical Mechanics and Its Applications*, vol. 314, no. 1–4, pp. 109–113, 2002.

[11] K. A. Seaton and L. M. Hackett, "Stations, trains and small-world networks," *Physica A: Statistical Mechanics and its Applications*, vol. 339, no. 3-4, pp. 635–644, 2004.

[12] P. Sen, S. Dasgupta, A. Chatterjee, P. A. Sreeram, G. Mukherjee, and S. S. Manna, "Small-world properties of the Indian railway network," *Physical Review E: Statistical, Nonlinear, and Soft Matter Physics*, vol. 67, no. 3, Article ID 036106, 2003.

[13] W. Li and X. Cai, "Statistical analysis of airport network of China," *Physical Review E*, vol. 69, no. 4, Article ID 046106, 2004.

[14] M. Kurant and P. Thiran, "Extraction and analysis of traffic and topologies of transportation networks," *Physical Review E*, vol. 74, no. 3, Article ID 036114, 2006.

[15] R. Guimerà and L. A. N. Amaral, "Modeling the world-wide airport network," *European Physical Journal B*, vol. 38, no. 2, pp. 381–385, 2004.

[16] R. Guimerà, S. Mossa, A. Turtschi, and L. N. Amaral, "The worldwide air transportation network: anomalous centrality, community structure, and cities' global roles," *Proceedings of the National Academy of Sciences of the United States of America*, vol. 102, no. 22, pp. 7794–7799, 2005.

[17] L.-P. Chi, R. Wang, H. Su et al., "Structural properties of US flight network," *Chinese Physics Letters*, vol. 20, no. 8, pp. 1393–1396, 2003.

[18] G. Bagler, "Analysis of the airport network of India as a complex weighted network," *Physica A: Statistical Mechanics and Its Applications*, vol. 387, no. 12, pp. 2972–2980, 2008.

[19] M. Guida and F. Maria, "Topology of the Italian airport network: a scale-free small-world network with a fractal structure?" *Chaos, Solitons & Fractals*, vol. 31, no. 3, pp. 527–536, 2007.

[20] W. Li and X. Cai, "Statistical analysis of airport network of China," *Physical Review E—Statistical, Nonlinear, and Soft Matter Physics*, vol. 69, no. 4, Article ID 046106, 2004.

[21] B. Jiang and C. Claramunt, "Topological analysis of urban street networks," *Environment and Planning B: Planning and Design*, vol. 31, no. 1, pp. 151–162, 2004.

[22] P. Holme, B. J. Kim, C. N. Yoon, and S. K. Han, "Attack vulnerability of complex networks," *Physical Review E: Statistical, Nonlinear, and Soft Matter Physics*, vol. 65, no. 5, Article ID 056109, 2002.

[23] A. W. Krings and A. Azadmanesh, "A graph based model for survivability applications," *European Journal of Operational Research*, vol. 164, no. 3, pp. 680–689, 2005.

[24] J. J. Wu, H. J. Sun, and Z. Y. Gao, "Cascading failures on weighted urban traffic equilibrium networks," *Physica A: Statistical Mechanics and Its Applications*, vol. 386, no. 1, pp. 407–413, 2007.

[25] Economy Statistics for Chemical Engineering in China, http://wms.mofcom.gov.cn/article/ztxx/nyhg/hyyxqk /201307/ 20130700190421.shtml.

[26] S. Porta, P. Crucitti, and V. Latora, "The network analysis of urban streets: A dual approach," *Physica A: Statistical Mechanics and Its Applications*, vol. 369, no. 2, pp. 853–866, 2006.

[27] A.-L. Barabási, R. Albert, and H. Jeong, "Mean-field theory for scale-free random networks," *Physica A: Statistical Mechanics and Its Applications*, vol. 272, no. 1, pp. 173–187, 1999.

[28] S. H. Yook, H. Jeong, A.-L. Barabási, and Y. Tu, "Weighted evolving networks," *Physical Review Letters*, vol. 86, no. 25, pp. 5835–5838, 2001.

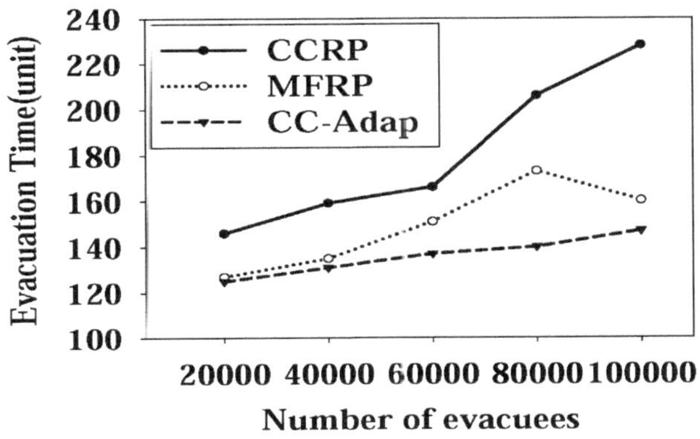

(a) $n = 256$; $m = 768$; $s = 26$

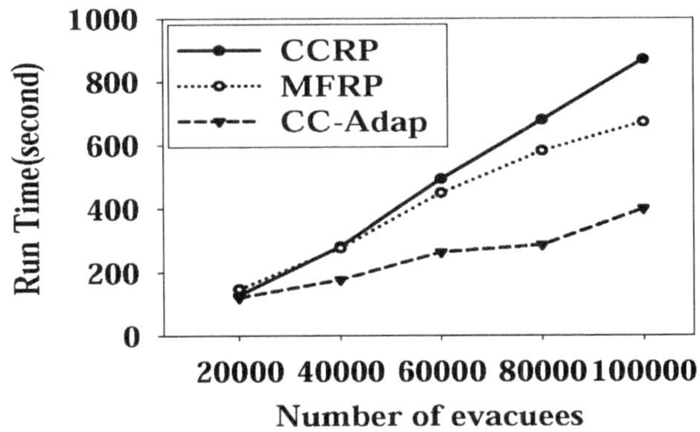

(b) $n = 256$; $m = 768$; $s = 26$

FIGURE 4: Evacuation quality in a small transportation network.

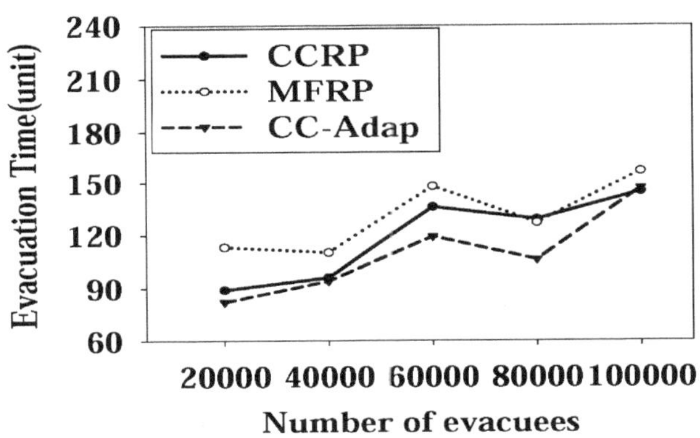

(a) $n = 1024$; $m = 5258$; $s = 102$

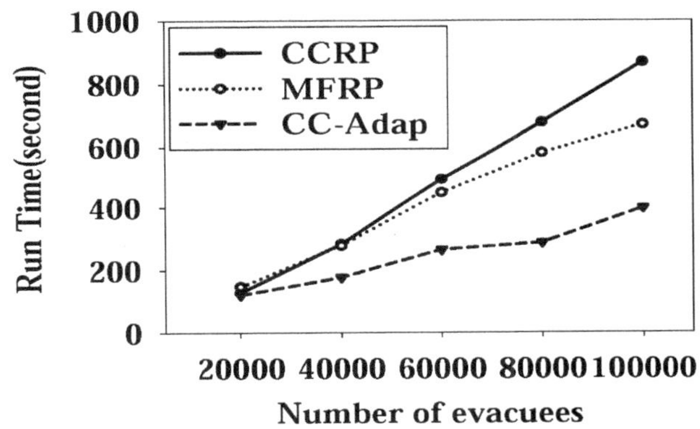

(b) $n = 1024$; $m = 5258$; $s = 102$

FIGURE 5: Evacuation quality in a large transportation network.

5.3. Experimental Result

5.3.1. Feasibility Verification. The purpose of this section is to evaluate the feasibility of using CC-Adap in evacuation planning. CC-Adap is compared with CCRP and MFRP to determine whether CC-Adap can be used for evacuation planning and to optimize evacuation plans by applying a contraflow-based method.

In this section, two experiments are carried out. Each experiment has five test groups and the number of vertices, edges, source, and sink vertices is fixed. We vary the number of evacuees from 20000 to 100000. Figures 4 and 5 depict the results on the evacuation time and run time of the three algorithms in small-scale and large-scale transportation networks, respectively. According to Figure 4(a), CC-Adap has the smallest and nearly linearly growing evacuation time with the increase of the number of evacuees, while CCRP and MFRP have steeply increasing evacuation time and the evacuation time of CCRP is almost double that of CC-Adap. The run time of CCRP and MFRP grows rapidly as the number of evacuees increases, while CC-Adap has a stable run time and the run time of CCRP and MFRP is more than twice that of CC-Adap when the number of evacuees is 100000 in Figure 4(b). As shown in Figure 5, CC-Adap has the shortest evacuation time with the increase in the number of evacuees in Figure 5(a), followed by CCRP, and MFRP has the longest evacuation time. In terms of the run time, CC-Adap, CCRP, and MFRP all show nearly linear growth with the increase in the number of evacuees, and CC-Adap has a run time that is nearly half those of CCRP and MFRP.

5.3.2. Scalability Verification. The purpose of this section is to evaluate the scalability of CC-Adap to complex evacuation situations and CC-Adap is compared with CCRP, MFRP, and Greedy algorithms.

Comparison on Networks with Different Number of Edges. Figure 6 plots the experimental results in terms of evacuation time and run time of the four algorithms against the number of edges. This experiment is carried out with five subexperiments. The number of vertices is fixed at 256, and

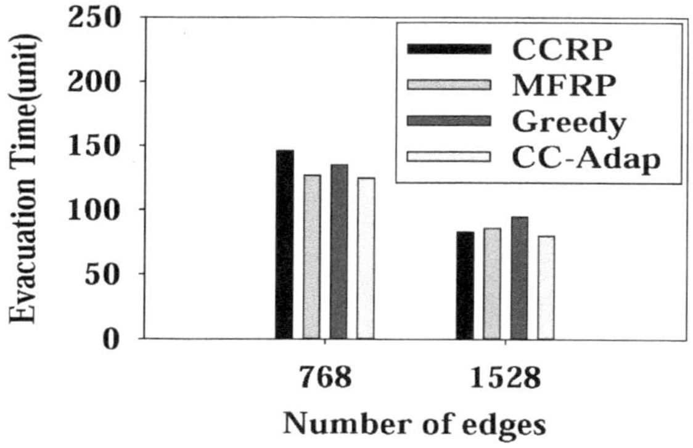

(a) $n = 256$; $s = 26$; $p = 20000$

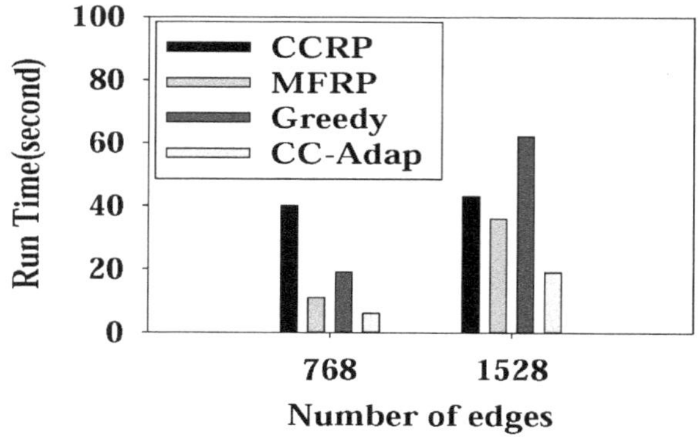

(b) $n = 256$; $s = 26$; $p = 20000$

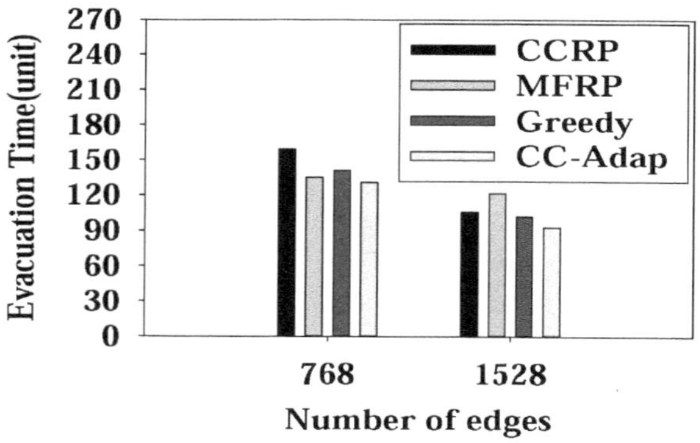

(c) $n = 256$; $s = 26$; $p = 40000$

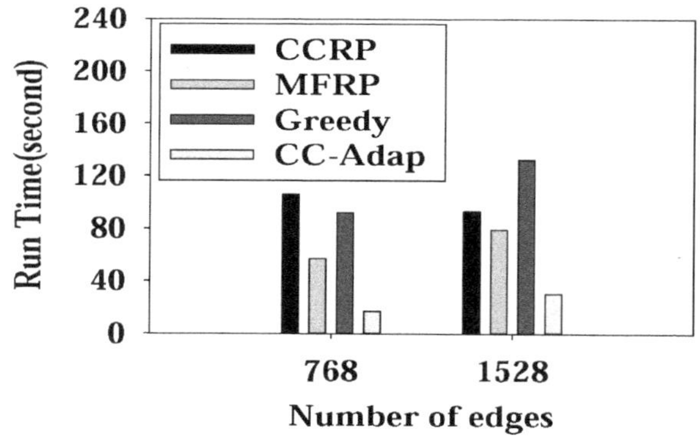

(d) $n = 256$; $s = 26$; $p = 40000$

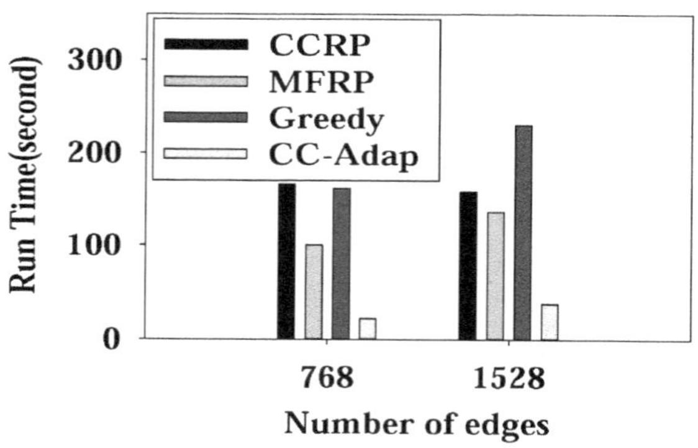

(e) $n = 256$; $s = 26$; $p = 60000$

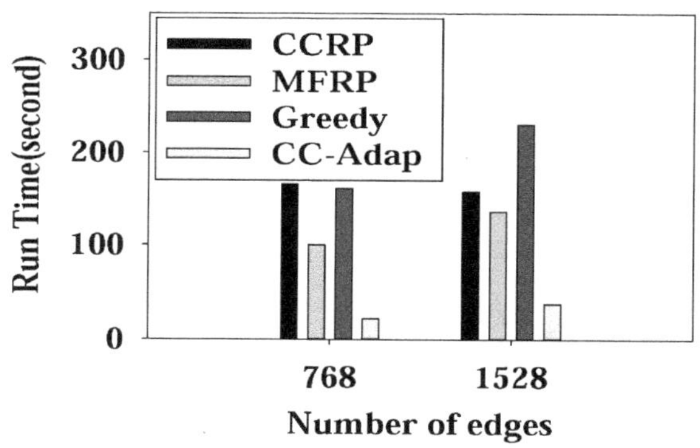

(f) $n = 256$; $s = 26$; $p = 60000$

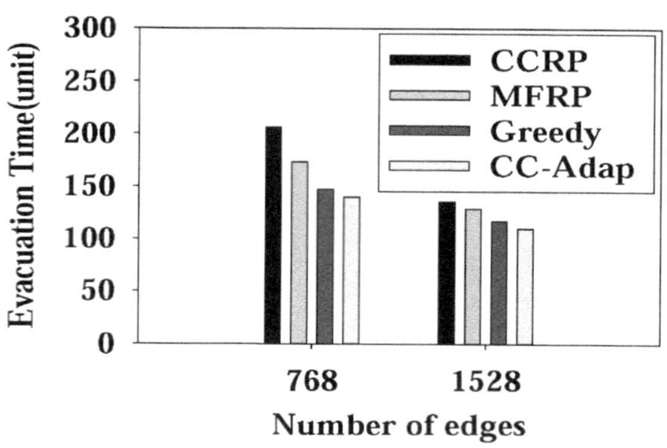

(g) $n = 256$; $s = 26$; $p = 80000$

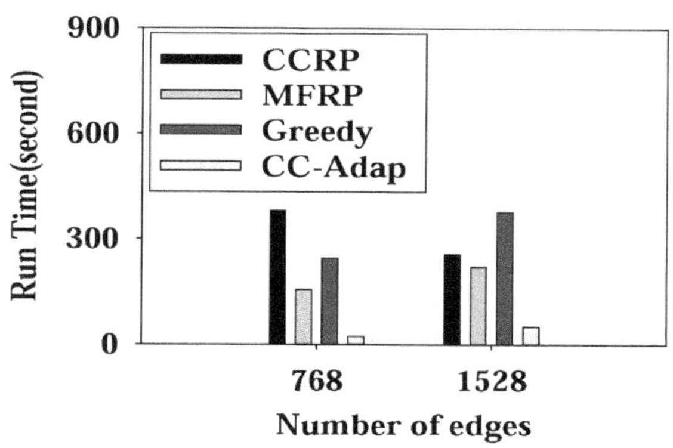

(h) $n = 256$; $s = 26$; $p = 80000$

Figure 6: Continued.

Permissions

All chapters in this book were first published in JAT, by Hindawi Publishing Corporation; hereby published with permission under the Creative Commons Attribution License or equivalent. Every chapter published in this book has been scrutinized by our experts. Their significance has been extensively debated. The topics covered herein carry significant findings which will fuel the growth of the discipline. They may even be implemented as practical applications or may be referred to as a beginning point for another development.

The contributors of this book come from diverse backgrounds, making this book a truly international effort. This book will bring forth new frontiers with its revolutionizing research information and detailed analysis of the nascent developments around the world.

We would like to thank all the contributing authors for lending their expertise to make the book truly unique. They have played a crucial role in the development of this book. Without their invaluable contributions this book wouldn't have been possible. They have made vital efforts to compile up to date information on the varied aspects of this subject to make this book a valuable addition to the collection of many professionals and students.

This book was conceptualized with the vision of imparting up-to-date information and advanced data in this field. To ensure the same, a matchless editorial board was set up. Every individual on the board went through rigorous rounds of assessment to prove their worth. After which they invested a large part of their time researching and compiling the most relevant data for our readers.

The editorial board has been involved in producing this book since its inception. They have spent rigorous hours researching and exploring the diverse topics which have resulted in the successful publishing of this book. They have passed on their knowledge of decades through this book. To expedite this challenging task, the publisher supported the team at every step. A small team of assistant editors was also appointed to further simplify the editing procedure and attain best results for the readers.

Apart from the editorial board, the designing team has also invested a significant amount of their time in understanding the subject and creating the most relevant covers. They scrutinized every image to scout for the most suitable representation of the subject and create an appropriate cover for the book.

The publishing team has been an ardent support to the editorial, designing and production team. Their endless efforts to recruit the best for this project, has resulted in the accomplishment of this book. They are a veteran in the field of academics and their pool of knowledge is as vast as their experience in printing. Their expertise and guidance has proved useful at every step. Their uncompromising quality standards have made this book an exceptional effort. Their encouragement from time to time has been an inspiration for everyone.

The publisher and the editorial board hope that this book will prove to be a valuable piece of knowledge for researchers, students, practitioners and scholars across the globe.

List of Contributors

Khaled Shaaban and Hassan Hamad
Department of Civil Engineering, Qatar University, Doha, Qatar

Yi Li and Yuren Chen
The Key Laboratory of Road and Traffic Engineering, Ministry of Education, College of Transportation Engineering, Tongji University, No. 4800 Cao'an Hwy., Shanghai 201804, China

Leyu Wei, Jinliang Xu, Xingli Jia, Xiaodong Zhang and Haoru Li
School of Highway, Chang'an University, Middle Section of South 2 Ring Rd., Beilin District, Xi'an, Shaanxi 710064, China

Marina Milenković and Dražnko Glavić
University of Belgrade, Faculty of Transport and Traffic Engineering, V. Stepe 305, 11000 Belgrade, Serbia

Miloš N. Mladenović
Aalto University, Department of Built Environment, Otakaari 4, 02150 Espoo, Finland

Gen Li
School of Transportation, Southeast University, Nanjing 210096, China

Lu Sun
Department of Operations Research and Financial Engineering, Princeton University, 229 Sherrerd Hall, Princeton, NJ 08544, USA

Wenyuan Wang, Yun Peng and Qi Tian
State Key Laboratory of Coastal and Offshore Engineering, Dalian University of Technology, Dalian 116024, China

Wenyuan Wang and Ying Jiang
Mobilities and Urban Policy Lab, Graduate School for International Development and Cooperation, Hiroshima University, Hiroshima, Japan

Yong Zhou
College of Transport & Communications, Shanghai Maritime University, Shanghai 201306, China

Mohammad Almoarawi and Essam Dabbour
Department of Civil Engineering, Abu Dhabi University, Abu Dhabi, UAE

Mariana Gonçalves de Carvalho Wolff and Marco Antonio Farah Caldas
Department of Industrial Engineering, Universidade Federal Fluminense, Niterói, RJ, Brazil

Shijie Shu, Jing Zhao and Yin Han
Department of Traffic Engineering, University of Shanghai for Science and Technology, Shanghai 200093, China

Dongmin Lee
Associate Professor, Department of Transportation Engineering, University of Seoul, 163 Seoulsiripdae-ro, Dongdaemun-gu, Seoul 02504, Republic of Korea

Sooncheon Hwang
Graduate Student, Department of Transportation Engineering,The University of Seoul, Korea Seoulsiripdaero 163, Dongdaemun-gu, Seoul 02504, Republic of Korea

Eunhan Ka
Researcher, Institute of Construction and Environmental Engineering, Seoul National University, 1 Gwanak-ro, Gwanak-gu, Seoul 08826, Republic of Korea

Chungwon Lee
Associate Professor, Department of Civil and Environmental Engineering, Seoul National University, Korea 1 Gwanak-ro, Gwanak-gu, Seoul 151-744, Republic of Korea

Qinrui Tang
Institute of Transportation Systems, German Aerospace Center, Berlin, Germany

Qinrui Tang and Bernhard Friedrich
Institute of Transportation and Urban Engineering, Technische Universität Braunschweig, Braunschweig, Germany

V. L. Knoop, M. de Baat, H. Taale and S. P. Hoogendoorn
Delft University of Technology, Netherlands

M. Keyvan Ekbatani
University of Canterbury, New Zealand

H. Taale
Rijkswaterstaat and Delft University of Technology, Netherlands

Giovanni Tesoriere, Tiziana Campisi and Antonino Canale
Faculty of Engineering and Architecture, University of Enna Kore, Italy

Tedi Zgrablić
Faculty of Engineering and Architecture, University of Maribor, Slovenia

Wu Ni, Chengxin Xiong and Dan Guo
School of Computer and Information, Hefei University of Technology, Hefei 230009, China

Wenbo Li and Hailei Wang
Institute of Technology Innovation, Hefei Institutes of Physical Science, Chinese Academy of Sciences, Hefei 230088, China

Yue Zheng and Wenquan Li
Jiangsu Key Laboratory of Urban ITS, Jiangsu Province Collaborative Innovation Center of Modern Urban Traffic Technologies, School of Transportation, Southeast University, Nanjing, China

Feng Qiu
Department of Computer Science, University of Victoria, Victoria, Canada

Xiaoyuan Wang, Fang Wang and Yaqi Liu
College of Electromechanical Engineering, Qingdao University of Science & Technology, Qingdao 266000, China

Xiaoyuan Wang
State Key Laboratory of Automotive Safety and Energy, Tsinghua University, Beijing 100084, China

Fang Wang, Dong Kong, Yaqi Liu, Liping Liu and Chen Chen
Institute of Intelligent Transportation, School of Transportation and Vehicle Engineering, Shandong University of Technology, Zibo 255049, China

Andriy Viktorovich Goncharenko
Aircraft Airworthiness Retaining Department, Academic and Research Aerospace Institute, National Aviation University, 1, Kosmonavta Komarova Avenue, Kyiv 03058, Ukraine

Chen Chen, Xiaohua Zhao, Ying Yao, Jian Rong and Xiaoming Liu
Beijing Key Laboratory of Traffic Engineering, Beijing Engineering Research Center of Urban Transport Operation Guarantee, Beijing University of Technology, Beijing 100124, China

Yunlong Zhang
Texas A&M University, College Station, TX 77843, USA

Index

www.ingramcontent.com/pod-product-compliance
Lightning Source LLC
LaVergne TN
LVHW061509220926
848485LV00017B/597